Emerging Drugs and Targets for Parkinson's Disease

RSC Drug Discovery Series

Editor-in-Chief:
Professor David Thurston, *King's College, London, UK*

Series Editors:
Dr David Fox, *Vulpine Science and Learning, UK*
Professor Ana Martinez, *Medicinal Chemistry Institute-CSIC, Madrid, Spain*
Professor David Rotella, *Montclair State University, USA*

Advisor to the Board:
Professor Robin Ganellin, *University College London, UK*

Titles in the Series:
1: Metabolism, Pharmacokinetics and Toxicity of Functional Groups
2: Emerging Drugs and Targets for Alzheimer's Disease; Volume 1
3: Emerging Drugs and Targets for Alzheimer's Disease; Volume 2
4: Accounts in Drug Discovery
5: New Frontiers in Chemical Biology
6: Animal Models for Neurodegenerative Disease
7: Neurodegeneration
8: G Protein-Coupled Receptors
9: Pharmaceutical Process Development
10: Extracellular and Intracellular Signaling
11: New Synthetic Technologies in Medicinal Chemistry
12: New Horizons in Predictive Toxicology
13: Drug Design Strategies: Quantitative Approaches
14: Neglected Diseases and Drug Discovery
15: Biomedical Imaging
16: Pharmaceutical Salts and Cocrystals
17: Polyamine Drug Discovery
18: Proteinases as Drug Targets
19: Kinase Drug Discovery
20: Drug Design Strategies: Computational Techniques and Applications
21: Designing Multi-Target Drugs
22: Nanostructured Biomaterials for Overcoming Biological Barriers
23: Physico-Chemical and Computational Approaches to Drug Discovery
24: Biomarkers for Traumatic Brain Injury
25: Drug Discovery from Natural Products
26: Anti-Inflammatory Drug Discovery
27: New Therapeutic Strategies for Type 2 Diabetes: Small Molecules
28: Drug Discovery for Psychiatric Disorders
29: Organic Chemistry of Drug Degradation
30: Computational Approaches to Nuclear Receptors
31: Traditional Chinese Medicine
32: Successful Strategies for the Discovery of Antiviral Drugs
33: Comprehensive Biomarker Discovery and Validation for Clinical Application
34: Emerging Drugs and Targets for Parkinson's Disease

How to obtain future titles on publication:
A standing order plan is available for this series. A standing order will bring delivery of each new volume immediately on publication.

For further information please contact:
Book Sales Department, Royal Society of Chemistry, Thomas Graham House, Science Park, Milton Road, Cambridge, CB4 0WF, UK
Telephone: +44 (0)1223 420066, Fax: +44 (0)1223 420247,
Email: booksales@rsc.org
Visit our website at www.rsc.org/books

Emerging Drugs and Targets for Parkinson's Disease

Edited by

Ana Martinez and Carmen Gil
Instituto de Química Médica-CSIC, Madrid, Spain
Email: amartinez@iqm.csic.es, cgil@iqm.csic.es

RSCPublishing

RSC Drug Discovery Series No. 34

ISBN: 978-1-84973-617-6
ISSN: 2041-3203

A catalogue record for this book is available from the British Library

© The Royal Society of Chemistry 2013

All rights reserved

Apart from fair dealing for the purposes of research for non-commercial purposes or for private study, criticism or review, as permitted under the Copyright, Designs and Patents Act 1988 and the Copyright and Related Rights Regulations 2003, this publication may not be reproduced, stored or transmitted, in any form or by any means, without the prior permission in writing of The Royal Society of Chemistry or the copyright owner, or in the case of reproduction in accordance with the terms of licences issued by the Copyright Licensing Agency in the UK, or in accordance with the terms of the licences issued by the appropriate Reproduction Rights Organization outside the UK. Enquiries concerning reproduction outside the terms stated here should be sent to The Royal Society of Chemistry at the address printed on this page.

The RSC is not responsible for individual opinions expressed in this work.

Published by The Royal Society of Chemistry,
Thomas Graham House, Science Park, Milton Road,
Cambridge CB4 0WF, UK

Registered Charity Number 207890

For further information see our web site at www.rsc.org

Preface

"Now this is not the end. It is not even the beginning of the end. But it is, perhaps, the end of the beginning."

Sir Winston Churchill

Despite the great goals achieved in our era, such as reaching the moon or finding the Higgs particle among others, human health remains fragile and our current therapeutic arsenal is completely insufficient to cure many severe diseases. Drug discovery today is fueled by the urgent need to find effective drugs for many unmet pathologies.

Parkinson's disease, the second most common neurodegenerative disorder, is one of the above mentioned pathologies. Following the death of dopamine-generating cells in the *substantia nigra*, it is characterized by progressive loss of muscle control, which leads to trembling of the limbs and head while at rest, stiffness, slowness, and impaired balance. As symptoms worsen, it may become difficult to walk, talk and complete simple tasks. The first descriptions of Parkinson's disease date back as far as 5000 BC. Around that time, an ancient Indian civilization called the disorder 'Kampavata' and treated it with the seeds of a plant containing therapeutic levels of what is today known as levodopa. The disease is named after the British doctor James Parkinson, who published its first detailed description in *An Essay on the Shaking Palsy* in 1817.

Although more than 5 million people worldwide are affected by Parkinson's disease, currently there is no treatment to cure this mid-brain neurodegenerative pathology. Several therapies are available to delay the onset of motor symptoms, and to ameliorate motor symptoms, thereby extending the patient's quality of life.

Recent research advances in molecular biology and technology have provided multiple credible hypotheses around which therapeutic agents can be developed. This book collects some of the most outstanding examples of new

---

RSC Drug Discovery Series No. 34
Emerging Drugs and Targets for Parkinson's Disease
Edited by Ana Martinez and Carmen Gil
© The Royal Society of Chemistry 2013
Published by the Royal Society of Chemistry, www.rsc.org

drugs currently under pharmaceutical development or new targets in the validation process that will reach the Parkinson's drugs market over the next few years as disease-modifying drugs. These new drugs will be able to provide effective treatment for motor and non-motor symptoms.

We wish to thank all of the contributors to the chapters in this book, firstly for their faith in the project, and we would like also to express our great appreciation to all of them for delivering clear, comprehensive reviews that will inform and enlighten readers on the state-of-the-art in their respective fields of research. We would also like to thank our families and students for their patience when we were immersed in editing, and the staff at the RSC, especially Gwen Jones and Cara Sutton, for their support in bringing the book to completion. It is very much hoped that this book will provide a useful resource to scientists, both in industry and academia, who are looking to find a solution for the many patients worldwide waiting for effective drugs.

Ana Martinez
Carmen Gil
Instituto de Química Médica-CSIC
Madrid, Spain

Contents

Introduction

Chapter 1 Parkinson's Disease: Symptoms, Unmet Needs and New Therapeutic Targets   3
Mónica M. Kurtis and Pablo Martínez-Martín

    1.1 Introduction   3
    1.2 Motor Signs and Symptoms   4
        1.2.1 Bradykinesia   4
        1.2.2 Rigidity   7
        1.2.3 Rest Tremor   7
        1.2.4 Gait Disturbances   7
        1.2.5 Motor Fluctuations and Dyskinesias   8
        1.2.6 Dystonia   8
    1.3 Non-Motor Symptoms   9
        1.3.1 Neuropsychiatric Symptoms   9
        1.3.2 Sleep Disturbances   12
        1.3.3 Dysautonomia   13
        1.3.4 Other Symptoms   15
    1.4 Clinical Parkinson's Disease Subtypes   16
        1.4.1 Empirically Driven Subtypes   17
        1.4.2 Data-Driven Subtypes   17
    1.5 Current Diagnosis   18
    1.6 Current Treatment   18
    1.7 Unmet Needs and New Therapeutic Targets   19
        1.7.1 Symptoms Beyond the Dopamine Scope   19
        1.7.2 Biomarkers   20
        1.7.3 The Cure: Cause-Directed Therapy   21
        1.7.4 Neuroprotection   21
    References   21

Chapter 2		**Molecular Pathogenesis and Pathophysiology of Parkinson's Disease: New Targets for New Therapies** José G. Castaño, Carmen González, José A. Obeso and Manuel Rodriguez	**26**
	2.1	Introduction	26
	2.2	Molecular Pathogenesis of Parkinson's Disease	27
		2.2.1 Genetic Channels	29
		2.2.2 Metabolic Disturbance Channel Toxins	31
		2.2.3 Inflammatory and Immunity Channels	32
		2.2.4 Reactive Oxygen and Nitrogen Species Channels	32
		2.2.5 Channels of Mitochondrial DNA Mutations	32
		2.2.6 Channels of Somatic Nuclear Mutations	33
		2.2.7 Channels of Cell Division Activation	34
		2.2.8 Connecting Channels to Define the Pathogenesis of Parkinson's Disease	34
		2.2.9 Transcriptional Re-Programming and Epigenetic Control	36
	2.3	The Pathophysiology of Parkinson's Disease	38
		2.3.1 Dopaminergic Neurons are Complex	38
		2.3.2 Dopamine Cells are Not the Only ones Involved in PD	40
		2.3.3 New Models of the Basal Ganglia are Necessary to Understand Parkinson's Disease	41
		2.3.4 New Structural Data for the Basal Ganglia Model	42
		2.3.5 Functional Data to be Included in the Basal Ganglia Model	45
	2.4	Targets for New Parkinson's Disease Therapies	45
		Acknowledgements	47
		References	48

L-DOPA and Dopaminergic Agents

Chapter 3		**Dopaminergic Treatments for Parkinson's Disease: Light and Shadows** Nicola Simola	**61**
	3.1	Introduction	61
	3.2	Overview of the Drugs used in Dopamine-Replacement Therapy	63
		3.2.1 L-DOPA	63
		3.2.2 Adjuncts to L-DOPA: COMT and MAO Inhibitors	63
		3.2.3 Dopaminergic Agonists	63

3.3	Effect of Dopamine-Replacement Therapy on the Motor Features of Parkinson's Disease	68
	3.3.1 Motor Impairment	68
	3.3.2 Motor Complications	69
3.4	Effect of Dopamine-Replacement Therapy on the Non-Motor Features of Parkinson's Disease	71
	3.4.1 Non-Motor Symptoms	71
	3.4.2 Addictive-Like Behavior Associated with Dopamine-Replacement Therapy	72
3.5	Effect of Dopamine-Replacement Therapy on Disease Progression	73
3.6	Future Directions in Dopamine-Replacement Therapy	76
	3.6.1 Continuous Drug Delivery	76
	3.6.2 Development of New Drugs	76
3.7	Final Remarks	77
Acknowledgements		77
References		77

Chapter 4 Catechol-O-Methyl-Transferase Inhibitors: Present Problems and Relevance of the New Ones   83
P. Nuno Palma, László E. Kiss and Patrício Soares-da-Silva

4.1	Introduction	83
4.2	The Role of COMT Inhibitors in the Symptomatic Treatment of Parkinson's Disease	84
4.3	Pyrogallol and Catechol Derivatives as COMT Inhibitors	85
4.4	Nitrocatechol COMT Inhibitors	88
4.5	Non-Clinical Pharmacology of COMT Inhibitors	92
4.6	Metabolic Profile of COMT Inhibitors	93
4.7	Human Pharmacology	95
4.8	New Insights into the Mechanism of COMT Inhibition	100
	4.8.1 *In Vitro* Potency	100
	4.8.2 *In Vivo* Duration of Action	103
4.9	Final Remarks	105
Acknowledgements		106
References		106

Chapter 5 Pharmacologic Management of Dopaminergic-Induced Dyskinesias in Parkinson's Disease   110
Mildred D. Gottwald and Michael J. Aminoff

5.1	Introduction	110
5.2	Why Do Dyskinesias Occur?	111

5.3	Scales Used for Assessing Dyskinesias	111
5.4	Medical Therapies	112
	5.4.1 Delaying the Onset of Dyskinesias with Dopamine Agonists	112
	5.4.2 Therapeutic Strategies for Existing Dyskinesias	114
	5.4.3 Parenteral Therapies	117
	5.4.4 New Compounds in Development	119
5.5	Conclusion	122
	References	122

Chapter 6 D_3 Receptor Agonists and Antagonists as Anti-Parkinsonian Therapeutic Agents — 126
Mark Johnson and Aloke Dutta

6.1	Introduction: The D_3 Receptor	126
	6.1.1 D_3 Localization and Distribution in the Brain	127
6.2	D_3 Receptor-Selective Ligands	128
	6.2.1 D_3 Receptor-Selective Agonists	129
	6.2.2 D_3 Receptor-Selective Antagonists	131
6.3	Role of D_3 Receptors in Levodopa-Induced Dyskinesias	134
	6.3.1 D_3 Receptor Modulation in the Treatment of Levodopa-Induced Dyskinesias	136
6.4	Neuroprotective Action of D_3 Receptor-Preferring Agonists	138
	6.4.1 D_3 Receptor-Independent Neuroprotection	139
6.5	Conclusion	142
	Acknowledgements	142
	References	142

Chapter 7 Protein Phosphatases in Parkinson's Disease — 149
Petr Heneberg

7.1	Introduction	149
7.2	Protein Tyrosine Phosphatases	149
	7.2.1 Receptor Protein Tyrosine Phosphatase β/ζ	150
	7.2.2 Protein Tyrosine Phosphatase PTP-PEST	151
	7.2.3 Striatum-Enriched Protein Tyrosine Phosphatase	151
	7.2.4 Src Homology 2 Domain-Containing Phosphatase 2	153

	7.2.5	Phosphatase and Tensin Homolog Deleted on	
		Chromosome 10	154
	7.2.6	Dual-Specificity Protein Phosphatase 1	156
7.3	Protein Serine/Threonine Phosphatases		157
	7.3.1	Protein Phosphatase 1	158
	7.3.2	Protein Phosphatase 2A	160
	7.3.3	Protein Phosphatase 3	163
	7.3.4	PH Domain Leucine-Rich Repeat Protein Phosphatase 1	163
7.4	Future Views		164
Acknowledgments			164
References			164

The α-Synuclein Hypothesis

Chapter 8 Synuclein and Parkinson's Disease: An Update   175
Kurt A. Jellinger

8.1	Introduction: α-Synuclein and Disease	175
8.2	The Synuclein Family	176
8.3	The Biochemistry of α-Synuclein	176
	8.3.1 Structure	176
	8.3.2 Localization and Regulation	177
	8.3.3 Physiological Functions	179
	8.3.4 Genetics	180
8.4	α-Synuclein and Neurodegeneration	183
	8.4.1 α-Synuclein Neurotoxicity and the Oligomer Toxicity Hypothesis	184
	8.4.2 Mitochondrial Involvement in Parkinson's Disease	187
	8.4.3 Lysosomal Dysfunction and Autophagy	188
	8.4.4 Oxidative and Nitrative Injuries	189
	8.4.5 α-Synuclein and Neuroinflammation	189
8.5	α-Synuclein Interaction with Other Proteins	190
8.6	α-Synuclein Spread and Disease Propagation	191
8.7	Neuropathology of Lewy Body Disorders	193
	8.7.1 Sporadic Parkinson's Disease	193
	8.7.2 Dementia with Lewy Bodies and Parkinson's Disease	196
8.8	Animal Models of Parkinson's Disease	197
8.9	α-Synuclein as a Biomarker for Synucleinopathies	197
8.10	Conclusions and Outlook for the Future	198
References		200

Neuroprotective Therapies

Chapter 9   New Approaches to Neuroprotection in Parkinson's Disease   219
María Angeles Mena, Juan Perucho, José Luis López-Sendón and Justo García de Yébenes

    9.1 Introduction 219
    9.2 Pathogenic Mechanisms in Parkinson's Disease due to Genetic Defects 221
        9.2.1 α-Synuclein 221
        9.2.2 PARKIN 223
        9.2.3 Other Genes Involved in Hereditary Parkinsonism 224
    9.3 Temporal Profile of Clinical Features and Pathological Changes Observed in Parkinson's Disease 225
        9.3.1 The Temporal Spectrum on Clinical Findings 225
        9.3.2 The Temporal Pattern of Pathological Changes 228
    9.4 Neuroprotective Therapies: Towards Neuroprotection Based on Pathogenesis in Parkinson's Disease 229
    Acknowledgements 232
    References 232

Chapter 10   Glutamate Receptor Modulators as Emergent Therapeutic Agents in the Treatment of Parkinson's Disease   237
Sylvain Célanire, Benjamin Perry, Robert Lutjens, Sonia Poli and Ian J. Reynolds

    10.1 Introduction 237
        10.1.1 Glutamate Receptors: Nomenclature and Links with Parkinson's Disease 238
        10.1.2 Allosteric Modulators *versus* Orthosteric Ligands 240
        10.1.3 Animal Models of Parkinson's Disease 240
    10.2 Recent Progress of Ionotropic Glutamate Receptor Modulators in Parkinson's Disease 241
        10.2.1 NMDA Receptor Blockers 241
        10.2.2 AMPA Receptor Modulators 244
    10.3 Recent Progress of Metabotropic Glutamate Receptor Modulators in Parkinson's Disease 246
        10.3.1 Group I mGluRs: Focus on mGluR5 Negative Allosteric Modulators 246
        10.3.2 Group II mGluR Modulators 248
        10.3.3 Group III mGluRs: Focus on mGluR4 and mGluR8 Activators 250

	10.4	Clinical Development of Novel Glutamate-Based Therapeutics in Parkinson's Disease	253
		10.4.1 NMDA Receptor Blockers	253
		10.4.2 AMPA Receptor Modulators	254
		10.4.3 mGluR5 Receptor Negative Allosteric Modulators	254
	10.5	Conclusion and Perspectives	255
	Acknowledgements		256
	References		256

Chapter 11 LRRK2 Kinase Inhibitors as New Drugs for Parkinson's Disease?   266
Sandra Schulz, Stefan Göring, Boris Schmidt and Carsten Hopf

11.1	Introduction	266
11.2	Insight into LRRK2 Inhibitor SARs from Structural Biology Studies and Molecular Modeling	267
11.3	Medicinal Chemistry for the Design of Potent and Selective LRRK2 Inhibitors	269
	11.3.1 Non-Selective LRRK2 Inhibitors	269
	11.3.2 Potent and Selective LRRK2 Inhibitors	273
	11.3.3 Examples from Recently Published Patent Applications	277
11.4	The Role of LRRK2 Outside the Brain and Implications for Potential Mechanism-Based Toxicity of LRRK2 Inhibitors as Drugs	277
11.5	Invertebrate and Vertebrate Animal Models for Pharmacological Evaluation of LRRK2 Inhibitors	282
11.6	Pharmacokinetics and Pharmacodynamics of LRRK2 Inhibitors: The Current State-of-the-Art	284
11.7	Will LRRK2 Kinase Inhibitors be Developed into Drugs (for the Treatment of Parkinson's Disease)?	286
	11.7.1 Patient Stratification	287
	11.7.2 Mechanism-Based Toxicity	288
	11.7.3 Utility in Other Therapy Areas Besides Parkinson's Disease	288
References		289

Chapter 12 Phosphodiesterase Inhibitors as a New Therapeutic Approach for the Treatment of Parkinson's Disease   294
Ana Martinez and Carmen Gil

12.1	Introduction	294
12.2	Dopamine and Cyclic Adenosine Monophosphate	295

12.3	Phosphodiesterases and their Role in Dopamine Signaling	296
12.4	Phosphodiesterases as Drug Targets Beyond Dopamine	298
12.5	Phosphodiesterase Inhibitors as New Drugs for Parkinson's Disease	298
	12.5.1 PDE1 Inhibitors	300
	12.5.2 PDE4 Inhibitors	300
	12.5.3 PDE7 Inhibitors	301
	12.5.4 PDE10 Inhibitors	302
12.6	Conclusions	303
Acknowledgements		303
References		303

Chapter 13 5-HT$_{1A}$ Receptors as a Therapeutic Target for Parkinson's Disease — 308
Saki Shimizu and Yukihiro Ohno

13.1	Introduction	308
13.2	5-HT$_{1A}$ Receptors	310
13.3	Role of 5-HT$_{1A}$ Receptors in the Treatment of Parkinson's Disease	312
	13.3.1 Treatment of Parkinsonian Symptoms	312
	13.3.2 Treatment of L-DOPA-Induced Dyskinesia	314
	13.3.3 Treatment of Non-Motor Symptoms in Parkinson's Disease	316
13.4	5-HT$_{1A}$ Receptor Ligands	320
13.5	Conclusion	320
References		322

Chapter 14 Tryptophan Metabolism in Parkinson's Disease: Future Therapeutic Possibilities — 327
Zsófia Majláth and László Vécsei

14.1	Introduction	327
14.2	Tryptophan Metabolism	328
	14.2.1 Serotonin Pathway	328
	14.2.2 Kynurenine Pathway	328
14.3	Pathogenesis of Parkinson's Disease	330
	14.3.1 Some of the Main Aspects of the Pathogenesis of Parkinson's Disease	330
	14.3.2 Altered Tryptophan Metabolism in Parkinson's Disease	331

14.4	Future Therapeutic Possibilities	333
14.5	Conclusions	334
Acknowledgements		335
References		335

Chapter 15 Role of P2X7 Receptor Signaling in the Treatment of Parkinson's Disease and Other Neurodegenerative Disorders   341

Takato Takenouchi, Kazunari Sekiyama, Masayo Fujita, Shuei Sugama, Yoshifumi Iwamaru, Hiroshi Kitani and Makoto Hashimoto

15.1	Introduction	341
15.2	Role of Neuroinflammation in the Progression of Parkinson's Disease	342
	15.2.1 Neuroinflammation in Parkinson's Disease Brains and Animal Models of Parkinson's Disease	342
	15.2.2 Role of IL-1β in Neuroinflammation in the Progression of Parkinson's Disease	343
15.3	Expression in the Central Nervous System and Drugs for the Modulation of P2X7R	345
	15.3.1 P2X7R Expression in the Central Nervous System	345
	15.3.2 The Dual Neuroprotective and Neurotoxic roles of P2X7R	346
	15.3.3 Modulators of P2X7R Function	347
15.4	Altered Expression and Function of P2X7R in Parkinson's and Other Neurodegenerative Conditions	349
15.5	Effects of P2X7R Modulators or Deficiency in Animal Models of Parkinson's and Other Neurodegenerative Diseases	351
	15.5.1 Effects of P2X7R Antagonists or Deficiency	351
	15.5.2 Possible Effects of Other P2X7R Modulators	353
15.6	Conclusion and Perspective	354
Acknowledgements		355
References		355

Neuroregenerative Strategies

Chapter 16 Carotid Body Transplants as a Therapy for Parkinson's Disease — 363
Javier Villadiego, Ana Belén Muñoz-Manchado, Simón Mendez-Ferrer, Juan José Toledo-Aral and José López-Barneo

16.1	Cell Therapy in Parkinson's Disease	363
16.2	Anatomical and Physiological Features of the Carotid Body	364
16.3	Carotid Body Cell Therapy for Parkinson's Disease	366
	16.3.1 Initial Preclinical Studies: The Carotid Body as a Source of Dopamine Cells	366
	16.3.2 Recent Preclinical Studies: The Carotid Body as a Biological Pump Releasing Dopaminotrophic Factors	367
16.4	Clinical Studies of Carotid Body Autotransplantation on Parkinson's Disease Patients	371
16.5	Conclusions and Perspectives	372
	Acknowledgements	373
	References	373

Chapter 17 Stem-Cell-Based Cell-Replacement Therapy in Parkinson's Disease — 376
Jan Tønnesen and Merab Kokaia

17.1	Parkinson's Disease and Cell-Replacement Therapy	376
17.2	Proof-of-Principle: Fetal-Cell-Replacement Therapy	377
17.3	Candidate Stem Cells for Parkinson's Disease Cell-Replacement Therapy	378
	17.3.1 Fetal Neural Stem Cells	378
	17.3.2 Embryonic Stem Cells	379
	17.3.3 Induced Pluripotent Stem Cells	379
	17.3.4 Directly Induced Neurons	382
17.4	Stem Cell Integration in a Host Tissue	382
17.5	Genetically Enhanced Stem Cells	384
17.6	Concluding Remarks and Future Perspectives	384
	References	385

Subject Index — 390

Introduction

CHAPTER 1

Parkinson's Disease: Symptoms, Unmet Needs and New Therapeutic Targets

MÓNICA M. KURTIS*[a] AND
PABLO MARTINEZ-MARTÍN[b,c]

[a] Movement Disorders Unit, Department of Neurology, Hospital Ruber Internacional, Madrid, Spain; [b] Area of Applied Epidemiology, National Centre of Epidemiology and CIBERNED, Carlos III Institute of Health, Madrid, Spain; [c] Alzheimer Disease Research Unit, CIEN Foundation, Carlos III Institute of Health, Alzheimer Center Reina Sofia Foundation, Madrid, Spain
*Email: mkurtis@ruberinternacional.es

1.1 Introduction

Since James Parkinson wrote the first systematic clinical description of six patients in 1817 in his essay titled "Paralysis Agitans",[1] Parkinson's disease (PD) has been considered a motor disorder, consisting of tremor, rigidity and gait difficulties. A few decades later, Jean Martin Charcot characterized the feature of bradykinesia and added other observations not pertaining to the motor domain, consisting of arthropathy, dysautonomia, and pain.[2] In the mid 1950s, pathological changes in the PD mid-brain described as "neuronal degeneration of the *substantia nigra*" were defined by Greenfield and Bosanquet.[3] The delineation of the nigrostriatal pathway in the 1960s and discovery by Arvid Carlsson and colleagues of the direct correlation between striatal dopamine loss and clinical Parkinsonian manifestations were a major breakthrough in the neurosciences and provided the opportunity for the development of effective therapies.[4] In 1967 George Cotzias and others demonstrated the benefits of oral levodopa in patients, paving the way for substitutive dopaminergic treatments.[5,6]

1.2 Motor Signs and Symptoms

Based on these early findings, the classic features that define the Parkinsonian syndrome are: bradykinesia, rigidity, tremor at rest, and gait disturbances (flexed posture, freezing, and loss of postural reflexes). At least two of these signs should be present before the diagnosis of Parkinsonism is put forth.[7] The etiology is widely variable, therefore multiple primary and secondary causes must be considered when evaluating a patient, considering that PD is the most prevalent of the primary causes. In the following section, the well-established PD motor symptoms are enumerated and defined (Table 1.1).

1.2.1 Bradykinesia

The terms akinesia, literally means absence; bradykinesia, meaning slowness; and hypokinesia, meaning decreased amplitude; are all used, often interchangeably, to describe the most prominent phenomena of Parkinsonism. Patients show poverty of automatic movements (*i.e.*, blinking, arm swing) and also present reduced speed when initiating and executing single and repetitive movements with progressive loss of amplitude. Characteristically, there is greater difficulty in moving with self-initiated cues than with externally triggered movements and this abnormal activation and slowness affects most body parts.

Pathophysiology of bradykinesia can generally be explained by the classic model of the basal ganglia–thalamocortical circuitry postulated in the 1980s. In the absence of dopamine, the main output nucleus of the basal

Table 1.1 The motor symptom and sign complex of PD.

Axial symptoms	*Limb symptoms*
Hypomimia	Micrographia
Blepharospasm	Loss of dexterity
Hypophonia	Asymmetric arm swing
Dysarthria	Slow gait
Dysphagia	Rest tremor of the hand or foot
Sialorrhea	Foot dystonia
Chin, lip and tongue tremor	
Vertical eye movement restriction and convergence insufficiency	
Freezing of gait	
Flexed posture	
Loss of postural reflexes	
Camptocormia	

ganglia, the *globus pallidus interna* (GPi) is abnormally active, thus inhibiting the ventroanterior and ventrolateral motor thalamus, and subsequently the primary motor cortex, resulting in slowness. Current findings add complexity to the model, attempting to explain the other features of akinesia. It is hypothesized that the main disturbances in PD are the non-generation of phasic neurons, and the time-locked inhibition of GPi neurons which cannot facilitate recruitment of cortical motor neurons that are appropriately adjusted to produce voluntary movement. The primary motor cortex is also altered and there is a functional uncoupling with premotor areas that is not well understood. The loss of automatic movements in PD is probably related to alterations of basal ganglia projections to the brainstem central pattern generators, with excessive inhibition being the net result.[8]

1.2.1.1 Hypomimia

Bradykinesia affecting the facial muscles results in decreased expression, sometimes called 'poker' face, alluding to card players that do not show any emotion during their game, and can be an initial sign of the disease. This sign can also be seen in depressed patients and the differential diagnosis must be kept in mind. With disease progression, the lips can remain open most of the time and blink rate becomes severely decreased, leading to ocular problems such as dry eye.

1.2.1.2 Hypophonia

Hypophonia, meaning soft voice, is an axial sign that can also be a first complaint and is generally noted by the patient's family and friends. The person with PD is usually unaware that he/she is speaking softly and tends to blame others for being 'hard of hearing'. Some patients complain that their tone of voice has changed and become monotonous (termed "aprosody").

1.2.1.3 Dysarthria

Difficulty in articulating language is a reflection of bradykinesia of the tongue, oral cavity and larynx musculature. Some patients may talk too fast, presenting tachyphemia, others may stutter, due to freezing of speech episodes, and in advanced stages of the disease, patients may develop progressively severe mumbling that can make language unintelligible.

1.2.1.4 Dysphagia

Difficulty swallowing secondary to neurological disease generally affects liquids more than solids. Patients complain of coughing during their meals due to minor choking episodes and with disease progression, dysphagia may be severe and lead to aspiration, causing pulmonary infections such as pneumonia. In order to avoid this, patients must be instructed to maintain proper posture when swallowing and avoid food textures, liquids and volumes they have difficulty managing. Ultimately, gastrostomy may be considered although the danger of saliva aspiration is not avoided.

1.2.1.5 Sialorrhea

Excessive salivation is probably secondary to decreased spontaneous swallowing although in some patients, saliva characteristics can differ from normal (becoming denser), and thus hypersalivation may also be considered a non-motor dysautonomic problem.

1.2.1.6 Eye Movement Abnormalities

In the past two decades, neuro-ophtalmologic symptoms in PD have been objectively measured and thus defined. Horizontal and vertical pursuit can show bradykinesia and decreased amplitude even in the early stages of the disease[9] and as in other body parts, slowness becomes increasingly marked with repetition.[10] It is not infrequent to find restriction of vertical eye movements. The saccade system is also altered, showing characteristically slow and hypometric saccades, and occasionally prolonged latencies.[11,12] Patients often complain of blurred or double vision, secondary to convergence insufficiency as Biousse et al. found in an early untreated PD cohort.[13] This study also showed that, when compared to controls, PD patients declared more local ocular symptoms (irritation, pain, conjunctival redness), eyelid problems (blepharospasm and decreased blinking) and dry eye.[13] Dry eye is probably multi-factorial, secondary to motor disturbances (decreased blinking), and dysautonomic changes of the lacrimal glands.[14]

1.2.1.7 Micrographia

Small handwriting can often be the first symptom noted when PD affects the dominant side of the body. Patients describe that their writing starts out normally but becomes increasingly smaller as they keep writing. Their handwriting can become illegible and their signature can change to the point of misunderstandings with banks and official documents.

1.2.1.8 Slow Gait

Some patients' initial complaint is that they walk more slowly. They describe walking as tremendously effortful, since their legs feel heavy, as if they had weights pulling them down. Often a person close to them will note that they have decreased arm swing, usually asymmetric. Recent studies show that there are marked alterations in the rhythmicity and timing of gait, even in the early stages of the disease when speed can be intact.[15]

1.2.2 Rigidity

Rigidity is defined as increased muscle tone at rest that can be palpated, reduced distension when the limb is passively moved, increased resistance when the limb is stretched, and facilitation of the shortening reaction.[8] Resistance is more noticeable when the limb is passively moved slowly, can manifest as cogwheeling since the limb gives way in a stepwise fashion, and is increased with voluntary movement of other body parts (Froment's maneuver). Flexor muscles are generally affected earlier than the extensors. Rigidity is not explained by the classic model of the Parkinsonian state, where overactivity of the basal ganglia's main output nucleus (GPi) leads to cortical inhibition. Projections to the brainstem and spinal mechanisms probably play an important role as experimental findings suggest that spinal cord motorneurons present a shift towards increased activity in response to peripheral stimulation.[8]

1.2.3 Rest Tremor

Parkinsonian tremor usually involves distal parts of the extremities (called "pill rolling" when the thumb and index are involved) or the lips and chin, and characteristically occurs at rest. About two thirds of patients with PD show the typical rest tremor with a frequency measured by motor neurophysiological testing (electromyography and accelometery) of 4–5 Hz. Some rest tremors re-emerge after a short latency period of a few seconds, thus appearing during some actions such as posture holding. Kinetic and postural tremors can also be seen, but are generally not significant and do not interfere with the patient's activities. The pathophysiological mechanism behind PD tremor is unclear. To date there is no proven model that explains the link between dopamine deficiency and abnormal oscillatory activity in an extensive motor network that involves the basal ganglia, the cerebellum, the thalamus and the motor cortex.[8]

1.2.4 Gait Disturbances

1.2.4.1 Freezing of Gait

When initiating gait or turning, the feet literally become stuck to the ground, so the patient feels he/she cannot take a step. Some patients may present freezing in the initial stages of the disease, although it is rare in the first three years. The problems begin when they want to initiate gait (start hesitation), when turning, in tight spaces, or in doorways. With disease progression, patients develop destination freezing (*i.e.*, stopping a few steps from the chair where they want to sit) and freezing may interrupt gait at any time, even in open spaces. Characteristically patients with freezing do not have trouble with other complex motor programs such as climbing stairs or riding a bicycle. To date, the physiopathology of freezing is not well understood and the best treatment for freezing is based on physiotherapy since most patients benefit from external visual or auditory cues.

1.2.4.2 Flexed Posture

As the disease advances, patients tend to walk with flexion at the neck, elbows, hips and knees, with the forearms placed in front of the body. When extreme, this flexed posture can lead to pronounced kyphoscoliosis.

1.2.4.3 Loss of Postural Reflexes

Balance is tested in the office by a gentle pull backwards on the shoulders by the examiner. In the early stages of the disease, patients may have to take a few steps backwards (up to two is considered normal) in order to regain their balance. As the disease progresses, patients will not be able to recuperate due to loss of postural reflexes and falls become a major problem.

1.2.4.4 Festination

The patient with a festinating gait walks progressively faster and faster, taking shorter and shorter steps as he/she tries to catch up with his/her axial center of gravity. It is a result of stooped posture and altered postural reflexes.

1.2.4.5 Falls

One of the symptoms with highest morbidity in PD is falling, conditioned by multiple factors. Freezing is one of the primary etiologies of falls, as the body moves forward or to the side when turning, but the feet do not follow. Festinating gait can also cause the patient to fall forward as eventually the lower limbs cannot catch up with the forward tilting trunk. Loss of postural reflexes means that any small obstacle or nudge will throw the patient off balance, generally leading to a fall backwards. Cognitive impairment also plays a role in patients' falls, since loss of insight and risk appraisal can be affected.

1.2.5 Motor Fluctuations and Dyskinesias

With disease progression, treated patients may develop motor fluctuations, signifying they present what is termed an "off" state, in which their motor symptoms re-appear as medication benefits disappear, and an "on" state, when medications are effective and symptoms are well controlled. In addition to this, they may develop dyskinesias or involuntary movements that resemble chorea or dancing movements, generally appearing in the areas most affected by parkinsonism and probably secondary to dopamine receptor hypersensitivity.

1.2.6 Dystonia

Abnormal posturing due to sustained muscle contractions can be the first sign of the disease, or it can develop years later as a consequence of dopaminergic treatment, representing a similar phenomenon to dyskinesias. Axial dystonia can affect the eyes, causing blepharospasm; the neck, usually producing antecollis; or the trunk, resulting in camptocormia or stooped posture that increasingly worsens with walking, or Pisa syndrome, as patients lean the trunk to the side. Axial dystonia characteristically normalizes when standing against a wall or lying down. Secondary dystonia of the limbs can be associated with tremor (dystonic tremor), which tends to be faster and more erratic than the typical 4–5 Hz rest tremor, and when affecting the lower extremities can lead to abnormal gait.

1.3 Non-Motor Symptoms

At present, PD is still defined clinically by the presence of two or more of the cardinal motor symptoms described above. However, in the past decade, research has expanded to the prevalent non-motor symptom complex that affects patients during all stages of the disease, even in the premotor phase. Non-motor symptoms including sleep, mood, cognition, pain, and autonomic disorders have been identified as important PD manifestations, which often remain undeclared unless specifically sought.[16,17] It is important to recognize and treat these symptoms since they are key determinants of patients' quality of life. Results of a recent study by our group showed that non-motor symptoms have, as a whole, a greater impact on health-related quality of life (HRQoL) than motor symptoms, and non-motor symptom progression contributes importantly to HRQoL decline in PD patients.[18]

The neurochemical and pathological substrates for most of the non-motor symptoms remain a puzzle. Key dopaminergic areas in the brain (the *substantia nigra pars compacta*, ventral tegmental area, and hypothalamus) project extensively to form four main circuits: the mesocortical, meso-limbic, nigrostriatal, and tuberoinfundibular pathways, which mediate several non-motor symptoms such as cognition, sleep, and pain. Other non-dopaminergic pathways depending on neurotransmitters such as serotonin, norepinephrine and acetylcholine also play a major role.[19] Therefore, the non-motor symptoms that are classified and described in the next section may be modulated by dopaminergic therapy (Table 1.2) while others rarely respond.

1.3.1 Neuropsychiatric Symptoms

1.3.1.1 Depression

Depression is very frequent in PD patients, although prevalent cases range from 2.7 to 70%, depending on the study,[20] possibly due to differing methodologies. In practice, about 40% of patients[21] show signs of depression which can be expressed as sadness, but often presents as irritability, hopelessness, pessimism or worry, more often than guilt or remorse, and contributes to insomnia and general slowness. Studies have suggested that depression is probably another premotor non-motor biomarker that can precede the development of clinical PD as it is currently defined.[22–24] Serotoninergic

Table 1.2 The non-motor symptom complex of Parkinson's disease. Adapted from Chaudhuri et al.[40]

Neuropsychiatric domain	Sleep domain
Depression[a]	Restless legs[a]
Anxiety[a]	Insomnia[ab]
Apathy[a]	Vivid dreaming
Impulse control disorder	Rapid eye movement (REM) sleep behavior disorder
Dopaminergic dysregulation syndrome	
Hallucinations	Excessive daytime sleepiness
Cognitive impairment	
Dementia	
Dysautonomic domain	*Neuro-opthalmological domain*
Urinary: urgency, frequency and nocturia[a]	Dry eye (xerostomia)[b]
Sexual dysfunction[a]	Blurred vision[b]
Gastrointestinal: sialorrhea[b], dysphagia[b], constipation[a], bloating	
Orthostatic hypotension	
Sweating	
Seborrhea	
Other	
Pain[a]	
Fatigue[a]	
Olfactory disturbance	
Neuro-opthalmologic disturbances	
Weight changes	

[a]Some modalities of these symptoms may respond to dopaminergic treatment.[19]
[b]Mixed symptoms, may also correlate with motor abnormalities.

neurotransmission disturbances as well as meso-limbic noradrenergic and dopaminergic mechanism alterations are a primary part of the disease process.[25] It is vital to recognize and address depression since repeated studies have shown that it is the most important symptom in determining patient HRQoL.

1.3.1.2 Anxiety

The prevalence of anxiety lies between 20 and 49% of patients.[26] Symptoms include apprehensiveness, nervousness, irritability, insomnia and feelings of impending disaster, palpitations, and hyperventilation that can lead to panic attacks. Anxiety often co-exists with depression. Both these symptoms can fluctuate with the motor state, appearing in the wearing-off period, thus reflecting a partial response to levodopa, or remain a constant underlying problem independent of the dopaminergic state.

1.3.1.3 Apathy

Apathy is a generalized lack of motivation accounting for a decrease in goal-orientated activities, loss of interest, and blunted emotional experience. It is seen in up to 60% of patients with PD[26] and can occur as the sole major neuropsychiatric manifestation of the disease, although it frequently co-exists with depression, dementia, or executive dysfunction. It can be distinguished from depression, as patients are hardly bothered by the symptom and it is generally the caregiver who reports the problem. Like fatigue, it is possibly related to disturbances in the areas that mediate goal-directed behavior and reward.

1.3.1.4 Impulse Control Disorders

Impulse control disorders such as hypersexuality, compulsive eating, pathological gambling, compulsive shopping, punding and dopamine dysregulation syndrome have been described in patients with PD and are related to dopaminergic medication. Large cross-sectional and case control studies show a prevalence of 13.6% and a recent epidemiological study including more than 3000 patients demonstrated a clear association with dopamine agonist use and a weaker association with higher levodopa dose. It also showed that increased sexuality and pathological gambling was more frequent in men, while compulsive shopping and eating was more prevalent in women.[27] In general, younger patients are more at risk of developing these disorders, which may be associated with depression, anxiety, novelty seeking, impulsivity, enhanced risk preference and obsessive symptoms. The pathophysiological explanation may be that in PD, ventral striatal dopamine is preserved and dopaminergic treatment used to treat the dorsal motor striatal deficiencies result in 'overdosing' the ventral *striatum* which projects to the cognitive and meso-limbic pathways. In this way, impulse control disorders may be the non-motor correlate of motor dyskinesias.[28]

1.3.1.5 Hallucinations and Psychosis

Hallucinations can be visual, auditory, olfactory or tactile. The pathophysiology is unclear and it has been suggested that: (1) brainstem nuclei degeneration (pedunculopontine nucleus, locus ceruleus and raphe nuclei) is involved, since rapid eye movement (REM) sleep behavior disorder is a possible risk factor;[29] (2) cortical Lewy body pathology is causative since patients with cognitive impairment are at risk for developing hallucinations, particularly with dopaminergic treatment; (3) decreased visual acuity plays a role; and (4) there may be a genetic predisposition as demonstrated by the existence of polymorphism in the cholecystokinin system.[30] As the disease progresses, delusions can develop, ranging from unfounded jealousy to paranoid ideation involving caregivers and family members. Delirium and confusional states are often catalyzed by concurrent systemic disease such as infections, or dopaminergic medication increases.

1.3.1.6 Cognitive Impairment

Cognitive difficulty in PD is characterized by a dysexecutive syndrome, defined by disturbance of strategic planning and problem solving, maintaining and shifting attention, and behavioral regulation which operates the ability to initiate, execute, inhibit, and monitor a sequence of actions, in addition to impairments in visuospatial and working memory. The depletion of dopamine in the nigrostriatal and mesocortical circuits affecting associative areas probably plays a role in early executive problems. The relative contribution of these two systems may vary during the course of the illness and depends on the nature of the specific executive process under investigation.[8]

1.3.1.7 Dementia

The prevalence of dementia has been estimated to be about 40%[31] but some studies measure it to be as high as 78%,[32] depending on disease duration. The classic pathological substrate of dementia in PD is the presence of cortical Lewy bodies and Lewy neurites although current data suggests that about 25% of patients have associated Alzheimer's disease pathology (senile plaques and neurofibrillary tangles)[33] and vascular pathology also plays a role. Based on the pathoanatomical PD model suggested by Braak, α-synuclein deposition in structures relevant to cognition progress in an ascending pattern from the central subnucleus of the amygdala to the anteromedial temporal mesocortex, followed by the insular and cingulate mesocortex and finally reach the neocortex, affecting association and primary fields in stages V and VI.[34] Cholinergic cell loss in the nucleus basalis of Meynert also plays a major role in PD and forms the basis of cholinergic treatment for dementia in the disease.

1.3.2 Sleep Disturbances

Almost two thirds of PD patients report some type of sleep disturbance. From a clinical point of view, they can be classified as parasomnias, insomnia, restless legs and excessive daytime sleepiness.

1.3.2.1 REM Sleep Behavior Disorder

A type of parasomnia that is characterized by loss of skeletal muscle atonia during REM sleep, thus enabling patients to act out their dreams. Bed partners report vocalizations (talking, laughing, shouting) and dramatic movements (leg or arm thrashing, falling out of bed, violent assaults) that can be dangerous to the patient or partner. REM sleep behavior disorder (RBD) may be a premotor symptom in up to 40% of PD patients as demonstrated by longitudinal data.[35] The pathophysiology remains unclear, however, degeneration of lower brainstem nuclei, such as the pedunculopontine and peri-ceruleal nuclei, which have connections with the dopaminergic ventral tegmental area in the mid-brain, are thought to be key areas related to the origin of this disorder.[19]

1.3.2.2 Vivid Dreams

Contrary to RBD, where the patient is asleep and undisturbed and it is the patient's bed partner that points out the symptoms, vivid dreams are bothersome to the patient. They may wake up in anguish or actually fear going to sleep because of the vivid nightmares. Classically thought of as a prodrome that progresses to hallucinations and cognitive impairment, a recent 10-year prospective longitudinal study challenges this view, showing that vivid dreaming did not predict the development of hallucinations in their cohort.[36]

1.3.2.3 Insomnia

Patients with PD frequently complain of insomnia, which can present as inability to fall asleep and/or maintain sleep due to numerous nighttime awakenings. The main etiology of insomnia in PD is probably due to truly altered circadian rhythms by the disease itself, however, other influencing factors must be considered including restless legs, periodic leg movements, mood disturbances, re-emergence of akinesia during the night, nocturia and medication side-effects.

1.3.2.4 Restless Legs Syndrome

Restless legs syndrome (RLS) is defined by sensory symptoms in the lower limbs described as uncomfortable paresthesias that appear when lying down or sitting, follow a circadian pattern (usually more prominent at sundown or bedtime) and are alleviated by voluntary movement. It can be a cause of insomnia as the person seeks relief by getting up and walking. The disease can spread to the arms (called "restless limbs") and to earlier parts of the day, beginning earlier and earlier. It occurs in up to 20% of patients with PD, having a two-fold increase in prevalence when compared to controls.[37] Frequently it is associated with periodic leg movements which consist of brief 1–2 second jerks of one or both legs producing dorsiflexion of the toe or foot. These occur in 20 second intervals for periods of minutes or hours and may fragment sleep.

1.3.2.5 Excessive Daytime Sleepiness

Involuntary dozing affects up to 50% of patients and may be a preclinical marker of the disease.[38,39] The etiology of excessive daytime somnolence in PD is multi-factorial, mediated by the disease process, medications, other sleep disorders, and possibly dysautonomia and cognitive impairment.

1.3.3 Dysautonomia

The pathophysiology underlying dysautonomic disorders affecting the genitourinary, gastrointestinal, endocrine and cardiac domains is complex and probably involves degeneration and dysfunction of the dorsal vagal nucleus, *nucleus ambigus*, and other medullary centers which exert differential control on the sympathetic and parasympathetic preganglionic neurons *via* descending pathways.[40] Although problematic dysautonomia usually develops in late stages of the disease, cardiac ^{123}I-*meta*-iodobenzylguanidine (MIBG) imaging challenges this view by showing early cardiac sympathetic denervation in PD.[41]

1.3.3.1 Urinary Bladder Symptoms

Urinary frequency, urgency and nocturia are the most prevalent dysautonomic symptoms in PD. Detrusor hyperactivity caused by basal ganglia dysfunction possibly due to a combination of underactive D_1 stimulation (normally inhibits the micturion reflex) and exacerbation of D_2 stimulation (activates micturion), leads to urinary urgency and frequency.[19] Premature uninhibited bladder contractions and severe motor difficulties when having the urge to micturate or retention with overflow can lead to urinary incontinence. In men, prostate enlargement causing obstruction should be evaluated and followed. True neurogenic incontinence is not characteristic of PD and when present should be investigated further, keeping in mind the differential diagnosis of other Parkinsonian syndromes (including multiple-system atrophy) and that voiding dysfunction in PD is correlated with disease severity, not with age or gender.[7]

1.3.3.2 Sexual Dysfunction

Sexual problems are a common complaint of both men and women with PD when specifically asked by their physician. However, studies are scarce and thus frequency and severity poorly understood. The problem is probably multi-factorial, depending on motor symptoms, medication side-effects, mood disorders, and dysautonomia. Dysautonomia may manifest as erectile dysfunction in men and reduced genital sensitivity and lubrication and difficulties reaching orgasm in women.[42]

1.3.3.3 Gastrointestinal Symptoms

Problems can be found along the entire gastrointestinal tract, from the oral masticatory phase to defecation. As discussed above, dysphagia can be problematic in advanced diseases and drooling is socially handicapping. Constipation is by far the most common problem, defined as less than three bowel movements per week, and may be secondary to Lewy body degeneration of the parasympathetic myenteric plexus as well as of central autonomic nuclei. Other factors that may aggravate constipation include dopaminergic drugs, immobility, and reduced fluid and fiber intake. Sometimes constipation is associated with effortful voiding, which may be a consequence of bradykinesia of the pelvic floor striated muscles and sphincter. Epidemiological data suggest that it may be a premotor symptom preceding the disease by many years. In a cohort including 7000 men followed for 24 years, results showed that those with initial constipation had a three-fold risk of developing PD after a mean interval of 10 years.[43] Thus, current longitudinal prospective cohort studies are evaluating whether this may be an early non-motor biomarker. Parasympathetic failure may also cause delayed esophageal and gastric emptying as well as prolonged colonic transit leading to gastric reflux and a sense of bloating, nausea and indigestion.

1.3.3.4 Orthostatic Hypotension

Orthostatic hypotension is defined as a symptomatic drop of 20 mmHg in systolic or 10 mmHg in diastolic blood pressure.[44] Patients report lightheadedness when moving from lying down or sitting to a standing position. Rarely do they syncope, but presyncopes can lead to falls. Often their complaints are non-specific as they report: dizziness, unsteadiness, "having their head in the clouds", being unable to think clearly or even headache or neck tightness (called "coat hanger pain"). Orthostatic hypotension may be caused by sympathetic failure to produce reactive vasoconstriction and handle intravascular volume adequately.

1.3.3.5 Excessive Sweating

Sudden drenching sweats can be very bothersome to the patient and also socially inhibiting. Sometimes sweating can fluctuate with the dopaminergic state, occurring in the "off" state, but can also appear in the on state with dyskinesias.[7]

1.3.3.6 Seborrhea

Excessive sebum accumulation may lead to greasy facial skin and is probably secondary to facial immobility more than sebum overproduction. When severe, it can lead to seborrheic dermatitis, dandruff and blepharitis.

1.3.4 Other Symptoms

1.3.4.1 Pain

Another symptom that greatly affects quality of life and is not uncommon in PD patients is pain. Pain can be primary and present as paresthesias or dysesthesias that involve the oral and genital areas, the lower limbs, or appear as visceral pain. Pain can also be secondary to akinesia and rigidity as seen in frozen shoulder syndrome or secondary to sustained dystonic posturing and therefore fluctuate with the dopaminergic state.

1.3.4.2 Fatigue

Fatigue is commonly reported by patients with PD. The etiology of fatigue in PD is unclear and may be multi-factorial and related to sleep problems, mood disorders and the primary disease process.

1.3.4.3 Olfactory Disturbance

Eventually, a decreased sense of smell affects up to 90% of patients with PD and is currently also being studied for its potential as a premotor clinical biomarker.[45] To date, epidemiological studies corroborate Braak's pathoanatomical theory, where stage I affects the olfactory bulb and anterior olfactory nucleus, thus leading to early olfactory disturbance long before nigral degeneration.[34]

1.3.4.4 Neuro-opthalmological Disturbances

Contrast sensitivity and color discrimination are also established signs of PD that worsen as the disease advances[46] and improve with dopaminergic treatment. The retina is rich in D_1 and D_2 receptors which are probably affected by systemic dopaminergic progressive loss. As described earlier, patients with visual impairment may be more at risk for developing visual hallucinations.[47]

1.3.4.5 Weight Changes

As seen in other neurodegenerative diseases, weight loss in PD patients is associated with a loss of fat mass that progresses as the disease advances. It may precede the formal diagnosis by years. The etiology of weight changes is probably multi-factorial and secondary to increased metabolic demand from motor symptoms, decreased caloric intake, medication side-effects, and more importantly, due to pathophysiological mechanisms underlying the disease itself that are not well understood. One of the recognized side-effects of deep brain stimulation surgery is weight gain. Recently, it has been postulated that noradrenergic transmission between the locus ceruleus, hypothalamus and subthalamic nucleus may play a key role in modulating homeostatic control centers and energy metabolism. Furthermore, subthalamic nucleus deep brain stimulation may benefit this dysfunction, directly or indirectly, leading to normalization of nutrient metabolism. Post surgical weight gain would be explained by the patients' tendency to maintain presurgery caloric intake with a normalized energy expenditure.[48]

1.4 Clinical Parkinson's Disease Subtypes

PD is a conspicuously variable condition, which has led to the possibly more appropriate term "Parkinson's diseases".[49] This evolving concept reflects the heterogeneity of symptom profiles, severity and progression encountered in the clinic, and is supported by the expanding list of genetic causes and diverse neuropathologic substrates.

The heterogeneous pathophysiology underlying the disease, compensatory mechanisms, therapeutic effects and complications probably explain clinical diversity. The search for clinical subtype identification is justified, since homogeneous patient groups likely reflect clinical, pathological, and genetic coherence, and thus may enhance our understanding of the disease process.[50]

1.4.1 Empirically Driven Subtypes

The classic method for patient classification is based on preset clinical criteria, such as motor symptomatology or age, and further analysis of the similarities within the group and differences between groups. Motorically, patients can be subdivided into phenotypes based on their most salient features and are thus described as: tremor predominant, rigid akinetic, mixed, or postural-instability-gait-disorder (PIGD). It is well established that the tremor predominant phenotype is generally more benign since these patients progress slower than the rigid–akinetic forms and have relatively less balance and cognition problems. The motor classification subtypes have been extensively examined and pathological and imaging data strongly support biological differences between the four phenotypes with respect to nigral cell loss, pallidal and striatal dopamine levels, striatal dopamine transporter binding and cerebral blood flow patterns,[51] thus feeding the debate about the existence of different types of "Parkinson's diseases". Distinctions according to age of onset can also be clinically helpful as young onset patients, defined as <45 years old, tend to have a slowly progressive disease, early motor fluctuations and less cognitive problems; while elderly patients tend to have more rigid–akinetic phenotypes with balance problems earlier on,[52] which may be related to non-dopaminergic aging-related processes.[53]

1.4.2 Data-Driven Subtypes

Another approach to classifying patients that may be more objective and rigorous is through a data-driven approach and cluster analysis which avoids the use of predefined empirical criteria (such as age of onset or predominant motor symptom). Using this method, a recent cross-sectional study examined two longitudinal PD cohorts (PROPARK and ELEP). The study identified four prevalent clusters or subpopulations at mean disease duration of 8–10 years. Cluster 1 was characterized by overall mild severity in all clinical domains and patients were relatively young. Cluster 2 had severe motor complications, important sleep and depressive symptoms and patients had the youngest age of onset. Cluster 3 was made up of patients with medium severity across all domains who were relatively old and had an older age at onset. Cluster 4 included patients who were severely affected in most domains (including non-dopaminergic domains), with prominent motor complications, and who were older.[50] Other cluster analysis studies have found a young onset subtype, a rapid disease progression subtype, a tremor dominant subtype and a non-tremor dominant subtype.[51] One of the advances of cluster subtyping classifications is that most of them have included a wide range of non-motor symptoms and have therefore contributed to a more holistic view of the disease. However, further investigation demonstrating pathological and metabolic differences between these clusters is necessary before applying them in clinical practice.

1.5 Current Diagnosis

To date, the diagnosis of PD is clinical, based on the cardinal motor features described above, with the presence of two of them being suggestive of parkinsonism. A thorough and meticulous clinical history is necessary to detect exposure to neuroleptics and generally basic blood tests (may include ceruloplasmin and thyroid hormones) and brain imaging are sufficient to rule out secondary forms of parkinsonism.

The differential diagnosis of PD and the other primary Parkinsonian syndromes, including multiple-system atrophy (MSA), progressive supranuclear palsy (PSP), cortico–basal degeneration (CBD), Lewy body dementia and frontotemporal dementia, can be challenging and is also based on clinical criteria. Brain magnetic resonance imaging (MRI) can sometimes be helpful since the non-PD entities can show characteristic changes (*i.e.*, the hot-cross-bun sign in MSA, the humming bird sign in PSP). Single-photon emission computerized tomography (SPECT) scans marked with radioactive ioflupane which binds to the dopaminergic transporter can help in distinguishing PD from other tremoric entities or drug-induced parkinsonism. Cardiac MIBG measurements can be useful in separating patients with MSA from PD,[54] as can ^{123}I-iodobenzamide SPECT scans and dysautonomic testing. Recent investigations with ultrasound have repeatedly demonstrated characteristic hyperechogenicity of the *substantia nigra* in PD patients, a finding that is not found in the other Parkinsonian syndromes.[55] Ongoing research with PET scans measuring glucose activity has shown promising results in detecting specific activity network patterns for Lewy body diseases including PD, MSA, PSP and CBD.[56] However, although these diagnostic tools may be helpful in a particular clinical context and for research purposes, their diagnostic accuracy is insufficient to recommend their generalized use.

1.6 Current Treatment

PD is, to this day, the only neurodegenerative disease with an effective treatment. For motor symptoms, dopaminergic therapy is uncontestedly beneficial for bradykinesia, rigidity and tremor. More than five decades after its discovery, levodopa remains the most potent, efficacious and well-tolerated treatment available. In addition, the development of dopaminergic agonists has offered an alternative or complementary substitutive treatment with increasingly longer half-life and easier dosing regimen. The advantage of agonist therapy over levodopa is that it reduces the risk of developing dyskinesias when used during the initial five years since disease onset.[57] This benefit is not sustained at longer follow-up however.[58] Add-on monaminoxidase inhibitors (MAOI-B) or catechol-*O*-methyl-transferase (COMT) inhibitors have demonstrated efficacy in prolonging levodopa half-life and therefore reducing motor fluctuations and "off" time.[59] Amantadine is the only oral treatment that has demonstrated efficacy in reducing dyskinesias.[60]

For the non-motor symptoms, few therapies have demonstrated efficacy in double-blind, randomized controlled studies.[61] Clozapine is useful in controlling psychosis,[62] rivastigmine is beneficial for dementia,[63] pramipexole has demonstrated a small but significant improvement of depression,[64] as have venlafaxine and paroxetine,[65] and long-acting dopaminergic therapy (for example, rotigotine) can help improve sleep and morning akinesia.[66]

For problematic motor fluctuations and dyskinesias that cannot be managed with oral medication, the recently developed continuous infusion therapies have proven effectiveness, although they are not free of side-effects. Continuous apomorphine subcutaneous infusion reduces "off" time by 60%, dyskinesias by 35% and oral medication by 47%.[67] The most common secondary effects are local, as subcutaneous nodules can develop at the injection site. The most problematic adverse effects are sleepiness, confusion and hallucinations. Continuous levodopa infusion directly into the duodenum has demonstrated significant decrease in motor fluctuations and improved quality of life. The most common adverse events are mechanical, due to pump and catheter obstructions, or tube displacement.[68] This therapy is probably effective for some of the dopamine-mediated non-motor symptoms.[69] Finally, stereotactic surgery (lesion or deep brain stimulation) of the subthalamic nucleus and the GPi has demonstrated benefit for all dopamine-responsive symptoms, and has the additional advantages of also being very effective in controlling motor fluctuations and tremors.[70–72]

1.7 Unmet Needs and New Therapeutic Targets

Despite the availability of efficacious treatments for motor dysfunction, there are basic aspects of PD still waiting for a definitive solution. The most intimate cause and initial pathophysiological mechanisms of the disorder remain unknown for the highest proportion of patients (the non-genetically bound disease). Current therapies are unable to stop progression of neurodegeneration and, in addition to neuroprotection, the restoration of structures and functions seems to be a far-off achievement. In medicine, specific and accurate diagnosis is always needed, but in clinical practice and for PD, only the passing of time can result in certainty of the diagnosis. Finally, management of the disease's manifestations and complications needs to be improved urgently. The long-term condition of PD patients is far from being acceptable. Complex advanced therapies only show partial palliative success with non-negligible morbidity. Evidence on the efficacy of treatments for many of the non-motor symptoms is lacking, but the severity and accumulation of motor and non-motor manifestations in advanced stages makes life very difficult for patients and caregivers.

1.7.1 Symptoms Beyond the Dopamine Scope

As discussed earlier, the currently available medical treatments for PD are based on dopaminergic substitutive therapy, which is highly effective for most motor symptoms and may improve a few of the non-motor symptoms. The therapeutic challenge faced today is that the predominant problems that affect quality of life, disability and mortality in PD are not responsive to levodopa therapy. The most prevalent motor symptoms in late-stage PD are axial, such as dysphagia, dysarthria, flexed posture, freezing of gait, and loss of postural reflexes, and these barely respond to oral medication, continuous infusions, or surgery.[73] As Hely and colleagues demonstrated in their Australian PD cohort, after 15 years of disease diagnosis, the most prevalent motor symptoms were falls 81%, dysphagia 50% and severe dysarthria 27%), none of which benefit from dopamine.[74]

Most of the non-motor symptoms also lack effective treatments, even though they can be the main determinants of quality of life at any stage of the disease. In early stages, mood and sleep disturbances may predominate while in late stages of the disease, the most frequent non-motor problems are in the neuropsychiatric and dysautonomic domains as shown in the Australian study (cognitive decline 84%, daytime sleepiness 79%, depression 50%, dementia 48%, urinary incontinence 41%, symptomatic postural hypotension 35% and hallucinations 21%).[74] These symptoms can be only partially mitigated by current treatments. The lack of randomized controlled clinical trials targeting the non-motor symptoms in the PD population is well recognized.[61] Further efforts are needed to provide evidence-based classification of the non-motor symptoms according to their response to levodopa, in order to ameliorate their management and better define therapeutic needs.

It is clearly manifest that the non-dopamine-dependent symptoms warrant further investigation and should be the object of further investigation. To facilitate this, a new 'working' pathophysiology PD model that attempts to explain all the possible motor and non-motor manifestations in early and later stages of the disease is necessary. It must incorporate the deficit of other neurotransmitters besides dopamine, such as serotonin, noradrenaline, acetylcholine, and the role of compensatory mechanisms in order to provide a more holistic therapeutic approach.[8]

1.7.2 Biomarkers

The development of biomarkers must address the need for trait markers that can evaluate the risk of disease in presymptomatic individuals, state markers that may assess the presence of established disease (even if preclinical) and rate markers that are capable of evaluating the progression of the disease and response to treatment.[75] The characterization of patient populations by precise and reliable clinical, biological and neuroimaging instruments, and the use of biomarkers to detect small changes in disease progression, would decrease the number of patients needed for intervention clinical trials and would thus represent an advancement in developing new therapies. Pathological findings would also be very informative if they could be correlated with clearly defined populations in all disease stages.

1.7.3 The Cure: Cause-Directed Therapy

PD therapy directed to the etiology or etiologies of the disease is a conspicuously unmet need. For putative disease-modifying therapies to be developed, investigative efforts should be directed towards: (1) characterization of the pathogenic mechanisms in one or more forms of typical PD; (2) specific therapeutic target validation in preclinical models in order to test new drugs; and (3) differentiation of typical PD clinical subtypes based on objective biomarkers. The process of stratification using biological markers plays a crucial role in matching a drug with its target in the appropriate patient cohort and this could have a profound effect on the success rate of cause-directed therapeutic trials.[76]

1.7.4 Neuroprotection

Currently, by the time motor symptoms develop and a diagnosis of PD is postulated, about 70–80% of dopaminergic nigral cells have degenerated. To date, none of the proposed neuroprotective molecules have shown efficacy, possibly due to late intervention. However, ongoing multi-center trials are in progress including co-enzyme Q10, inosine, isradipine and trophic therapies *via* viral vectors or by direct application.[75] In the future, with the aid of premotor biomarkers, it may be possible to identify patients in the earliest stages of neurodegeneration, and this may represent a crucial advancement in the quest for developing effective neuroprotective therapy.

References

1. J. Parkinson, *Medical Classics*, 1817, **10**, 964.
2. J. Charcot, in *Lectures on Diseases of the Nervous System*, H. C. Lea, Philadelphia, 1879.
3. J. G. Greenfield and F. D. Bosanquet, *J. Neurol. Neurosurg. Psychiatry*, 1953, **16**, 213.
4. A. Carlsson, *Pharmacol Rev.*, 1959, **11**, 300.
5. W. Birkmayer and O. Hornikewicz, *Arch. Psychiatr. Nervenkr.*, 1962, **203**, 560.
6. G. C. Cotzias, M. H. Van Woert and L. M. Schiffer, *N. Engl. J. Med.*, 1967, **276**, 374.
7. S. Fahn and J. Jankovic, *Principles and Practice of Movement Disorders*, Elsevier, Philadelphia, 2007.
8. M. C. Rodriguez-Oroz, M. Jahanshahi, P. Krack, I. Litvan, R. Macias, E. Bezard and J. A. Obeso, *Lancet Neurol.*, 2009, **8**, 1128.
9. S. Marino, E. Sessa, G. Di Lorenzo, P. Lanzafame, G. Scullica, A. Bramanti, F. La Rosa, G. Iannizzotto, P. Bramanti and P. Di Bella, *Eur. Neurol.*, 2007, **58**, 193.
10. G. U. Lekwuwa, G. R. Barnes, C. J. Collins and P. Limousin, *J. Neurol. Neurosurg. Psychiatry*, 1999, **66**, 746.

11. D. Zee, in *Walsh and Hoyt's Clinical Neuro-ophthalmology*, N. N. Miller, Williams & Wilkins, Baltimore, 5th edn, 1985, p. 1316.
12. T. J. Crawford, L. Henderson and C. Kennard, *Brain*, 1989, **112**(6), 1573.
13. V. Biousse, B. C. Skibell, R. L. Watts, D. N. Loupe, C. Drews-Botsch and N. J. Newman, *Neurology*, 2004, **62**, 177.
14. C. Tamer, I. M. Melek, T. Duman and H. Oksuz, *Ophthalmology*, 2005, **112**, 1795.
15. R. Baltadjieva, N. Giladi, L. Gruendlinger, C. Peretz and J. M. Hausdorff, *Eur J. Neurosci.*, 2006, **24**, 1815.
16. L. M. Shulman, R. L. Taback, A. A. Rabinstein and W. J. Weiner, *Parkinsonism Relat. Disord.*, 2002, **8**, 193.
17. K. R. Chaudhuri, C. Prieto-Jurcynska, Y. Naidu, T. Mitra, B. Frades-Payo, S. Tluk, A. Ruessmann, P. Odin, G. Macphee, F. Stocchi, W. Ondo, K. Sethi, A. H. Schapira, J. C. Martinez Castrillo and P. Martinez-Martin, *Mov. Disord.*, 2010, **25**, 704.
18. P. Martinez-Martin, C. Rodriguez-Blazquez, M. M. Kurtis and K. R. Chaudhuri, *Mov. Disord.*, 2011, **26**, 399.
19. K. R. Chaudhuri and A. H. Schapira, *Lancet Neurol.*, 2009, **8**, 464.
20. D. J. Burn, *Mov. Disord.*, 2002, **17**, 445.
21. J. S. Reijnders, U. Ehrt, W. E. Weber, D. Aarsland and A. F. Leentjens, *Mov. Disord.*, 2008, **23**, 183.
22. E. C. Lauterbach, A. Freeman and R. L. Vogel, *J. Neuropsychiatry Clin. Neurosci.*, 2004, **16**, 29.
23. F. M. Nilsson, L. V. Kessing and T. G. Bolwig, *Acta Psychiatr. Scand.*, 2001, **104**, 380.
24. A. G. Schuurman, M. van den Akker, K. T. Ensinck, J. F. Metsemakers, J. A. Knottnerus, A. F. Leentjens and F. Buntinx, *Neurology*, 2002, **58**, 1501.
25. P. Remy, M. Doder, A. Lees, N. Turjanski and D. Brooks, *Brain*, 2005, **128**, 1314.
26. D. A. Gallagher and A. Schrag, *Neurobiol. Dis.*, 2012, **46**, 581.
27. D. Weintraub, J. Koester, M. N. Potenza, A. D. Siderowf, M. Stacy, V. Voon, J. Whetteckey, G. R. Wunderlich and A. E. Lang, *Arch. Neurol.*, 2010, **67**, 589.
28. V. Voon, A. R. Mehta and M. Hallett, *Curr. Opin. Neurol.*, 2011, **24**, 324.
29. M. Onofrj, A. Thomas, G. D'Andreamatteo, D. Iacono, A. L. Luciano, A. Di Rollo, R. Di Mascio, E. Ballone and A. Di Iorio, *Neurol. Sci.*, 2002, **23**(2), S91.
30. J. G. Goldman, C. G. Goetz, E. Berry-Kravis, S. Leurgans and L. Zhou, *Arch. Neurol.*, 2004, **61**, 1280.
31. M. Emre, *Lancet Neurol.*, 2003, **2**, 229.
32. D. Aarsland, K. Andersen, J. P. Larsen, A. Lolk and P. Kragh-Sorensen, *Arch. Neurol.*, 2003, **60**, 387.
33. D. J. Irwin, M. T. White, J. B. Toledo, S. X. Xie, J. L. Robinson, V. Van Deerlin, V. M. Lee, J. B. Leverenz, T. J. Montine, J. E. Duda, H. I. Hurtig and J. Q. Trojanowski, *Ann. Neurol.*, 2012, **72**, 587.

34. H. Braak, K. Del Tredici, U. Rub, R. A. de Vos, E. N. Jansen Steur and E. Braak, *Neurobiol. Aging*, 2003, **24**, 197.
35. C. H. Schenck, S. R. Bundlie and M. W. Mahowald, *Neurology*, 1996, **46**, 388.
36. C. G. Goetz, B. Ouyang, A. Negron and G. T. Stebbins, *Neurology*, 2010, **75**, 1773.
37. W. G. Ondo, K. D. Vuong and J. Jankovic, *Arch. Neurol.*, 2002, **59**, 421.
38. D. B. Rye and J. Jankovic, *Neurology*, 2002, **58**, 341.
39. D. Garcia-Borreguero, O. Larrosa and M. Bravo, *Sleep Med. Rev.*, 2003, **7**, 115.
40. K. R. Chaudhuri, D. G. Healy and A. H. Schapira, *Lancet Neurol.*, 2006, **5**, 235.
41. D. S. Goldstein, C. Holmes, S. T. Li, S. Bruce, L. V. Metman and R. O. Cannon, III, *Ann. Intern. Med.*, 2000, **133**, 338.
42. G. Meco, A. Rubino, N. Caravona and M. Valente, *Parkinsonism Relat. Disord.*, 2008, **14**, 451.
43. R. D. Abbott, H. Petrovitch, L. R. White, K. H. Masaki, C. M. Tanner, J. D. Curb, A. Grandinetti, P. L. Blanchette, J. S. Popper and G. W. Ross, *Neurology*, 2001, **57**, 456.
44. H. Lahrmann, P. Cortelli, M. Hilz, C. J. Mathias, W. Struhal and M. Tassinari, *Eur J. Neurol.*, 2006, **13**, 930.
45. M. M. Ponsen, D. Stoffers, J. Booij, B. L. van Eck-Smit, E. Wolters and H. W. Berendse, *Ann. Neurol.*, 2004, **56**, 173.
46. N. J. Diederich, R. Raman, S. Leurgans and C. G. Goetz, *Arch. Neurol.*, 2002, **59**, 1249.
47. N. J. Diederich, C. G. Goetz, R. Raman, E. J. Pappert, S. Leurgans and V. Piery, *Clin. Neuropharmacol.*, 1998, **21**, 289.
48. J. Guimaraes, E. Moura, M. A. Vieira-Coelho and C. Garrett, *Mov. Disord.*, 2012, **27**, 1078.
49. W. R. Galpern and A. E. Lang, *Ann. Neurol.*, 2006, **59**, 449.
50. S. M. van Rooden, F. Colas, P. Martinez-Martin, M. Visser, D. Verbaan, J. Marinus, R. K. Chaudhuri, J. N. Kok and J. J. van Hilten, *Mov. Disord.*, 2011, **26**, 51.
51. C. Marras and A. Lang, *J. Neurol. Neurosurg. Psychiatry*, 2013, **84**, 409.
52. N. J. Diederich, C. G. Moore, S. E. Leurgans, T. A. Chmura and C. G. Goetz, *Arch. Neurol.*, 2003, **60**, 529.
53. G. Levy, E. D. Louis, L. Cote, M. Perez, H. Mejia-Santana, H. Andrews, J. Harris, C. Waters, B. Ford, S. Frucht, S. Fahn and K. Marder, *Arch. Neurol.*, 2005, **62**, 467.
54. H. Takatsu, K. Nagashima, M. Murase, H. Fujiwara, H. Nishida, H. Matsuo, S. Watanabe and K. Satomi, *JAMA, J. Am. Med. Assoc.*, 2000, **284**, 44.
55. A. Gaenslen, B. Unmuth, J. Godau, I. Liepelt, A. Di Santo, K. J. Schweitzer, T. Gasser, H. J. Machulla, M. Reimold, K. Marek and D. Berg, *Lancet Neurol.*, 2008, **7**, 417.

56. S. Hellwig, F. Amtage, A. Kreft, R. Buchert, O. H. Winz, W. Vach, T. S. Spehl, M. Rijntjes, B. Hellwig, C. Weiller, C. Winkler, W. A. Weber, O. Tuscher and P. T. Meyer, *Neurology*, 2012, **79**, 1314.
57. Parkinson Study Group, *J. Am. Med. Assoc.*, 2000, **284**, 1931.
58. R. Katzenschlager, J. Head, A. Schrag, Y. Ben-Shlomo, A. Evans and A. J. Lees, *Neurology*, 2008, **71**, 474.
59. U. K. Rinne, J. P. Larsen, A. Siden and J. Worm-Petersen, *Neurology*, 1998, **51**, 1309.
60. P. Del Dotto, N. Pavese, G. Gambaccini, S. Bernardini, L. V. Metman, T. N. Chase and U. Bonuccelli, *Mov. Disord.*, 2001, **16**, 515.
61. K. Seppi, D. Weintraub, M. Coelho, S. Perez-Lloret, S. H. Fox, R. Katzenschlager, E. M. Hametner, W. Poewe, O. Rascol, C. G. Goetz and C. Sampaio, *Mov. Disord.*, 2011, **26**(3), S42.
62. Parkinson Study Group, *N. Engl J. Med.*, 1999, **340**, 757.
63. M. Emre, D. Aarsland, A. Albanese, E. J. Byrne, G. Deuschl, P. P. De Deyn, F. Durif, J. Kulisevsky, T. van Laar, A. Lees, W. Poewe, A. Robillard, M. M. Rosa, E. Wolters, P. Quarg, S. Tekin and R. Lane, *N. Engl. J. Med.*, 2004, **351**, 2509.
64. P. Barone, W. Poewe, S. Albrecht, C. Debieuvre, D. Massey, O. Rascol, E. Tolosa and D. Weintraub, *Lancet Neurol.*, 2010, **9**, 573.
65. I. H. Richard, M. P. McDermott, R. Kurlan, J. M. Lyness, P. G. Como, N. Pearson, S. A. Factor, J. Juncos, C. Serrano Ramos, M. Brodsky, C. Manning, L. Marsh, L. Shulman, H. H. Fernandez, K. J. Black, M. Panisset, C. W. Christine, W. Jiang, C. Singer, S. Horn, R. Pfeiffer, D. Rottenberg, J. Slevin, L. Elmer, D. Press, H. C. Hyson and W. McDonald, *Neurology*, 2012, **78**, 1229.
66. C. Trenkwalder, B. Kies, M. Rudzinska, J. Fine, J. Nikl, K. Honczarenko, P. Dioszeghy, D. Hill, T. Anderson, V. Myllyla, J. Kassubek, M. Steiger, M. Zucconi, E. Tolosa, W. Poewe, E. Surmann, J. Whitesides, B. Boroojerdi and K. R. Chaudhuri, *Mov. Disord.*, 2011, **26**, 90.
67. P. J. Garcia Ruiz, A. Sesar Ignacio, B. Ares Pensado, A. Castro Garcia, F. Alonso Frech, M. Alvarez Lopez, J. Arbelo Gonzalez, J. Baiges Octavio, J. A. Burguera Hernandez, M. Calopa Garriga, D. Campos Blanco, B. Castano Garcia, M. Carballo Cordero, J. Chacon Pena, A. Espino Ibanez, A. Gorospe Onisalde, S. Gimenez-Roldan, P. Granes Ibanez, J. Hernandez Vara, R. Ibanez Alonso, F. J. Jimenez Jimenez, J. Krupinski, J. Kulisevsky Bojarsky, I. Legarda Ramirez, E. Lezcano Garcia, J. C. Martinez-Castrillo, D. Mateo Gonzalez, F. Miquel Rodriguez, P. Mir, E. Munoz Fargas, J. Obeso Inchausti, J. Olivares Romero, J. Olive Plana, P. Otermin Vallejo, B. Pascual Sedano, V. Perez de Colosia Rama, I. Perez Lopez-Fraile, A. Planas Comes, V. Puente Periz, M. C. Rodriguez Oroz, D. Sevillano Garcia, P. Solis Perez, J. Suarez Munoz, J. Vaamonde Gamo, C. Valero Merino, F. Valldeoriola Serra, J. M. Velazquez Perez, R. Yanez Bana and I. Zamarbide Capdepon, *Mov. Disord.*, 2008, **23**, 1130.
68. D. Devos, *Mov. Disord.*, 2009, **24**, 993.

69. H. Honig, A. Antonini, P. Martinez-Martin, I. Forgacs, G. C. Faye, T. Fox, K. Fox, F. Mancini, M. Canesi, P. Odin and K. R. Chaudhuri, *Mov. Disord.*, 2009, **24**, 1468.
70. P. Limousin, P. Pollak, A. Benazzouz, D. Hoffmann, E. Broussolle, J. E. Perret and A. L. Benabid, *Mov. Disord.*, 1995, **10**, 672.
71. P. Krack, A. Benazzouz, P. Pollak, P. Limousin, B. Piallat, D. Hoffmann, J. Xie and A. L. Benabid, *Mov. Disord.*, 1998, **13**, 907.
72. K. A. Follett, F. M. Weaver, M. Stern, K. Hur, C. L. Harris, P. Luo, W. J. Marks, Jr., J. Rothlind, O. Sagher, C. Moy, R. Pahwa, K. Burchiel, P. Hogarth, E. C. Lai, J. E. Duda, K. Holloway, A. Samii, S. Horn, J. M. Bronstein, G. Stoner, P. A. Starr, R. Simpson, G. Baltuch, A. De Salles, G. D. Huang and D. J. Reda, *N. Engl. J. Med.*, 2010, **362**, 2077.
73. J. A. Obeso and W. Olanow, *Mov. Disord.*, 2011, **26**, 2303.
74. M. A. Hely, J. G. Morris, W. G. Reid and R. Trafficante, *Mov. Disord.*, 2005, **20**, 190.
75. C. Klein, D. Krainc, M. G. Schlossmacher and A. E. Lang, *Arch. Neurol.*, 2011, **68**, 709.
76. M. G. Schlossmacher and B. Mollenhauer, *Biomark Med.*, 2010, **4**, 647.

CHAPTER 2

Molecular Pathogenesis and Pathophysiology of Parkinson's Disease: New Targets for New Therapies

JOSÉ G. CASTAÑO,*[a,b] CARMEN GONZÁLEZ,[c]
JOSÉ A. OBESO[b,d] AND MANUEL RODRIGUEZ[b,e]

[a] Departamento de Bioquímica, Instituto de Investigaciones Biomédicas "Alberto Sols", Facultad de Medicina, Universidad Autónoma de Madrid, Spain; [b] Centro de Investigación Biomédica en Red sobre Enfermedades Neurodegenerativas, Madrid, Spain; [c] Departamento de Farmacologia, Facultad de Medicina, Universidad de Castilla-La Mancha, Albacete, Spain; [d] Laboratorio de Trastornos del Movimiento, Centro de Investigación Médica Aplicada, University of Navarra, Pamplona, Spain; [e] Laboratory of Neurobiology and Experimental Neurology, Department of Physiology, Faculty of Medicine, University of La Laguna, Tenerife, Canary Islands
*Email: joseg.castano@uam.es

2.1 Introduction

Parkinson's disease (PD) is a clinical entity of unknown etiology. Motor manifestations such as bradykinesia or akinesia, resting tremor, and rigidity, are the cardinal features of PD which begins at adulthood and has a progressive course. PD patients have a severe loss of nigrostriatal dopamine neurons leading to a depletion of dopamine (DA) and its metabolites in the *striatum*.

The brain pathology of PD patients shows the presence of eosinophilic intracytoplasmic inclusions in neurons; those inclusions are immunoreactive to α-synuclein (Snca) and ubiquitin, the so-called "Lewy bodies", as well as Snca immunoreactive neuritic pathology.[1] The aggregation of Snca has been observed in other neurological diseases: dementia with Lewy bodies, multiple-system atrophy, pure autonomic failure, Hallervorden–Spatz disease, REM sleep behavior disorder, *etc.* that together with PD are known globally as "synucleinopathies".[2]

The etiology of PD is still elusive and it clearly is a complex multi-factorial disease. Patients with PD may have prodromal features such as constipation, anosmia, REM sleep disorder and depression.[3] The typical asymmetric motor features of the disease become noticeable when some 50% of the *substantia nigra* DA neurons and 70–80% of striatal DA have been lost. As the disease duration increases, the neurodegenerative process becomes more widespread and extends to the limbic system, the amigdala, the hippocampus, the cortical multi-modal association areas and the cortical primary receptive areas. As a consequence, a plethora of non-dopaminergic manifestations dominate the clinical picture at those stages; major problems are gait and equilibrium abnormalities, dysautonomia and dementia.[4]

We review here the molecular mechanisms that may be implicated in the pathogenesis of PD, the anatomo-physiological complexity of the basal ganglia (BG) that will illustrate how far we are from understanding the molecular mechanism at an adequate system level and finally we will suggest possible new targets for the treatment of PD.

2.2 Molecular Pathogenesis of Parkinson's Disease

Any disease process may be described as a "failure of homeostasis" whose particular clinical manifestation are determined individually by the interplay of genetic and epigenetic mechanisms in response to environmental changes and manifested in an organ- and tissue-specific manner.[5] Complex chronic diseases (diabetes, atherosclerosis, cancer, neurodegeneration) have common channels or pathways that contribute to their development. Cell-autonomous, neuronal, non-cell-autonomous, glial cells and vasculature mechanisms may contribute significantly to compensate or accelerate the process of brain degeneration, as clearly illustrated in the case of amyotrophic lateral sclerosis.[6]

A conceptual framework with the main channels operating in the pathogenesis of PD is presented in Figure 2.1, based on the scheme used by Garraway *et al.*[7] to illustrate another chronic disease, cancer and adapted for PD pathology.

The conceptual framework is that PD is a 'developmental' process resulting in stressed dopamine (DA) neurons that progressively become dysfunctional and eventually die (upper part of Figure 2.1). The pathological developmental process is governed by transcriptional re-programming of the cells, including epigenetic control mechanisms, resulting in compromised cell and organ function. The transcriptional re-programming of DA neurons along PD

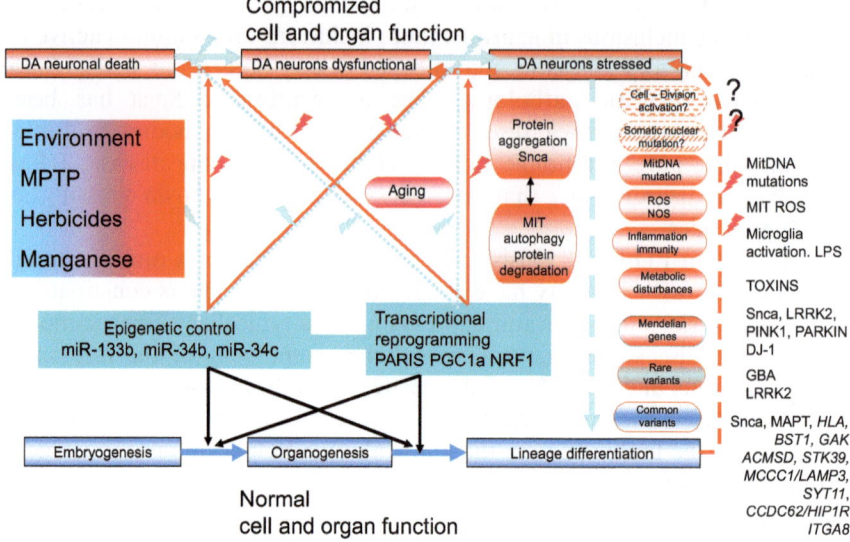

Figure 2.1 Conceptual framework of the pathogenetic mechanisms of Parkinson's disease. Normal cell and organ function develops by a transcriptional programme in different steps from embryogenesis to late differentiated states of the cells in the adult that is also controlled by epigenetic mechanisms (middle and bottom part of the figure). A similar program is postulated to occur, but in reverse, during neurodegeneration (upper part of the figure) and the epigenetic players (mainly miRNAs) and transcription factors implicated in the reprogramming are indicated. These events are promoted by aging, and the environment and postulated to be a developmental program. The main pathogenetic channels contributing to degeneration are indicated by red boxes. Two of them are postulated (dotted boxes), activation of cell division (partial evidence) and somatic nuclear gene mutations (no evidence yet available). The other channels going from normal to the disease state are indicated on the right-hand side. Protein aggregation (Snca) is indicated as a channel that may be 'unique' for brain degenerative diseases including PD, that cross-talks with the protein degradation and mitoautophagy channel. Blue lines represent reversibility by compensatory mechanisms or by the action of therapeutic intervention. For a detailed description see the main text. Based on the figure used by Garraway et al.[7] to illustrate cancer development.

pathological development would be the 'opposite' of the DA neuronal differentiation program taking place at the end of embryogenesis and organogenesis that results in normal cell and organ function (lower part of Figure 2.1). The transcriptional re-programming may vary between individuals and the particular clinical manifestations are also determined individually by the interplay of genetic and epigenetic mechanisms in response to the environment.

The major factor in brain degeneration (and other chronic diseases) is age; the increased longevity of the human population is the major factor contributing to increased incidence and prevalence of brain degenerative disorders, including PD. The process of aging is not well understood. Certainly, a key factor in aging is the accumulation of structural and functional damage of cell components, as a result of sustained stress during the lifespan of an organism. Nevertheless, it is conceivable that aging is another phase of development of an organism; a genetic (and epigenetic) program of aging may exist and can be accelerated, as demonstrated by the progeria syndromes.[8] This developmental aging program may also include, in the case of the nervous system, a re-programming of cells for active growth and dedifferentiation similar to what happens in cancer cells, but without cell division.[9]

Protein aggregation is a hallmark of brain degenerative diseases; the aggregated proteins are expressed almost ubiquitously, but the aggregation within brain cells can be considered as cause, or consequence, of the brain degenerative process. Accordingly, protein homeostasis (proteostasis) in the cell is a critical factor in brain degeneration, with Snca being the protein that aggregates in PD. The other main organelle implicated in PD is the mitochondria. Mitochondrial dysfunction can also be considered both as a consequence and as a trigger of the process of brain degeneration occurring in PD.

The pathway from normal DA cell function to stressed DA neurons, that eventually results in DA dysfunction and death, has several channels. We will review those channels and afterwards consider the connections between those channels and their contribution to the understanding of the pathogenesis of PD.

2.2.1 Genetic Channels

The classical genetic contribution to disease is based on the fact that familial forms of diseases can be attributed to a 'single' malfunctioning gene (in a reduced percentage of cases) in an appropriate genetic background or due to the influence of multi-factorial gene expression profiles and interactions.

Genome-wide association studies (GWAS) are widely used nowadays to provide first-line evidence to unravel the genetic contribution in multi-factorial diseases. Common single nucleotide polymorphisms (SNPs) have been associated with several multi-factorial diseases in cases and controls studies. Due to their frequency in the population (<5%), the contribution of SNPs to the development of a specific disease cannot be very significant. Each disease-associated SNP will produce a minor increase in the odds ratio (OR) for the development of the disease. Having the "pathogenic-associated" SNPs in several of the associated loci will increase the OR and will contribute significantly to the development of disease in those individuals. Most of the non-coding SNPs found to be associated with different diseases lie in regulatory DNA sequences marked by deoxyribonuclease I hypersensitive sites and seem to systematically perturb transcription factor recognition sequences.[10] There are several GWAS association studies in PD and recently they were compiled with the appropriate meta-analysis in the PD gene database (www.pdgene.org).[11] SNPs in twelve loci showed genome-wide significant association (BST1, CCDC62/HIP1R, DGKQ/GAK, GBA, LRRK2, MAPT, MCCC1/LAMP3, PARK16, SNCA, STK39, SYT11/RAB25 and ITGA8) with the risk of developing PD and two loci: HLA and ACMSD requiring further study.[11]

While included in the GWAS analysis, the SNPs of LRRK2 (OR 2.23 in Asians) and GBA (OR 3.51 in Caucasians), are based on candidate-gene approaches (still a valid approach in the 'GWAS era') since these SNPs are not explored in the usual GWAS platforms. Their frequency in the general population is very low (<0.05%) and as a consequence they can be classified as rare variants (see Figure 2.1).

Mendelian gene mutations are also associated with familial forms of PD. An increasing number of Mendelian gene mutations have been linked to PD, but not always clearly validated.[12] The genes whose mutations are clearly linked to PD are listed in Figure 2.1, Snca, LRRK2, PINK1, PARKIN, DJ-1 and a complete database can be consulted in PDmutDB (http://www.molgen.ua.ac.be/PDmutDB/default.cfm?MT=0&ML=0&Page=Home).

A total of three missense mutations of the Snca/PARK1 gene on chromosome 4 have been linked to dominant forms of PD, A53T,[13] A30P[14] and E46K mutations.[15] In addition to point mutations, Snca locus duplication[16,17] and triplication[18] have been described in certain familial forms of PD and hypomethylation of Snca intron 1 is found in sporadic PD patients.[19] Genetic studies of the Snca gene indicate that either a point mutation or an increased expression of Snca are directly involved in the pathogenesis of PD.

Mutations in DJ-1/PAK7 are linked with autosomal recessive early-onset PD. Pathogenic mutations identified in the DJ-1/PARK7 gene include CNVs (exonic deletions and truncations), homozygous (L10P, M26I, E64D, E163K and L166P) and heterozygous (A39S, A104T and D149A) missense mutations. Rare polymorphisms (R98Q, A171S) have also been identified in normal individuals and are not associated with PD.[20]

PARKIN is a member of the E3 ubiquitin ligase family,[21] and mutations found in this gene include: deletion of exons, point mutations, insertion and deletions, including changes in the open reading frame and internal stop codons, most of them resulting in a loss of function of the protein as an E3 ligase (by no expression or expression of mutated or truncated protein). Mutations of PARK2 are the most frequent cause of recessive Parkinsonism worldwide. PINK1/PARK6 is a serine–threonine protein kinase located in the mitochondria outer membrane and most of the mutations found in familial PD patients are missense or truncation, but gene and exonic re-arrangements also have been described correlated with null activity of PINK1.[22] Mutations of the LRRK2/PARK8 gene (leucine-rich repeat kinase 2 or dardarin) were identified as a cause in dominant familial PD[23,24] and account for roughly 5% of cases of familial PD and 1–2% of non-familial PD. Mutations in LRRK2 found in familial PD patients either increase[25] or decrease[26] its protein kinase activity. Surprisingly even the kinase activity of LRRK2 may not be an absolute requirement for inducing DA neuron dysfunction. A double inducible transgenic mouse of LRRK2 and A53T Snca shows acceleration of the 'toxic' effect of A53T Snca, but similar effects were found with a kinase-death LRRK2 mutant.[27] These results indicate that other domains of LRRK2 apart from the kinase domain are important for developing pathological features. The GTPase domain of LRRK2 is a good candidate. In fact, in yeast[28] and *C. elegans*[29] it has been shown that the GTPase domain of LRRK2 is responsible for its effect on Golgi distribution, vesicle trafficking and autophagy and more recently it has been shown that the ADP-ribosylation factor GTPase-activating protein 1 (ArfGAP1) is a GTPase-activating protein for LRRK2 and activates its kinase activity. Both ArfGAP1 and LRRK2 co-operate to produce neurite shortening indicating a crosstalk between LRRK2 protein kinase activity and its GTPase activity.[30] Clearly more studies are needed to actually understand the molecular basis of the cellular function of LRRK2.

2.2.2 Metabolic Disturbance Channel Toxins

From an initial description of the epidemic of PD produced in California by contamination of meperidine with 1-methyl-4-phenyl-1,2,3,6-tetrahydropyridine (MPTP) for recreational use,[31] many studies have shown that MPTP is metabolized by MAO-B producing 1-methyl-4-phenylpyridinium (MPP$^+$), the active toxin that is actively taken up by the active dopamine transporter (Dat). MPP$^+$ gets concentrated in DA neurons within vesicles by the vesicle monoamine transporter (Vmat2) and explains the exquisite sensitivity of the DA neurons to this compound. MPP$^+$ binds to complex 1 of the mitochondrial respiratory chain, blocks it irreversibly and produces cell death.[32,33] MPTP intoxication has been used as a model to produce "Toxic-Parkinsonism" in monkeys and mice with different regimes of administration (acute or chronic) and the dose is dependent on the animal species under study.[34,35] Herbicides like paraquat, an analogue of MPP$^+$ which is used widely in agriculture, was considered for a few years as a putative causative agent in patients with idiopathic PD, especially for those exposed by living in rural areas. However, in recent years, it has become clear that paraquat is unlikely to play a role in human PD, for two reasons: it does not cross the blood–brain barrier and most patients severely exposed die of pulmonary edema before having the opportunity to develop Parkinsonism.[36] Manganese is a known cause of Parkinsonism in people exposed to high concentrations, especially Chilean miners[37] and induces levodopa-unresponsive Parkinsonism in monkeys due to the lesion of the *globus pallidus* and the *substantia nigra pars reticularis* (SNr), having little resemblance with PD neither clinically nor pathologically.[38] Manganese has recently attracted new attention as PARK9/ATP13A2 (a vacuolar ATPase in yeast) seems to protect cells from manganese toxicity[39] and patients carrying a PARK9/ATP13A2 mutation have increased iron content in the caudate and putamen.[40] Accumulation of iron in the *substantia nigra* is known to be associated with PD and with Parkinsonism.[41] Iron accumulation is also found in neurodegeneration with brain iron accumulation, an hereditary disorder related to mutations of the panthotenate kinase 2 [ref. 42] that has been reproduced in *Drosophila*.[43] At present, there is no clear-cut evidence from epidemiological studies that a 'common' or 'specific' toxin is associated with the development of PD; however the exposure to pesticides does confer a greater risk.

2.2.3 Inflammatory and Immunity Channels

Inflammatory mechanisms are a stress response that is commonly associated with chronic diseases and particularly with brain degeneration.[44] Neuroinflammation has been increasingly associated with the characteristic selective and gradual death of DA neurons from the *substantia nigra* and the loss of their projections to the *striatum*, participating in the basis of the pathogenesis of PD.[45] Slight microenvironmental changes within the brain elicit pro-neuroinflammatory responses by activating glial cells (primarily, microglia and, to a lesser extent, astroglia). Inflammatory and neurotoxic mediators such as cytokines, lipids, free radicals, reactive oxygen species (ROS) and reactive nitrogen species (RNS) are produced and their accumulation has a high impact on the survival of the extremely vulnerable nigral DA neurons, enhancing their progressive death.[45] The main evidence in favor of neuroinflammatory mechanisms in PD comes from postmortem brain studies[46,47] and *in vivo* direct imaging of the microglia reaction with [^{11}C]-(R) PK11195 by positron emission tomography,[48,49] a ligand of the mitochondrial peripheral benzodiazepine receptor. Finally, the role of microglia activation in the initiation (or extension) of PD pathology has also been suggested by the extensive microgliosis observed in postmortem studies of PD patients that have received treatment with grafted embryonic neurons.[50–53]

2.2.4 Reactive Oxygen and Nitrogen Species Channels

The production of ROS and RNS is a consequence of normal mitochondrial respiration and is especially increased when mitochondrial respiratory complexes are inhibited and electrons get directly transferred to molecular oxygen. A mitochondrial dysfunction is probably operating in many, if not all, neurodegenerative diseases. The mitochondrial dysfunctions that have been associated with PD are: decreased functioning of the mitochondrial electron transport in the respiratory chain affecting oxidative phosphorylation; mitochondrial DNA damage; impaired calcium buffering; and abnormalities in mitochondrial morphology and dynamics.[54] Decreased activity of the mitochondrial respiratory complexes has been described in postmortem brain studies of PD patients due to reduced activity of Complex I in the *substantia nigra* and also in the frontal cortex. Similar Complex I deficits have been described in skeletal muscle and platelets from PD patients.[55] In contrast, none of the hereditary dysfunctions of Complex I have clinical features of PD,[56] PINK1-deficient, but not PARKIN-deficient, cells show a decrease in Complex I activity that can be rescued by overexpression of yeast NADH dehydrogenase, Ndi1p.[57]

2.2.5 Channels of Mitochondrial DNA Mutations

Mitochondrial DNA deletions have been shown to accumulate in DA neurons of the *substantia nigra* and not in other brain regions and cell types, from both

PD patients and normal age-matched controls.[58,59] Those somatic mutations have been attributed, but not demonstrated, to be a consequence of ROS damage, while they are more likely due to the mitochondrial DNApol γ mutational rate.[60] Rare DNApol γ mutations have been demonstrated in early-onset PD without ophthalmoplegia.[61]

The mtDNA Mutator and the MitoPark mouse are two main models developed to study the role of autonomous mitochondrial deficits in general pathology and PD. The mtDNA mutator mice are a knock-in mouse harboring the D257A mutation in DNA POLG leading to defective proof-reading activity of the mitochondrial polymerase.[62] Homozygous knock-in mice are normal until 25 weeks of age and then, they develop a premature aging phenotype greying of hair and alopecia, weight loss, hearing loss, heart enlargement, osteoporosis, etc. Apart from an increase number of point mutations accumulated with increasing age, the mitochondrial DNA of these animals also contains extended deletions both in linear and circular forms of mitDNA.[62] The mitochondrial respiratory deficit in these mice occurs with a normal rate of translation of mitochondrial-encoded respiratory proteins. Accordingly, a defect in the assembly of the respiratory complex due to point mutations is more likely responsible for this phenotype.[63] Other authors have claimed that the phenotype is better correlated with mitDNA deletions,[64] in spite of the fact that mice with big deletions of mitDNA do not show a progeroid syndrome.[64,65] The MitoPark mouse was produced by inactivation in DA cells of the nuclear-encoded mitochondrial transcription factor A (TFAM), these mice show some clinical features of PD with adult-onset and progressive degeneration of nigrostriatal DA circuitry; motor deficit sensitive to L-3,4-dihydroxyphenylalanine and altered response to treatment as the disease progresses.[66] Notably, protein inclusion also developed in these DAT-driven TFAM mice that are immunoreactive with h116–131 anti-Snca antibodies but not with other anti-Snca antibodies. Those protein inclusions were demonstrated not to contain Snca, as they also appear (and react with h116–131 antibodies) in double transgenic Snca/TFAM null mice.[66] Furthermore, TFAM null mice have extensive cell death by apoptosis, but without ROS generation.[67]

2.2.6 Channels of Somatic Nuclear Mutations

Another channel listed in Figure 2.1 is nuclear somatic mutations, but at present this is undemonstrated in brain degenerative diseases. If these nuclear somatic mutations were found in brain degenerative diseases there would be a change of paradigm in the study of brain degenerative disease, as happened with cancer. After initial methodological discussion and proof-of-concept projects, it could be interesting to launch a massive whole genome sequencing of samples from different regions of the brain from normal and diseased individuals with different brain degenerative pathologies including PD co-ordinately with the study of epigenetic modifications of the somatic genome and analyzing samples from men and women at 30, 50, 65 and 75 years of age.

2.2.7 Channels of Cell Division Activation

Affected neurons and/or non-neuronal cells in brain degeneration are probably being re-programmed through transcriptional regulation to an active program of growth and dedifferentiation (a reverse of the normal development) along the aging process, similar to what happens in cancer cells. This aging re-programming may lead to activation of cell division in brain cells (not yet demonstrated), but the consequence in the brain would be cell death. Note that neurons are postmitotic cells, but they have a high activity of the mitotic E3 ligases, APC/C-Cdh1 and –Cdc20 complexes that participate in axon and dendrite morphogenesis, remodeling and synapse differentiation.[68] The possible dedifferentiation program is illustrated by the phenotypes of EN-1/2 and NURR-1 heterozygous and the conditional GDNF mouse and the activation of cell cycle by the effect of the pathogenic LRRK2 on overproduction of E2F1/DP (see below under transcriptional reprogramming).

2.2.8 Connecting Channels to Define the Pathogenesis of Parkinson's Disease

The above-presented channels involved in PD may look like a collection of unconnected channels that somehow co-operate to produce PD. Nevertheless, it has emerged that some connections between the different channels may eventually produce a clear picture of PD pathogenesis. Certainly PD pathology shows the aggregation of Snca and mitochondrial dysfunction (Figure 2.1). Genetic studies of the Mendelian forms of PD[69] point out that Snca point mutation or overexpression is a genetic determinant. While overexpression can be 'understood' as a bigger burden for the cell to clear out the protein produced in excess, perhaps promoting its aggregation, the mechanisms responsible for the dominant phenotype of Snca point mutants (A30P, A53T and E46K) are unclear. The role of Snca aggregation in progression of PD has direct evidence from postmortem studies of PD patients that have been transplanted with embryonic neurons. Surprisingly enough, those patients developed PD pathology with Lewy bodies (LB) inclusions in the grafted neurons[50–53] concomitant with extensive microgliosis. These results suggest that Snca belongs to the transmissible family of proteins, behaving like a prion-like protein, prionoid.[70] Recently clear-cut experiments have been published showing that aggregated Snca enters neurons, promoting aggregation of soluble endogenous Snca; this aggregation starts in axons and then propagates into perikarya. The accumulation of aggregated Snca led to impairments in the excitability and connectivity of neurons and final neuronal death.[71] Furthermore, intracerebral injections of aggregated Snca induce the formation of intracellular LB-like inclusions and the onset of neurological symptoms in A53T Snca transgenic animals, but have no effect in Snca-null mice.[72] There are numerous questions still pending to validate the prionoid hypothesis for PD. It is not known what factors trigger protein mis-folding (aging, environmental stress, genetic predisposition, *etc.*) of Snca (and the other prionoid proteins Tau, Abeta, *etc.* implicated in other neurodegenerative diseases), how aggregated Snca get access to the interneuronal space (liberated by cell death, secretion, multi-vesicular bodies), how aggregated Snca is captured by neurons (endocytosis, cytoplasmic penetration) or can be transmitted by tunneling nanotubes. What are the routes of propagation of the misfolded protein? The mis-folding process begins somewhere else (blood, intestine) and then gets to the central nervous system through the autonomous nervous system like prions? Or there are alternative routes? Finally, there should be evidence of disease transmission between individuals to categorize PD as a *bonafide* prion disease. Another central issue is to characterize how neurons, and specifically DA neurons, respond to aggregated Snca.

In the case of mitochondrial dysfunction, the evidence from environmental exposure to toxins is clear for MPTP, but there is still no convincing epidemiological study that relates PD with exposure to one or several toxic substances. In contrast, the studies of the Mendelian forms of PD have a possible connection between stressed (oxidative) mitochondria and disease (mitoautophagy, Figure 2.1). Mitochondria are dynamic organelles, stressed mitochondria fused with healthy mitochondria to be repaired. When the mitochondrial damage is more severe with loss of the transmembrane potential, PINK1 is no longer imported from the mitochondrial outer membrane and degraded; its accumulation promotes PARKIN translocation to the mitochondrial outer membrane leading to the ubiquitylation of mitochondrial outer membrane proteins, like mitofusins 1 and 2 that are degraded by the proteasome preventing mitochondrial fusion. Other ubiquitylated outer membrane mitochondrial proteins are the signal for promotion of mitoautophagy and as a consequence the damaged mitochondria are eliminated.[73] Other genes like LRRK2, GBA and ATP13A2 may also regulate the cellular dynamics of these processes, but their connecting channels are still unclear.[74] Accordingly, mitochondrial quality control will be impaired in patients with mutations in their PARKIN and PINK1 genes. Is this mechanism operating in idiopathic PD where no mutation of PARKIN and PINK1 genes is found? One of the consequences of oxidative stress is the activation of protein tyrosine kinase c-Abl and this kinase has been shown to phosphorylate PARKIN, resulting in the inactivation of its E3 ligase activity.[75,76] As a consequence, either by gene mutation or oxidative stress, regulation of PARKIN activity is critical in PD development.

Another connecting channel relates PARKIN E3 ligase activity with transcriptional reprogramming. PARIS (ZNF746) is a transcriptional repressor of the transcriptional co-activator PGC-1α and its protein levels are regulated by ubiquitylation mediated by PARKIN and proteasomal degradation. Inactivation of PARKIN by gene mutation or after phosphorylation by c-Abl in response to oxidative stress, leads to increased levels of PARIS with concomitant decrease expression of the PGC-1α co-activator and a decrease in Nrf-1 expression.[77] This transcriptional re-programming makes the cell more vulnerable to oxidative stress as the enzymatic transcriptional network that copes with ROS and RNS is orchestrated by Nrf1[78] and also Nrf2[79] transcription factors and with a decrease in the expression of Nrf1 the cellular response to oxidative stress would be compromised.

2.2.9 Transcriptional Re-Programming and Epigenetic Control

The vulnerability of DA neurons to oxidative stress in PD is likely to be related and due to some specific 'mechanisms' or features of DA neurons. The pacemaker activity of DA neurons relies on L-type Ca(v)1.3 Ca^{2+} channels and renders them vulnerable to stressors.[80] The pacemaker activity of DA neurons creates by itself an oxidative stress that induces transient, mild mitochondrial depolarization or uncoupling that is increased by deletion of DJ-1/PARK7.[81] Similarly, the cholinergic neurons of the dorsal motor nucleus of the vagus, affected early in PD patients, show oxidative stress as a consequence of its pacemaker activity due to poorly buffered calcium entry and the oxidative stress is also increased in DJ-1/PARK7 null mice.[82] The role of pacemaker activity related to calcium homeostasis and oxidative stress as a vulnerability factor is a recently opened avenue that has not been explored in detail.

Another way to account for the selectivity of DA vulnerability is the interference with growth factors or transcription factors that are responsible for the maintenance of the DA phenotype. We have described above a case of transcriptional re-programming in the case of PARIS and its regulation by PARKIN participating in the oxidative transcriptional response to oxidative stress that shows some specificity for DA neurons. More generally, the re-programming can be DA-specific if it does affect the expression of DA-specific transcription factors. As already mentioned in the Introduction, it is conceivable that a similar program may exist for aging (another phase of development) and there are indications that the same transcription factors involved in early development can play a role in the aging process, as shown in *C. elegans*.[83] During mice embryogenesis, mesDA neurons are derived from the ventral mid-line of the mesencephalon through the intersection of Shh expressed along the ventral neural tube and Fgf8, locally released at the mid-hind–brain boundary and in the rostral forebrain,[84] and by Wnt1[85] signaling. MesDA neurons originate in the floor plate of the mesencephalon; the pathway Otx2-Limx1a-Msh1 is also involved in specification and generation of mesDA identity.[86] Foxa1 and 2 mediate the effect of Shh and also participate in the above process both up-stream and down-stream of Lmx1a.[87] Afterwards, mesDA neurons acquire the expression of transcription factors selective for mesDA neurons including Nurr1 (Nr4a2), Lmx1b, Pitx3, En1/2 and Foxa1/2, activating the expression of specific DA neuronal markers (TH, Dat, Vmar2) and promoting the terminal differentiation and maintenance of mesDA neurons.[88] Other growth factors also participate in the process of growth control (TGF α/β) and in the survival and maintenance of mesDA neurons like GDNF and MANF/CDBF families of growth factors.[86,89] At present, further studies are needed to demonstrate a significant association of genetic variations in LMX1A, LMX1B, EN1/2 and PITX3 and its expression with idiophatic PD. Furthermore the null animal models for those transcription factors show no or variable generation of DA neurons and as a consequence can not be considered 'good' models for PD.[86,89] Nurr1 has been extensively studied because of its central role in DA neurons. Several mutations in the 5'-region and 5'-UTR of the NURR1 gene have been correlated with decreased expression of NURR1 and PD development.[90,91] A missense mutation (S125C) in NURR1 has also been described in a PD patient.[92] Furthermore, NURR1 expression is reduced in neurons with pathological signs in the brains of PD patients[93] and a decrease in NURR1 activity is observed in the peripheral blood lymphocytes of PD patients.[94] Traditional knock-out mice with DA-specific transcription factors show only defective numbers of DA neurons and are not very useful. There are exceptions, with some heterozygous knock-out mice. $EN1^{+/-}$; $EN2^{+/-}$ and $EN1^{+/-}$; $EN2^{-/-}$ mice exhibit slow and progressive loss of nigral DA neurons which leads to decreases in DA levels in the *striatum*.[95,96] Importantly, $EN1^{+/-}$; $EN2^{-/-}$ adult mice show motor deficits such as akinesia and bradykinesia, a phenotype that resembles key pathological features of PD,[95] and old $NURR1^{+/-}$ mice display an age-dependent DA dysfunction associated with motor impairment that is analogous to Parkinsonism, showing decreases in rotarod performance and locomotor activities.[97] The best approach to studying the possible effects of any transcription factor in animals is to obtain conditional knock-out mice that can be induced in the adult stage (somehow mimicking what may be happen in humans), using this approach, a mouse strain with conditional targeting of the NURR1 gene has been generated.[98] In those mice when NURR1 is ablated at late stages of mesDA neuron development there is a rapid loss of striatal DA, loss of DA neuronal markers and neuron degeneration. The ablation in the adult brain results in a slower progressive loss as in PD, with DA neurons at the *substantia nigra* being more vulnerable in this model.[98] Recently it has also been reported that Snca overexpression in adult mice results in an accelerated degradation of NURR1 and as a consequence it may explain the selective vulnerability of DA neurons.[99] While this latter study connects Snca effects with the specificity of its effect for DA neurons, in our own studies of the mechanism of NURR1 degradation (Alvarez-Castelao et al.)[206] we looked for the possible effects of Snca overexpression in NURR1 degradation, but we failed to find any significant effect. As a consequence, this specific connection needs to be replicated by some other groups. A similar approach with conditional null mice is the tamoxifen inducible knock-out of the glial cell line-derived neurotrophic factor (GDNF) in adult (two-month-old) mice that promotes a dramatic cell death of catecholaminergic neurons affecting the *locus coeruleus*, the *substantia nigra* and the Ventral Tegmental Area (VTA), resulting in an hypokinetic phenotype.[100]

Very little is known with respect to the epigenetic control of gene expression related to DA neurons and PD and mainly centered in miRNAs.[101] The first report was the identification of miR-133b, specifically expressed in DA neurons and whose expression is reduced in PD patients. miR-133b regulates the maturation and function of mid-brain DA. Actually, Pitx3 specifically induces transcription of miR-133b and Pitx3 activity is down-regulated by miR-133b within a negative feedback circuit that increases the robustness and speed response time of DA neurons and that circuit would be lost in PD patients.[102] More recently, it has been shown that pathogenic LRRK2 mutations interact with microRNAs to regulate protein synthesis reducing let-7 and miR-184 and resulting in an increase of E2F1 and DP expression levels. E2F1 and DP are known as heteromeric transcription factors that mediate G1/S transition in the regulation of a cell cycle. In postmitotic neurons, increased expression of E2F1 and DP1 will promote a catastrophe by activating G1S transition and promoting cell death.[103] A decreased expression of miR-34b and miR-34c in brains from PD patients and down-regulation of those miRNAs in SH-SY5Y resulted in down-regulation of DJ1 and PARKIN expression, similar results were observed in PD brain samples displaying strong depletion of miR-34/bc expression.[104]

2.3 The Pathophysiology of Parkinson's Disease

While the general overview presented above on the molecular pathogenesis of PD has allowed us to understand some of the molecular pathways implicated, there is still a tremendous gap between such molecular and cellular studies and the understanding of the real mechanisms at the system level implicated in the pathophysiology of PD. The most clearly unanswered question is why dopaminergic (DA-ergic) neurons are the main targets in PD, some hints were described above, as was how the whole basal ganglia (BG) physiology adapts to the disease process and progression to other brain structures. This section deals first with the properties of DA-ergic neurons and then with the first system level, the anatomo-physiological circuits that are pathologically implicated in the clinical manifestations of PD.

2.3.1 Dopaminergic Neurons are Complex

During the 1980s, morphological, odological, neurochemical and electro-physiological data were grouped in a single ("classical") model of the BG[105,106] which was later widely used to explain the clinical expression of PD and to develop new therapies for their symptomatic control. Although the classical model of BG has been useful for understanding clinical data in the first stages of PD, its explanatory value decreases with the progression of the illness. There is a large amount of new data which could contribute to improving the explanatory value of the classical model of the BG, thereby facilitating the development of new therapies for PD.

In the earliest description of DA-ergic cells,[107,108] nigrostriatal neurons seemed to be a homogeneous cell population with the somata located in the *substantia nigra pars compacta* (SNc) and with the synaptic projections only innervating the *striatum*. This initial impression has progressively changed with the incorporation of an enormous amount of data. There are non-nigral mesencephalic DA-ergic cells which also project to the *striatum* (often referred to as "meso-estriatal cells"). In addition, DA-ergic cells of the nigra can also project (directly or as collateral of meso-striatal axons) to other brain centers apart from the *striatum*, such as the subthalamic nucleus, several thalamic nuclei, the *globus pallidus* and the brain cortex.[109–114] The nigral DA-ergic neurons are far from being a uniform cell population with the same neurochemical characteristics. Thus, there are DA-cell groups in the nigra with a different expression of calbindin, calretinin and colecystokinin[115] and with a different accumulation of melanin (a product of the self-oxidation of DA to quinine/semi-quinone).[116–118] Even the expression of proteins needed for the synaptic management of DA is not the same in all the DA-ergic cells of the nigra, as in the case of the expression of DA transporters of the cell membrane (Dat) and synaptic vesicles. (Vmat2).[119,120] These differences could justify not only the different functional activity observed between nigral DA-ergic cells but also their different vulnerability to external (glutamate, toxins *etc*.) and internal (DA oxidation) aggression.[121,122] The study of the neurochemical differences between DA-ergic cells associated with their differential vulnerability could be useful for understanding the mechanisms involved in the origin and progression of PD.

Initially, the DA-cell role was considered to be that of maintaining a stable concentration of DA in the extracellular medium of the *striatum* (DA-pump model), whereby the striatal DA could be replaced by drugs able to directly stimulate postsynaptic DA receptors (volume transmitters). Morphological and electrophysiological evidence showed that in addition to behaving as a volume transmitter, DA could also behave as a neurotransmitter with a short and local action which cannot be replaced by drugs which cannot be handled by DA-ergic cells in the same way as they handle DA.[123,124] In addition, the DA-pump model does not explain the marked self-regulation of DA-ergic cells and their sensitivity to information coming from different brain centers (visual information, pain *etc*.). The DA-synapse has several regulatory mechanisms able to compensate for the partial degeneration of the DA-ergic system and which involve the regulation of DA-synthesis (modulated by presynaptic receptors, intracellular calcium, firing rate of DA-cells, end-product inhibition of tyrosine hydroxylase by DA), DA-release (presynaptic receptors), DA-uptake (firing rate) and postsynaptic DA actions (denervation hypersensitivity). These regulatory mechanisms are probably different for DA released in the *striatum* and the *substantia nigra*. The DA release is, at least partially, dependent on the firing rate of DA-cells which is accurately modulated by DA-cell inputs.[124–126]

Data about the high structural, electrophysiological and chemical diversity of DA-neurons could provide new lines of work for the development of more efficient symptomatic control of PD and for the future prevention of PD progression. The pharmacological control of DA-cell activity must consider the adaptation of these neurons to degeneration and to the effects of drugs. These adaptive mechanisms are necessary for levodopa and DA-receptor agonists to be able to selectively act on the partially degenerated DA-ergic system. Levodopa is mainly metabolized in cells with increased DA turn-over and DA-ergic agonists are more efficient in denervated cells which have developed denervation hypersensitivity.[127] If a drug modifies these adaptive mechanisms, its selectivity will be compromised in the medium term. The DA-ergic diversity should also be taken into account when trying to control the progression of DA-cell degeneration. Although the etiology of PD is unknown, present evidence suggests that the DA-cell degeneration involves the synergic actions of many different factors (multi-factorial theory). Not all DA-ergic neurons present the same vulnerability to each of these factors and the understanding of which DA-cell type is more vulnerable to each of the factors could facilitate the development of efficient etio-pathogenic treatments. Therefore, the complexity and diversity of DA-cells should be considered as an opportunity to understand their physiological role, to study the etiology of PD and develop new and more efficient therapies for controlling the symptomatic expression of PD and the progression of neurodegeneration.

2.3.2 Dopamine Cells are Not the Only ones Involved in PD

Since the 1960s, the degeneration of DA-ergic nigrostriatal neurons has been considered the hallmark of PD. Although this consideration is generally accepted nowadays, evidence accumulated over the last 10 years shows that DA-cells are not the only ones which degenerate in PD. The pathology of non-DA-neurons in PD has also been observed in different regions, some of which are in the peripheral nervous system (glossopharyngeal and vagal nerves) and in the anterior olfactory nucleus which show LB and Lewy neuritis, but do not exhibit significant cell loss.[128,129] Other cell populations with an early degeneration are those of the presupplementary cortex,[129] cholinergic neurons of the pedunculo-pontine nucleus,[130,131] and noradreneric neurons of the *nucleus accumbens*.[132] The pedunculo-pontine nuclei have cholinergic-nitrinergic-glutamatergic neurons which project to the *substantia nigra*[133] and that can degenerate before DA-cells. There is evidence suggesting that cholinergic activation (*e.g.*, in smokers) decreases the probability of developing PD,[134] a prevention that could be induced by the stimulation of the cholinergic receptors of DA-cells induced by pedunculo-pontine neurons. The noradrenergic neurons of the *nucleus accumbens* also degenerate in PD, an effect which is also observed after the experimental degeneration of the noradrenergic neurons of that nucleus in animals.[132] Some non-DA-cells which degenerate in PD could be associated with the progression of the illness more than with its onset. As an example, the GLU-ergic neurons of the intralaminar nuclei of the thalamus projecting to the *striatum* and whose degeneration could be secondary to the previous DA-ergic denervation of the *striatum* or of the thalamus itself.[135,136]

In summary, the degeneration of non-DA-ergic cells could play a relevant role in some clinical manifestations of PD and particularly in its manifestation during progression. The study of these other cell populations could be necessary for developing new treatments for controlling the expression of the high diversity of symptoms which appear in advanced PD and for preventing the onset and progression of DA-cell degeneration. Currently, particular attention is being paid to identifying early indicators of PD which can be used to diagnose the onset of the illness when at its presymptomatic motor stage.[137] This early diagnosis could be useful for studying the origin of the illness and to develop an efficient treatment for preventing the progression of cell degeneration. Progression is more difficult to stop when most of the cells have disappeared and the cells that are still alive are suffering the consequences of a degenerative process that started years beforehand. The study of the degeneration of non-DA-ergic cells and of the earliest stages of the disease will be hot issues in coming years.

2.3.3 New Models of the Basal Ganglia are Necessary to Understand Parkinson's Disease

The basal ganglia involves a group of subcortical nuclei whose main components are the *striatum*, *globus pallidus*, subthalamic nucleus (STN) and the *substantia nigra*. Each of these nuclei have been divided into different parts, the *striatum* into the putamen and caudate nucleus, the *globus pallidus* into external (GPe), internal (GPi) and ventral (VP) segments and the *substantia nigra* (SN) into SNr and SNc. The classical model of the BG proposes the *striatum* and STN as being the main input stations of the basal ganglia, the *striatum* receiving glutamatergic (GLU-ergic) inputs from the entire cerebral cortex (except the primary visual cortex) and the STN receiving GLU-ergic inputs mainly from the frontal cortex. About 95% of striatal neurons are GABA-ergic medium-sized spiny neurons (MsSN), 50% of which express the DA1 dopamine receptor and *substantia* P and project directly to the GPi and SNr ("direct pathway") and the other 50% express the DA2 dopamine receptor and enkephalin and project polysynaptically to the GPi and SNr by way of the GPe and STN ("indirect pathways"). The GLU-ergic neurons of the STN transfer the excitatory inputs of the cortex to the GPi and SNr, a transfer which is faster than that of the direct and indirect pathways ("hyperdirect pathway"). The GPi and SNr are the main output stations of the basal ganglia, sending GABA-ergic projections to the ventral and rostral nuclei of the thalamus. The GPi/SNr projections to the thalamus induce a tonic inhibition of the thalamo-cortical GLU-ergic pathway, thus inhibiting the cortical areas innervated by these thalamic nuclei. Therefore, the main flow of information within the BG crosses the cortico–subcortico–cortical loop, returning (after being processed in a way which is not still clear) to the cortical region of origin. This loop has been segregated into four loops; motor, oculomotor, associative and limbico, whose anatomical organization is similar, but whose cortical and subcortical distribution is parallel and segregated.[138] All these loops are controlled by DA-ergic striatal afferents which modulate the excitatory action of cortical inputs on MsSN. The modulatory action of DA is different for the MsSN of the "direct pathway" (activated after stimulating its DA1 receptors) and for the MsSN of the "indirect pathway" (inhibited after stimulating its DA2 receptor). In this way, the striatal DA controls the balance between the direct and indirect pathways, decreasing the tonic GABA-ergic inhibition of GPe and SNr cells and promoting the tonic thalamic activation of the brain cortex. The BG model described above has proved to be useful in explaining the pathophysiology of different clinical disturbances in BG disorders and in developing new pharmacological and neurosurgical methods for controlling the clinical expression of PD. The basic idea is that the decrease of striatal DA induced by the degeneration of nigrostriatal cells reduces the inhibitory activity of the "indirect pathway" and increases the excitatory activity of the "direct pathways" on SNr/GPi neurons. In this way, the tonic GABA-ergic inhibition of SNr/GPi projections to the thalamus induces a subsequent tonic inhibition of the excitatory thalamo-cortical projections and of the cortical activity. The pharmacological control of PD symptoms basically consists of promoting the DA-ergic transmission in the *striatum* by administering DA precursors (levodopa) or DA-receptor agonists, procedures which restore the basal equilibrium between the direct and the indirect pathways. Neurosurgical treatments are aimed at restoring this equilibrium by preventing the hyperactivity of subthalamo-nigral projections (with the lesions or high-rate stimulation of the STN) or by preventing the consequences of the direct/indirect pathway imbalance on the thalamic activity by lesioning the GPi. SNr lesions are ruled out as a possible treatment, at present, by the possible collateral lesion to the DA-ergic cells of the SNc.

This model allows explanation of some of the cardinal features of PD, the pharmacological response to dopaminergic drugs and the impact of neurosurgery on PD control, but its value decreases for explaining disease progression. Many data not included in the original model of the BG have been published during the last 10 years. The illness has gone beyond the DA-ergic cells and the clinical expression of the illness has extended from motor disturbances to many different sensitive, cognitive and emotional functions, transforming the illness into a complex puzzle which requires an updating of the original BG model.

2.3.4 New Structural Data for the Basal Ganglia Model

Structural data reported during the last few years should be incorporated into the BG model. First of all, probably not all the information originating from different cortical areas or different somatotopic regions flow in parallel within the BG. Two opposing views have been proposed.[139] One is the "parallel processing hypothesis"[140–142] which suggests information from different cortical areas is always processed independently in different parts of the BG. Another view is that the BG and particularly the *striatum*, perform an integrative convergence of the cortical information.[143] Both views are possible to operate, with the integration of information being particularly relevant when it comes from cortical areas with similar functions. Similarly, convergence could also be produced inside particular loops where projections come from cortical areas with a functional relationship. For instance, those projections arriving from the primary motor cortex and supplementary motor area converge in the same striatal regions; whereas cortical areas with different functions project to different striatal regions. The cortical projections from the supplementary motor area could innervate the same striatal region as those innervated by the primary motor cortex but not the same as those innervated by the presupplementary motor area. This convergence should preserve the somatotopic distribution of the original areas in the cortex, with the same parts of the homunculus of different cortical areas being in nearby regions of the *striatum*.

Changes in the somatotopy suggest that the parallel *vs.* convergence of information within the BG loops can be altered in PD. For instance, GPe and GPi neurons only change their activity when a specific movement of a single joint in a specific direction is performed by a contralateral member. This selectivity changes in the Parkinsonian state where neurons begin to respond to multiple movements of multiple joints of the upper and lower limbs of both sides,[144,145] a cross-talk which has been related to the DA-ergic denervation of the *striatum*.[146] A similar disintegration of the homunculus has been reported in dystonia where patients trying to move one body part also move ("motor overflow") other nearby regions of the body.[147,148]

Another structural characteristic to be included in the BG model is the short loop. There are a number of short loops in the long cortico–basal loop involving only a portion of the basal ganglia. Three examples of these loops, which were already known in the 1980s, are the SNc–*striatum*–SNc loop (DA-ergic nigrostriatal cells activating GABA-ergic nigrostriatal cells which retrogradely inhibit the firing rate of the DA-ergic nigral neurons), the *striatum*-SNr-intralaminar nuclei of the thalamus–*striatum* loop and the thalamo–cortex–thalamo GLU-ergic loop.[149] The study of the functional relevance of these short loops (and of others such as the thalamus–*striatum*–GPi-thalamus, thalamus–STN–GPi-thalamus and STN–GPi–STN loops) has attracted attention during the past decade.[150,151]

Another structural characteristic to be considered is the interhemispheric connections of the BG and particularly the bilateral cortico–striatal projections. Ipsilateral cortico–striatal projections arise in the rat from two different populations of cortico–fugal neurons, known as intratelencephalic (IT) and pyramidal-tract (PT) neurons. The contralateral cortico–striatal projections only originate from PT neurons. IT neurons mainly innervate the MsSN of the direct pathway, whereas PT neurons preferentially target the MsSN of the indirect pathway. Nigrostriatal DA-cells send a small percentage of projections to the contralateral *striatum* (4%), which do not prevent the DA-ergic cells of each brain side from being able to present a marked regulatory interaction with the same cell type of the contralateral mesencephalon. The interhemispheric regulation is probably mediated by connections between the right and left deep mesencephalic nucleus (DmN), and between thalamic nuclei of both brain sides, which could facilitate the interhemispheric regulation of the electrophysiological behavior[152] and the neurochemical characteristics of nigrostriatal cells.[153,154]

There are a number of other structural details which need to be addressed. For instance, not all nigrostriatal cells are DA-ergic. About 10–50% of these cells are GABA-ergic or present a GABA-DA cotransmission.[155,156]

The functional significance of these cells is presently unknown, but evidence showing their change after DA-ergic degeneration suggests their involvement in PD. There are new candidates to be included in the BG loops. An example is the DmN which receives massive striatal inputs and sends projections to the thalamus[157] and whose activity changes after DA-cell degeneration.[158] In addition, not all the cortical information arriving at the *striatum* returns to the cortex. Centers such as the colliculus and pedunculo-pontine nucleus (PPN) receive BG inputs and do not return information to the cortex.

The classical model of the BG did not take into account the role of interneurons, particularly in the *striatum* where the cholinergic tonically active interneurons and the GABA-ergic fast-spiking interneurons are the most prominent.[159–161] The tonically active interneurons modulate MsSN excitability by presynaptic inhibition of the cortical GLU-ergic input, thus mediating the DA-ergic mechanisms involved in long-term depression. Although less abundant, the fast-spiking interneurons are involved in the feed-forward inhibition which, together with collateral of MsSN axons, provide intrastriatal inhibition. These mechanisms could promote the intense activation of some micro-regions of the *striatum* while inhibiting collateral regions, which could be involved in the selection of motor patterns.

Glial cells were not included in the classical model of the BG. Until recently, glia was thought to be involved in supporting activities but not in the information processing that was specifically attributed to neurons. In this view, glial cells of the BG were mainly considered as possible substrates for the mechanisms involved in the neurodegenerative process. This could also be the case of microglia, whose activation in the *substantia nigra* of PD could promote DA-cell degeneration and of astrocytes, cells suitable for preventing the excitotoxic effect of GLU and for providing energy to close-by neurons which are threatened by the degenerative process.[162–165] The passive role of glia in information processing is now being re-considered. An example is the role of astrocytes in GLU-ergic neurotransmission, the most common synapse in the BG.[166–169]

Finally, there is recent evidence showing that many structural characteristics of the BG are not permanent and can be modified by experience and by the action of many different factors. The dendritic spines of MsINs are an example of these changes. Each MsSN may have more than 5000 spines which, as mentioned above, receive massive glutamatergic inputs from the cerebral cortex and thalamus, together with modulatory DA-ergic inputs from the SNc.[170,171] The cortico–striatal synapse shows a marked plasticity with short-term and long-term synaptic changes modulated by glutamate and DA and which has been associated with a re-configuration of motor abilities.[172–175]

Both studies in PD patients and in animal models of the disease have reported that the decrease of DA can induce a marked reduction in the number of spines of MsSN neurons (30–50%), suggesting that these structural modifications could be relevant for motor disturbances of PD.[176,177] These studies have built a bridge between the structural and functional abnormalities of BG in PD.

2.3.5 Functional Data to be Included in the Basal Ganglia Model

The classical model considers the neuronal firing rate as the biological substrate for the information processing in the BG. Although this assumption was not an explicit statement, it was the basis of many explanations provided by the model, e.g., explaining the relationship between BG centers as an activation/inhibition and explaining motor disturbances as the result of changes in the firing rates of neurons of a particular center or in the inhibition/excitation between particular BG. However, the electrophysiological behavior of BG cells seems to be much more complex than the simple increase/decrease of their firing rate. An example of this complexity is provided by the electrophysiological ultra-stability shown by DA-ergic nigrostriatal cells. Each DA-cell shows a stable firing rate (around 4.5 Hz and always between 0.3 and 10 Hz) which is actively maintained over time. Transient modifications of the firing stability will induce an opposite reaction during the following seconds. As a consequence, a transient firing increase over the basal firing activity will be followed by a transient decrease which will persist until the average firing rate is recovered.[178] Another example of the complex firing code is the fractal pattern of firing observed in GABA-ergic cells of the SNr.[179] The time-invariance of fractal firing could be useful in complex systems which, as BG, use the massive arrival of information to organize complex motor patterns with different cadences in parallel, for example walking, saluting and speaking simultaneously.

Another characteristic of BG cells is their propensity to fire as groups and following oscillatory patterns, which has been suggested as being a determining factor in the confection of the motor state. Changes in this oscillatory pattern (hypersynchrony) have been associated with motor disturbance in PD. Thus, recordings of local field potentials through deep-brain stimulation macro-electrodes implanted in the STN have shown enhanced beta-frequency (≈ 18 Hz) oscillations in the "off" motor state. Such beta activity is reduced in the "on" medication state while γ-band (>60 Hz) activity is increased. Similar oscillations also occur in the GPi and cortex of PD patients.[180] Abnormal oscillations in the theta band (≈ 6 Hz) have been associated with dyskinesias and impulsivity induced by levodopa and DA-ergic agonists.[181,182] Pallidotomy has been suggested to eliminate dyskinesia by interrupting the hypersynchrony in output centers of the BG.[183]

Accordingly, the code for neuronal communication in the BG may be much more complex than a simple excitation/inhibition relationship. The task execution probably involves a spatio-temporal assembly of BG neurons whose synchronization could be necessary for the generation of adequate motor programs.[184–186] The development of new methods capable of monitoring hundreds of neurons in animals will allow the study of the likely complexity of BG neurons.

2.4 Targets for New Parkinson's Disease Therapies

The blue lines in Figure 2.1 indicate compensatory mechanisms or therapeutic interventions that may lead to the recovery of cell and organ function.

The therapy of chronic complex diseases may have common strategies to tackle and reduce cellular stress. Some treatments used for a specific chronic disease may also produce beneficial outcomes, even if small, for other chronic diseases. Therefore, it is not surprising that the main recommendations for prevention of brain degenerative disorders and other chronic diseases are similar: healthy nutritional habits and moderate mental and physical exercise.[187] Some drugs used for the treatment of other complex diseases may have application for finding clues to the pathogenesis of PD or even its treatment. Let us illustrate this point with some examples.

Metformin is a Complex I inhibitor used worldwide for the treatment of type 2 diabetes and epidemiological studies suggest it may also protect against different types of cancer.[188] Metformin's mechanism of action is not completely understood, as its anti-gluconeogenic liver effects are not due to the activation of the LKB1/AMPK pathway but by a 'simple' decrease in hepatic energy state.[189] A simple question: does metformin treatment have any epidemiological relationship with PD? No report exists in Pubmed on metformin and PD and it may be interesting to do an epidemiological study if one has not already been done. Clearly it has not been reported that metformin treatment increases the risk of suffering PD, even though it crosses the blood–brain barrier and seems to increase the production of Abeta.[190] Similarly, there is a clinical trial of pioglitazone (NINDS NET-PD, http://parkinsontrial.ninds.nih.gov/), another drug used in type 2 diabetes, for the treatment of early PD patients.

Blocking L-type $Ca(v)1.3$ Ca^{2+} channels may produce the 're-juvenation' of DA-ergic neurons.[80] The buffering of cytoplasmic calcium levels by Endoplasmic Reticulum (ER) and mitochondria is more critical for normal function of DA-ergic neurons in the *substantia nigra* than in other neurons.[191] Accordingly, specific calcium blockers are predicted as neuroprotectors for PD. Epidemiological evidence in cohorts and case-control studies of patients that are treated with calcium blockers for hypertension have shown no correlation with PD, while a recent case-control study of the use of specific L-type dihydropyridine calcium blockers in Denmark showed a reduced risk of PD [ref. 192 and references therein).

The most precise delineation of those common channels not unexpectedly comes from simple organisms that are 'less sophisticated'. Expression of Snca in yeast cells at low levels does not produce any significant effect on yeast viability, but its overexpression kills yeast cells and there are several pathways that can modify Snca toxicity both positively and negatively.[39,193–196] Neutral lipids accumulate in lipid droplets in yeast cells overexpressing Snca.[193] The inhibition with statins of the mevalonate-ergosterol biosynthesis pathway as well as inhibition with rapamycin of the mTOR pathway increases Snca toxicity in yeast cells.[39] In contrast to yeast results, mice treated with rapamycin[197] or simvastatin are protected against MPTP-induced nigral degeneration[198] and treatment with lovastatin reduces Snca accumulation and oxidation in Snca transgenic mice.[199] In humans, the macrolide rapamycin seems not to cross the blood–brain barrier and we could not find any epidemiological studies on PD and rapamycin. There are conflicting results on the brain accessibility of the different statins used to control cholesterol metabolism for prevention of atherosclerosis, clearly more epidemiological data are needed to clarify the possible relationship between therapies with different statins and PD.[200]

Proteasome inhibitors have been in use as a treatment for multiple myeloma and mantle cell lymphoma for more than a decade and are now indicated for other types of cancer.[201] An epidemiological study on those treated patients in relation to PD and other brain degenerative diseases may give us some clues to understand the conflicting results of the bench experiments with proteasome inhibitors.

The unexpected finding that copper(II) diacetylbis(N(4)-methylthiosemicarbazonato) used for the PET imaging of hypoxia as a potent scavenger of NOS and its protective role in animal models of PD, suggests that this compound or its derivatives could be used in clinical trials in PD patients.[202]

Specific drug treatments for PD based on molecular studies of its pathogenesis have revealed two clear targets, c-Abl and LRRK2 and possibly autophagy as discussed above. The use of c-Abl inhibitors that cross the blood–brain barrier may be effective for prevention or to stop progression of PD, as those inhibitors, like dasatinib, will prevent the inactivation of PARKIN by oxidative stress.[75,76] The use of LRRK2 inhibitors[203] or other ways to tackle with the effect of mutations in the LRRK2 kinase in the development of PD are worthy of clinical trials in patients with this dominant form of PD.[204] Increasing autophagy may be considered a good therapeutic goal to treat neurodegenerative diseases, as it would promote the removal of protein aggregates and damaged mitochondria. Nevertheless, use of autophagic inducers has to be carefully evaluated before going to clinical trials, as they may have potential deleterious effects.[205]

The combined research of molecular channels and their connections in cell regulation, together with better knowledge of the physiology of the basal ganglia will generate a better understanding of PD pathogenesis and pathophysiology. Those studies may provide useful biomarkers for early detection, to monitor progression and the effectiveness of new therapies. Treatment of PD will benefit from those studies, as well as through exploring therapeutics already available for other chronic diseases. Hopefully the bidirectional dialogue, bench-to-bed and bed-to-bench, will produce effective therapies to stop PD progression or even its prevention, ameliorating the daily life of PD patients.

Acknowledgements

This work was supported by grants from MINECO SAF-2012-34556, Comunidad de Madrid P2010_BMD-2331 & CIBERNED to JGC, by MICYT SAF2008-03746 and CIBERNED to MR and by MICYT (SAF2005-08416; SAF2008-04276) and CIBERNED to JAO.

References

1. C. M. Muller, R. A. De Vos, C. A. Maurage, D. R. Thal, M. Tolnay and H. Braak, *J. Neuropathol. Exp. Neurol.*, 2005, **64**, 623.
2. K. A. Jellinger, *Mov. Disord.*, 2003, **18**(6), S2.
3. L. Ishihara and C. Brayne, *Acta Neurol. Scand.*, 2006, **113**, 211.
4. H. Braak, J. R. Bohl, C. M. Muller, U. Rub, R. A. De Vos and K. Del Tredici, *Mov. Disord.*, 2006, **21**, 2042.
5. H. A. Hirsch, D. Iliopoulos, A. Joshi, Y. Zhang, S. A. Jaeger, M. Bulyk, P. N. Tsichlis, L. Shirley and X and K. Struhl, *Cancer Cell*, 2010, **17**, 348.
6. H. Ilieva, M. Polymenidou and D. W. Cleveland, *J. Cell. Biol.*, 2009, **187**, 761.
7. L. A. Garraway and W. R. Sellers, *Nat. Rev. Cancer*, 2006, **6**, 593.
8. B. A. Kudlow, B. K. Kennedy and R. J. Monnat, *Nat. Rev. Mol. Cell. Biol.*, 2007, **8**, 394.
9. M. V. Blagosklonny and M. N. Hall, *Aging*, 2009, **1**, 357.
10. M. T. Maurano, R. Humbert, E. Rynes, R. E. Thurman, E. Haugen, H. Wang, A. P. Reynolds, R. Sandstrom, H. Qu, J. Brody, A. Shafer, F. Neri, K. Lee, T. Kutyavin, S. Stehling-Sun, A. K. Johnson, T. K. Canfield, E. Giste, M. Diegel, D. Bates, R. S. Hansen, S. Neph, P. J. Sabo, S. Heimfeld, A. Raubitschek, S. Ziegler, C. Cotsapas, N. Sotoodehnia, I. Glass, S. R. Sunyaev, R. Kaul and J. A. Stamatoyannopoulos, *Science*, 2012, **337**, 1190.
11. C. M. Lill, J. T. Roehr, M. B. McQueen, F. K. Kavvoura, S. Bagade, B. M. Schjeide, L. M. Schjeide, E. Meissner, U. Zauft, N. C. Allen, T. Liu, M. Schilling, K. J. Anderson, G. Beecham, D. Berg, J. M. Biernacka, A. Brice, A. L. DeStefano, C. B. Do, N. Eriksson, S. A. Factor, M. J. Farrer, T. Foroud, T. Gasser, T. Hamza, J.A. Hardy, P. Heutink, E. M. Hill-Burns, C. Klein, J. C. Latourelle, D. M. Maraganore, E. R. Martin, M. Martinez, R. H. Myers, M. A. Nalls, N. Pankratz, H. Payami, W. Satake, W. K. Scott, M. Sharma, A. B. Singleton, K. Stefansson, T. Toda, J. Y. Tung, J. Vance, N. W. Wood, C. P. Zabetian, P. Young, R. E. Tanzi, M. J. Khoury, F. Zipp, H. Lehrach, J. P. Ioannidis and L. Bertram, *PLoS Genet.*, 2012, **8**, e1002548.
12. S. Lesage and A. Brice, *Hum. Mol. Genet.*, 2009, **18**, R48.
13. M. H. Polymeropoulos, C. Lavedan, E. Leroy, S. E. Ide, A. Dehejia, A. Dutra, B. Pike, H. Root, J. Rubenstein, R. Boyer, E. S. Stenroos, S. Chandrasekharappa, A. Athanassiadou, T. Papapetropoulos, W. G. Johnson, A. M. Lazzarini, R. C. Duvoisin, G. Di Iorio, L. I. Golbe and R. L. Nussbaum, *Science*, 1997, **276**, 2045.
14. R. Kruger, W. Kuhn, T. Muller, D. Woitalla, M. Graeber, S. Kosel, H. Przuntek, J. T. Epplen, L. Schols and O. Riess, *Nat. Genet.*, 1998, **18**, 106.
15. J. J. Zarranz, J. Alegre, J. C. Gomez-Esteban, E. Lezcano, R. Ros, I. Ampuero, L. Vidal, J. Hoenicka, O. Rodriguez, B. Atares, V. Llorens,

T. E. Gomez, T. Del Ser, D. G. Munoz and J. G. de Yebenes, *Ann. Neurol.*, 2004, **55**, 164.
16. M. C. Chartier-Harlin, J. Kachergus, C. Roumier, V. Mouroux, X. Douay, S. Lincoln, C. Levecque, L. Larvor, J. Andrieux, M. Hulihan, N. Waucquier, L. Defebvre, P. Amouyel, M. Farrer and A. Destee, *Lancet*, 2004, **364**, 1167.
17. P. Ibanez, A. M. Bonnet, B. Debarges, E. Lohmann, F. Tison, P. Pollak, Y. Agid, A. Durr and A. Brice, *Lancet*, 2004, **364**, 1169.
18. A. B. Singleton, M. Farrer, J. Johnson, A. Singleton, S. Hague, J. Kachergus, M. Hulihan, T. Peuralinna, A. Dutra, R. Nussbaum, S. Lincoln, A. Crawley, M. Hanson, D. Maraganore, C. Adler, M. R. Cookson, M. Muenter, M. Baptista, D. Miller, J. Blancato, J. Hardy and K. Gwinn-Hardy, *Science*, 2003, **302**, 841.
19. A. Jowaed, I. Schmitt, O. Kaut and U. Wullner, *J. Neurosci.*, 2010, **30**, 6355.
20. K. Nuytemans, J. Theuns, M. Cruts and C. van Broeckhoven, *Hum. Mutat.*, 2010, **31**, 763.
21. T. Kitada, S. Asakawa, N. Hattori, H. Matsumine, Y. Yamamura, S. Minoshima, M. Yokochi, Y. Mizuno and N. Shimizu, *Nature*, 1998, **392**, 605.
22. L. Samaranch, O. Lorenzo-Betancor, J. M. Arbelo, I. Ferrer, E. Lorenzo, J. Irigoyen, M. A. Pastor, C. Marrero, C. Isla, J. Herrera-Henriquez and P. Pastor, *Brain*, 2010, **133**, 1128.
23. C. Paisan-Ruiz, S. Jain, E. W. Evans, W. P. Gilks, J. Simon, B. M. van der, d. M. Lopez, S. Aparicio, A. M. Gil, N. Khan, J. Johnson, J. R. Martinez, D. Nicholl, I. M. Carrera, A. S. Pena, R. de Silva, A. Lees, J. F. Marti-Masso, J. Perez-Tur, N. W. Wood and A. B. Singleton, *Neuron*, 2004, **44**, 595.
24. A. Zimprich, S. Biskup, P. Leitner, P. Lichtner, M. Farrer, S. Lincoln, J. Kachergus, M. Hulihan, R. J. Uitti, D. B. Calne, A. J. Stoessl, R. F. Pfeiffer, N. Patenge, I. C. Carbajal, P. Vieregge, F. Asmus, B. Muller-Myhsok, D. W. Dickson, T. Meitinger, T. M. Strom, Z. K. Wszolek and T. Gasser, *Neuron*, 2004, **44**, 601.
25. M. R. Cookson, *Nat. Rev. Neurosci.*, 2010, **11**, 791.
26. I. N. Rudenko, A. Kaganovich, D. N. Hauser, A. Beylina, R. Chia, J. Ding, D. Maric, H. Jaffe and M. R. Cookson, *Biochem. J.*, 2012, **446**, 99.
27. X. Lin, L. Parisiadou, X. L. Gu, L. Wang, H. Shim, L. Sun, C. Xie, C. X. Long, W. J. Yang, J. Ding, Z. Z. Chen, P. E. Gallant, J. H. Tao-Cheng, G. Rudow, J. C. Troncoso, Z. Liu, Z. Li and H. Cai, *Neuron*, 2009, **64**, 807.
28. Y. Xiong, C. E. Coombes, A. Kilaru, X. Li, A. D. Gitler, W. J. Bowers, V. L. Dawson, T. M. Dawson and D. J. Moore, *PLoS. Genet.*, 2010, **6**, e1000902.
29. C. Yao, R. El Khoury, W. Wang, T. A. Byrd, E. A. Pehek, C. Thacker, X. Zhu, M. A. Smith, A. L. Wilson-Delfosse and S. G. Chen, *Neurobiol. Dis.*, 2010, **40**, 73.

30. K. Stafa, A. Trancikova, P. J. Webber, L. Glauser, A. B. West and D. J. Moore, *PLoS. Genet.*, 2012, **8**, e1002526.
31. J. W. Langston, *Neurology*, 1996, **47**, S153.
32. A. Schober, *Cell Tissue Res.*, 2004, **318**, 215.
33. J. Bove, D. Prou, C. Perier and S. Przedborski, *NeuroRx.*, 2005, **2**, 484.
34. M. E. Emborg, *ILAR J.*, 2007, **48**, 339.
35. N. M. Filipov, A. B. Norwood and S. C. Sistrunk, *Neuroreport*, 2009, **20**, 713.
36. R. J. Dinis-Oliveira, J. A. Duarte, A. Sanchez-Navarro, F. Remiao, M. L. Bastos and F. Carvalho, *Crit. Rev. Toxicol.*, 2008, **38**, 13.
37. I. Mena, *Ann. Clin. Lab. Sci.*, 1974, **4**, 487.
38. C. W. Olanow, P. F. Good, H. Shinotoh, K. A. Hewitt, F. Vingerhoets, B. J. Snow, M. F. Beal, D. B. Calne and D. P. Perl, *Neurology*, 1996, **46**, 492.
39. A. D. Gitler, A. Chesi, M. L. Geddie, K. E. Strathearn, S. Hamamichi, K. J. Hill, K. A. Caldwell, G. A. Caldwell, A. A. Cooper, J. C. Rochet and S. Lindquist, *Nat. Genet.*, 2009, **41**, 308.
40. S. A. Schneider, C. Paisan-Ruiz, N. P. Quinn, A. J. Lees, H. Houlden, J. Hardy and K. P. Bhatia, *Mov. Disord.*, 2010, **25**, 979.
41. W. R. Martin, M. Wieler and M. Gee, *Neurology*, 2008, **70**, 1411.
42. A. Gregory, B. J. Polster and S. J. Hayflick, *J. Med. Genet.*, 2009, **46**, 73.
43. Z. Wu, C. Li, S. Lv and B. Zhou, *Hum. Mol. Genet.*, 2009, **18**, 3659.
44. C. K. Glass, K. Saijo, B. Winner, M. C. Marchetto and F. H. Gage, *Cell*, 2010, **140**, 918.
45. M. G. Tansey and M. S. Goldberg, *Neurobiol. Dis.*, 2010, **37**, 510.
46. P. L. McGeer, S. Itagaki, B. E. Boyes and E. G. McGeer, *Neurology*, 1988, **38**, 1285.
47. E. Croisier, L. B. Moran, D. T. Dexter, R. K. Pearce and M. B. Graeber, *J. Neuroinflammation*, 2005, **2**, 14.
48. Y. Ouchi, E. Yoshikawa, Y. Sekine, M. Futatsubashi, T. Kanno, T. Ogusu and T. Torizuka, *Ann. Neurol.*, 2005, **57**, 168.
49. Y. Ouchi, S. Yagi, M. Yokokura and M. Sakamoto, *Parkinsonism Relat. Disord.*, 2009, **15**(3), S200.
50. J. H. Kordower, Y. Chu, R. A. Hauser, C. W. Olanow and T. B. Freeman, *Mov. Disord.*, 2008, **23**, 2303.
51. J. H. Kordower, Y. Chu, R. A. Hauser, T. B. Freeman and C. W. Olanow, *Nat. Med.*, 2008, **14**, 504.
52. J. Y. Li, E. Englund, J. L. Holton, D. Soulet, P. Hagell, A. J. Lees, T. Lashley, N. P. Quinn, S. Rehncrona, A. Bjorklund, H. Widner, T. Revesz, O. Lindvall and P. Brundin, *Nat. Med.*, 2008, **14**, 501.
53. I. Mendez, A. Vinuela, A. Astradsson, K. Mukhida, P. Hallett, H. Robertson, T. Tierney, R. Holness, A. Dagher, J. Q. Trojanowski and O. Isacson, *Nat. Med.*, 2008, **14**, 507.
54. R. Banerjee, A. A. Starkov, M. F. Beal and B. Thomas, *Biochim. Biophys. Acta*, 2009, **1792**, 651.

55. A. H. Schapira, *Lancet Neurol.*, 2008, **7**, 97.
56. S. Papa, V. Petruzzella, S. Scacco, A. M. Sardanelli, A. Iuso, D. Panelli, R. Vitale, R. Trentadue, D. De Rasmo, N. Capitanio, C. Piccoli, F. Papa, M. Scivetti, E. Bertini, T. Rizza and G. De Michele, *Biochim. Biophys. Acta*, 2009, **1787**, 502.
57. S. Vilain, G. Esposito, D. Haddad, O. Schaap, M. P. Dobreva, M. Vos, M. S. Van, V. A. Morais, S. B. De and P. Verstreken, *PLoS. Genet.*, 2012, **8**, e1002456.
58. A. Bender, K. J. Krishnan, C. M. Morris, G. A. Taylor, A. K. Reeve, R. H. Perry, E. Jaros, J. S. Hersheson, J. Betts, T. Klopstock, R. W. Taylor and D. M. Turnbull, *Nat. Genet.*, 2006, **38**, 515.
59. Y. Kraytsberg, E. Kudryavtseva, A. C. McKee, C. Geula, N. W. Kowall and K. Khrapko, *Nat. Genet.*, 2006, **38**, 518.
60. N. G. Larsson, *Ann. Rev. Biochem.*, 2010, **79**, 683.
61. G. Davidzon, P. Greene, M. Mancuso, K. J. Klos, J. E. Ahlskog, M. Hirano and S. DiMauro, *Ann. Neurol.*, 2006, **59**, 859.
62. A. Trifunovic, A. Wredenberg, M. Falkenberg, J. N. Spelbrink, A. T. Rovio, C. E. Bruder, Y. Bohlooly, S. Gidlof, A. Oldfors, R. Wibom, J. Tornell, H. T. Jacobs and N. G. Larsson, *Nature*, 2004, **429**, 417.
63. A. Trifunovic, A. Hansson, A. Wredenberg, A. T. Rovio, E. Dufour, I. Khvorostov, J. N. Spelbrink, R. Wibom, H. T. Jacobs and N. G. Larsson, *Proc. Natl. Acad. Sci. U. S. A.*, 2005, **102**, 17993.
64. M. Vermulst, J. Wanagat, G. C. Kujoth, J. H. Bielas, P. S. Rabinovitch, T. A. Prolla and L. A. Loeb, *Nat. Genet.*, 2008, **40**, 392.
65. K. Inoue, K. Nakada, A. Ogura, K. Isobe, Y. Goto, I. Nonaka and J. I. Hayashi, *Nat. Genet.*, 2000, **26**, 176.
66. M. I. Ekstrand, M. Terzioglu, D. Galter, S. Zhu, C. Hofstetter, E. Lindqvist, S. Thams, A. Bergstrand, F. S. Hansson, A. Trifunovic, B. Hoffer, S. Cullheim, A. H. Mohammed, L. Olson and N. G. Larsson, *Proc. Natl. Acad. Sci. U. S. A.*, 2007, **104**, 1325.
67. J. Wang, J. P. Silva, C. M. Gustafsson, P. Rustin and N. G. Larsson, *Proc. Natl. Acad. Sci. U. S. A.*, 2001, **98**, 4038.
68. Y. Yang, A. H. Kim and A. Bonni, *Curr. Opin. Neurobiol.*, 2010, **20**, 92.
69. O. Corti, S. Lesage and A. Brice, *Physiol. Rev.*, 2011, **91**, 1161.
70. C. Soto, *Cell*, 2012, **149**, 968.
71. L. A. Volpicelli-Daley, K. C. Luk, T. P. Patel, S. A. Tanik, D. M. Riddle, A. Stieber, D. F. Meaney, J. Q. Trojanowski and V. M. Lee, *Neuron*, 2011, **72**, 57.
72. K. C. Luk, V. M. Kehm, B. Zhang, P. O'Brien, J. Q. Trojanowski and V. M. Lee, *J. Exp. Med*, 2012, **209**, 975.
73. R. J. Youle and A. M. van der Bliek, *Science*, 2012, **337**, 1062.
74. G. K. Tofaris, *Mov. Disord.*, 2012.
75. H. S. Ko, Y. Lee, J. H. Shin, S. S. Karuppagounder, B. S. Gadad, A. J. Koleske, O. Pletnikova, J. C. Troncoso, V. L. Dawson and T. M. Dawson, *Proc. Natl. Acad. Sci. U. S. A.*, 2010, **107**, 16691.

76. S. Z. Imam, Q. Zhou, A. Yamamoto, A. J. Valente, S. F. Ali, M. Bains, J. L. Roberts, P. J. Kahle, R. A. Clark and S. Li, *J. Neurosci.*, 2011, **31**, 157.
77. J. H. Shin, H. S. Ko, H. Kang, Y. Lee, Y. I. Lee, O. Pletinkova, J. C. Troconso, V. L. Dawson and T. M. Dawson, *Cell*, 2011, **144**, 689.
78. M. Biswas and J. Y. Chan, *Toxicol. Appl. Pharmacol.*, 2010, **244**, 16.
79. T. W. Kensler, N. Wakabayashi and S. Biswal, *Annu. Rev. Pharmacol. Toxicol.*, 2007, **47**, 89.
80. C. S. Chan, J. N. Guzman, E. Ilijic, J. N. Mercer, C. Rick, T. Tkatch, G. E. Meredith and D. J. Surmeier, *Nature*, 2007, **447**, 1081.
81. J. N. Guzman, J. Sanchez-Padilla, D. Wokosin, J. Kondapalli, E. Ilijic, P. T. Schumacker and D. J. Surmeier, *Nature*, 2010, **468**, 696.
82. J. A. Goldberg, J. N. Guzman, C. M. Estep, E. Ilijic, J. Kondapalli, J. Sanchez-Padilla and D. J. Surmeier, *Nat. Neurosci.*, 2012, **15**, 1414–1421.
83. Y. V. Budovskaya, K. Wu, L. K. Southworth, M. Jiang, P. Tedesco, T. E. Johnson and S. K. Kim, *Cell*, 2008, **134**, 291.
84. W. Ye, K. Shimamura, J. L. Rubenstein, M. A. Hynes and A. Rosenthal, *Cell*, 1998, **93**, 755.
85. N. Prakash, C. Brodski, T. Naserke, E. Puelles, R. Gogoi, A. Hall, M. Panhuysen, D. Echevarria, L. Sussel, D. M. Weisenhorn, S. Martinez, E. Arenas, A. Simeone and W. Wurst, *Development*, 2006, **133**, 89.
86. M. P. Smidt and J. P. Burbach, *Nat. Rev. Neurosci.*, 2007, **8**, 21.
87. W. Lin, E. Metzakopian, Y. E. Mavromatakis, N. Gao, N. Balaskas, H. Sasaki, J. Briscoe, J. A. Whitsett, M. Goulding, K. H. Kaestner and S. L. Ang, *Dev. Biol.*, 2009, **333**, 386.
88. N. Flames and O. Hobert, *Nature*, 2009, **458**, 885.
89. J. O. Andressoo and M. Saarma, *Curr. Opin. Neurobiol.*, 2008, **18**, 297.
90. W. D. Le, P. Xu, J. Jankovic, H. Jiang, S. H. Appel, R. G. Smith and D. K. Vassilatis, *Nat. Genet.*, 2003, **33**, 85.
91. P. M. Sleiman, D. G. Healy, M. M. Muqit, Y. X. Yang, B. M. van der, J. L. Holton, T. Revesz, N. P. Quinn, K. Bhatia, J. K. Diss, A. J. Lees, M. R. Cookson, D. S. Latchman and N. W. Wood, *Neurosci. Lett.*, 2009, **457**, 75.
92. D. A. Grimes, F. Han, M. Panisset, L. Racacho, F. Xiao, R. Zou, K. Westaff and D. E. Bulman, *Mov. Disord.*, 2006, **21**, 906.
93. Y. Chu, W. Le, K. Kompoliti, J. Jankovic, E. J. Mufson and J. H. Kordower, *J. Comp. Neurol.*, 2006, **494**, 495.
94. W. Le, T. Pan, M. Huang, P. Xu, W. Xie, W. Zhu, X. Zhang, H. Deng and J. Jankovic, *J. Neurol. Sci.*, 2008, **273**, 29.
95. P. Sgado, L. Alberi, D. Gherbassi, S. L. Galasso, G. M. Ramakers, K. N. Alavian, M. P. Smidt, R. H. Dyck and H. H. Simon, *Proc. Natl. Acad. Sci. U. S. A.*, 2006, **103**, 15242.
96. P. Sgado, C. Viaggi, C. Fantacci and G. U. Corsini, *Parkinsonism Relat. Disord.*, 2008, **14**(2), S103.
97. C. Jiang, X. Wan, Y. He, T. Pan, J. Jankovic and W. Le, *Exp. Neurol.*, 2005, **191**, 154.

98. B. Kadkhodaei, T. Ito, E. Joodmardi, B. Mattsson, C. Rouillard, M. Carta, S. Muramatsu, C. Sumi-Ichinose, T. Nomura, D. Metzger, P. Chambon, E. Lindqvist, N. G. Larsson, L. Olson, A. Bjorklund, H. Ichinose and T. Perlmann, *J. Neurosci.*, 2009, **29**, 15923.
99. X. Lin, L. Parisiadou, C. Sgobio, G. Liu, J. Yu, L. Sun, H. Shim, X. L. Gu, J. Luo, C. X. Long, J. Ding, Y. Mateo, P. H. Sullivan, L. G. Wu, D. S. Goldstein, D. Lovinger and H. Cai, *J. Neurosci.*, 2012, **32**, 9248.
100. A. Pascual, M. Hidalgo-Figueroa, J. I. Piruat, C. O. Pintado, R. Gomez-Diaz and J. Lopez-Barneo, *Nat. Neurosci.*, 2008, **11**, 755.
101. M. M. Harraz, T. M. Dawson and V. L. Dawson, *J. Chem. Neuroanat.*, 2011, **42**, 127.
102. J. Kim, K. Inoue, J. Ishii, W. B. Vanti, S. V. Voronov, E. Murchison, G. Hannon and A. Abeliovich, *Science*, 2007, **317**, 1220.
103. S. Gehrke, Y. Imai, N. Sokol and B. Lu, *Nature*, 2010, **466**, 637.
104. E. Minones-Moyano, S. Porta, G. Escaramis, R. Rabionet, S. Iraola, B. Kagerbauer, Y. Espinosa-Parrilla, I. Ferrer, X. Estivill and E. Marti, *Hum. Mol. Genet.*, 2011, **20**, 3067.
105. R. L. Albin, A. B. Young and J. B. Penney, *Trends Neurosci.*, 1989, **12**, 366.
106. M. R. DeLong, *Trends Neurosci.*, 1990, **13**, 281.
107. A. Dahlstroem and K. Fuxe, *Acta Physiol. Scand.*, 1964, **232**, 1–55.
108. U. Ungerstedt, *Acta Physiol. Scand.*, 1971, **367**, 1.
109. A. Parent and Y. Smith, *Brain Res.*, 1987, **426**, 397.
110. M. Cossette, M. Levesque and A. Parent, *Neurosci. Res.*, 1999, **34**, 51.
111. B. Lavoie, Y. Smith and A. Parent, *J. Comp. Neurol.*, 1989, **289**, 36.
112. J. Marcusson and K. Eriksson, *Brain Res.*, 1988, **457**, 122.
113. L. Prensa, M. Cossette and A. Parent, *J. Chem. Neuroanat.*, 2000, **20**, 207.
114. K. S. Rommelfanger and T. Wichmann, *Front. Neuroanat.*, 2010, **4**, 139.
115. T. Gonzalez-Hernandez and M. Rodriguez, *J. Comp. Neurol.*, 2000, **421**, 107.
116. E. Hirsch, A. M. Graybiel and Y. A. Agid, *Nature*, 1988, **334**, 345.
117. T. A. Newcomer, A. M. Palmer, P. A. Rosenberg and E. Aizenman, *J. Neurochem.*, 1993, **61**, 911.
118. A. Rescigno, A. C. Rinaldi and E. Sanjust, *Biochem. Pharmacol.*, 1998, **56**, 1089.
119. I. Cruz-Muros, D. Afonso-Oramas, P. Abreu, P. Barroso-Chinea, M. Rodriguez, M. C. Gonzalez and T. G. Hernandez, *Exp. Neurol.*, 2007, **204**, 147.
120. I. Cruz-Muros, D. Afonso-Oramas, P. Abreu, M. Rodriguez, M. C. Gonzalez and T. Gonzalez-Hernandez, *Neurobiol. Aging*, 2008, **29**, 1702.
121. P. Damier, E. C. Hirsch, Y. Agid and A. M. Graybiel, *Brain*, 1999, **122**(8), 1421.
122. M. Rodriguez, P. Barroso-Chinea, P. Abdala, J. Obeso and T. Gonzalez-Hernandez, *Exp. Neurol.*, 2001, **169**, 163.

123. W. Schultz, *J. Neurophysiol.*, 1998, **80**, 1.
124. M. Rodriguez, I. Morales, I. Gomez, S. Gonzalez, T. Gonzalez-Hernandez and J. L. Gonzalez-Mora, *J. Pharmacol. Exp. Ther.*, 2006, **319**, 31.
125. F. G. Gonon, *Neuroscience*, 1988, **24**, 19.
126. R. M. Wightman and J. B. Zimmerman, *Brain Res. Rev.*, 1990, **15**, 135.
127. E. Cubo, K. Kompoliti, S. E. Leurgans and R. Raman, *Clin. Neuropharmacol.*, 2004, **27**, 30.
128. H. Braak, T. K. Del, U. Rub, R. A. de Vos, E. N. Jansen Steur and E. Braak, *Neurobiol. Aging*, 2003, **24**, 197.
129. G. M. Halliday, T. K. Del and H. Braak, *J. Neural Transm.*, 1971, **367**, 49–68.
130. W. P. Gai, G. M. Halliday, P. C. Blumbergs, L. B. Geffen and W. W. Blessing, *Brain*, 1991, **114**(5), 2253.
131. E. C. Hirsch, A. M. Graybiel, C. Duyckaerts and F. Javoy-Agid, *Proc. Natl. Acad. Sci. U. S. A.*, 1987, **84**, 5976.
132. C. Delaville, P. D. Deurwaerdere and A. Benazzouz, *Front. Syst. Neurosci.*, 2011, **5**, 31.
133. E. Scarnati, F. Hajdu, C. Pacitti and T. Tombol, *J. Hirnforsch.*, 1988, **29**, 95.
134. K. M. Powers, D. M. Kay, S. A. Factor, C. P. Zabetian, D. S. Higgins, A. Samii, J. G. Nutt, A. Griffith, B. Leis, J. W. Roberts, E. D. Martinez, J. S. Montimurro, H. Checkoway and H. Payami, *Mov. Disord.*, 2008, **23**, 88.
135. M. S. Aymerich, P. Barroso-Chinea, M. Perez-Manso, A. M. Munoz-Patino, M. Moreno-Igoa, T. Gonzalez-Hernandez and J. L. Lanciego, *Eur. J. Neurosci.*, 2006, **23**, 2099.
136. K. Sedaghat, D. I. Finkelstein and A. L. Gundlach, *Brain Res.*, 2009, **1271**, 83.
137. Y. Wu, W. Le and J. Jankovic, *Arch. Neurol.*, 2011, **68**, 22.
138. G. E. Alexander, M. R. DeLong and P. L. Strick, *Annu. Rev Neurosci.*, 1986, **9**, 357.
139. A. Parent and L. N. Hazrati, *Brain Res. Rev.*, 1995, **20**, 91.
140. G. E. Alexander, M. R. DeLong and P. L. Strick, *Annu. Rev. Neurosci.*, 1986, **9**, 357.
141. J. E. Hoover and P. L. Strick, *Science*, 1993, **259**, 819.
142. J. E. Hoover and P. L. Strick, *J. Neurosci.*, 1999, **19**, 1446.
143. G. Percheron and M. Filion, *Trends Neurosci.*, 1991, **14**, 55.
144. M. Filion, L. Tremblay and P. J. Bedard, *Brain Res.*, 1988, **444**, 165.
145. M. Pessiglione, D. Guehl, A. S. Rolland, C. Francois, E. C. Hirsch, J. Feger and L. Tremblay, *J. Neurosci.*, 2005, **25**, 1523.
146. H. Bergman, A. Feingold, A. Nini, A. Raz, H. Slovin, M. Abeles and E. Vaadia, *Trends Neurosci.*, 1998, **21**, 32.
147. J. L. Vitek, V. Chockkan, J. Y. Zhang, Y. Kaneoke, M. Evatt, M. R. DeLong, S. Triche, K. Mewes, T. Hashimoto and R. A. Bakay, *Ann. Neurol.*, 1999, **46**, 22.
148. S. Chiken, P. Shashidharan and A. Nambu, *J. Neurosci.*, 2008, **28**, 13967.

149. M. Castle, M. S. Aymerich, C. Sanchez-Escobar, N. Gonzalo, J. A. Obeso and J. L. Lanciego, *J. Comp. Neurol.*, 2005, **483**, 143.
150. J. A. Obeso, M. C. Rodriguez-Oroz, M. Rodriguez, J. L. Lanciego, J. Artieda, N. Gonzalo and C. W. Olanow, *Trends Neurosci.*, 2000, **23**, S8.
151. J. G. McHaffie, T. R. Stanford, B. E. Stein, V. Coizet and P. Redgrave, *Trends Neurosci.*, 2005, **28**, 401.
152. M. A. Castellano and D. M. Rodriguez, *Brain Res. Bull.*, 1991, **27**, 213.
153. M. Rodriguez, M. A. Castellano and M. D. Palarea, *Life Sci.*, 1990, **47**, 377.
154. T. Gonzalez-Hernandez, P. Barroso-Chinea and M. Rodriguez, *Mov. Disord.*, 2004, **19**, 1029.
155. M. Rodriguez and T. Gonzalez-Hernandez, *J. Neurosci.*, 1999, **19**, 4682.
156. T. Gonzalez-Hernandez, P. Barroso-Chinea, A. Acevedo, E. Salido and M. Rodriguez, *Eur. J. Neurosci.*, 2001, **13**, 57.
157. M. Rodriguez, P. Abdala, P. Barroso-Chinea and T. Gonzalez-Hernandez, *J. Comp. Neurol.*, 2001, **438**, 12.
158. T. Gonzalez-Hernandez, P. Barroso-Chinea, MA Perez de la Cruz, P. Valera, J. G. Dopico and M. Rodriguez, *Neuroscience*, 2002, **113**, 311.
159. E. Bracci, D. Centonze, G. Bernardi and P. Calabresi, *J. Neurophysiol.*, 2002, **87**, 2190.
160. A. Klaus, H. Planert, J. J. Hjorth, J. D. Berke, G. Silberberg and J. H. Kotaleski, *Front. Syst. Neurosci.*, 2011, **5**, 57.
161. T. Koos and J. M. Tepper, *J. Neurosci.*, 2002, **22**, 529.
162. G. M. Halliday and C. H. Stevens, *Mov. Disord.*, 2011, **26**, 6.
163. P. Marcaggi and D. Attwell, *Glia*, 2004, **47**, 217.
164. P. M. Rappold and K. Tieu, *Neurotherapeutics*, 2010, **7**, 413.
165. P. Teismann and J. B. Schulz, *Cell Tissue Res.*, 2004, **318**, 149.
166. A. Araque, V. Parpura, R. P. Sanzgiri and P. G. Haydon, *Trends Neurosci.*, 1999, **22**, 208.
167. P. G. Haydon and G. Carmignoto, *Physiol. Rev.*, 2006, **86**, 1009.
168. L. S. Overstreet, *Trends Neurosci.*, 2005, **28**, 59.
169. R. M. Villalba and Y. Smith, *Front. Syst. Neurosci.*, 2011, **5**, 68.
170. J. M. Kemp and T. P. Powell, *Philos. Trans. R. Soc., B*, 1971, **262**, 429.
171. Y. Smith, D. V. Raju, J. F. Pare and M. Sidibe, *Trends Neurosci.*, 2004, **27**, 520.
172. J. Ding, J. D. Peterson and D. J. Surmeier, *J. Neurosci.*, 2008, **28**, 6483.
173. M. Garcia-Munoz, L. Carrillo-Reid and G. W. Arbuthnott, *Front. Neuroanat.*, 2010, **4**, 144.
174. F. Tecuapetla, L. Carrillo-Reid, J. Bargas and E. Galarraga, *Proc. Natl. Acad. Sci. U. S. A.*, 2007, **104**, 10258.
175. J. R. Wickens, A. J. Begg and G. W. Arbuthnott, *Neuroscience*, 1996, **70**, 1.
176. C. A. Ingham, S. H. Hood and G. W. Arbuthnott, *Brain Res.*, 1989, **503**, 334.
177. Y. Smith, D. V. Raju, J. F. Pare and M. Sidibe, *Trends Neurosci.*, 2004, **27**, 520.

178. M. Rodriguez, E. Pereda, J. Gonzalez, P. Abdala and J. A. Obeso, *Synapse*, 2003, **49**, 216.
179. M. Rodriguez, E. Pereda, J. Gonzalez, P. Abdala and J. A. Obeso, *Exp. Brain. Res.*, 2003, **151**, 167.
180. P. Brown, A. Oliviero, P. Mazzone, A. Insola, P. Tonali and L. Di, *J. Neurosci.*, 2001, **21**, 1033.
181. F. Alonso-Frech, I. Zamarbide, M. Alegre, M. C. Rodriguez-Oroz, J. Guridi, M. Manrique, M. Valencia, J. Artieda and J. A. Obeso, *Brain*, 2006, **129**, 1748.
182. M. C. Rodriguez-Oroz, J. Lopez-Azcarate, D. Garcia-Garcia, M. Alegre, J. Toledo, M. Valencia, J. Guridi, J. Artieda and J. A. Obeso, *Brain*, 2011, **134**, 36.
183. P. Brown and A. Eusebio, *Mov. Disord.*, 2008, **23**, 12.
184. L. Carrillo-Reid, F. Tecuapetla, D. Tapia, A. Hernandez-Cruz, E. Galarraga, R. Drucker-Colin and J. Bargas, *J. Neurophysiol.*, 2008, **99**, 1435.
185. M. Garcia-Munoz, L. Carrillo-Reid and G. W. Arbuthnott, *Front. Neuroanat.*, 2010, **4**, 144.
186. O. Jaidar, L. Carrillo-Reid, A. Hernandez, R. Drucker-Colin, J. Bargas and A. Hernandez-Cruz, *J. Neurosci.*, 2010, **30**, 11326.
187. S. Y. Angell, I. Danel and K. M. DeCock, *Science*, 2012, **337**, 1456.
188. N. Papanas, E. Maltezos and D. P. Mikhailidis, *Expert Opin. Investig. Drugs*, 2010, **19**, 913.
189. M. Foretz, S. Hebrard, J. Leclerc, E. Zarrinpashneh, M. Soty, G. Mithieux, K. Sakamoto, F. Andreelli and B. Viollet, *J. Clin. Invest.*, 2010, **120**, 2355.
190. Y. Chen, K. Zhou, R. Wang, Y. Liu, Y. D. Kwak, T. Ma, R. C. Thompson, Y. Zhao, L. Smith, L. Gasparini, Z. Luo, H. Xu and F. F. Liao, *Proc. Natl. Acad. Sci. U. S. A.*, 2009, **106**, 3907.
191. C. S. Chan, T. S. Gertler and D. J. Surmeier, *Trends Neurosci.*, 2009, **32**, 249.
192. B. Ritz, S. L. Rhodes, L. Qian, E. Schernhammer, J. H. Olsen and S. Friis, *Ann. Neurol.*, 2010, **67**, 600.
193. T. F. Outeiro and S. Lindquist, *Science*, 2003, **302**, 1772.
194. A. A. Cooper, A. D. Gitler, A. Cashikar, C. M. Haynes, K. J. Hill, B. Bhullar, K. Liu, K. Xu, K. E. Strathearn, F. Liu, S. Cao, K. A. Caldwell, G. A. Caldwell, G. Marsischky, R. D. Kolodner, J. Labaer, J. C. Rochet, N. M. Bonini and S. Lindquist, *Science*, 2006, **313**, 324.
195. A. D. Gitler, B. J. Bevis, J. Shorter, K. E. Strathearn, S. Hamamichi, L. J. Su, K. A. Caldwell, G. A. Caldwell, J. C. Rochet, J. M. McCaffery, C. Barlowe and S. Lindquist, *Proc. Natl. Acad. Sci. U. S. A.*, 2008, **105**, 145.
196. E. Yeger-Lotem, L. Riva, L. J. Su, A. D. Gitler, A. G. Cashikar, O. D. King, P. K. Auluck, M. L. Geddie, J. S. Valastyan, D. R. Karger, S. Lindquist and E. Fraenkel, *Nat. Genet.*, 2009, **41**, 316.

197. C. Malagelada, Z. H. Jin, V. Jackson-Lewis, S. Przedborski and L. A. Greene, *J. Neurosci.*, 2010, **30**, 1166.
198. A. Ghosh, A. Roy, J. Matras, S. Brahmachari, H. E. Gendelman and K. Pahan, *J. Neurosci.*, 2009, **29**, 13543.
199. A. O. Koob, K. Ubhi, J. F. Paulsson, J. Kelly, E. Rockenstein, M. Mante, A. Adame and E. Masliah, *Exp. Neurol.*, 2010, **221**, 267.
200. C. Becker and C. R. Meier, *Expert Opin. Drug Saf.*, 2009, **8**, 261.
201. R. Z. Orlowski and D. J. Kuhn, *Clin. Cancer Res.*, 2008, **14**, 1649.
202. L. W. Hung, V. L. Villemagne, L. Cheng, N. A. Sherratt, S. Ayton, A. R. White, P. J. Crouch, S. Lim, S. L. Leong, S. Wilkins, J. George, B. R. Roberts, C. L. Pham, X. Liu, F. C. Chiu, D. M. Shackleford, A. K. Powell, C. L. Masters, A. I. Bush, G. O'Keefe, J. G. Culvenor, R. Cappai, R. A. Cherny, P. S. Donnelly, A. F. Hill, D. I. Finkelstein and K. J. Barnham, *J. Exp. Med.*, 2012, **209**, 837.
203. B. D. Lee, V. L. Dawson and T. M. Dawson, *Trends Pharmacol. Sci.*, 2012, **33**, 365.
204. I. N. Rudenko, R. Chia and M. R. Cookson, *BMC Med.*, 2012, **10**, 20.
205. H. Harris and D. C. Rubinsztein, *Nat. Rev. Neurol.*, 2012, **8**, 108.
206. B. Alvarez-Castelao, F. Losada, Patrícia Ahicart and J. G. Castaño, *PLoS One*, 2013, **8**, e55999.

L-DOPA and Dopaminergic Agents

CHAPTER 3
Dopaminergic Treatments for Parkinson's Disease: Light and Shadows

NICOLA SIMOLA

Department of Biomedical Sciences, Section of Neuropsychopharmacology, University of Cagliari, Cagliari, Italy
Email: nicola.simola@unica.it

3.1 Introduction

Parkinson's disease (PD) is the second most common neurodegenerative disease after Alzheimer's disease,[1] and is considered the paradigmatic motor disorder. The pathological hallmark of PD is the degeneration of neuromelanin-containing dopaminergic neurons, which are located in the *Substantia Nigra pars compacta* (SNc) and project to the *striatum*. This degeneration dramatically attenuates the striatal dopaminergic tone, leading to a reduced stimulation of both dopamine (DA) D_1 and D_2 receptors, and eventually to motor impairment (Figure 3.1). On these bases, the restoration of dopaminergic transmission with so-called "dopamine-replacement therapy" (DRT) is the leading strategy used for the pharmacological management of motor impairment associated with PD.

DRT relies on the DA precursor L-3,4-dihydroxyphenylalanine (L-DOPA) and direct dopaminergic agonists. At times, L-DOPA can be associated with inhibitors of either catechol-*O*-methyl-transferase (COMT) or monoamine oxidase (MAO), the enzymes governing DA catabolism. These drugs are

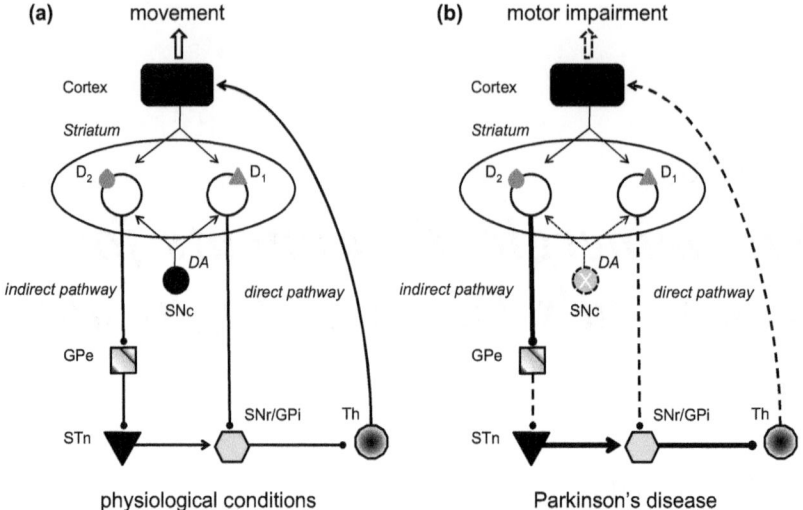

Figure 3.1 Modifications in basal ganglia circuits during PD. Under physiological conditions (a), the SNc sends dopaminergic input to striatal neurons. Endogenous DA activates the neurons belonging to the nigrostriatal, or direct, pathway. Those neurons send GABA-ergic projections to the *substantia nigra pars reticulata/globus pallidus pars interna* (SNr/GPi), and express stimulatory D_1 receptors. At the same time, endogenous DA inhibits the neurons belonging to the striatopallidal, or indirect, pathway. Those neurons send GABA-ergic projections to the SNr/GPi *via* the *globus pallidus pars externa* (GPe) and subthalamic nucleus (STN), and express inhibitory D_2 receptors. The balanced activity of the two striatal efferent pathways underlies the correct execution of movement. Degeneration of SNc neurons during PD removes the DA input to the *striatum* (b). This causes the disinhibition of the neurons in the indirect pathway, and boosts the inhibitory influence they exert on the GPe, in turn leading to an overactivation of the STN. The reduction in striatal DA input from the SNc also causes a decreased activation of neurons in the direct pathway. Taken together, these modifications in basal ganglia circuits result in an imbalanced activity of the striatal efferent pathways, and in an increased inhibitory output from the SNr–GPi complex. As a consequence, excessive inhibition of thalamo-cortical neurons (Th) causes the motor deficits associated with PD. ──▶ excitatory transmission (glutamatergic); ──• inhibitory transmission (GABA-ergic).

utilized to amplify the therapeutic effects of L-DOPA, and/or to attenuate the motor complications it may induce. DRT can be highly effective on motor impairment, particularly in early-stage PD.[2–4] However, DRT can also be associated with untoward effects, either motor or non-motor, which emerge with its prolonged use.[3,5–7] Furthermore, it has to be remarked that DRT is scarcely effective on the non-motor symptoms of PD, which are mostly non-dopaminergic in nature.[8,9]

This chapter will provide an overview of the light and shadows of DRT used in PD, encompassing preclinical investigations in experimental animals and clinical studies in patients. New avenues currently explored in DRT are also briefly discussed.

3.2 Overview of the Drugs used in Dopamine-Replacement Therapy

3.2.1 L-DOPA

L-DOPA has a long-standing clinical history as the most effective antiparkinsonian medication, and is still considered the gold standard therapy for PD.[3,10] L-DOPA is an intermediate in DA biosynthesis (Figure 3.2), and the drug needs to be converted into DA to exert its beneficial effects.[10] L-DOPA is highly effective on several features of motor impairment associated with PD, and is usually well tolerated.[3] However, chronic therapy with L-DOPA is often associated with a progressive reduction in drug efficacy, the so-called "wearing-off" phenomenon, as well as emergence of motor fluctuations and motor complications.[11,12]

3.2.2 Adjuncts to L-DOPA: COMT and MAO Inhibitors

COMT and MAO are the major enzymes involved in DA catabolism (Figure 3.2).[13,14] Not only do COMT and MAO metabolize endogenous DA, they also degrade exogenous DA derived from L-DOPA, therefore reducing the efficacy of the drug. On these bases, inhibitors of either COMT or MAO (more specifically of the MAO-B isoform) can be added to L-DOPA to stabilize its plasma levels.[15,16] Thus, COMT and MAO-B inhibitors may amplify the therapeutic effects of L-DOPA, and attenuate the motor complications it may induce. As of today, entacapone and tolcapone are the best characterized COMT inhibitors, whereas selegiline and rasagiline are the best characterized MAO-B inhibitors. New molecules of both classes are continuously being developed and investigated.

3.2.3 Dopaminergic Agonists

Dopaminergic agonists can be used either in combination with, or as alternatives to, L-DOPA. In contrast to L-DOPA, these drugs stimulate dopaminergic receptors without the need to be converted into DA. Dopaminergic agonists are a heterogeneous group of molecules, and are classically subdivided in ergolinic and non-ergolinic derivatives, based on their chemical structure (Figure 3.3).

In spite of their structural differences, all the dopaminergic agonists currently in use act chiefly by stimulating the D_2 family of receptors.[17] Nevertheless,

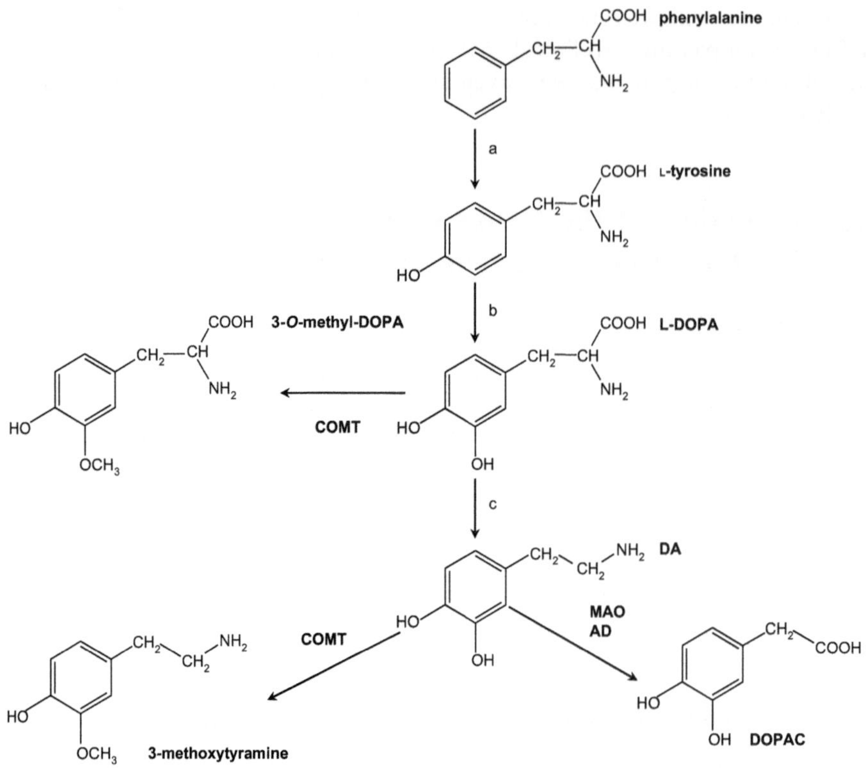

Figure 3.2 Biosynthesis and catabolism of DA. DA is synthesized from the amino acid phenylalanine, which is converted into L-tyrosine by the enzyme phenylalanine hydroxylase (a). L-tyrosine is then hydroxylated to L-DOPA by the enzyme tyrosine hydroxylase (b). Finally, L-DOPA is converted into dopamine by the enzyme L-amino acid decarboxylase (c). Both DA and L-DOPA are subjected to degradation by COMT, being converted to 3-methoxytyramine and 3-O-methyl-DOPA, respectively. DA can also be catabolized to 3,4-dihydroxyphenylacetic acid (DOPAC) by MAO and aldehyde dehydrogenase (AD) enzymes.

some of these drugs may differ in their affinities for the receptor subtypes within the D_2 family, and/or may bind non-dopaminergic receptors.[18–21] Taken together, these properties shape the therapeutic profiles of the various dopaminergic agonists. Dopaminergic agonists often possess more favorable pharmacokinetics than those of L-DOPA, and this is thought to be one of the factors underlying the delayed onset of motor complications observed during DRT with these drugs. However, the use of dopaminergic agonists may also be associated with untoward effects, some of which may be severe.[6,7,22,23] An overview of the pharmacological profile of some of the dopaminergic agonists used in PD therapy is reported in (Table 3.1).

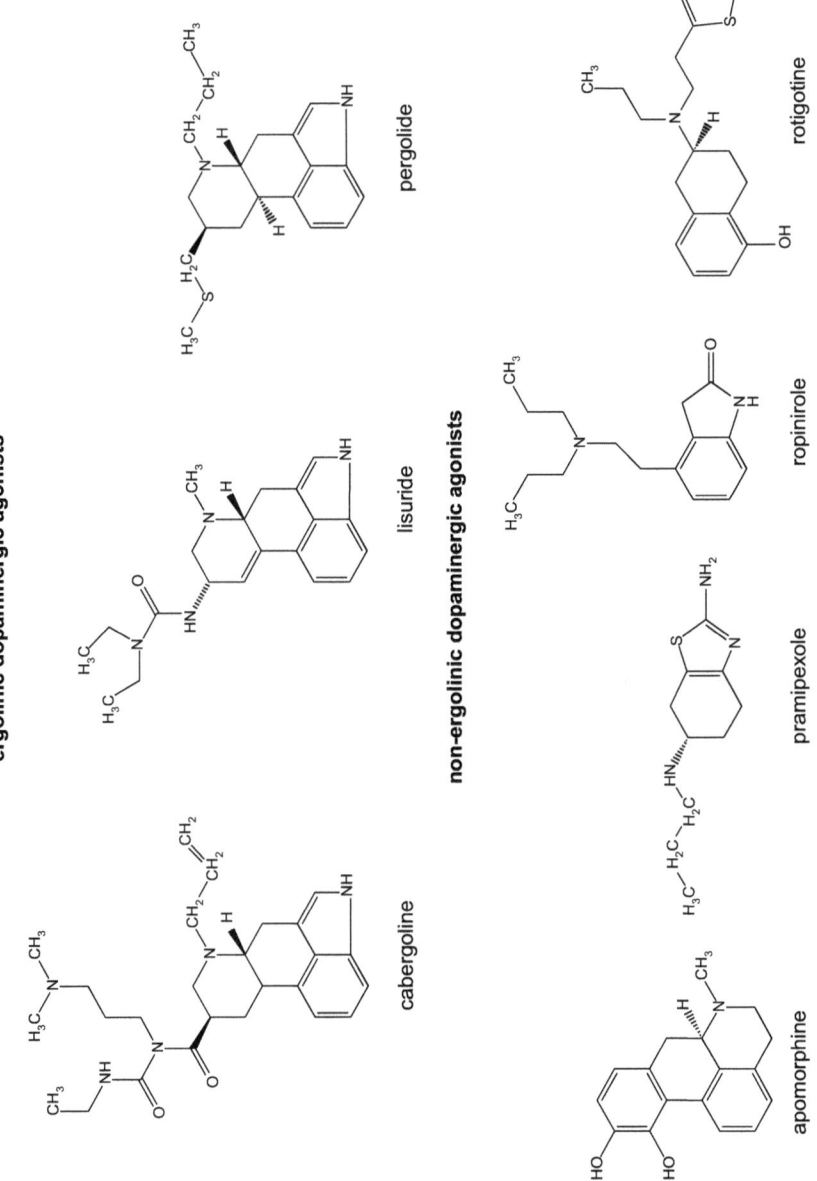

Figure 3.3 Chemical structures of some of the dopaminergic agonists used in DRT of PD.

Table 3.1 Pharmacological profiles of some of the dopaminergic agonists used in DRT of PD.

Agonist	Class	Pharmacodynamics	Pharmacokinetics	Beneficial effects	Untoward effects
Apomorphine	Non-ergolinic	Stimulates D_1 and D_2 receptors	Requires subcutaneous administration $t_{1/2} \sim 30$ min	Relieves motor disability efficiently Useful as a 'rescue' therapy in patients with severe freezing	Hypersexuality Hypotension Nausea and vomiting Subcutaneous nodules
Cabergoline	Ergolinic	Agonist at D_2 receptors Binds norepinephrine and serotonin receptors	Absorbed orally $t_{1/2} \sim 65$ h	Monotherapy in early PD Adjunct therapy in late-stage PD to manage motor complications induced by L-DOPA	Hypotension Leg edema Nausea and vomiting Psychiatric problems Retroperitoneal and pulmonary fibrosis
Dihydroergocriptine	Ergolinic	Agonist at D_2 receptors Partial agonist at D_1 receptors	Absorbed orally, but with low bioavailability ($\sim 5\%$) $t_{1/2}$ ranges from 12 to 25 h	Effective in monotherapy, or as an adjunct to L-DOPA May delay the onset of motor complications	Similar to other ergolinic agonists (see cabergoline)
Lisuride	Ergolinic	Agonist at D_2 receptors Partial agonist at $5HT_{2A}$ and $5HT_{2C}$ serotonin receptors	Low oral bioavailability (~ 10 to 20%) Can be delivered with transdermal patches $t_{1/2} \sim 2$ h after oral administration	May be useful in monotherapy	Similar to other ergolinic agonists
Pergolide	Ergolinic	Agonist at D_1 and D_2 receptors	Absorbed orally $T_{1/2} \sim 21$ h	Effective in monotherapy, or as an adjunct to L-DOPA in	Similar to other ergolinic derivatives -withdrawn from the

Drug	Type	Mechanism	Pharmacokinetics	Clinical use	Adverse effects
Piribedil	Non-ergolinic	Agonist at D_2 receptors Blocks α-2 norepinephrine receptors Binds serotonin receptors	Absorbed orally $t_{1/2} \sim 20$ h	Effective in monotherapy, or in combination with L-DOPA Can be used in both early and advanced PD patients with motor complications	market in the U.S.A. due to risk of valvular hearth damage Similar to other non-ergolinic agonists (see apomorphine and pramipexole)
Pramipexole	Non-ergolinic	Agonist at D_2 receptors preferential D_3 agonist Very low affinity for norepinephrine and serotonin receptors	Very high oral bioavailability (90% or more) $t_{1/2} \sim 10$ h	Effective in both early and advanced PD Can be used in monotherapy, or as an adjunct to L-DOPA Useful in patients who experience motor complications	Gastrointestinal problems Neuropsychiatric troubles, hypersexuality, pathological gambling Orthostatic hypotension Sudden sleep attacks
Ropinirole	Non-ergolinic	D_2 agonist, binds D_3 receptors Low affinity for α-2 norepinephrine and $5HT_2$ serotonin receptors	Absorbed orally, but with low bioavailability (46%) $t_{1/2} \sim 10$ h	Effective in monotherapy Can be administered in combination with L-DOPA in patients experiencing motor complications	Similar to other non-ergolinic agonists
Rotigotine	Non-ergolinic	Agonist at D_2 receptors Preferential D_3 agonist Binds norepinephrine and serotonin receptors	Delivered by transdermal patches $t_{1/2} \sim 24$ h	Effective in early PD	Can induce skin reactions at the site of application Other untoward effects are similar to those elicited by other non-ergolinic agonists

3.3 Effect of Dopamine-Replacement Therapy on the Motor Features of Parkinson's Disease

3.3.1 Motor Impairment

The ability of DRT to counteract several features of motor impairment is a firmly established concept in the pharmacotherapy of PD. In this regard, it is worth recalling that the availability of suitable preclinical models which adequately reproduce different aspects of motor deficits associated with PD has dramatically contributed to the development, and effectiveness, of DRT. Briefly, experimental models of PD can be subdivided in pharmacological, toxin-based and genetic.[24,25] Pharmacological models usually involve a drug-induced, reversible, impairment of DA transmission, the function of which returns to baseline levels when the drug ceases its effects. On the other hand, toxin-based models rely on a permanent degeneration of DA neurons induced by substances which selectively, or preferentially, damage those neurons. Finally, genetic models study the behavioral and neurochemical modifications in mutant animals lacking, or overexpressing, specific genes thought to play a role in PD. Pharmacological and toxin-based models are the most popular, and are suited to both acute and chronic studies with dopaminergic drugs.

Acute administration of either the catecholamine-depleting agent reserpine, or haloperidol, an antagonist of D_2 receptors, are the most common pharmacological models of PD.[25] When administered to rodents, either drug induces akinesia and rigidity resembling those occurring in PD. Even though these models do not rely on DA neuron neurodegeneration, they have predictive validity, and are often used in the preliminary screening of new anti-parkinsonian agents.[25] Several drugs currently used in DRT have been found to be effective in these models, as they reduce the duration and severity of catalepsy, and stimulate motor activity.[26–29]

A large number of agents, including metals, neurotoxins and pesticides, have been reported to induce PD-like neurodegeneration and motor impairment, when administered to experimental animals.[30–34] As of today, the best characterized and most used experimental paradigms of PD employ the neurotoxins 6-hydroxydopamine (6-OHDA) and 1-methyl-1-4-phenyl-1,2,3,6-tetrahydropyridne (MPTP).[35,36]

The classical model of 6-OHDA-induced parkinsonism relies on the unilateral infusion of the neurotoxin in the medial forebrain bundle (which projects from the SNc to the *striatum*), or less frequently in the SNc, causing the degeneration of DA neurons.[36] This model usually employs rats, though other species can be used.[36] At the behavioral level, rats intracerebrally infused with 6-OHDA display akinesia and motor asymmetry, the latter evident as turning behavior ipsilateral to the side of toxin infusion.[36] The ability of a drug to stimulate turning behavior contralateral to the side of 6-OHDA infusion is deemed indicative of anti-parkinsonian activity; therefore the intensity of drug-induced contralateral turning is a major parameter evaluated in the 6-OHDA model.[36] Nevertheless, this paradigm can also be used to reproduce more subtle motor deficits, like impairments in gait and sensory–motor integration.[37,38] Differently from the 6-OHDA paradigm, the MPTP model employs a systemic neurotoxin administration, and is usually performed in mice and non-human primates, where it induces overt motor impairment evident as akinesia and hypokinesia.[25]

DRT has proven highly effective when evaluated in either the 6-OHDA or the MPTP model. Both L-DOPA and dopaminergic agonists reduce motor impairment, as indicated by the induction of contralateral turning behavior and the stimulation of motor activity, ameliorate forelimb akinesia, and counteract sensory-motor integration deficits.[39,40] Clinical studies have confirmed and extended results obtained in experimental animals. Thus, administration of DRT to patients counteracts motor disability, reduces the duration of the "off" phase, namely the period when motor impairment is overtly manifested, and improves several features of motor impairment measured according to the Unified Parkinson's Disease Rating Scale (UPDRS).[2–4,41,42] In this regard, a number of investigations have demonstrated that monotherapy with dopaminergic agonists may efficiently counteract motor disability, particularly in early-stage PD.[2,4,41,42] However, it has to be acknowledged that the majority of clinical studies indicate that L-DOPA has a higher efficacy than dopaminergic agonists, and that L-DOPA still remains the gold standard drug to manage PD motor symptoms.[3,21,41,42] This suggests that the therapeutic protocol with DRT should be carefully designed, based on both the drugs' properties and the patients' individual needs.

3.3.2 Motor Complications

Notwithstanding the beneficial effects on motor disability, chronic use of DRT is often associated with the emergence of motor complications, the most common being dose deterioration, best known as wearing-off, and dyskinesias.[11,12] Wearing-off is a progressive reduction in the duration of the beneficial effect DRT exerts on motor disability, whereas dyskinesia is abnormal and purposeless movements of the head, trunk and limbs. In addition, motor fluctuations like the "on–off" phenomenon may be associated with chronic DRT. During "on–off", patients may rapidly pass from a state in which therapy reverses motor disability while at the same time eliciting dyskinesias, to one in which DRT is not effective at all. Motor complications are classically envisaged as something peculiar to late-stage patients who have gone through several years of chronic L-DOPA. Recent studies, however, show that a significant percentage of patients may experience motor complications even within 1–2 years from the beginning of DRT.[11,12,43] Furthermore, motor complications seem not to be confined to DRT with L-DOPA, since the use of dopaminergic agonists may be associated with wearing-off and dyskinesias as well.[44,45]

Experimental paradigms of PD have greatly contributed to the preclinical modeling of both behavioral and neurochemical features of motor complications associated with DRT. Wearing-off has been reproduced in rats infused with 6-OHDA and chronically treated with L-DOPA, in which a progressive reduction in the duration of contralateral turning stimulated by L-DOPA has been reported.[46,47] With regard to dyskinesias, several investigations have shown that rats infused with 6-OHDA and repeatedly treated with either L-DOPA or dopaminergic agonists display a series of behavioral modifications involving a progressive increase in the intensity of contralateral turning, associated with Abnormal Involuntary Movements (AIMs) of the trunk, limbs, and head which mimic human dyskinesias.[48–51] In addition, a behavioral syndrome resembling human dyskinesias can be reproduced in primates rendered parkinsonian with MPTP and treated with DRT.[11,35] Behavioral modifications in MPTP-treated primates are considered closer to dyskinesias than the syndrome elicited by DRT in rats infused with 6-OHDA; therefore the primate treated with MPTP is considered the best preclinical model for the study of dyskinesias.[35] It has to be acknowledged that both the rat 6-OHDA and the primate MPTP models of motor complications have been criticized with regard to their clinical relevance.[12,25,52] Nevertheless, the results obtained in either model have often been predictive of findings later obtained in patients.[25] Thus, studies in parkinsonian rats and primates have demonstrated that repeated administration of dopaminergic agonists triggers a behavioral syndrome of mild severity, compared with that produced by chronic L-DOPA in the same animals.[53–56] Notably, this is consistent with clinical investigations concerned with the dyskinetic potential of the drugs used in DRT, which indicate that L-DOPA has a higher propensity to induce dyskinesias than dopaminergic agonists.[2–4,41,42]

Studies performed in the rat 6-OHDA and primate MPTP models which focused on dyskinesias induced by L-DOPA have shown that this motor complication is associated with several molecular and functional changes in the basal ganglia. In more detail, those investigations found modifications in receptors, peptides and early genes, and have indicated that the signal cascade of the D_1 receptor has a major role in motor complications.[57–60] Even though the mechanisms underlying the motor complications associated with DRT have not been elucidated yet, both preclinical and clinical studies converge on the important role played by phasic, intermittent, stimulation of DA receptors.[11,12] Under physiological conditions, DA is released at a low and continuous rate, leading to a near-permanent occupation of its receptors. Conversely, dopaminergic tone in the dopamine-denervated *striatum* exposed to DRT is not continuous, and is chiefly dependent on the plasma half-life and bioavailability of the drugs used. Several lines of evidence suggest that intermittent stimulation of DA receptors may favor maladaptive phenomena in the basal ganglia, and eventually motor complications.[11,12] Moreover, this could be a factor underlying the different propensity of L-DOPA and dopaminergic agonists to induce motor complications observed in clinical studies.[2–4,41,42] In fact, L-DOPA has a very short plasma half-life (about 1–2 h), which is associated with a rapid fluctuation of DA levels in the brain. On the other hand, dopaminergic agonists possess significantly longer plasma half-lives (Table 3.1), resulting in an extended occupancy of DA receptors.[21]

The complex pathophysiology of motor complications associated with DRT is the basis for the substantial lack of adequate therapies for these untoward effects. The addition of a COMT inhibitor to L-DOPA may help in the management of wearing-off, whereas dyskinesias can be somewhat controlled with amantadine, a weak antagonist of *N*-methyl-D-aspartate (NMDA) glutamate receptors.

However, a number of clinical studies indicate that the efficacy of those drugs is limited.[61,62] The management of motor complications associated with DRT is one of the major challenges in PD, and the search for new therapies is ongoing.

3.4 Effect of Dopamine-Replacement Therapy on the Non-Motor Features of Parkinson's Disease

3.4.1 Non-Motor Symptoms

In addition to motor impairment, PD may feature several non-motor symptoms which can be of different severity and involve a considerable percentage of patients (Table 3.2).[8]

Table 3.2 Summary of the non-motor symptoms most frequently encountered in PD. Symptoms in italics can be produced and/or exacerbated by DRT.

Non-motor symptom	Subcategory (when present)
Anxiety	
Apathy	
Autonomic dysfunction	Gastrointestinal problems Sexual dysfunction Sialorrhea Sweating Urinary problems
Cognitive dysfunction	Memory impairment Dementia
Depression and other mood disorders	
Dopamine dysregulation syndrome	
Fatigue	
Impulse control disorders	*Binge eating* *Compulsive shopping* *Hypersexuality* *Kleptomania* *Pathological gambling*
Psychosis	*Delusions* *Hallucinations*
Punding	*Hobbyism*
Sleep–wakefulness disorders	Hypersomnia and daytime sleepiness Insomnia REM sleep behavior disorder Sleep fragmentation *Sudden onset of sleep*
Walkabout	

Several lines of clinical evidence suggest that DRT does not significantly ameliorate non-motor features of PD.[8] This lack of effect is classically justified in the light of the evidence indicating that DA is negligibly involved in non-motor symptoms of PD, which seem mostly dependent on neurotransmitters like acetylcholine, norepinephrine and serotonin.[8,9] Nevertheless, recent clinical trials indicate that DRT may counteract at least some non-motor features of PD. Thus, apomorphine and cabergoline may be useful for managing sexual dysfunction, since they induce penile erection.[63,64] Moreover, pramipexole and ropinirole have been found to ameliorate depressive symptoms.[65–67] On these bases, more detailed investigations on the potential effects of DRT on non-motor symptoms of PD are warranted.

3.4.2 Addictive-Like Behavior Associated with Dopamine-Replacement Therapy

Chronic DRT may be associated with a pattern of behavioral disturbances sharing many features with addictive behavior (Table 3.2). These disturbances may involve, on the one hand, modifications in the intake of dopaminergic medications, which are taken at high dosages, and more frequently than requested by the therapeutic protocol.[6,68,69] This pattern of DRT consumption goes under the name "dopamine dysregulation syndrome" (DDS), is compulsive, displays several psychiatric features of drug addiction (*e.g.*, agitation, irritability, paranoia), and is carried on notwithstanding the development of severe dyskinesias and intoxication. On the other hand, DRT can trigger behavior that is not directed to dopaminergic medications, like punding, hobbyism, walkabout, and the so-called "impulse control disorders" (ICDs) (Table 3.2).[5,7] Briefly, punding is a complex and stereotyped behavior that can also be observed in individuals addicted to pyschostimulants.[70] It is characterized by a fascination with meaningless activities, like handling or sorting objects, and/or equipment. Hobbyism, is a more complex form of punding, where the patients focus on a specific hobby, for example repairing devices or writing.[71] Both punding and hobbyism are usually carried on in a continuous and purposeless fashion. Walkabout is defined as an aimless wandering, and can involve either walking or driving.[6] Finally, ICDs are a series of behavior types all characterized by "the failure to resist to an impulse, drive or temptation to perform an act that is harmful to the person or others".[72]

Several clinical studies have demonstrated the important role played by DRT in the emergence of ICDs by showing that: (1) only PD patients under DRT display ICDs; (2) ICDs emerge only after DRT has begun; and (3) ICDs can be observed also in patients not affected by PD who take DA agonists to treat fibromyalgia or restless leg syndrome.[73,74] It is worth recalling that DA plays a crucial role in reward and addiction.[75] Therefore, it can be hypothesized that the non-physiological stimulation of DA receptors by DRT may lead to sensitization of reward neurocircuits, and to the instatement of addictive-like behavior.[7] However, it is still not clear whether the dose of DRT could influence the manifestation of ICDs.[76–79] In addition, no solid data exist which indicate modifications to the risk of developing ICDs according to the specific drug used.[76–79] Finally, although DRT appears to be crucial for ICDs to emerge, epidemiological studies have demonstrated that factors others than DRT do also play a role in ICDs.[76] Thus, ICDs have been linked with young age at PD onset, history of ethanol and drug abuse and precedents for anxiety and depression. All things considered, it is worth emphasizing that only a percentage of PD patients under DRT display ICDs. This is estimated to range between 2 and 10% of all patients, depending on the specific behavior considered, although the actual prevalence of ICDs is not known, since many of these behavior types have not been thoroughly investigated in the clinical setting.[76]

The management of DDS and ICDs is largely unsatisfactory at present. Remission of these symptoms can be achieved by modifying the therapeutic protocol, for instance by introducing long-acting dopaminergic agonists or switching to a different dopaminergic agonist.[76] However, this approach is not always feasible, since it may result in severe motor deterioration. Moreover, no specific therapies are available for the management of DDS and ICDs, and the development of effective drugs is greatly limited by the substantial lack of adequate experimental models for the study of those non-motor features of PD.

3.5 Effect of Dopamine-Replacement Therapy on Disease Progression

Counteracting the degeneration of DA neurons is an unmet need in PD management. Preclinical investigations have characterized several putative neuroprotective agents, but no drugs have been licensed with this indication yet. It has to be remarked that the study of neurodegeneration/neuroprotection in PD is complicated by a number of factors. Thus, it is still unknown why DA neurons degenerate during PD.[80] Moreover, the majority of the available experimental models often reproduce a neurodegeneration which is too rapid, and/or chiefly dependent on a specific mechanism of toxicity (Table 3.3). This is strikingly different from idiopathic PD, where the demise of DA neurons is progressive, occurs over an extended period of time, and likely results from the combination of various mechanisms (e.g., oxidative stress, mitochondrial failure, neuroinflammation, proteolytic damage).[80,81] Finally, clinical trials performed so far to evaluate putative neuroprotectants for PD have shown major limitations, the most important being a substantial lack of end-points adequately reflecting disease progression.[80,82] Whether drugs used in DRT may impact PD progression is still up for debate, since preclinical studies have shown that they can have either neurotoxic or neuroprotective effects.[83,84]

Several *in vitro* studies suggest that high doses of L-DOPA may be toxic to cultured neurons,[85–87] raising some concern over the possibility that L-DOPA may hasten PD progression. Investigations in experimental animals, however, failed to observe degeneration of DA neurons after the administration of L-DOPA at doses comparable to those used in DRT.[88] The effects of L-DOPA

Table 3.3 Summary of the models of toxin-induced neurodegeneration most frequently used in the preclinical study of neuronal damage and neuroprotection in PD.

Model	Main toxic effects	Main features of the model	Effect of DRT drugs
Cell cultures	Decrease in cell viability	Acute *in vitro* model May allow the rapid identification of protective agents and their mechanisms of action May employ different cell types and toxins	Protective effects have been reported for apomorphine, cabergoline, lisuride, pergolide, pramipexole, rasagiline, ropinirole and selegiline Some studies indicate that high doses of L-DOPA may be toxic
Intracerebral infusion of 6-OHDA in rats and mice	Demise of DA neurons in the SNc Reduction of DA levels in the striatum	Acute model Can be used to specifically target selected brain regions Neurodegeneration is rapid and driven by oxidative stress	Protective effects have been reported for cabergoline, pergolide and ropinirole
Intracerebral or systemic lipopolysaccharide in rats and mice	Activation of glial cells Degeneration of DA neurons in the SNc Increase in the levels of inflammatory mediators	Can be either acute or chronic The model is selective for DA neurons in SNc The neurodegeneration is progressive, delayed, and driven by inflammatory mechanisms	Protective effects have been reported for pramipexole
Systemic administration of MPTP in mice and primates	Decrease in the content of striatal DA Degeneration of dopaminergic terminals in the striatum Glial activation Loss of DA neurons in the SNc	Acute or chronic model Is clinically relevant The extent of DA lesion can be adjusted with the protocol of toxin administration The toxin acts mainly by impairing mitochondrial function	Protective effects have been reported for apomorphine, pramipexole, rasagiline and selegiline
Systemic administration of rotenone in rats and mice	Degeneration of DA neurons in the SNc Reduction of dopaminergic terminals and DA levels in the striatum	Chronic model Is clinically relevant Is associated with inclusions in DA neurons resembling those observed in human PD Is associated with systemic toxicity and non-specific brain damage, high variability in DA neuronal loss, and mortality	Protective effects have been reported for pramipexole and selegiline
Systemic administration of paraquat and Systemic administration of paraquat + maneb in mice	Degeneration of striatal DA terminals Loss of DA neurons in the SNc	Chronic model DA degeneration is variable, and the model is associated with high mortality Is clinically relevant	Protective effects have been reported for selegiline

on PD progression have also been evaluated in patients by means of the ELLDOPA study.[89] In this trial, patients received a 40 week treatment with L-DOPA at three different dosages (low, medium and high) or placebo, followed by a two-week wash-out period, after which changes in UPDRS scores were evaluated. Moreover, a subpopulation of patients underwent single photon emission computerized tomography (SPECT) with 2-β-carboxymethoxy-3-β-(4-iodophenyl)-tropane (β-CIT, which binds the DA transporter) to evaluate the integrity of the nigrostriatal dopaminergic system. The results obtained in the ELLDOPA study, however, have proven somewhat contradictory.[90] Analysis of UPDRS scores showed that patients in the L-DOPA group displayed a marked improvement of motor symptoms compared with the placebo group. This seems to suggest, on the one hand, that L-DOPA does not hasten PD progression, and could even indicate that L-DOPA protects DA neurons from degeneration. However, it has to be considered that the study could not rule out an extension of the therapeutic effects of L-DOPA beyond the two-week wash-out period.[90] On the other hand, the SPECT evaluation showed a significantly lower β-CIT binding in the L-DOPA group, compared with placebo. This result would indicate nigrostriatal damage, and therefore of L-DOPA-induced neurotoxicity. However, it cannot be completely ruled out that L-DOPA could have exerted some non-specific effects on the uptake of β-CIT, which can justify the apparently discrepant results obtained in this trial.

Neuroprotection by dopaminergic agonists has been reported by several *in vitro* studies (Table 3.3).[91–98] Moreover, some dopaminergic agonists have been found to protect the nigrostriatal system from the noxious effects of 6-OHDA, MPTP, and other toxins *in vivo* (Table 3.3).[99–110] Notably, clinical trials with dopaminergic agonists have yielded results in line with those from preclinical studies. The CALM-PD-CIT study evaluated the effects of pramipexole on PD progression, by means of β-CIT SPECT.[111] Moreover, the REAL-PET study was undertaken to evaluate the neuroprotective properties of ropinirole, by measuring the uptake of fluorodopa with positron emission tomography (PET).[112] Both CALM-PD-CIT and REAL-PET trials compared the effects of the specific dopaminergic agonist investigated with those of L-DOPA, and found a significantly lower decline in the imaging biomarker studied in patients treated with the dopaminergic agonist. Even though these results appear intriguing, it is worth highlighting that both CALM-PD-CIT and REAL-PET trials lacked a placebo group, hence a direct appreciation of the effects of the dopaminergic agonists investigated cannot be made on the basis of the data available. Therefore, the results of these studies, on the one hand, might be indicative of neuroprotection by dopaminergic agonists but, on the other, could suggest the existence of neurotoxicity by L-DOPA.[83]

Putative neuroprotective properties have also been suggested for MAO-B inhibitors. Preclinical studies, both *in vitro* and *in vivo*, have demonstrated that selegiline and rasagiline protect DA neurons.[113,114] Clinical trials, however, failed to obtain direct evidence demonstrating that MAO-B inhibitors can slow down PD progression. Neuroprotection by selegiline was evaluated in the DATATOP and SINDEPAR studies.[115,116] Both trials reported an improvement in UPDRS score in patients treated with selegiline, compared with placebo. However, neither study could isolate the influence of the symptomatic effects of selegiline on the result obtained. Similar considerations may also apply to the TEMPO and ADAGIO clinical trials, which evaluated the effects of rasagiline on PD progression.[82,117–119]

Taken together, the clinical trials performed so far have not proven the existence of neuroprotection by DRT, but at the same time they do not rule it out. Should this be demonstrated in future trials, at least for some of the drugs used in DRT, it would greatly benefit the strategies aimed at contrasting PD progression. In fact, DRT drugs are already licensed for PD, and their use would overcome all the issues related to the approval of new neuroprotective agents.

3.6 Future Directions in Dopamine-Replacement Therapy

As of today, the most promising new strategies in DRT are continuous drug delivery and the development of new drugs.

3.6.1 Continuous Drug Delivery

The idea of continuous drug delivery stems from the concept that extended, non-intermittent, stimulation of DA receptors may be associated with a reduced incidence of motor complications.[11,12] This is in agreement with the wealth of preclinical and clinical studies showing that long-acting dopaminergic agonists are less likely to induce motor complications than L-DOPA, which has a very short half-life.[11,12,21] However, it should be considered that discrepancies in plasma half-life do not appear sufficient to justify the different propensity of DRT drugs to induce motor complications, and that the protocol of administration has emerged as a crucial factor.[11,12,120] In particular, continuous delivery of DRT has been shown to counteract motor impairment, being at the same time associated with a reduced incidence of motor complications, compared with classical protocols of administration.[12,120] Continuous delivery of DRT can be performed in different ways, depending on the specific drug considered. Thus, dopaminergic agonists can be administered by continuous subcutaneous infusion, subcutaneous slow-release systems, or transdermal patches, whereas L-DOPA can be administered by intravenous or intrajejunal continuous infusions.[120–122]

3.6.2 Development of New Drugs

New molecules are continuously being developed and evaluated for their effects on both the motor and non-motor symptoms of PD.[123] Some of them target neurotransmitters other than DA (e.g., antagonists of adenosine A_{2A} receptors, antagonists of metabotropic and ionotropic glutamate receptors, antagonists of histamine H_3 receptors, agonists of serotonin receptors), whereas the most innovative derivatives combine agonist activity at DA receptors with agonist/ antagonist actions on non-dopaminergic systems.[123] The idea of targeting neurotransmitters other than DA is based on the growing evidence indicating an important role for non-dopaminergic systems in both motor and non-motor symptoms of PD, as well as in motor complications associated with DRT. Several newly developed molecules have proven effective in experimental models of PD, alone or in combination with DRT, although in many cases clinical trials have failed to confirm, or have only partially replicated, preclinical results.[123]

3.7 Final Remarks

DRT still stands as the most effective pharmacological strategy for the management of motor impairment associated with PD. New avenues in DRT have been continuously explored in recent years; however, important limitations still accompany the use of DRT, such as the substantial ineffectiveness on non-motor symptoms, and the emergence of neurobehavioral disturbances. Finally, it is advisable for new strategies in DRT to look not only at drug efficacy, but also at pharmacoeconomic matters, the importance of which is often underestimated. Thus, the incidence of PD is increasingly expected to grow, in the light of the continuous rise in life expectancy, and so will healthcare costs. Therefore, new strategies in DRT should not only bear an increased efficacy compared with the current ones, but also try to reach a favorable cost-to-benefit ratio, in order to make therapies accessible to all patients.

Acknowledgements

Dr. Nicola Simola gratefully acknowledges Sardinia Regional Government for the financial support (P.O.R. Sardegna F.S.E. Operational Programme of the Autonomous Region of Sardinia, European Social Fund 2007–2013 - Axis IV Human Resources, Objective I.3, Line of Activity I.3.1 "Avviso di chiamata per il finanziamento di Assegni di Ricerca")

References

1. M. C. de Rijk, L. J. Launer, K. Berger, M. M. Breteler, J. F. Dartigues, M. Baldereschi, L. Fratiglioni, A. Lobo, J. Martinez-Lage, C. Trenkwalder and A. Hofman, *Neurology*, 2000, **54**, S21.
2. R. G. Holloway, I. Shoulson, S. Fahn, K. Kieburtz, A. Lang, K. Marek, M. McDermott, J. Seibyl, W. Weiner, B. Musch, C. Kamp, M. Welsh, A. Shinaman, R. Pahwa, L. Barclay, J. Hubble, P. LeWitt, J. Miyasaki, O. Suchowersky, M. Stacy, D. S. Russell, B. Ford, J. Hammerstad, D. Riley, D. Standaert, F. Wooten, S. Factor, J. Jankovic, F. Atassi, R. Kurlan, M. Panisset, A. Rajput, R. Rodnitzky, C. Shults, G. Petsinger,

C. Waters, R. Pfeiffer, K. Biglan, L. Borchert, A. Montgomery, L. Sutherland, C. Weeks, M. DeAngelis, E. Sime, S. Wood, C. Pantella, M. Harrigan, B. Fussell, S. Dillon, B. Alexander-Brown, P. Rainey, M. Tennis, E. Rost-Ruffner, D. Brown, S. Evans, D. Berry, J. Hall, T. Shirley, J. Dobson, D. Fontaine, B. Pfeiffer, A. Brocht, S. Bennett, S. Daigneault, K. Hodgeman, C. O'Connell, T. Ross, K. Richard and A. Watts, *Arch. Neurol.*, 2004, **61**, 1044.
3. W. Poewe, A. Antonini, J. C. Zijlmans, P. R. Burkhard and F. Vingerhoets, *Clin. Interv. Aging*, 2010, **5**, 229.
4. O. Rascol, D. J. Brooks, A. D. Korczyn, P. P. De Deyn, C. E. Clarke and A. E. Lang, *N. Engl. J. Med.*, 2000, **342**, 1484.
5. P. Ambermoon, A. Carter, D. W. Hall, N. N. Dissanayaka and J. D. O'Sullivan, *Addiction*, 2011, **106**, 283.
6. G. Giovannoni, J. D. O'Sullivan, K. Turner, A. J. Manson and A. J. Lees, *J. Neurol. Neurosurg. Psychiatry*, 2000, **68**, 423.
7. V. Voon, P. O. Fernagut, J. Wickens, C. Baunez, M. Rodriguez, N. Pavon, J. L. Juncos, J. A. Obeso and E. Bezard, *Lancet Neurol.*, 2009, **8**, 1140.
8. O. Bernal-Pacheco, N. Limotai, C. L. Go and Y. Fernandez, *Neurologist*, 2012, **18**, 1.
9. K. L. Eskow Jaunarajs, M. Angoa-Perez, D. M. Kuhn and C. Bishop, *Neurosci. Biobehav. Rev.*, 2011, **35**, 556.
10. O. Hornykiewicz, *J. Neurol.*, 2010, **257**, S249.
11. P. Jenner, *Mov. Disord.*, 2008, **23**, S585.
12. F. Stocchi, P. Jenner and J. A. Obeso, *Eur. Neurol.*, 2010, **63**, 257.
13. E. E. Billett, *Neurotoxicology*, 2004, **25**, 139.
14. M. J. Bonifácio, P. N. Palma, L. Almeida and P. Soares-da-Silva, *CNS Drug Rev.*, 2007, **13**, 352.
15. L. W. Elmer and J. M. Bertoni, *Expert Opin. Pharmacother.*, 2008, **9**, 2759.
16. H. Heikkinen, J. G. Nutt, P. A. LeWitt, W. C. Koller and A. Gordin, *Clin. Neuropharmacol.*, 2001, **24**, 150.
17. P. Jenner, *Curr. Opin. Neurol.*, 2003, **16**, S3.
18. D. Deleu, M. G. Northway and Y. Hanssens, *Clin. Pharmacokinet.*, 2002, **41**, 261.
19. T. Kvernmo, S. Härtter and E. Burger, *Clin. Ther.*, 2006, **28**, 1065.
20. Y. W. Lam, *Pharmacotherapy*, 2000, **20**, 17S.
21. D. Nyholm, *Clin. Pharmacokinet.*, 2006, **45**, 109.
22. H. Reichmann, A. Bilsing, R. Ehret, W. Greulich, J.B. Schulz, A. Schwartz and O. Rascol, *J. Neurol.*, 2006, **253**, IV36.
23. V. G. Rasmussen, K. Østergaard, E. Dupont and S. H. Poulsen, *Mov. Disord*, 2011, **26**, 801.
24. M. F. Chesselet and F. Richter, *Lancet Neurol.*, 2011, **10**, 1108.
25. S. Duty and P. Jenner, *Br. J. Pharmacol.*, 2011, **164**, 1357.
26. J. P. Finberg and M. B. Youdim, *Neuropharmacology*, 2002, **43**, 1110.

27. T. Kobayashi, T. Araki, Y. Itoyama, M. Takeshita, T. Ohta and Y. Oshima, *Life Sci.*, 1997, **61**, 2529.
28. J. Maj, Z. Rogóz, G. Skuza and K. Kołodziejczyk, *Eur. J. Pharmacol.*, 1997, **324**, 31.
29. M. Miyagi, N. Arai, F. Taya, F. Itoh, Y. Komatsu, M. Kojima and M. Isaji, *Biol. Pharm. Bull.*, 1996, **19**, 1499.
30. J. R. Cannon and J. T. Greenamyre, *Prog. Brain Res.*, 2010, **184**, 17.
31. R. E. Heikkila and P. K. Sonsalla, *Neurochem. Int.*, 1992, **20**, 299S.
32. M. M. Iravani, C. C. Leung, M. Sadeghian, C. O. Haddon, S. Rose and P. Jenner, *Eur. J. Neurosci.*, 2005, **22**, 317.
33. M. Parenti, C. Flauto, E. Parati, A. Vescovi and A. Groppetti, *Brain Res.*, 1986, **367**, 8.
34. K. Ossowska, M. Smiałowska, K. Kuter, J. Wierońska, B. Zieba, J. Wardas, P. Nowak, J. Dabrowska, A. Bortel, I. Biedka, G. Schulze and H. Rommelspacher, *Neuroscience*, 2006, **141**, 2155.
35. P. Jenner, *Parkinsonism Relat. Disord.*, 2009, **15**, S18.
36. N. Simola, M. Morelli and A. R. Carta, *Neurotox. Res.*, 2007, **11**, 151.
37. M. Olsson, G. Nikkhah, C. Bentlage and A. Björklund, *J. Neurosci.*, 1995, **15**, 3863.
38. T. Schallert, S. M. Fleming, J. L. Leasure, J. L. Tillerson and S. T. Bland, *Neuropharmacology*, 2000, **39**, 777.
39. B. Henry, A. R. Crossman and J. M. Brotchie, *Exp. Neurol.*, 1998, **151**, 334.
40. A. Pinna, S. Pontis, F. Borsini and M. Morelli, *Synapse*, 2007, **61**, 606.
41. P. Martinez-Martin and M. M. Kurtis, *Parkinsonism. Relat. Disord.*, 2009, **15**, S58.
42. W. H. Poewe, O. Rascol, N. Quinn, E. Tolosa, W. H. Oertel, E. Martignoni, M. Rupp and B. Boroojerdi, *Lancet Neurol.*, 2007, **6**, 513.
43. R. A. Hauser, M. P. McDermott and S. Messing, *Arch. Neurol.*, 2006, **63**, 1756.
44. R. A. Hauser, O. Rascol, A. D. Korczyn, A. J. Stoessl, R. L. Watts, W. Poewe, P. P. De Deyn and A. E. Lang, *Mov. Disord.*, 2007, **22**, 2409.
45. A. Thomas, L. Bonanni, A. Di Iorio, S. Varanese, F. Anzellotti, A. D'Andreagiovanni, F. Stocchi and M. Onofrj, *J. Neurol.*, 2006, **253**, 1633.
46. F. Bibbiani, J. D. Oh, A. Kielaite, M. A. Collins, C. Smith and T. N. Chase, *Exp. Neurol.*, 2005, **196**, 422.
47. C. Marin, E. Aguilar, M. Bonastre, E. Tolosa and J. A. Obeso, *Exp. Neurol.*, 2005, **192**, 184.
48. M. A. Cenci, C. S. Lee and A. Björklund, *Eur. J. Neurosci.*, 1998, **10**, 2694.
49. H. S. Lindgren, D. Rylander, K. E. Ohlin, M. Lundblad and M. A. Cenci, *Behav. Brain Res.*, 2007, **177**, 150.
50. S. Konitsiotis and C. Tsironis, *Behav. Brain Res.*, 2006, **170**, 337.
51. A. Pinna, S. Pontis and M. Morelli, *Behav. Brain Res.*, 2006, **171**, 175.
52. C. Marin, M. C. Rodriguez-Oroz and J. A. Obeso, *Exp. Neurol.*, 2006, **197**, 269.

53. R. Grondin, M. Goulet, T. Di Paolo and P. J. Bédard, *Brain Res.*, 1996, **735**, 298.
54. M. J. Jackson, L. A. Smith, G. Al-Barghouthy, S. Rose and P. Jenner, *Exp. Neurol.*, 2007, **204**, 162.
55. M. Lundblad, M. Andersson, C. Winkler, D. Kirik, N. Wierup and M. A. Cenci, *Eur. J. Neurosci.*, 2002, **15**, 120.
56. K. A. Tayarani-Binazir, M. J. Jackson, S. Rose, C. W. Olanow and P. Jenner, *Mov. Disord.*, 2010, **25**, 377.
57. E. Bezard, J. M. Brotchie and C. E. Gross, *Nat. Rev. Neurosci.*, 2001, **2**, 577.
58. P. Calabresi, M. Di Filippo, V. Ghiglieri and B. Picconi, *Mov. Disord.*, 2008, **23**, S570.
59. M. A. Cenci, *Trends Neurosci.*, 2007, **30**, 236.
60. P. Jenner, *Nat. Rev. Neurosci.*, 2008, **9**, 665.
61. N. Crosby, K. H. Deane and C. E. Clarke, *Cochrane Database Syst. Rev.*, 2003, **1**, CD003468.
62. S. Kaakkola, *Int. Rev. Neurobiol.*, 2010, **95**, 207.
63. J. D. O'Sullivan and A. J. Hughes, *Mov. Disord.*, 1998, **13**, 536.
64. M. Wittstock, R. Benecke and D. Dressler, *Neurology*, 2002, **58**, 831.
65. P. Barone, W. Poewe, S. Albrecht, C. Debieuvre, D. Massey, O. Rascol, E. Tolosa and D. Weintraub, *Lancet Neurol.*, 2010, **9**, 573–80.
66. P. Barone, L. Scarzella, R. Marconi, A. Antonini, L. Morgante, F. Bracco, M. Zappia and B. Musch, *J. Neurol.*, 2006, **253**, 601.
67. I. Rektorova, M. Balaz, J. Svatova, K. Zarubova, I. Honig, V. Dostal, S. Sedlackova, I. Nestrasil, J. Mastik, M. Bares, J. Veliskova and L. Dusek, *Clin. Neuropharmacol.*, 2008, **31**, 261.
68. D. Merims and N. Giladi, *Parkinsonism Relat. Disord.*, 2008, **14**, 273.
69. S. S. O'Sullivan, A. H. Evans and A. J. Lees, *CNS Drugs*, 2009, **23**, 157.
70. G. Rylander, *Psychiatr. Neurol. Neurochir*, 1972, **75**, 203.
71. V. Voon and S.H Fox, *Arch. Neurol.*, 2007, **64**, 1089.
72. American Psychiatric Association, *Diagnostic and Statistical Manual of Mental Disorders*, American Psychiatric Association, Washington DC, 4th edn, 1994.
73. E. D. Driver-Dunckley, B. N. Noble, J. G. Hentz, V. G. Evidente, J. N. Caviness, J. Parish, L. Krahn and C. H. Adler, *Clin. Neuropharmacol.*, 2007, **30**, 249.
74. A. J. Holman, *J. Gambl. Stud.*, 2009, **25**, 425.
75. G. Di Chiara and V. Bassareo, *Curr. Opin. Pharmacol.*, 2007, **7**, 69.
76. R. Ceravolo, D. Frosini, C. Rossi and U. Bonuccelli, *J. Neurol.*, 2010, **257**, S276.
77. E. Driver-Dunckley, J. Samanta and M. Stacy, *Neurology*, 2003, **61**, 422.
78. D. A. Gallagher, S. S. O'Sullivan, A. H. Evans, A. J. Lees and A. Schrag, *Mov. Disord*, 2007, **22**, 1757.
79. C. Lu, A. Bharmal and O. Suchowersky, *Arch. Neurol.*, 2006, **63**, 298.
80. C. W. Olanow, K. Kieburtz and A. H. Schapira, *Ann. Neurol.*, 2008, **64**, S101.

81. G. M. Halliday and C. H. Stevens, *Mov. Disord.*, 2011, **26**, 6.
82. R. de la Fuente-Fernández, M. Schulzer, E. Mak and V. Sossi, *Parkinsonism Relat. Disord.*, 2010, **16**, 365.
83. C. W. Olanow, *Neurology*, 2009, **72**, S59.
84. A. H. Schapira, *Eur. J. Neurol.*, 2002, **9**, 7.
85. T. Maeda, N. Cheng, T. Kume, S. Kaneko, H. Kouchiyama, A. Akaike, M. Ueda, M. Satoh, Y. Goshima and Y. Misu, *Brain Res.*, 1997, **771**, 159.
86. C. Mytilineou, R. H. Walker, R. Baptiste and C. W. Olanow, *J. Pharmacol. Exp. Ther.*, 2003, **304**, 792.
87. I. Ziv, R. Zilkha-Falb, D. Offen, A. Shirvan, A. Barzilai and E. Melamed, *Mov. Disord.*, 1997, **12**, 17.
88. E. Melamed, D. Offen, A. Shirvan, R. Djaldetti, A. Barzilai and I. Ziv, *Ann. Neurol.*, 1998, **44**, S149.
89. S. Fahn, D. Oakes, I. Shoulson, K. Kieburtz, A. Rudolph, A. Lang, C. W. Olanow, C. Tanner and K. Marek, *N. Engl. J. Med.*, 2004, **351**, 2498.
90. S. Fahn, *J. Neural Transm. Suppl.*, 2006, **70**, 419.
91. S. Chen, X. Zhang, D. Yang, Y. Du, L. Li, X. Li, M. Ming and W. Le, *FEBS Lett.*, 2008, **582**, 603.
92. M. Gassen, A. Gross and M. B. Youdim, *Mov. Disord*, 1998, **13**, 661.
93. G. Gille, W. D. Rausch, S. T. Hung, R. Moldzio, A. Ngyuen, B. Janetzky, A. Engfer and H. Reichmann, *J. Neural Transm.*, 2002, **109**, 157.
94. G. Lombardi, F. Varsaldi, G. Miglio, M. G. Papini, A. Battaglia and P. L. Canonico, *Eur. J. Pharmacol.*, 2002, **457**, 95.
95. W. Maruyama, T. Takahashi, M. Youdim and M. Naoi, *J. Neural Transm.*, 2002, **109**, 467.
96. C. Mytilineou, E. K. Leonardi, P. Radcliffe, E. H. Heinonen, S. K. Han, P. Werner, G. Cohen and C. W. Olanow, *J. Pharmacol. Exp. Ther.*, 1998, **284**, 700.
97. D. Uberti, L. Piccioni, A. Colzi, D. Bravi, P. L. Canonico and M. Memo, *Eur. J. Pharmacol.*, 2002, **434**, 17.
98. L. Zou, J. Jankovic, D. B. Rowe, W. Xie, S. H. Appel and W. Le, *Life Sci.*, 1999, **64**, 1275.
99. M. Asanuma, N. Ogawa, S. Nishibayashi, M. Kawai, Y. Kondo and E. Iwata, *Arch. Int. Pharmacodyn. Ther.*, 1995, **329**, 221.
100. E. Grünblatt, S. Mandel, G. Maor and M. B. Youdim, *J. Neurochem.*, 2001, **77**, 146.
101. M. Iida, I. Miyazaki, K. Tanaka, H. Kabuto, E. Iwata-Ichikawa and N. Ogawa, *Brain Res.*, 1999, **838**, 51.
102. M. Inden, Y. Kitamura, A. Tamaki, T. Yanagida, T. Shibaike, A. Yamamoto, K. Takata, H. Yasui, T. Taira, H. Ariga and T. Taniguchi, *Neurochem. Int.*, 2009, **55**, 760.
103. M. M. Iravani, C. O. Haddon, J. M. Cooper, P. Jenner and A. H. Schapira, *J. Neurochem.*, 2006, **96**, 1315.
104. M. M. Iravani, M. Sadeghian, C. C. Leung, B. C. Tel, S. Rose, A. H. Schapira and P. Jenner, *Exp. Neurol.*, 2008, **212**, 522.

105. H. H. Liou, R. C. Chen, T. H. Chen, Y. F. Tsai and M. C. Tsai, *Toxicol. Appl. Pharmacol.*, 2001, **172**, 37.
106. D. Muralikrishnan, S. Samantaray and K. P. Mohanakumar, *Synapse*, 2003, **50**, 7.
107. Y. Sagi, S. Mandel, T. Amit and M. B. Youdim, *Neurobiol. Dis.*, 2007, **25**, 35.
108. K. S. Saravanan, K. M. Sindhu, K. S. Senthilkumar and K. P. Mohanakumar, *Neurochem. Int.*, 2006, **49**, 28.
109. M. Yoshioka, K. Tanaka, I. Miyazaki, N. Fujita, Y. Higashi, M. Asanuma and N. Ogawa, *Neurosci. Res.*, 2002, **43**, 259.
110. L. Zou, J. Xu, J. Jankovic, Y. He, S. H. Appel and W. Le, *Neurosci. Lett.*, 2000, **281**, 67.
111. Parkinson Study Group, *JAMA, J. Am. Med. Assoc.*, 2002, **287**, 1653.
112. A. L. Whone, R. L. Watts, A. J. Stoessl, M. Davis, S. Reske, C. Nahmias, A. E. Lang, O. Rascol, M. J. Ribeiro, P. Remy, W. H. Poewe, R. A. Hauser and D. J. Brooks, *Ann. Neurol.*, 2003, **54**, 93.
113. F. Blandini, *CNS Drug Rev.*, 2005, **11**, 183.
114. K. Magyar, M. Pálfi, V. Jenei and E. Szöko, *J. Neural Transm. Suppl.*, 2006, **71**, 143.
115. C. W. Olanow, R. A. Hauser, L. Gauger, T. Malapira, W. Koller, J. Hubble, K. Bushenbark, D. Lilienfeld and J. Esterlitz, *Ann. Neurol.*, 1995, **38**, 771.
116. Parkinson Study Group, *N. Engl. J. Med.* 1989, **321**, 1364.
117. C. W. Olanow, O. Rascol, R. Hauser, P. D. Feigin, J. Jankovic, A. Lang, W. Langston, E. Melamed, W. Poewe, F. Stocchi and E. Tolosa, *N. Engl. J. Med.*, 2009, **361**, 1268.
118. S. M. Hoy and G. M. Keating, *Drugs*, 2012, **72**, 643.
119. Parkinson Study Group, *Arch. Neurol.*, 2002, **59**, 1937.
120. A. Antonini, G. Ursino, D. Calandrella, L. Bernardi and M. Plebani, *J. Neurol.*, 2010, **257**, S305.
121. D. Nyholm and S. M. Aquilonius, *Clin. Neuropharmacol.*, 2004, **27**, 245.
122. M. Sanford and L. J. Scott, *CNS Drugs*, 2011, **25**, 699.
123. N. Simola, A. Pinna and S. Fenu, *Recent Pat. CNS Drug Discov.*, 2010, **5**, 221.

CHAPTER 4

Catechol-O-Methyl-Transferase Inhibitors: Present Problems and Relevance of the New Ones

P. NUNO PALMA,[a] LÁSZLÓ E. KISS[a] AND PATRÍCIO SOARES-DA-SILVA*[a,b]

[a] Department of Research & Development, BIAL – Portela & C[a], S.A., À Avenida da Siderurgia Nacional, Mamede do Coronado, Portugal;
[b] Department of Pharmacology & Therapeutics, Faculty of Medicine, University Porto, Porto, Portugal
*Email: psoares.silva@bial.com

4.1 Introduction

Catechol-O-methyl-transferase (COMT)[1] is a magnesium-dependent enzyme found in both the periphery (liver, kidney and intestinal mucosa) and brain that catalyzes the transfer of a methyl group from the ubiquitous cosubstrate S-adenosyl-L-methionine (AdoMet) to substrates containing a catechol motif. The reaction results in the formation of a mono-O-methylated product and S-adenosylhomocysteine (AdoHcy) (Figure 4.1).

COMT is responsible for the metabolic inactivation of endogenous catechol neurotransmitters[2] and xenobiotics.[3,4] COMT inhibitors are used in the symptomatic treatment of Parkinson's disease (PD)[5,6] which is a neurodegenerative, slowly progressive disorder characterized by a severe decline in dopamine levels. The neuronal loss in specific regions of the brain (mainly in the *substantia nigra*) results in a reduction of the dopamine levels in the

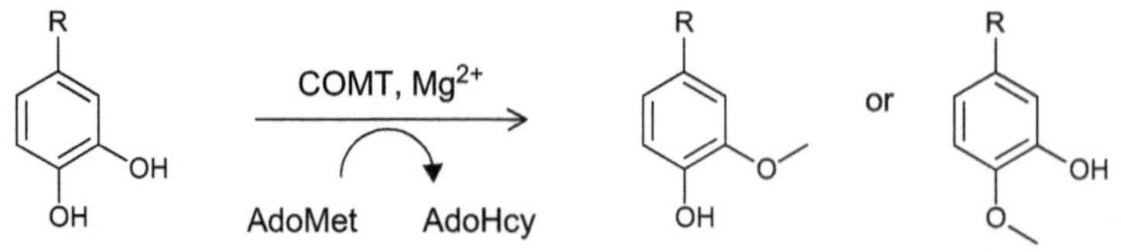

Figure 4.1 COMT-Catalyzed transfer of a methyl group to a catecholic substrate.

striatum, which becomes symptomatic over a certain threshold. The major symptoms include bradykinesia, resting tremor and postural reflex impairment.

The chapter will focus on the clinically most advanced COMT inhibitors (entacapone, tolcapone and opicapone) discussing their main pharmacokinetic and pharmacodynamic properties as well as their clinical relevance in the treatment of PD.

4.2 The Role of COMT Inhibitors in the Symptomatic Treatment of Parkinson's Disease

Dopamine itself is unable to permeate across the blood–brain barrier (BBB) and therefore it cannot be used as a drug. However, the dopamine precursor levodopa (also known as L-DOPA) readily gains access to the brain, where it is converted to dopamine by L-amino acid decarboxylase (AADC, also known as "dopa decarboxylase"). As a result, levodopa can be employed clinically as an artificial source of dopamine and currently remains the gold standard for the symptomatic treatment of PD. However, when levodopa is orally administered, it is extensively metabolized to dopamine in the peripheral tissues by AADC, so that only a limited amount of the administered levodopa reaches the brain (Figure 4.2).

In the absence of an AADC inhibitor, high doses of levodopa are required to have significant pharmacological effects. This repeated administration is often accompanied by adverse effects such as nausea, vomiting and hypotension. Generally, levodopa treatment is most effective in the early stages of PD due to the progressive degeneration of the dopamine system. To avoid the premature degradation of levodopa, a peripherally selective AADC inhibitor such as carbidopa or benserazide[7,8] is administered concomitantly. Administration of an AADC inhibitor results in an approximately 70–80% reduction in the clinically effective dose of levodopa. Subsequently, it was discovered that by blocking the decarboxylation route, the COMT enzyme becomes the major metabolic pathway for levodopa involving *O*-methylation to 3-*O*-methyl-levodopa (3-OMD). 3-OMD exhibits a long elimination half-life and may compete with levodopa for transport across the BBB. Therefore, accumulation of 3-OMD may decrease the bioavailability of levodopa in the brain. Concomitant administration of a COMT inhibitor with a peripherally selective AADC inhibitor prolongs the plasma half-life of levodopa, thereby more levodopa is delivered to the brain for dopamine biosynthesis. The combined administration of a COMT inhibitor plus levodopa plus an AADC inhibitor may allow lowering the oral daily dose of levodopa and also improve its tolerability and safety.

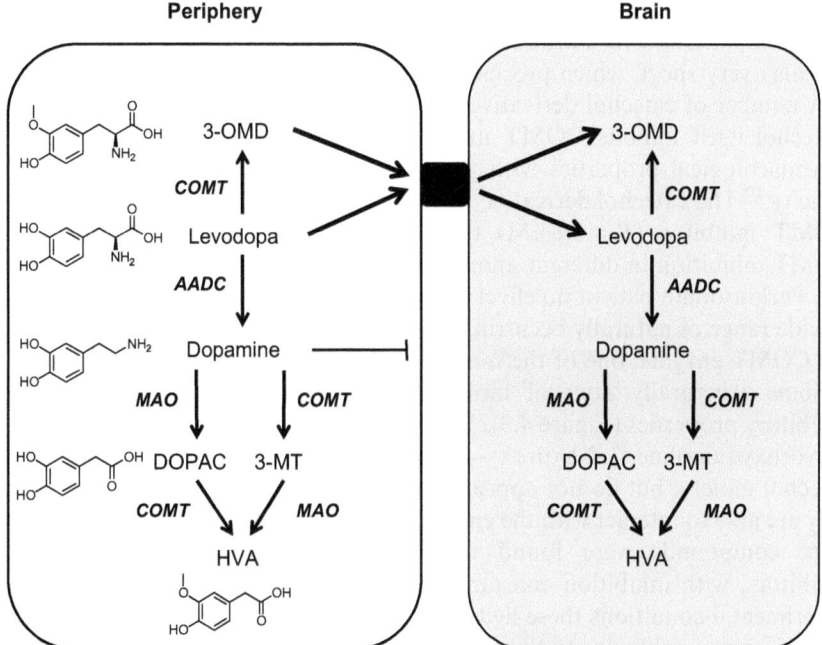

Figure 4.2 Levodopa active transport and metabolism.

4.3 Pyrogallol and Catechol Derivatives as COMT Inhibitors

The interest in developing clinically effective COMT inhibitors goes back at least four decades. During the early period (1960s and 1970s) a number of structurally simple compounds were identified and studied as COMT inhibitors. Most of these early inhibitors bear vicinal-dihydroxybenzene (catechol), trihydroxybenzene (pyrogallol) motifs or isosteric structures such as tropolones, hydroxypyrones and hydroxyquinolines (Figure 4.3).[2,9] The binding mode of catechol and pyrogallol ligands is fully reversible through the 1,2-dihydroxy moiety, which is compatible with their mode of action as competitive inhibitors. Pyrogallol was one of the first reported compounds to inhibit COMT enzyme under *in vitro* conditions with an inhibition constant (K_i) of 13 μM.[10,11] Its *in vivo* half-life was found to be very short (<1 h), and repeated administration every 30 min was required to sustain prolonged COMT inhibition. Moreover, pyrogallol inhibits cerebral COMT, which may result in undesirable CNS-related adverse effects. Gallic acid is a less potent *in vitro* inhibitor ($K_i = 70$ μM) of COMT than pyrogallol and is endowed with considerably higher cell toxicity.[12] The rationale for using butyl gallate (GPA 1714) as a COMT inhibitor was to reduce the toxicity risk of COMT inhibitors, considering that propyl and lauryl gallates were used in the past as safe food additives. Indeed, butyl gallate was found to be better tolerated than gallic acid, but unfortunately at least a daily dose of 750 mg was required to have beneficial effects in patients with Parkinsonism. The duration of action of butyl gallate was also very short, which precluded its clinical use.[9]

A number of catechol derivatives are known to inhibit COMT (Figure 4.3). Catechol itself inhibits COMT in a competitive manner and shares similar pharmacological properties with pyrogallol, such as low efficacy *in vivo* and toxicity.[10] The catechol derivative U-0521,[13] which was identified as a stronger COMT inhibitor ($K_i = 7.8\,\mu M$) than pyrogallol, demonstrated remarkable COMT inhibition in different animal species, but when administered orally to one Parkinsonian patient no effect on erythrocyte COMT activity was found.[14] A wide range of naturally occurring flavonoids has also been claimed to inhibit the COMT enzyme, one of the most potent being quercetin ($K_i = 8.4\,\mu M$).[15]

Some structurally 'atypical' molecules have been reported to show COMT inhibitory properties (Figure 4.3). Tropolone,[16] 3-hydroxypyridin-4(1*H*)-one,[17] 8-hydroxyquinoline,[18] 3-hydroxy-4-pyrone[17] are isosteric compounds with a catechol moiety, but do not appear to be substrates for COMT. Nevertheless, they are able to interact with the enzyme through α-keto–enol tautomerism. All these compounds were found to be weakly active competitive COMT inhibitors, with inhibition constants in the micromolar range. Under *in vivo* experimental conditions these ligands displayed low efficacy with short plasma half-life, presumably due to the easily reversible binding mode to the enzyme.[19]

Some COMT inhibitors incorporating duplicate catechol functions have been known in the literature since the 1960s, but until the late 1990s no interest was shown in studying their bifunctional nature.[4,20] A homologous series of symmetrical diamide type compounds incorporating two catechol structures were prepared and evaluated *in vitro* by Brevitt and Tan.[21] The authors reported, that no evidence was found for direct correlation between potency and bifunctional nature.[21] Nevertheless, the size of the linker between the two catechol nucleus was found to be important. One of the most potent bifunctional inhibitors within the catechol series has a spacer of three methylene groups (compound **1**, Figure 4.3; $K_i = 0.3\,\mu M$).

The design of bisubstrate inhibitors of COMT, incorporating a catechol moiety covalently attached to the C(5')-OH group of the adenosine fragment of AdoMet, through an appropriate linker has also originated potent inhibitors *in vitro* and is exemplified by compound **2** ($IC_{50} = 9\,nM$; Figure 4.3).[22–24] Generally, bisubstrate inhibitors are quite polar compounds with high molecular weights. Although no experimental *in vivo* data has been published to date in the literature, their *in vivo* efficacy may be questionable.

Until recently, relatively little attention has been devoted to the early 'atypical' COMT inhibitors, such as the 3-hydroxypyridin-4(1*H*)-one structures, since they demonstrated low efficacy *in vivo* and possessed poor target selectivity.[17] However, in the last few years much interest has been shown in development of potentially clinically effective 'atypical' COMT inhibitors. Substituted hydroxypyridinone[25,26] and hydroxypyrimidinone[27] derivatives have been claimed to demonstrate superior *in vitro* COMT activity over their earlier predecessors. The most interesting compounds include

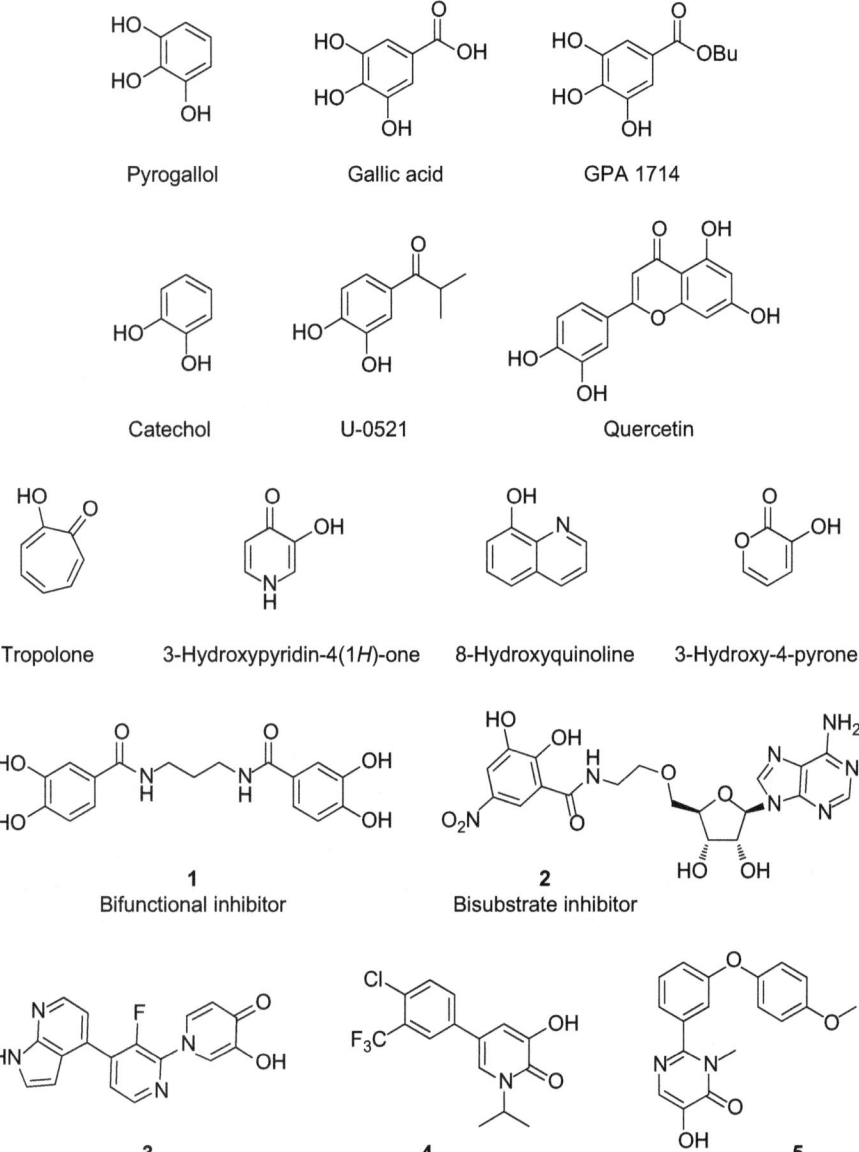

Figure 4.3 Representative examples of pyrogallol- and catechol-based inhibitors.

3-hydroxypyridin-4(1H)-one compound **3**, 3-hydroxypyridin-2(1H)-one compound **4** and 5-hydroxypyrimidin-4(3H)-one compound **5**, which were shown to be highly potent inhibitors of human membrane-bound COMT (MB-COMT) with IC_{50} values of 8 nM, 30 nM and 93 nM respectively (Figure 4.3). However, it is unknown whether these new analogues overcome the pharmacokinetic and pharmacodynamic problems of other previously disclosed

COMT inhibitors. Compounds **3**, **4** and **5** are thought to be useful in the symptomatic treatment of certain psychiatric disorders, such as schizophrenia, due to their potent inhibitory effect upon brain COMT.[25–27] The structure of the molecule chosen for clinical evaluation has not been publicly revealed to date.

4.4 Nitrocatechol COMT Inhibitors

In the late 1980s research groups at Orion Corp. and Hoffman–La Roche Ltd independently developed new series of disubstituted catechols incorporating an electron-withdrawing group (EWG) at the *ortho*-position to one of the hydroxyl groups. These new COMT inhibitors were endowed with significantly enhanced potency and selectivity over the previously studied pyrogallol and catechol derivatives. One of the 'first-generation' nitrocatechol inhibitors was 3,5-dinitrocatechol (OR-486) compound **6**, which displayed inhibitory activity ($IC_{50} = 12$ nM) 2–3 orders of magnitude higher than the most typical pyrogallol and catechol inhibitors (Table 4.1).[28] OR-486 is a poor substrate of the COMT enzyme and characterized as a fully reversible tight-binding inhibitor. Aside from the fact that OR-486 displays favorable pharmacodynamic properties, concerns have arisen regarding its toxicity and safety hazards, which seriously question its future clinical development. Therefore OR-486 serves solely as a biological tool for studying COMT and as a starting point in medicinal chemistry optimization programs.

Thus, a number of first-generation derivatives were synthesized, incorporating different EWGs (compounds **6–12**, Table 4.1). The substituent effect on COMT inhibitory ability was evaluated *in vitro* in rat brain homogenates.[28,29] Replacing the nitro substituent located furthest from the catechol hydroxyls with other electronegative groups such as cyano (**7**), formyl (**8**), or acetyl (**9**) gave compounds which were slightly less potent than compound **6**. In contrast, replacement of the nitro group at the *ortho*-position to the hydroxyl group with cyano (**10**), trifluoromethyl (**11**) or methylsulfonyl (**12**) resulted in a

Table 4.1 *In vitro* COMT inhibition by first-generation nitrocatechols in rat brain homogenate.

Compound	R_1	R_2	IC_{50}/nM
6	$-NO_2$	$-NO_2$	12
7	$-NO_2$	$-CN$	30
8	$-NO_2$	$-CHO$	24
9	$-NO_2$	$-(CO)CH_3$	16
10	$-CN$	$-CHO$	160
11	$-CF_3$	$-CHO$	2300
12	$-SO_2CH_3$	$-CHO$	20 000

remarkable reduction in COMT inhibitory activity. It was concluded that the presence of a nitro group at the *ortho*-position to the hydroxyl group was crucial for exhibiting high *in vitro* potency. Therefore, subsequent structure–activity relationship (SAR) studies focused on replacement of the nitro group at position 5 of the catechol ring in order to modulate the pharmacokinetic properties of nitrocatechol compounds. Several of these early 'second-generation' nitrocatechol COMT inhibitors were found to display relatively weak *in vivo* COMT inhibitory activity, which was attributed to their poor absorption and metabolic instability. However, a series of aryl ketones[29] and vinylic derivatives[28] presented favorable *in vivo* pharmacological and toxicological properties. Further optimization of the aryl ketone and vinylic side-chains led to the discovery of nitecapone (OR-462),[30] entacapone (OR-611)[31] and tolcapone (Ro 40-7592)[32] (Figure 4.4). Nitecapone, entacapone and tolcapone are fully reversible, tight-binding inhibitors of COMT with subnanomolar inhibition constants (K_i), in rat liver COMT.[33] Moreover, they do not interfere with other enzymes involved in the metabolism of catecholamines including tyrosine hydroxylase, dopamine-β-hydroxylase (DβH) and monoamine oxidase (MAO).[28] Nitecapone, which was the early clinical candidate of Orion demonstrated moderate *in vivo* efficacy in different animal species such as mouse, rat and monkey.[34] Nitecapone was found to be well tolerated and its inhibition was confined to the periphery.[35] In clinical studies

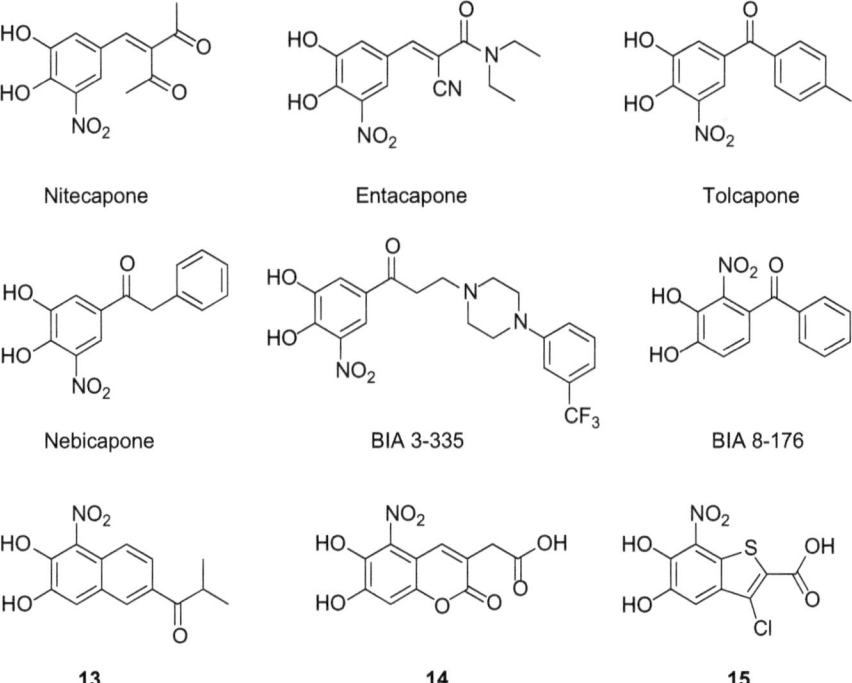

Figure 4.4 Representative examples of second-generation nitrocatechol inhibitors.

nitecapone had a beneficial effect on both levodopa and 3-OMD pharmacokinetics, but eventually ceded its place in the clinical development to entacapone, which was judged to be a better candidate. Although entacapone is a more potent compound than nitecapone, it has a very short duration of action[36] and requires high dosages, repeated up to eight times daily. Its clinical effectiveness in the treatment of PD has been questioned.[37] Entacapone is a purely peripherally acting COMT inhibitor at doses of up to 30 mg kg^{-1}.[38] Conversely tolcapone, the clinical candidate of Hoffman–La Roche is an equipotent inhibitor of both brain and peripheral COMT, endowed with a longer duration of action than entacapone. Both entacapone and tolcapone were introduced into clinical practice in the late 1990s as adjuncts to levodopa therapy for PD. However, shortly after its launch, tolcapone was withdrawn from the market due to liver toxicity; several cases of fatal fulminant hepatitis[39] were reported. Although entacapone is generally regarded as a safe and well-tolerated drug, its principal drawback is its short plasma half-life and limited oral bioavailability.

Undoubtedly, there was still clearly a definite need for improved COMT inhibitors to address the unmet medical needs of patients with PD. An ideal COMT inhibitor would be peripherally selective (more selective than tolcapone), and endowed with sufficiently long duration of action (longer than entacapone) that would make it suitable for a once-daily administration regime, which would be particularly convenient for PD patients.

In the last fifteen years several new classes of second-generation nitrocatechol COMT inhibitors have been designed in an attempt to eliminate the drawbacks of entacapone and tolcapone. Nebicapone[40] (Figure 4.4) was purposely designed to marry the positive attributes of tolcapone and entacapone, which is to possess a longer duration of COMT inhibition in the periphery than entacapone, while simultaneously having more limited access to the brain than tolcapone. However, the clinical development of nebicapone was discontinued because its safety profile was not considered sufficiently improved over that of tolcapone. Further modification of the phenylacetyl side-chain of nebicapone resulted in a novel series of COMT inhibitors containing an *N*-phenylpiperazine side-chain.[41] SAR optimization around the *N*-phenylpiperazine core led to the identification of BIA 3-335, which displays a longer duration of action than entacapone in mice at a dose of 30 mg kg^{-1} (Figure 4.4). Additionally, BIA 3-335 exhibits limited access to the brain, but shows lower *in vivo* efficacy when compared to that of tolcapone.

In a further study, several regioisomerically nitrated ("*ortho*-nitro") derivatives of COMT inhibitors were prepared and their ability to inhibit COMT enzyme under *in vitro* conditions was assessed. One of the most potent *ortho*-nitrated compounds is BIA 8-176, which inhibits COMT with IC$_{50}$ values of 3 and 130 nM for rat brain and liver, respectively (Figure 4.4).[42] *In vivo* evaluation in mice revealed that BIA 8-176 provided a fast onset of inhibition ($T_{max} \leq 0.5$ h) but showed relatively short duration of action. The latter could be explained by differences in the metabolic stability of *meta*-nitrated regioisomers. Additionally, the extent of peripheral and central COMT inhibition by BIA 8-176 was almost identical which might have been problematic during clinical development.

Several new second-generation nitrocatechol compounds bearing fused-bicyclic structures have been developed by Orion. These new classes include naphthalenes,[43] coumarins[44] and more recently benzo[b]thiophenes[45] (Figure 4.4). Naphthalene, compound **13** and coumarin, compound **14** are potent inhibitors of COMT under tight-binding conditions with a IC_{50}s of 25 and 10 nM, respectively. To the best of our knowledge no compounds from the naphthalene or coumarin series made the transition from laboratory to clinical trials. The more recent newcomer, benzothiophene, compound **15** has a quite impressive inhibition constant ($K_i < 1$ nM). Despite its outstanding pharmacodynamic interaction with the COMT enzyme under *in vitro* conditions, compound **15** was found to possess poor oral bioavailability and high systemic clearance.[46] It was later reported by Orion that the development of the lead molecule was discontinued due to the conclusion that its pharmacological properties would not outperform entacapone or tolcapone.

Shortly after Orion's disclosure on benzo[b]thiophene derivatives,[45] a novel class of non-condensed heterocycle-based nitrocatechol COMT inhibitors, structurally unrelated to their classical second-generation predecessors, were reported by a Portuguese research group.[47,48] It was found that the prototype unsubstituted oxadiazolylpyridine-*N*-oxide derivative presented a reduced propensity to cause toxicity and retained activity comparable to tolcapone.[48] Extensive SAR studies were carried out in order to improve intestinal absorption and the duration of COMT inhibition whilst simultaneously maintaining the low toxicity risk. Pyridine-*N*-oxide derivatives substituted with small lipophilic groups (Figure 4.5) were generally endowed with an enhanced inhibitory profile over second-generation inhibitors. The most promising of those compounds, BIA 9-1067 (INN: opicapone), is represented in Figure 4.5. Opicapone was found to be an extremely potent and purely peripheral inhibitor of COMT and endowed with a truly unprecedented duration of action. Furthermore, opicapone presents an atypical, yet extremely favorable pharmacodynamic interaction with levodopa, leading to stable and sustained plasma levodopa levels over a 24 h period.

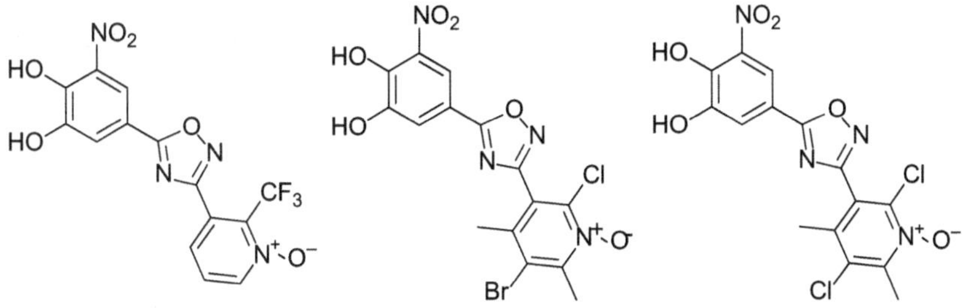

Figure 4.5 Representative examples of third-generation nitrocatechol inhibitors.

These characteristics clearly differentiate this class of compounds as having considerably improved biological properties over established COMT inhibitors. Therefore, opicapone is presented as a structurally novel, 'third-generation' nitrocatechol COMT inhibitor, which is currently under clinical evaluation as an adjunct to L-DOPA therapy for the treatment of PD.[48] Within this chapter, several different aspects of opicapone's enzymological mechanism of action, as well as of its non-clinical and clinical pharmacology will be discussed.

4.5 Non-Clinical Pharmacology of COMT Inhibitors

In the non-clinical pharmacology phase, the most relevant second-generation COMT inhibitors have been studied in different animal species for their ability to inhibit both central and peripheral COMT enzyme.

The *in vivo* timecourse inhibition profile of entacapone, tolcapone and opicapone was evaluated in the rat at a dose of 3 mg kg^{-1}.[48] While entacapone and tolcapone displayed a rapid onset of action in the liver with a maximum inhibitory effect at 0.5 h and 1 h respectively, opicapone was found to be slowly absorbed achieving maximal inhibition (>99%) at only 3 h post dose. However, opicapone was endowed with a much longer duration of action than that of tolcapone and entacapone, sustaining notable peripheral COMT inhibition (below 50% of initial levels) up to 24 h post dose.[48,49]

Considering that tolcapone has a quite lipophilic 4′-methylbenzoyl substituent at the *meta*-position relative to the nitro group, it is not surprising that tolcapone is able to penetrate the BBB. Conversely, due to the reduced overall lipophilicity and favorable partition coefficient (log$P$), entacapone and opicapone have no measurable COMT inhibition in the brain at any timepoint. The purely peripheral nature of opicapone effects may have an advantage over tolcapone, since inhibition of brain COMT is believed to be less important and may potentially cause undesired adverse effects. The *in vivo* inhibitory potency of the most important COMT inhibitors was assessed in the rat.[48] Entacapone, tolcapone and opicapone were found to inhibit rat liver COMT in a dose-dependent manner. The results show that, at 3 h post dose, opicapone is more potent than tolcapone with ED$_{50}$s (in mg kg^{-1}) of 1.05 ± 0.04 and 1.77 ± 0.10, respectively. On the other hand, entacapone was much less potent with an ED$_{50}$ of 7.80 ± 0.70 mg kg^{-1}. In the rat, entacapone was found to achieve maximum inhibitory potency at 1 h post dose (ED$_{50}$ = 1.90 ± 0.20 mg kg^{-1}).[40]

As mentioned above, the rationale for using COMT inhibitors as an adjunct to levodopa treatment PD is to simultaneously increase plasma levels of levodopa and decrease the formation of 3-OMD. For this reason, the pharmacodynamic interaction between COMT inhibitors and levodopa was evaluated in the rat after single-dose administration.[48] Entacapone, tolcapone and opicapone were administered to rats (3 mg kg^{-1} each) and at defined intervals (2, 7 and 24 h), levodopa (12 mg kg^{-1}) was co-administered with benserazide (3 mg kg^{-1}). One hour later, levodopa and 3-OMD were measured.

Levodopa levels in plasma were raised by entacapone (70% increase) at 2 h, but not at the subsequent timepoints, at 7 and 24 h post dose. Tolcapone caused an abrupt, three-fold, increase in plasma levodopa levels at 2 h, which subsequently declined to a two-fold increase at 7 h, and no effect was observed at 24 h post dose. The implications of this are that the relatively short-acting effects of entacapone and tolcapone require multiple periodic dosing of the inhibitor in order to sustain levodopa levels at a therapeutically useful threshold. Conversely, opicapone provided a steady sustained two-fold increase in plasma levodopa over the entire evaluation period, from 2 to 24 h post dose. The same trend, but in the opposite direction, was observed for 3-OMD levels in plasma. Entacapone markedly reduced 3-OMD plasma levels at the 2 h timepoint, but not at the subsequent timepoints at 7 and 24 h post dose. Tolcapone markedly decreased 3-OMD plasma levels at 2 h post dose, but this effect was reduced by 50% at the 7 h timepoint, and then returned to baseline at 24 h post dose. Opicapone, however, provided a steady sustained 50% increase in plasma levodopa over the entire evaluation period, from 2 to 24 h post dose. As a result of the increase in levodopa levels in plasma, an increased levodopa brain exposure was observed when opicapone and levodopa plus benserazide were administered to rats.[49]

The effect of opicapone upon the systemic and brain bioavailability of levodopa and related metabolites (3-OMD, HVA, DOPAC) was recently evaluated in cynomolgus monkeys.[50] Animals were randomized into two groups that received, in a cross-over design, vehicle or 100 mg kg^{-1} opicapone for 14 days. The wash-out period between the two phases was five days. Twenty-three hours after the last administration in each treatment period, animals were given levodopa/benserazide (12/3 mg kg^{-1}). The animals were implanted with indwelling guiding cannulas for the insertion of microdialysis probes in the *substantia nigra*, dorsal *striatum* and prefrontal cortex. Extracellular dialysate and blood samples were collected over 30 min periods before and up to 360 min after levodopa/benserazide. Opicapone increased levodopa exposure in the dorsal *striatum* dialysate without changes in the levodopa C_{max} values and reduced exposure to 3-OMD with marked decreases in 3-OMD C_{max} values. Similar findings were observed in the *substantia nigra* and prefrontal cortex. Opicapone also increased levodopa systemic exposure without changes in levodopa C_{max} values and reduced exposure to 3-OMD with marked decreases in 3-OMD C_{max} values. These changes were accompanied by a *ca.* 80–85% reduction in erythrocyte COMT activity.

4.6 Metabolic Profile of COMT Inhibitors

Typical metabolic pathways of second-generation COMT inhibitors are shown in Figure 4.6. Nitrocatechols are extensively metabolized in the liver mostly involving Phase II reactions such as *O*-glucuronidation, *O*-sulfation and *O*-methylation. These bio-transformations predominantly occur at the *meta*-hydroxy group relative to the nitro substituent.

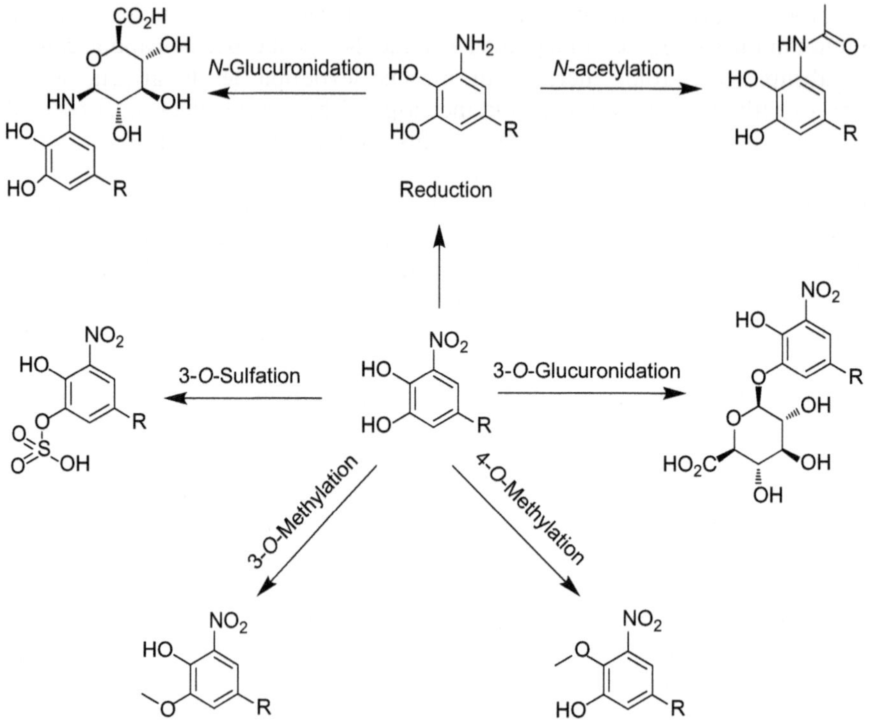

Figure 4.6 *In vivo* metabolism of nitrocatechols.

First-pass metabolism by sulfotransferases (SULTs) and UDP-glucuronosyltransferases (UGTs) at the hydroxyl group adjacent to the nitro substituent is much less likely. This hydroxyl group is only susceptible towards O-methylation by the COMT enzyme. The resultant 4-O-methyl metabolite can generally be detected at lesser amounts than the 3-O-methyl product. Phase I reduction of the nitro group to the catecholic aniline followed by N-acetyl conjugation or N-glucuronidation generally represent minor metabolic pathways.

In humans, entacapone,[51] tolcapone[52] and nebicapone,[53,54] are almost completely metabolized prior to excretion, with less than 1% of an orally administered dose of the parent compound excreted unchanged. Although entacapone and tolcapone incorporate the same nitrocatechol pharmacophore, notable differences in their conjugation profile can be observed. They are mainly metabolized through 3-O-glucuronidation, but tolcapone and nebicapone are also methylated by COMT to 3-O-methyl-tolcapone and 3-O-methyl-nebicapone, respectively. Nebicapone presents a similar metabolism and excretion profile to that of tolcapone. Nebicapone and BIA 3-476 (nebicapone 3-O-glucuronide) account for most early-phase circulating nebicapone-derived moieties, have limited circulating cell association, peak concentrations shortly after dosing, and short body residence.

3-O-Methyl-nebicapone was found to be endowed with a long residence time, but at a low exposure level.[53] After administration of [^{14}C]-nebicapone to healthy human volunteers, the urinary excretion (mainly BIA 3-476) accounted for $\sim$79% recovered radioactivity or 70.1% administered dose, with excretion of unchanged nebicapone <1%.[54] Faecal excretion accounted for approximately 21.4% of recovered radioactivity or 19% of the administered dose.[54] Formation of small amounts of amino catechol, N-glucuronide and N-acetyl derivatives could also be observed. O-Sulfation of entacapone to the corresponding 3-O-sulfate appears to be a minor metabolic pathway.[51] Side-chain-specific metabolism of COMT inhibitors may result in pharmacologically active metabolites. Indeed, entacapone undergoes isomerization from the (E)- to (Z)-isomer, which was found to be active against COMT (Figure 4.7).

The metabolism of tolcapone *via* CYPs to a primary alcohol followed by oxidation to the corresponding carboxylic acid has also been identified. The pharmacokinetic properties of opicapone in healthy human subjects have recently been assessed[55,56] and its metabolic profile does not follow the customary pathways for the metabolism of nitrocatechol derivatives. In fact, opicapone is mostly metabolized by O-conjugation with sulfate. The resultant 3-O-sulfate BIA 9-1103 is the major circulating entity.[56] This pharmacologically inactive metabolite has an elimination half-life of 39–67 h post dose. Opicapone also presents a side-chain metabolic reaction. The pyridine-N-oxide ring undergoes metabolic reduction to a pyridine derivative, known as BIA 9-1079, (Figure 4.7), which is endowed with inhibitory properties against COMT. The concentration of this reduced metabolite was found to be lower than the inactive O-sulfate metabolite. Plasma levels of other common metabolites such as O-methylated and O-glucuronide derivatives remained below the limit of quantification.

4.7 Human Pharmacology

In a double-blind, randomized, placebo-controlled, cross-over study with four single-dose treatment periods in 16 healthy male subjects the pharmacokinetic–pharmacodynamic profile of three different single doses of opicapone was assessed. The wash-out period between doses was at least 14 days; subjects were discharged 72 h post dose. Subjects had to attend four treatment periods and received a different dose of opicapone (25, 50 and 100 mg) or placebo during each of these treatment periods. Blood samples for the assay of plasma opicapone and its active metabolite BIA 9-1079 were taken at the following times: pre dose, 0.5, 1, 1.5, 2, 3, 4, 6, 8, 12, 16, 24, 48 and 72 h post dose. The same blood samples taken for the pharmacokinetic assessments were used for the preparation of washed erythrocytes for the assay of COMT activity in erythrocytes.

Figure 4.8 displays the plasma concentration–time profiles and Table 4.2 presents the plasma pharmacokinetic parameters of opicapone and its active metabolite BIA 9-1079, following single oral doses of opicapone ranging from 25 to 100 mg. Opicapone C_{max} and AUC showed a dose-dependent increase.

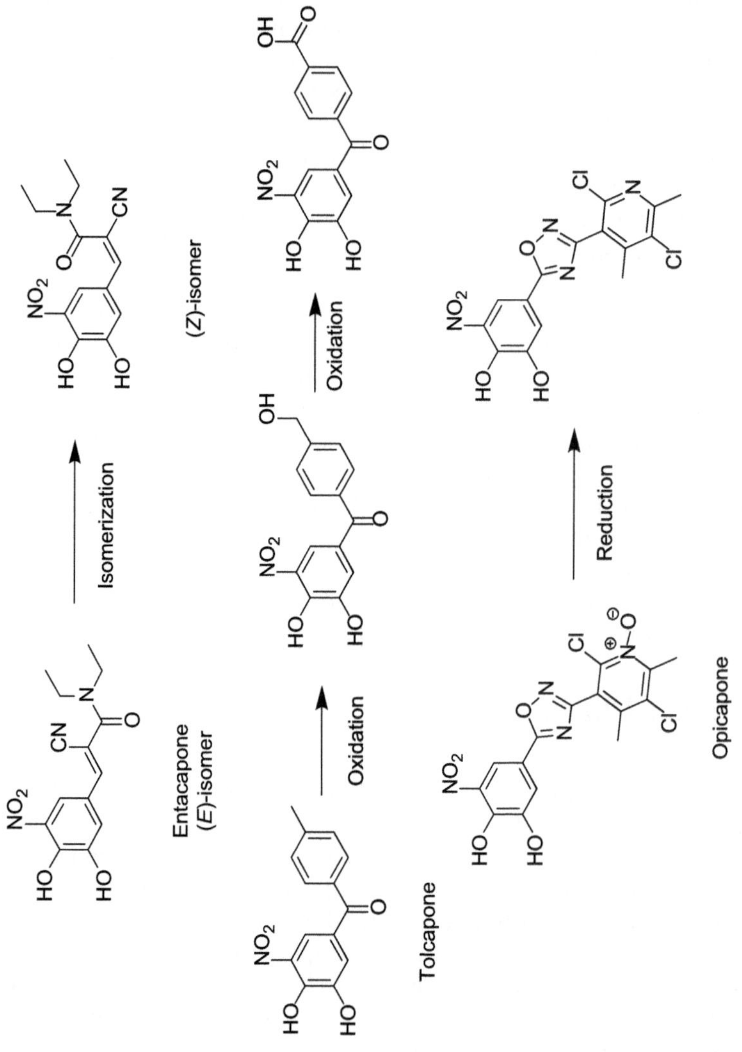

Figure 4.7 Side-chain-specific metabolism of entacapone, tolcapone and opicapone.

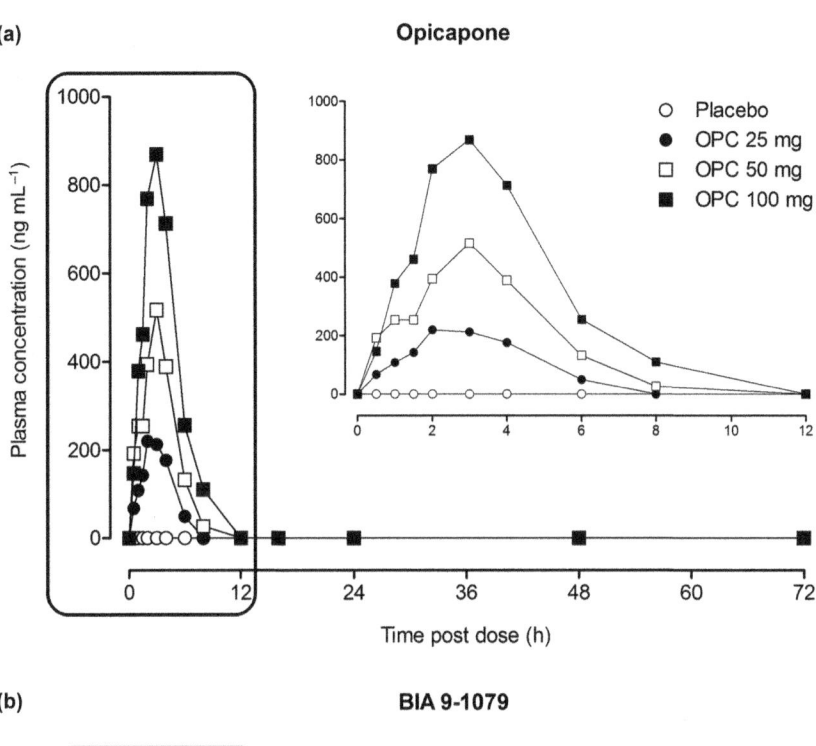
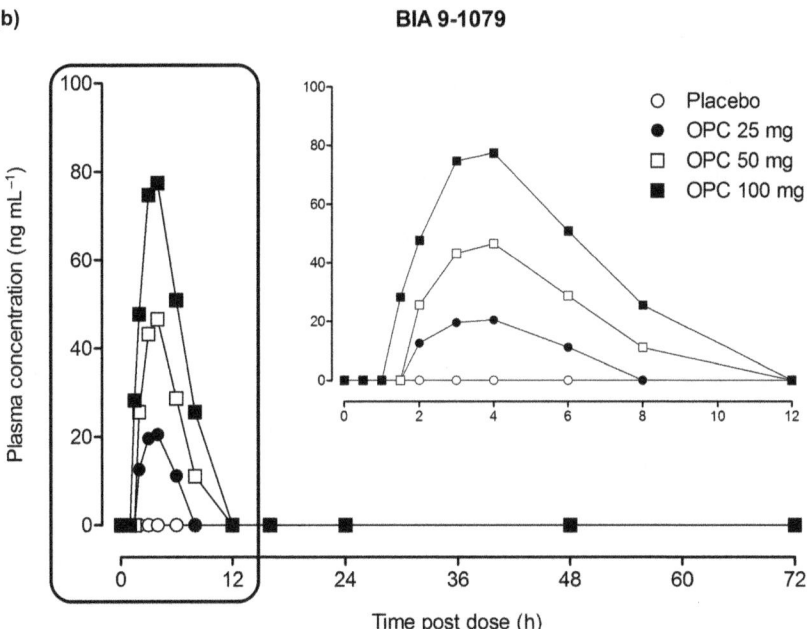

Figure 4.8 Mean plasma concentration–time profiles of (a) opicapone and (b) BIA 9-1079, the active metabolite of opicapone, following single oral doses of opicapone (OPC) ranging from 25 to 100 mg ($n=16$ per group). The insets show concentration–time profiles of opicapone and BIA 9-1079 up to 12 h post dose.

Table 4.2 Mean (coefficient of variation, %) pharmacokinetic parameters of opicapone and BIA 9-1079 following single oral administration of placebo and opicapone (OPC) 25, 50 and 100 mg ($n=16$). t_{max} values are median with range values in parentheses; NA = not available; NC = not calculated.

Opicapone	C_{max}/ng mL^{-1}	t_{max}/h	AUC_{0-t}/ng h mL^{-1}	$AUC_{0-\infty}$/ng h mL^{-1}	$t_{1/2}$/h
Placebo	NA	NA	NA	NC	NC
OPC 25 mg	327 (42.9)	2.00 (0.5–4.0)	881 (42.8)	1007 (31.2)	1.04 (27.7)
OPC 50 mg	642 (28.9)	3.00 (1.0–6.0)	2048 (39.5)	2200 (34.0)	1.01 (29.7)
OPC 100 mg	1110 (38.8)	2.00 (1.0–6.0)	3841 (51.1)	3888 (50.6)	1.03 (30.5)
BIA 9-1079					
Placebo	NA	NA	NA	NC	NC
OPC 25 mg	3.79 (36.2)	1.50 (0.5–3.0)	86.3 (45.6)	155 (24.1)	2.33 (16.5)
OPC 50 mg	52.4 (23.0)	4.00 (3.0–6.0)	213 (44.0)	296 (33.5)	2.33 (34.3)
OPC 100 mg	89.3 (35.8)	4.00 (2.0–6.0)	418 (57.1)	519 (49.1)	2.24 (30.5)

Interindividual variability ranged from 29 to 43% for C_{max} and 31 to 51% for $AUC_{0-\infty}$. t_{max} ranged from 0.4 to 6.0 h and $t_{1/2}$ was relatively short, ranging from 1.04 h (25 mg) to 1.03 h (100 mg). BIA 9-1079 C_{max} and $AUC_{0-\infty}$ also showed a dose-dependent increase and a relatively short $t_{1/2}$ (2.24 to 2.33 h). Both rate (as assessed by C_{max}) and extent (as assessed by AUC) of systemic exposure to opicapone and BIA 9-1079 increased in a dose-proportional manner.

Inhibition of soluble-COMT (S-COMT) activity in erythrocytes (by assaying erythrocyte S-COMT activity) was evaluated in a dose–escalation study. Figure 4.9 depicts S-COMT activity over time in the placebo and opicapone dose groups, and Table 4.3 presents the ANOVA of the main pharmacodynamic parameters *versus* placebo. A dose-dependent and long-lasting inhibitory effect of S-COMT activity was observed. At 24 h post dose, S-COMT inhibition ranged from 36.6% (25 mg) to 57.2% (100 mg); at 72 h post dose, it ranged from 24.1% (25 mg) to 31.5% (100 mg). The maximum S-COMT inhibition (E_{max}) occurred between 2.9 h (100 mg) and 6.1 h (25 mg) post dose ($t_{E_{max}}$), and ranged from 70.0% (25 mg) to 95.7% (100 mg). All opicapone treatments significantly inhibited both the peak and extent of S-COMT activity in relation to placebo. The peak and extent of S-COMT inhibition were dose-dependent. A statistical difference was found for $t_{E_{max}}$ between all doses of opicapone and placebo ($p=0.0034$ for 25 mg opicapone, $p=0.0004$ for 50 mg opicapone and $p=0.0005$ for 100 mg opicapone). Although the same dose-dependent tendency was observed for

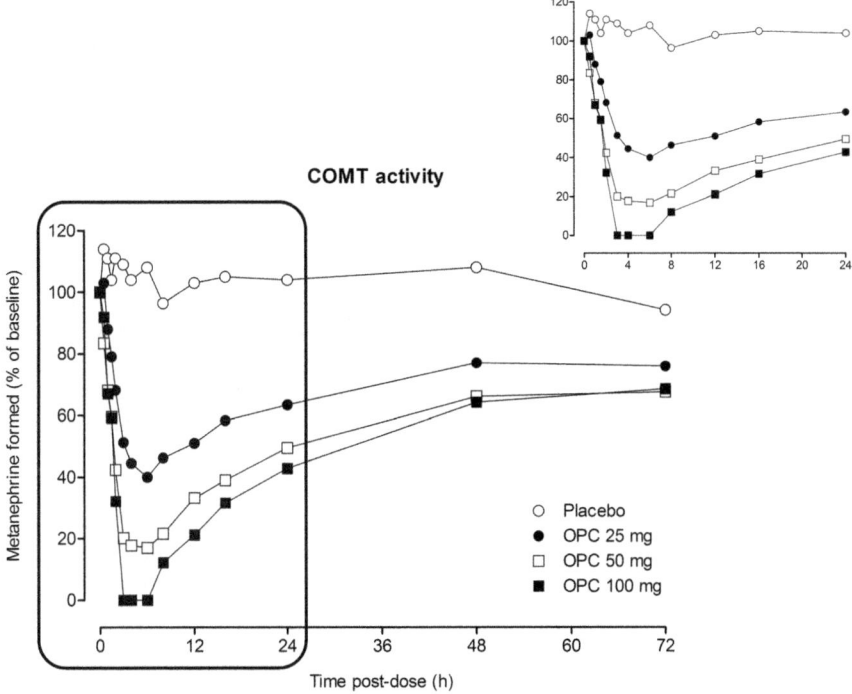

Figure 4.9 S-COMT activity (% of baseline) *versus* time curve over 0–72 h post dose following single oral doses of opicapone (OPC) ranging from 25 to 100 mg or placebo ($n = 16$ per dose group of opicapone; $n = 16$ for placebo). The inset shows S-COMT activity (% of baseline) *versus* time up to 24 h post dose.

Table 4.3 Mean ($\pm$SD) pharmacodynamic parameters for inhibition of S-COMT following single oral administration of placebo and opicapone 25, 50 and 100 mg ($n = 16$).

Dose	E_0 (pmol mg Hb^{-1} h^{-1})	E_{max} (pmol mg Hb^{-1} h^{-1})	$t_{E_{max}}$ (h)	$(E_0 - E_{max})/E_0$ ($\times 100\%$)	$AUEC_{0-24}$ (pmol mg Hb^{-1} h^{-1} h)[a]	$AUEC_{0-72}$ (pmol mg Hb^{-1} h^{-1} h)[a]
Placebo	35.7(9.88)	30.8(10.3)	27.1(26.9)	14.1(10.8)	891(253)	2664(796)
OPC 25 mg	39.7(14.0)	14.0(14.7)	6.13(3.80)	70.0(23.3)	538(341)	1935(987)
OPC 50 mg	37.1(13.3)	4.36(4.99)	3.94(1.44)	89.6(10.0)	324(156)	1433(554)
OPC 100 mg	37.2(12.6)	2.03(3.29)	2.87(1.68)	95.7(6.47)	255(142)	1325(502)

[a]AUEC = area under the effect curve.

opicapone-induced E_{max} changes, it was not possible to determine the exact time of return to baseline enzyme activity because at 72 h post dose (last timepoint of the profile) there was still inhibition at all opicapone doses.

A Phase I clinical trial aimed to evaluate the tolerability, pharmacokinetics and pharmacodynamics (S-COMT activity inhibition) of single oral doses of opicapone in the dose range of 25 to 100 mg. Dose-proportionality in C_{max} and $AUC_{0-\infty}$ was observed for opicapone and BIA 9-1079 (which results from the reduction of opicapone). BIA 9-1079 was found to be a minor metabolite, representing less than 15% of systemic exposure to opicapone; thus, its contribution to the therapeutic effect following administration of opicapone is expected to be of minor relevance.

The inhibitory effect of COMT inhibitors is usually evaluated by assaying the erythrocyte S-COMT activity.[57] In a dose-escalation Phase I clinical trial, dose-dependent and long-lasting S-COMT inhibition was observed following single oral doses of opicapone in the tested dose range. E_{max} occurred between 2.9 h (100 mg) and 6.1 h (25 mg) post dose and ranged from 70.0% (25 mg) to 95.7% (100 mg). E_{max} reached 89.6% with a dose of 50 mg.

The inhibitory effect on S-COMT by opicapone reported here is much stronger and more sustained than that reported for tolcapone, entacapone and nebicapone in healthy subjects. E_{max} was 72 and 80% for tolcapone 100 and 200 mg, respectively[58] 65% for entacapone 200 mg,[36] and 69 and 80% for nebicapone 75 and 150 mg, respectively.[59] Whereas S-COMT activity returned to baseline approximately 18 h after tolcapone administration,[58] 8 h after entacapone administration[36] and 16 h after nebicapone administration,[59] S-COMT inhibition with opicapone, ranged from 36.6% (25 mg) to 57.2% (100 mg) at 24 h post dose, and from 24.1% (25 mg) to 31.5% (100 mg) at 72 h post dose. The inhibition of S-COMT activity reached statistical significance at all doses of opicapone tested, in relation to placebo.

Overall, the results presented here suggest that opicapone is endowed with a much stronger and more sustained effect than the other COMT inhibitors and presents a pharmacodynamic profile adequate for a once-daily regimen; this represents an advantage over entacapone and tolcapone, which are recommended in several doses per day. Since COMT inhibitions of 37, 51 and 57% still remain 24 h after administration of 25, 50 and 100 mg of opicapone, respectively and it is expected that such inhibitory effect will increase after repeated administration, therapeutic doses are expected to be 50 mg or below.

4.8 New Insights into the Mechanism of COMT Inhibition

Since the three-dimensional structure of recombinant rat soluble COMT was unveiled, for the first time, by Vidgren and collaborators,[60,61] a number of other crystallographic structures of rat and human S-COMT, in complexes with various inhibitors, have enabled a detailed understanding of the mechanism of action of COMT inhibitors. These aspects have been reviewed in the recent literature (see for example[62–64] and references therein).

4.8.1 *In Vitro* Potency

Nitrocatecholic inhibitors are kinetically characterized as reversible tight-binding COMT inhibitors and behave competitively, with respect to the catechol substrate, while showing uncompetitive behavior with respect to the cosubstrate AdoMet.[33,65,66] The kinetic analysis of tight-binding inhibitors is complicated, as the determination of inhibition parameters is highly dependent on the enzyme preparations and assay conditions. Therefore some discrepancies among reported values may be observed. Tolcapone, the most potent of the two marketed COMT inhibitors was shown to inhibit MB-COMT *in vitro* with an observed inhibition constant (K_i) of 0.27 nM.[33] More recently, the observed K_i for opicapone, against recombinant human S-COMT, was reported to be 16 pM, one order of magnitude lower, when compared to that of tolcapone (160 pM), assayed under the same experimental conditions.[49] In the next paragraph the main factors that contribute to the difference in potency observed between the two COMT inhibitors are discussed.

Unlike most typical non-covalent and reversible enzyme inhibitors, nitrocatechol COMT inhibitors are, themselves, competitive substrates (though very poor ones), as they undergo reaction at the enzyme catalytic site to be released in the form of *O*-methylated products (Scheme 4.1).[49,67]

Hence, the experimentally observed inhibition constant (K_i) for these inhibitors is represented by the expression

$$K_i = \frac{k_{off} + k_{cat}}{k_{on}} = K_d + \frac{k_{cat}}{k_{on}} \tag{4.1}$$

where k_{on} and k_{off} are, respectively, the pseudo-first-order association and the first-order dissociation rate constants of the enzyme–inhibitor Michaelis complex (*EI*). The catalytic constant k_{cat} is characteristic of the rate at which the bound inhibitor is transformed and released in the form of a product *P* (the *O*-methylated derivative), thereby regenerating the uninhibited form of the enzyme. The rate of the enzyme–inhibitor association (k_{on}) is only dependent on the rates of diffusion in solution and in the case of these COMT–ligand complexes, is estimated to be approximately $10^7 \, M^{-1} \, s^{-1}$.[68] The inhibition constant, therefore, depends not only on the dissociation equilibrium of the *EI* complex ($K_d = k_{off}/k_{on}$), but also on the rate of the methylation reaction (k_{cat}) to release *E* and *P*. The assessment and comparison of the individual rate constants in eqn (4.1) is crucial to understand the relative contribution of each of the two independent processes to the observed inhibition constants of nitrocatechol COMT inhibitors.

The catalytic rate constants (k_{cat}) for the *O*-methylation of tolcapone and opicapone by human recombinant S-COMT have been determined and are, respectively, $10^{-3} \, s^{-1}$ and $10^{-4} \, s^{-1}$ (Table 4.4).[68,69] It is evident that, even

$$E + I \underset{k_{off}}{\overset{k_{on}}{\rightleftharpoons}} EI \xrightarrow{k_{cat}} E + P$$

Scheme 4.1 Simple mechanism of COMT inhibition by nitrocatecholic compounds. *E, I, EI* and *P* are the enzyme, inhibitor, enzyme–inhibitor complex, and the *O*-methylated product of the inhibitor, respectively.

though these two inhibitors are still substrates of the enzyme, the rates at which they are O-methylated show a remarkable reduction of three and four orders of magnitude, respectively, when compared to that of catechol or pyrogallol ($k_{cat} \sim 1\,s^{-1}$).[33,67] It can be shown that the catechol substituents found in the structures of tolcapone and, in particular, opicapone, may substantially enhance the electronic effects exerted upon the reacting hydroxyl, thereby contributing directly to the potency.

Electrophilic super-delocalizabilities, $S_r(E)$, are dynamic reactivity indices derived from molecular orbital theory, which may be related to the degree of nucleophilicity, or the relative ability of a given atom to participate in a charge-transfer process with a second, electrophilic reagent.[70,71] Therefore, the greater the electrophilic super-delocalizability of the catechol hydroxyl oxygen, the more likely that oxygen is to be methylated by COMT, provided that it is accessible to the electrophilic methyl donor, in an adequate reactive geometry. In contrast, a lower value of $S_r(E)$ indicates a relatively unreactive atom. A practical relative scale of reactivities can be circumscribed from the non-substituted catechol (which is easily O-methylated) and the doubly substituted 3,5-dinitrocatechol. It is illustrated, in Figure 4.10(a), that the nitro group and the dichlorodimethylpyridine-N-oxide moiety, in opicapone, exert a conjugated effect on the reactivity of the catechol, which is greater than that observed with the nitro group alone or with tolcapone, and is only comparable to that of the disubstituted 3,5-dinitrocatechol molecule. This is interpreted as a basis for the markedly low O-methylation rate (k_{cat}) observed with opicapone, as compared to that of catechol or even of potent nitrocatecholic COMT inhibitors, such as tolcapone.[69]

The second term that may affect the observed inhibition constants of a COMT inhibitor is the equilibrium dissociation constant (K_d) of the enzyme–inhibitor complex. However, this quantity cannot be readily determined by wet experiments, as it would require decoupling of the association/dissociation process from that of the catalyzed reaction. Alternatively, the binding affinities between human catechol-O-methyl-transferase and several substrates and inhibitors have been recently estimated by means of computer simulations.[68] In this study, the authors utilized high accuracy statistical mechanics methods to calculate the free energies of binding and the thermodynamic dissociation constants of tolcapone and opicapone with human COMT, in aqueous solvent. The results obtained indicate that opicapone forms an extremely high-affinity complex with human COMT, with a subpicomolar dissociation constant of 0.19 pM, two orders of magnitude lower than that estimated for tolcapone (Table 4.4). It is notable, that the estimated dissociation constants (K_d) are significantly lower, in both cases, than the observed inhibition constants (K_i). Considering the various terms that affect the value of the inhibition constant in eqn (4.1), it becomes obvious that, for such high-affinity inhibitors, K_i is largely dominated by the catalytic term k_{cat}/k_{on} ($K_i \approx k_{cat}/k_{on}$), whose values are significantly larger than those of K_d.

Nitrocatechol inhibitors are typically tight-binding inhibitors. Indeed, it has been shown that opicapone forms an extremely tight complex with human

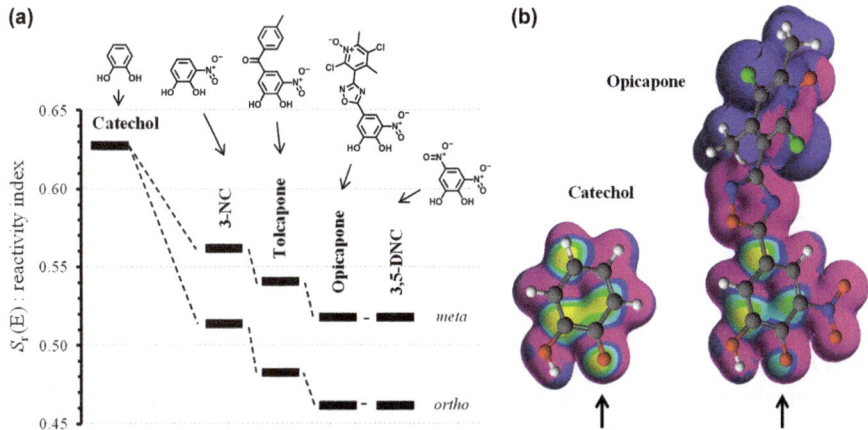

Figure 4.10 (a) Electrophilic superdelocalizability indices, $S_r(E)$, computed for each of the ionized catecholate oxygens (*ortho-* or *meta-* relative to the nitro group), while keeping the other hydroxyl protonated. (b) Catechol and opicapone are shown with their electron density surfaces color-coded according to susceptibility of attack by an electrophile (in decreasing order: purple < magenta < blue < cyan < green < yellow). The reacting hydroxylate anion (arrows) in catechol shows a higher nucleophilic character than the equivalent atom in opicapone. Molecular orbital calculations were computed using the PM5 semi-empirical Hamiltonian with the program Scigress Explorer v.7.7 (Fujitsu Limited, Tokyo, Japan).

Table 4.4 Equilibrium and kinetic constants associated with the mechanism of COMT inhibition by tolcapone and opicapone.

	K_i^a	K_d^b	k_{off}^b	k_{cat}^b	k_{cat}/k_{on}^b
Tolcapone	160 pM	14 pM	1.4×10^{-4} s^{-1}	10^{-3} s^{-1}	114 pM
Opicapone	16 pM	0.19 pM	1.9×10^{-6} s^{-1}	10^{-4} s^{-1}	10 pM

[a] Constants from ref. 49.
[b] Constants from ref. 68, $k_{on} = 10^7$ M^{-1} s^{-1} (see text).

COMT.[68] However, it is concluded that the natures of the high *in vitro* potencies shown by opicapone and tolcapone, are primarily dependent on the rates of formation of their *O*-methylated products, not on the relative binding affinities of their complexes with COMT. The chemical nature of the dichlorodimethylpyridine-*N*-oxide substituent certainly plays a role, not only in modulating the pharmacokinetic properties of the nitrocatechol,[48] but also in enhancing its potency.

4.8.2 *In Vivo* Duration of Action

Although inhibition potencies may be effectively characterized, in the test tube, using thermodynamic equilibrium constants (*e.g.*, K_d or K_i), the clinical efficacies of the drug candidates, on the other hand, are also critically dependent on the timescales of the onset and duration of pharmacological effects, after a drug enters a living organism. In such an open and dynamic system, the concentration of the drug, that is available to modulate the activity of a target receptor, is no longer constant, but varies continuously with time after dosing. For instance, the duration of action is dependent not only on the rate of clearance of the drug from its site of action, but also on the half-life of the drug–receptor complex, itself.[72–74]

Both animal experiments and clinical trials have shown that opicapone is an exceptionally long-acting COMT inhibitor. This molecule shows a low plasma exposure in humans, with a clearance half-life of about 1 h after reaching its maximum concentration at about 2 h post dose (Figure 4.8, Table 4.2). Yet, the levels of erythrocyte COMT inhibition are sustained far beyond the observable point of drug clearance (8–12 h) for all doses tested (Figure 4.9), The timecourse of the enzyme inhibition shows that, after the initial onset of pharmacodynamic effects, the level of COMT inhibition follows a two-stage exponential decay.[75] A faster rate of recovery is observed during the initial 24 h after dose, followed by a much slower, dose-independent recovery phase, with $t_{\frac{1}{2}}$ in excess of 50 h.

We propose that the dose-independent long duration of opicapone-induced COMT inhibition is a consequence of the very long residence time of the COMT–inhibitor complex. In pharmacological terms, the half-life of the enzyme inhibition is given by $t_{\frac{1}{2}} = 0.693/k_{off}$,[72] where k_{off} is the effective rate of the enzyme–inhibitor dissociation process.

As the inhibitor reaches the site of action, it may form a tight complex with COMT (*EI* in Scheme 4.2), with a subpicomolar dissociation constant. As discussed above, the rate-limiting step for the breakdown of this complex is the rate constant of the *O*-methylation of the inhibitor ($k_{cat} = 10^{-4}\,s^{-1}$), since this is significantly faster than the dissociation rate constant of the Michaelis complex ($k_{off} = 10^{-6}\,s^{-1}$). It should be noted, however, that the form of the enzyme thus released (*E**) is not a functional one. Upon methyl transfer to the catecholic inhibitor, the cosubstrate becomes demethylated, S-adenosylhomocysteine

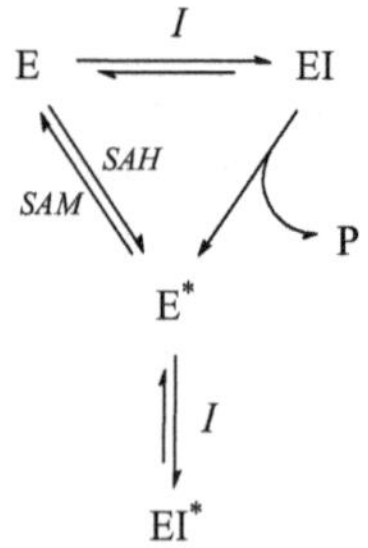

Scheme 4.2 Mechanism of COMT inhibition by nitrocatecholic compounds. *E*, *I*, *EI* and *P* are COMT (containing AdoMet), the inhibitor, enzyme–inhibitor complex and *O*-methylated product, respectively. *E** is the inactive binary complex between COMT and AdoHCy and *EI** is a nonproductive ternary complex between COMT, AdoHCy and the inhibitor.

(AdoHCy), and must be exchanged for a new molecule of S-adenosyl-L-methionine (AdoMet), before the full catalytic cycle of the enzyme is complete. This process may require extensive conformational re-arrangement of the enzyme's three-dimensional structure, as the cosubstrate is deeply buried inside the protein[60] and is, therefore, likely to be relatively slow. We hypothesize that, while the concentration of free opicapone in circulation is still significant, a second molecule of the inhibitor may bind to the transient complex COMT–AdoHCy (E^*), effectively competing with the slower AdoHcy/AdoMet exchange process. This would lead to a non-productive complex COMT–AdoHcy–OPC (EI^* in Scheme 4.2), lacking the ability to catalyze the O-methylation of the bound inhibitor, but possessing a long residence time.

As a consequence, the kinetics of the recovery of baseline COMT activity, from this point onwards, would no longer depend on the O-methylation rate of the inhibitor, or the exposure to free drug in circulation. Instead, it would become limited by the much slower dissociation rate of the complex ($k_{off} \ll k_{cat}$). Indeed, the observed half-life of opicapone-induced COMT inhibition, in human erythrocytes (>50 h), translates into an underlying kinetic process with a rate constant of approximately 3.6×10^{-6} s^{-1}. This value is consistent with the estimated dissociation rate constant of the COMT–opicapone complex ($k_{off} = 1.9 \times 10^{-6}$ s^{-1}).

Based on the above observations, we propose that the sustained enzyme inhibition, far beyond the observable point of clearance of circulating drug, is a direct consequence of the long residence time of the reversible COMT–opicapone complex. While the inhibition constant (K_i) may well reflect the differences in activity of different inhibitors, *in vitro*, the enzyme–inhibitor dissociation rate constant (k_{off}) is a key factor in distinguishing the inhibitor's pharmacological profile *in vivo*.

4.9 Final Remarks

Close partnering and interaction between different disciplines (chemistry, pharmacology, bioinformatics, biochemistry and clinical research) is making the development of new COMT inhibitors, with improved pharmacodynamics, pharmacokinetics, safety and tolerability profiles possible. One such example is opicapone that made its way through a unique path of drug development in which molecular modeling provided the design of high potency, low toxicity, and long-acting COMT inhibition. These characteristics are expected to change the way COMT inhibitors are perceived for their clinical use, namely on efficacy, safety and posology. Opicapone's drug development is, at present, in progress through Phase III clinical trials as once-daily dosing as an adjunct to levodopa therapy in PD patients with motor fluctuations.[76,77] In addition, as a pharmacological tool, the availability of opicapone is of particular importance, since it will allow for a better understanding of the inhibitor–enzyme complex, namely with regard to the functional relevance of the low dissociation rate constant (k_{off}) of the inhibitor from the enzyme–inhibitor complex, as a key determinant in defining opicapone's *in vivo* pharmacological profile and clinical use in PD.

Acknowledgements

The authors are particularly thankful to Drs Maria João Bonifácio, Ana I. Loureiro, José Francisco Rocha, Teresa Nunes (Department of Research and Development, BIAL – Portela & Cª, S.A.) and Professors Amílcar Falcão (Faculty of Pharmacy, University Coimbra, Portugal) and Luís Almeida (Health Science Section, University Aveiro, Portugal) for their contribution to the opicapone development program.

References

1. J. Axelrod and R. Tomchick, *J. Biol. Chem.*, 1958, **233**, 702.
2. H. C. Guldberg and C. A. Marsden, *Pharmacol. Rev.*, 1975, **27**, 135.
3. B. Zhu, E. Ezell and J. Liehr, *J. Biol. Chem.*, 1994, **269**, 292.
4. J. V. Burba and G. C. Becking, *Arch. Int. Pharmacodyn. Ther.*, 1969, **180**, 323.
5. P. T. Männistö and S. Kaakkola, *Trends Pharmacol. Sci.*, 1989, **10**, 54.
6. D. B. Calne, *Arch. Int. Pharmacodyn. Ther.*, 1993, **329**, 1021.
7. F. S. Messiha, T. H. Hsu and J. R. Bianchine, *J. Clin. Invest.*, 1972, **51**, 452.
8. M. Da Prada, R. Kettler, G. Zurcher, R. Schaffner and W. Haefely, *Eur. Neurol. Suppl.*, 1987, **27**, 9.
9. A. D. Ericsson, *J. Neurol. Sci.*, 1971, **14**, 193.
10. Z. Bacq, L. Gosselin, A. Dresse and J. Renson, *Science*, 1959, **130**, 453.
11. J. R. Crout, *Biochem. Pharmacol.*, 1961, **6**, 47.
12. A. N. Booth, M. Masri, D. J. Robbins, O. H. Emerson, F. T. Jones and F. Deeds, *J. Biol. Chem.*, 1959, **234**, 3014.
13. R. E. Giles and J. W. Miller, *J. Pharmacol. Exp. Ther.*, 1967, **158**, 189.
14. A. Reches and S. Fahn, *Adv. Neurol.*, 1984, **40**, 171.
15. R. Gugler and H. J. Dengler, *Naunyn-Schmiedeberg's Arch. Exp. Pathol. Pharmakol.*, 1973, **201**, 353.
16. B. Belleau and J. Burba, *Biochim. Biophys. Acta*, 1961, **54**, 195.
17. R. T. Borchardt, *J. Med. Chem.*, 1973, **16**, 581.
18. R. T. Borchardt, *J. Med. Chem.*, 1973, **16**, 382.
19. O. J. Broch Jr., *Acta Pharmacol. Toxicol.*, 1973, **33**, 417.
20. J. V. Burba and M. F. Murnaghan, *Biochem. Pharmacol.*, 1965, **14**, 823.
21. S. E. Brevitt and E. W. Tan, *J. Med. Chem.*, 1997, **40**, 2035.
22. B. Masjost, P. Ballmer, E. Borroni, G. Zurcher, F. K. Winkler, R. Jakob-Roetne and F. Diederich, *Chem.–Eur. J.*, 2000, **6**, 971.
23. C. Lerner, A. Ruf, V. Gramlich, B. Masjost, G. Zurcher, R. Jakob-Roetne, E. Borroni and F. Diederich, *Angew. Chem., Int. Ed.*, 2001, **40**, 4040.
24. C. Lerner, B. Masjost, A. Ruf, V. Gramlich, R. Jakob-Roetne, G. Zurcher, E. Borroni and F. Diederich, *Org. Biomol. Chem.*, 2003, **1**, 42.

25. S. Wolkenberg, J. C. Barrow, S. T. Poslusney, S. T. Harrison, B. W. Trotter, J. Mulhearn, K. K. Nanda, P. J. Manley, Z. Zhao, J. W. Scubert, N. Kett and A. Zartman, *Pat.*, WO20111/09254 A1, 2011.
26. S. Wolkenberg, J. C. Barrow, S. T. Harrison, B. W. Trotter, K. K. Nanda, P. J. Manley and Z. Zhao, *Pat.*, WO2011/109261 A1, 2011.
27. S. Wolkenberg, S. T. Harrison, J. C. Barrow, Z. Zhao, N. Kett and A. Zartman, *Pat.*, WO2011/109267 A1, 2011.
28. R. Bäckström, E. Honkanen, A. Pippuri, P. Kairisalo, J. Pystynen, K. Heinola, E. Nissinen, I.-B. Lindén, P. T. Männistö, S. Kaakkola and P. Pohto, *J. Med. Chem.*, 1989, **32**, 841.
29. J. Borgulya, H. Bruderer, K. Bernauer, G. Zürcher and M. Prada, *Helv. Chim. Acta*, 1989, **72**, 952.
30. I. Linden, E. Nissinen, E. Etemadzadeh, S. Kaakkola, P. Mannistö and P. Pohto, *J. Pharmacol. Exp. Ther.*, 1988, **247**, 289.
31. J. Prous, X. Rabasseda and J. Castañer, *Drugs Future*, 1994, **19**, 641.
32. J. Borgulya, M. Da Prada, J. Dingemanse, R. Scherschlicht, B. Schlappi and G. Zürcher, *Drugs Future*, 1991, **16**, 719.
33. T. Lotta, J. Vidgren, C. Tilgmann, I. Ulmanen, K. Melén, I. Julkunen and J. Taskinen, *Biochemistry*, 1995, **34**, 4202.
34. J. Cedarbaum, G. Leger, A. Reches and M. Guttman, *Clin. Neuropharmacol.*, 1990, **13**, 544–552.
35. M. Tornwall and P. T. Mannistö, *Pharmacol. Toxicol.*, 1991, **69**, 64.
36. T. Keranen, A. Gordin, M. Karlsson, K. Korpela, P. J. Pentikainen, H. Rita, E. Schultz, L. Seppala and T. Wikberg, *Eur. J. Clin. Pharmacol.*, 1994, **46**, 151–157.
37. S. A. Parashos, C. L. Wielinski and J. A. Kern, *Clin. Neuropharmacol.*, 2004, **27**, 119.
38. E. Nissinen, I.-B. Lindén, E. Schultz and P. Pohto, *Naunyn-Schmiedeberg's Arch. Pharmacol.*, 1992, **346**, 262.
39. F. Assal, L. Spahr, A. Hadengue, L. Rubbici-Brandt and P. R. Burkhard, *Lancet*, 1998, **352**, 958.
40. D. A. Learmonth, M. A. Vieira-Coelho, J. Benes, P. C. Alves, N. Borges, A. P. Freitas and P. Soares-da-Silva, *J. Med. Chem.*, 2002, **45**, 685.
41. D. A. Learmonth, P. N. Palma, M. A. Vieira-Coelho and P. Soares-da-Silva, *J. Med. Chem.*, 2004, **47**, 6207.
42. D. A. Learmonth, M. J. Bonifácio and P. Soares-da-Silva, *J. Med. Chem.*, 2005, **48**, 8070.
43. R. Bäckström, J. Pystynen, T. Lotta, M. Ovaska and J. Taskinen, *Pat.*, WO2002/22551 A1, 2002.
44. J. Pystynen, M. Ovaska, J. Vidgren, T. Lotta and M. Yliperttula-ikonen, *Pat.*, WO2002/02548 A1, 2002.
45. M. Ahlmark, R. Bäckström, A. Luiro, J. Pystynen and E. Tiainen, *Pat.*, WO2007/010085 A2, 2007.
46. J. Rautio, J. Leppänen, M. Lehtonen, K. Laine, M. Koskinen, J. Pystynen, J. Savolainen and M. Sairanen, *Bioorg. Med. Chem. Lett.*, 2010, **20**, 2614.

47. D. A. Learmonth, L. E. Kiss, P. N. Palma, H. S. Ferreira and P. Soares-da-Silva, *Pat.*, WO2007/013830 A1, 2007.
48. L. E. Kiss, H. S. Ferreira, L. Torrão, M. J. Bonifácio, P. N. Palma, P. Soares-da-Silva and D. A. Learmonth, *J. Med. Chem.*, 2010, **53**, 3396.
49. M. J. Bonifácio, L. Torrão, A. I. Loureiro, L. C. Wright and P. Soares-da-silva, *Parkinsonism Relat. Disord.*, 2012, **18**, S125.
50. M. J. Bonifácio, J. S. Sutcliffe, L. Torrão, L. C. Wright and P. Soares-da-Silva, *Parkinsonism Relat. Disord.*, 2012, **18**, S125.
51. T. Wikberg, A. Vuorela, P. Ottoila and J. Taskinen, *Drug Metab. Dispos.*, 1993, **21**, 81.
52. K. M. Jorga, B. Fotteler, P. Heizmann and R. Gasser, *Br. J. Clin. Pharmacol.*, 1999, **48**, 513.
53. L. C. Wright, J. Maia, A. I. Loureiro, L. Almeida and P. Soares-da-Silva, *Drug Metab. Lett.*, 2010, **4**, 149.
54. A. I. Loureiro, M. J. Bonifacio, C. Fernandes-Lopes, L. Almeida, L. C. Wright and P. Soares-Da-Silva, *Drug Metab. Dispos.*, 2006, **34**, 1856.
55. T. Nunes, J. F. Rocha, R. Pinto, R. Machado, L. C. Wright, A. Falcão, L. Almeida and P. Soares-da-silva, *Parkinsonism Relat. Disord.*, 2012, **18**, S126.
56. J. F. Rocha, T. Nunes, M. Vaz-da-silva, R. Machado, L. C. Wright, A. Falcão, L. Almeida and P. Soares-da-silva, *Parkinsonism Relat. Disord.*, 2012, **18**, S126.
57. J. J. Ferreira, L. Almeida, L. Cunha, M. Ticmeanu, M. M. Rosa, C. Januário, C.-E. Mitu, M. Coelho, L. Correia-Guedes, A. Morgadinho, T. Nunes, L. C. Wright, A. Falcão, C. Sampaio and P. Soares-da-silva, *Clin. Neuropharmacol.*, 2008, **31**, 2.
58. J. Dingemanse, K. M. Jorga, M. Schmitt, R. Gieschke, B. Fotteler, G. Zürcher, M. Da Prada and P. van Brummelen, *Clin. Pharmacol. Ther.*, 1995, **57**, 508.
59. P. Silveira, M. Vaz-da-Silva, L. Almeida, J. Maia, A. Falcão, A. Loureiro, L. Torrão, R. Machado, L. Wright and P. Soares-da-Silva, *Eur. J. Clin. Pharmacol.*, 2003, **59**, 603.
60. J. Vidgren, L. A. Svensson and A. Liljas, *Nature*, 1994, **368**, 354.
61. J. Vidgren, C. Tilgmann, K. Lundström and A. Liljas, *Proteins*, 1991, **11**, 233.
62. M. J. Bonifácio, P. N. Palma, L. Almeida and P. Soares-da-Silva, *CNS Drug Rev.*, 2007, **13**, 352.
63. P. N. Palma, M. J. Bonifácio, L. Almeida and P. Soares-da-Silva, in *Restoring Dopamine Levels*, eds. H. John Smith, Claire Simons and Robert D. E. Sewell, CRC Press: Boca Raton, 2007, 415–445.
64. E. Nissinen, in *Catechol-O-Methyltransferase Inhibition - An Innovative Approach To Enhance L-Dopa Therapy In Parkinson's Disease With Dual Enzyme Inhibition*, eds. Ronald J. Bradley, R. Adron Harris and Peter Jenner, Academic Press: Amsterdam, 2010, **95**, 1–5.
65. E. Schultz and E. Nissinen, *Biochem. Pharmacol.*, 1989, **38**, 3953.

66. N. Borges, M. A. Vieira-Coelho, A. Parada and P. Soares-da-silva, *J. Pharmacol. Exp. Ther.*, 1997, **282**, 812.
67. P. Lautala, I. Ulmanen and J. Taskinen, *Mol. Pharmacol.*, 2001, **59**, 393.
68. P. N. Palma, M. J. Bonifácio, A. I. Loureiro and P. Soares-da-Silva, *J. Comput. Chem.*, 2012, **33**, 970.
69. P. N. Palma, M. J. Bonifácio, A. I. Loureiro and P. Soares-da-silva, *Parkinsonism Relat. Disord.*, 2012, **18**, S125.
70. M. Karelson, V. S. Lobanov and A. R. Katritzky, *Chem. Rev.*, 1996, **96**, 1027.
71. K. Fukui, T. Yonezawa and C. Nagata, *Bull. Chem. Soc.*, 1954, **27**, 423.
72. R. A. Copeland and D. Pompliano, *Nat. Rev. Drug Discovery*, 2006, **1**.
73. P. J. Tummino and R. A. Copeland, *Biochemistry*, 2008, **47**, 5481.
74. G. Vauquelin and I. Van Liefde, *Trends Pharmacol. Sci.*, 2006, **27**, 356.
75. L. Almeida, J. F. Rocha, A. Falcão, P. N. Palma, A. I. Loureiro, R. Pinto, M. J. Bonifácio, L. C. Wright and P. Soares-da-Silva, *Clin. Pharmacokinet.*, 2013, **52**, 139–151.
76. A. Lees, R. Costa, C. Oliveira, N. Lopes, T. Nunes and P. Soares-da-Silva, *Movement Disord.*, 2012, **27**, S127.
77. J. J. Ferreira, J. F. Rocha, A. Santos, T. Nunes and P. Soares-da-Silva, *Movement Disord.*, 2012, **27**, S118.

CHAPTER 5

Pharmacologic Management of Dopaminergic-Induced Dyskinesias in Parkinson's Disease

MILDRED D. GOTTWALD*[a,b] AND
MICHAEL J. AMINOFF[c]

[a] Department of Clinical Pharmacy, University of California, San Francisco, USA; [b] Gilead Sciences Inc., Foster City, USA; [c] Department of Neurology, University of California, San Francisco, USA
*Email: mgottwald@gilead.com

5.1 Introduction

Levodopa is the most efficacious symptomatic treatment for Parkinson's disease (PD). In combination with carbidopa or benserazide, it is generally well-tolerated and improves motor function and quality of life. However, the response time to each dose eventually shortens, and levodopa-induced dyskinesias (LIDs) and motor fluctuations develop.[1–3] In the present era, patients treated with levodopa for 4–6 years have a risk of developing dyskinesias that is just under 40%, with the number increasing progressively thereafter.[1] The severity of LIDs correlates with levodopa dose, duration and stage of disease, and younger age at disease onset.[4–9] The occurrence of dyskinesias impacts the treatment regimen, affects functional disability, and impacts quality of life.[1,5,8] Dyskinesias are associated with response fluctuations to dopaminergic medication, and are often disabling and may necessitate surgical intervention.[8] Disease-modifying medical treatments that prevent the onset of dyskinesias do not currently exist; most strategies only reduce the severity of dyskinesias. In discussing the pharmacologic management of LIDs in this chapter, we have built on and extended our earlier report published elsewhere.[10]

5.2 Why Do Dyskinesias Occur?

The development of LIDs may relate to an abnormal distribution of striatal dopamine because of the disease or its treatment.[11,12] Presynaptic dysfunction of dopaminergic neurons may result in loss of the buffering capacity of surviving neurons or altered dopamine release. Suggested postsynaptic mechanisms for the induction of dyskinesias involve dopaminergic denervation coupled with chronic pulsatile stimulation of dopamine receptors with levodopa.[13,14] A lower incidence of dyskinesias occurs with longer acting dopamine agonists, continuously administered levodopa, and drug-delivery systems that enable a more continuous delivery of dopaminergic medication.[15,16] In rodents, degeneration of the nigrostriatal dopaminergic pathways leads to a switch in the regulation of ERK1/2/MAP signaling kinases in direct-pathway neurons and thus to increased sensitivity of dopaminergic D_1 receptor agonists.[17,18] A variety of other changes occur with pulsatile dopaminergic stimulation of the denervated striatum and involve glutamate and N-desmethyl-D-aspartate (NMDA) receptors, calcium channel activity, synaptic plasticity, loss of depotentiation, induction of immediate early genes, altered protein–DNA binding, and abnormal levels of phosphorylated DARPP-32, a signaling protein.[11,17–20]

5.3 Scales Used for Assessing Dyskinesias

Due to the lack of an accepted gold standard, clinical trials of anti-dyskinetic agents have used different scales, making it difficult to compare drugs or the results of different studies. The Movement Disorders Society (MDS) has examined this issue.[21] Of eight rating scales that were evaluated, only two met the criteria for recommended use in a PD population. These were the Abnormal Involuntary Movement Scale (AIMS) and the Rush dyskinesia rating scale. The AIMS does not evaluate type, duration, or pattern of dyskinesias. The Rush scale assesses functional disability, but only at specific timepoints rather than over the course of a waking day. An additional two scales required testing by investigators other than the original authors before they could be regarded as acceptable.

The most widely used scales are the AIMS (originally developed for assessment of tardive dyskinesias; less specific for a specific type of dyskinesia) and the Unified Parkinson's Disease Rating Scale, Part IV (UPDRS-IV), which assesses duration and functional impact of dyskinesias, but is part of a scale that assesses motor disability and has had minimal independent clinimetric testing.

5.4 Medical Therapies

Medical management strategies involve delaying the introduction of levodopa therapy, treating with an anti-dyskinetic agent, providing continuous dopaminergic stimulation, or using novel agents that target receptors implicated in the mechanisms underlying LIDs (Figure 5.1). Treatment with dopamine agonists (*e.g.*, pramipexole or ropinirole) allows introduction of levodopa to be delayed—once levodopa is added, however, the usual course of onset of dyskinesias is unchanged.[22,23]

Amantadine, an NMDA antagonist, is the only approved compound that provides a sustained anti-dyskinetic benefit without unacceptable side-effects, supporting a role for glutamate overactivity in LID development.[24–28] Levitiracetam, an anti-seizure agent, may also have a mild anti-dyskinetic effect.[29–32] Therapies providing relatively stable plasma levodopa concentrations, such as by intraintestinal or subcutaneous delivery, are promising but are invasive and associated with injection site reactions.[33,34] Genetic studies and elucidation of the role of neurotransmitters in the pathophysiology of LIDs have identified new compounds that modulate the direct and indirect striatal output pathways; some are in the early stages of development or undergoing proof-of-concept evaluation as anti-dyskinetic agents (Table 5.1).

5.4.1 Delaying the Onset of Dyskinesias with Dopamine Agonists

Although dopamine agonists do not prevent dyskinesias, they usually delay the need for up-titration of levodopa doses and reduce the risk for early onset of troublesome dyskinesias.[22,23,35–39] The decreased incidence of LIDs may be due in part also to the selectivity of certain agonists (*e.g.*, pramipexole and

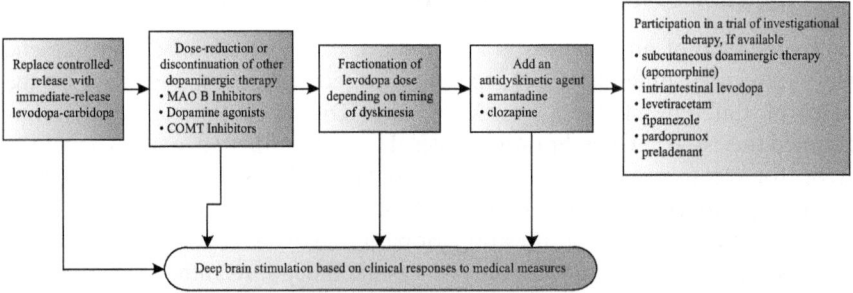

Stepwise treatment algorithm for managing levodopa-induced dyskinesias. Replacement of controlled-release with immediate levodopa-carbidopa, followed by possible dose-reduction of other dopaminergic therapies, and fractionation of levodopa doses are recommended medical intervention strategies. Addition of an antidyskinetic therapy for established dyskinesias may attenuate LIDs. Participation in a trial of an investigational agent may also be considered.

COMT = Catechol-*O*-methyltransferase
MAO = Monoamine Oxidase

Figure 5.1 Treatment algorithm for managing levodopa-induced dyskinesias.

Table 5.1 Medications for managing levodopa-induced dyskinesias. ER = extended release, PR = prolonged release, SV2A = synaptic vesicle protein 2A.

Agents that delay the onset of LIDs	
Pramipexole, pramipexole ER	D_2 selective dopamine agonist
Ropinirole, ropinirole PR	D_2 selective dopamine agonist
Anti-dyskinetic agents	
Amantadine	NMDA-receptor antagonist
Clozapine	$5HT_{2A}$ serotonin, D_4 dopamine receptor antagonist
Levetiracetam (possibly)	Binds to SV2A (anti-seizure agent)
Investigational parenteral therapies	
Subcutaneous apomorphine	$D_2/D_3/D_4$ dopamine; serotonin receptors
Intraintestinal levodopa	D_1/D_2 receptor agonist
Other investigational oral therapies	
Fipamezole	$\alpha 2$ adrenergic receptor antagonist
Pardoprunox	Partial D_2/D_3 agonist; full $5HT_1$ serotonin agonist
Preladenant	Adenosine A_{2A} receptor antagonist
IPX066 (Rytary™)	D_1/D_2 receptor agonist (carbidopa–levodopa)

ropinirole) for D_2 receptors and the relatively longer half-lives of these agonists compared to pulsatile levodopa delivery.

5.4.1.1 Pramipexole

In the randomized, controlled CALM-PD trial (Comparison of the Agonist Pramipexole *versus* Levodopa on Motor Complications of Parkinson's Disease), the risk of developing dyskinesias was evaluated in 301 patients with early PD initially started on either pramipexole or levodopa–carbidopa, followed by a maintenance phase during which open-label levodopa–carbidopa was permitted if required.[35–37] After four years, pramipexole-treated patients were receiving a mean daily dose of 2.78 mg pramipexole plus a mean daily dose of 434 mg levodopa, compared with 702 mg levodopa for those receiving only levodopa–carbidopa.[35] Patients in the pramipexole group also had a significantly lower risk for developing dyskinesias (24.5 *vs.* 54%, hazard ratio (HR) = 0.37; 95% confidence interval (CI), 0.25 to 0.56, $p<0.001$). In an open-label six-year follow-up study, most patients eventually received levodopa therapy regardless of initial treatment assignment, and additional levodopa use was needed in 72% of the patients originally started on pramipexole compared with 59% of those in the levodopa monotherapy arm ($p=0.001$).[35] Compared to those taking pramipexole, patients initially treated with levodopa–carbidopa had significantly more dyskinesias (20.4 *vs.* 36.8%, $p=0.004$), but no difference existed in the incidence of disabling or painful dyskinesias.[36] There was no significant difference in the Lang–Fahn ADL dyskinesia score.

Extended-release (ER) pramipexole is now available, and it also delays the need for initiating levodopa therapy.[38] However, somnolence, nausea, constipation and fatigue were more common adverse effects with both pramipexole formulations than with placebo.

These studies do not support a treatment effect of pramipexole on dyskinesias, but suggest it may delay their onset by limiting the need for initiation of levodopa.

5.4.1.2 Ropinirole

In patients receiving ropinirole monotherapy, the risk of developing dyskinesias was less than with levodopa–benserazide (HR = 2.82; [1.78, 4.44]; $p<0.001$) in a randomized, double-blind five-year study.[39] With the eventual addition of levodopa, the risk for developing dyskinesias increased and then was the same as that associated with levodopa alone (HR = 0.80; [0.48, 1.33]; $p=0.39$).[20] The use of ropinirole as monotherapy, with the addition of levodopa only later, delayed the onset of dyskinesias by up to three years.[23,39]

The risk for developing dyskinesias in the patients originally treated with ropinirole remained lower than for those initiated on therapy with levodopa in a 10-year open-label extension period (52.4%, 22 patients ropinirole vs. 77.8%, 21 patients levodopa—adjusted odds ratio (OR) = 0.3; 95% CI 0.1, 1.0; $p=0.046$).[40] Median time to development of dyskinesias was 1.6 years longer in the ropinirole group (8.6 vs. 7; adjusted HR = 0.4; 95% CI 0.2, 0.8; $p=0.007$).

Prolonged-release (PR) ropinirole was studied in a 107-week multi-center, randomized double-dummy, flexible-dose study of 104 patients with PD not optimally controlled with levodopa (<600 mg daily) compared to those receiving flexible doses of levodopa up to three times daily.[41] The primary endpoint was onset of dyskinesia as assessed by patient history. Patients in the ropinirole PR group (mean daily dose 10 mg) experienced dyskinesias less frequently than those in the levodopa group who required an additional 284 mg per day of levodopa (3% vs. 17%) and also had a delayed onset of dyskinesias without a significant change in UPDRS III motor scores. In another study in patients with motor fluctuations exceeding three hours daily, patients randomized to ropinirole PR had a reduced "off" time accompanied by a reduced daily need for levodopa (−278 mg with ropinirole vs. −164 mg with placebo) and an increase over placebo in the mean "on" time without troublesome dyskinesia (53.2 to 66% vs. 52.7 to 56% for placebo).[42]

Such results support the belief that D_2-selective dopamine agonists delay the onset of dyskinesias by delaying the introduction of levodopa; however, the addition of levodopa is inevitable, and consideration of the known neuropsychiatric side-effects of the agonists must be factored into the choice between levodopa and a dopamine agonist.

5.4.2 Therapeutic Strategies for Existing Dyskinesias

"Peak-dose" dyskinesias are the most common type of LIDs, occur shortly after the peak pharmacodynamic effect of levodopa is reached, and decline as the end of the dosing interval is reached.[5]

Strategies for the prevention or management of peak-dose dyskinesias (Figure 5.1) involve the following: (1) substitution of immediate- for controlled-release levodopa. The former is easier to adjust as its effects occur sooner and are shorter in duration. Both preparations produce pulsatile dopaminergic stimulation. Agents that prolong the half-life of levodopa, such as entacapone, also should be discontinued; (2) discontinuation of other dopaminergic therapy that may exacerbate dyskinesias (*e.g.*, selegiline, rasagiline, or, in some instances, dopamine agonists); (3) administration of levodopa more frequently but in lower doses; and (4) addition of an anti-dyskinetic agent such as amantadine.

Agents or delivery systems that provide more continuous (rather than pulsatile) dopaminergic stimulation may come to have a role in minimizing dyskinesias. However, initiating levodopa with entacapone failed to delay the onset or reduce the frequency of LIDs in one recent study, perhaps because of an inadequate dosing schedule so that stimulation was not continuous or because the treatment group had more severe disease and thus received more dopaminergic medication.[43,44]

Diphasic dyskinesias, which manifest at the beginning and end of a dosing cycle, may also occur with disease progression. They are managed in general with more frequent doses of levodopa, but therapy is individualized.[5]

5.4.2.1 Amantadine

In 1969, the anti-parkinsonian effects of amantadine were discovered. Amantadine's effects on the motor symptoms (bradykinesia, akinesia) are relatively weak.[24] A meta-analysis found inadequate evidence to support its continued use for managing the motor symptoms of PD. However, there are data supporting its utility in the treatment of LIDs.[45] Amantadine is a relative weak and non-selective antagonist of the NMDA-type glutamate receptor that also increases dopamine release and blocks dopamine re-uptake.[46] The finding that sensitization of NMDA glutamatergic receptors may increase cortical excitatory input suggested that an NMDA antagonist such as amantadine has an anti-dyskinetic effect. Amantadine was therefore studied as an adjuvant treatment in patients experiencing levodopa-induced motor complications, including dyskinesias, with the aim of reducing these effects without worsening parkinsonian symptoms.[24–28]

Three short-term, randomized placebo-controlled cross-over clinical studies in a small group ($n = 53$) of patients with PD assessed the severity of LIDs after acute levodopa administration.[24,25,27,28] Amantadine reduced the severity of LIDs in all three studies (up to 60% reduction), without affecting the beneficial effects of levodopa on motor function. However, these studies did not involve a wash-out and failed to evaluate or consistently demonstrate long-term anti-dyskinetic benefits. The long-term effect of amantadine on LIDs has been studied in a randomized, double-blind, placebo-controlled, parallel group study of 32 PD patients (14 on amantadine, mean daily dose 298 mg; 18 on placebo).[26] Eligible patients had received stable doses of amantadine for one year and stable anti-parkinsonian therapy for one month. They were then randomized to receive amantadine at their prestudy dose or matched placebo. After the three-week treatment period, patients were assessed for dyskinesia duration and level of disability using the total score from the UPDRS-IV (items 32 and 33). In patients switched to placebo from amantadine, there was an increase in dyskinesias compared to those maintained on amantadine (0.4; 95% CI −0.6, 1.3) or in the placebo group (1.2; 95% CI +0.2, 2.2). After three weeks, there was no difference between baseline and follow-up in patients continuing on amantadine. Additional time in the "on" stage with dyskinesias was reported more frequently in the placebo group. These results supported the long-term anti-dyskinetic effect of amantadine.

A meta-analysis pooled 11 randomized, placebo-controlled clinical studies involving 253 PD patients with LIDs mainly receiving amantadine or other NMDA receptor antagonists (dextromethorphan, memantine, or remacemide).[45] The analysis of the six studies (159 patients) that utilized the UPDRS-IV and treatment with amantadine showed a significant reduction in the standard mean difference for the UPDRS-IV (SMD −1.10; 95% CI −1.92, −0.28) and UPDRS-III (7 studies, standard mean difference (SMD) −0.35; 95% CI −0.60, −0.09), but no significant treatment effect for either measure with the other NMDA receptor antagonists studied. Treatment with amantadine reduced the short-term severity of LIDs, without adversely affecting motor function.

5.4.2.2 Clozapine

Clozapine binds with high affinity to D_4 receptors and serotonergic $5HT_{2A}$ and with low affinity to D_1, D_2 and D_3 receptors.[46,47] The effect of clozapine on LIDs was demonstrated in an early randomized double-blind, placebo-controlled study of 50 patients with severe PD and disabling dyskinesias who were receiving optimized and stable dopaminergic therapy (dopamine agonists, levodopa, or both) and clozapine at a mean dose of 39.4 ($\pm$ 4.5) mg daily ($n = 26$) or placebo ($n = 24$).[48] The primary end-point was change in "on" time with and without dyskinesia and "off" time between baseline and the end of the study. Secondary outcomes were response to an acute levodopa challenge before and after 70 days, with *ad hoc* scoring using a dyskinesia rating scale. After 10 weeks, the clozapine-treated patients had a reduced duration of "on" periods with LIDs compared to patients receiving placebo (−1.7 hours *vs.* −0.74 hours for placebo, $p = 0.003$), with a two-hour decrease in duration of LIDs without an increase in the duration of "off" periods. For the 38 patients who completed the acute levodopa challenge with *post hoc* video scoring, there was reduced severity of dyskinesias at rest, but without significant change during activity. These beneficial effects may be due to the serotonergic and striatal effects of clozapine on D_2 receptors, but clozapine does not benefit motor function. The side-effects of clozapine include agranulocytosis (0.7%), requiring mandatory hematology assessments and limiting its utility as a long-term anti-dyskinetic therapy.

5.4.3 Parenteral Therapies

The efficacy of continuous intravenous infusions of levodopa for the management of severe motor fluctuations, including troublesome dyskinesias, supports the notion that continuous dopaminergic stimulation may enable stable plasma concentrations of levodopa, thereby minimizing end-of-dose and peak plasma concentrations that may result in motor fluctuations and dyskinesias. Administration by this route is impractical, however, as an outpatient procedure.

Subcutaneous administration with a pump for delivery of apomorphine or the direct intraintestinal delivery of levodopa methods are more practical methods to achieve sustained delivery of dopaminergic therapy. Although primarily available in Europe and Canada, they are also being studied in the USA. Subcutaneous lisuride has also been evaluated, but is no longer being studied due to an unfavorable benefit-to-risk profile and is therefore not discussed here.

5.4.3.1 Subcutaneous Apomorphine

Apomorphine, a non-ergoline dopaminergic agent, interacts with D_2, D_3 and D_4 receptors, adrenergic $\alpha 1D$, $\alpha 2B$, and $\alpha 2C$ receptors, and to a lesser degree with D_1 receptors or serotonin $5HT_{1A}$, $5HT_{2A}$, $5HT_{2B}$ and $5HT_{2B}$ receptors.[49] Its anti-parkinsonian action relates to its interaction with postsynaptic D_2 receptors in the caudate–putamen. In addition to its use for managing "on–off" motor fluctuations, apomorphine has been studied as an anti-dyskinetic add-on or single replacement therapy.[50–53]

The results of 11 long-term open-label uncontrolled studies evaluating the efficacy of subcutaneous apomorphine as monotherapy or as an add-on to levodopa therapy have been published. These studies supported the effect of apomorphine in aborting "off" periods, reducing dyskinesias, improving motor scores, and reducing the need for levodopa.[33] In these studies (233 patients), continuous apomorphine at an average daily dose of 88 mg provided an approximately 50–80% reduction in time spent "off" after an average follow-up time of 32 months, and provided a 49% reduction in levodopa requirement. Six studies demonstrated reduced duration of biphasic and peak-dose dyskinesias (−50%) and severity (−45%) and these improvements in dyskinesias were two-fold greater when apomorphine was given as monotherapy rather than add-on therapy. The mean time to achieve maximal dyskinesia improvement was 12.1 months. One prospective study compared single-dose levodopa and apomorphine challenges before and after continuous daytime apomorphine infusion given over a six-month period in 12 patients on ongoing dopaminergic therapy.[54] After six months, apomorphine reduced dyskinesias by at least 36% (by the AIM and Goetz scales, both $p < 0.01$); and after single-dose levodopa challenge by at least 40% compared to baseline values. Additionally, patients were able to decrease their daily levodopa intake by 55% ($p < 0.01$), and four of the 12 patients (33%) were reportedly able to completely discontinue their other oral dopaminergic medications including short-term intermittent treatment with levodopa. After six months, patients who were able to discontinue other oral therapies and remain on apomorphine as monotherapy had significantly lower dyskinesia severity and duration as assessed by visual analogue scales ($p < 0.05$). The most commonly reported adverse events included panniculitis; skin inflammation prevented by rotation of injection site; and mental or behavioral disturbances. Therefore, this alternative may best be reserved for patients with severe motor complications that are unresponsive to oral medication, especially those with an adequate support system at home.

5.4.3.2 Intraintestinal Levodopa

Levodopa can be administered by direct infusion into the duodenum or upper jejunum. Intraduodenal infusion of levodopa–carbidopa using a suspension of micronized levodopa–carbidopa ($20:5\,\text{mg}\,\text{mL}^{-1}$) in a methylcellulose gel (Duodopa®) using a portable infusion pump (CADD-Legacy DUODOPA pump) is an approved therapy in Europe and Canada for the treatment of advanced levodopa-responsive PD when standard treatments have failed.[55] A "Fast Track" status has been granted by the US FDA, but it is not yet available in the USA as of writing, and long-term safety and efficacy studies are ongoing. A permanent access tube is inserted in the abdominal wall by percutaneous endoscopic gastrostomy (PEG) in patients with a positive clinical response to levodopa–carbidopa gel administered through a nasoduodenal tube. Dosing is usually accomplished by a morning bolus dose (100–300 mg levodopa), a continuous maintenance dose (40–120 mg levodopa per hour) and supplemental bolus doses as needed for hypokinesis.

In a double-blind, placebo-controlled, cross-over study of intraduodenal levodopa compared to intermittent oral levodopa–carbidopa in 10 PD patients, intraduodenal levodopa significantly improved functional "on" time by 1.1 hour ($p = 0.03$) and half of the patients remained on this treatment for up to 20 months.[56] Intraduodenal levodopa provided a stable response for up to 83 months (median, 44 months) in one open-label study.[57] In another, continuous jejunal infusion in 13 patients was supplemented with bolus doses as needed; in all patients "off" time was reduced immediately after jejunal levodopa administration (50% of awake-time before infusion and 11% after six months).[58] "On" time with disabling dyskinesias was also reduced (17% before infusion *vs.* 3% after). In France, a multi-center retrospective review of patients treated with Duodopa for advanced PD complicated by dyskinesias showed an improvement in dyskinesia (measured on a three-point scale) in 94.7% ($n = 71$).[34]

When therapy has had to be discontinued, it is usually because of technical problems with the infusion device, complications of the gastrostomy procedure, or stomach pain; therefore, careful consideration must be given to the selection of appropriate patients.[34,56-58] Further evaluation of Duodopa in placebo-controlled studies in levodopa-responsive patients is continuing to determine its effect on "on" time without dyskinesias.

5.4.4 New Compounds in Development

Various target receptors, if activated or blocked, prevent dyskinesias or have an anti-dyskinetic effect.[46] NMDA receptor antagonists (*e.g.*, amantadine) and 2-amino-3-(3-hydroxy-5-methyl isoxazol-4-yl) propionic acid (AMPA) receptor antagonists (*e.g.*, talampanel, perampanel) target overactivity of glutamate and excitatory pathways in the direct striatal pathway.[21] The α2 adrenergic receptor antagonists (*e.g.*, fipamezole, idazoxan) may be involved in modulating direct/indirect pathways through GABA-ergic and adrenergic mechanisms. Serotonergic ($5HT_{1A}$) agonists (*e.g.*, sarizotan) or $5HT_{2B}$ antagonists (*e.g.*, clozapine) affect striatal dopaminergic transmission and γ-amino butyric acid (GABA) and glutamate output regions of the basal ganglia. In animal models, when $5HT_{1A}$ and $5HT_{1B}$ agonists were co-administered, there was a synergistic effect on development and severity of dyskinesias; however, human studies are currently unavailable.[59] Adenosine A_{2A} receptor antagonists (*e.g.*, istradefylline, preladenant) colocalize with D_2 receptors and may restore balance in the D_1/D_2 pathway.[60] Although istradefylline has moderate affinity for the adenosine A_{2A} receptor, preladenant has a 1000-fold higher affinity compared to other adenosine subtypes. Cannabinoid CB1 receptors are concentrated in the basal ganglia, and cannabinoid receptor agonists (*e.g.*, nabilone) can potentially increase γ-aminobutyric acid transmission in the lateral segment of the globus pallidus and reduce glutamate release in the striatum.[46] Large cohort studies have revealed a reduced incidence of PD with sustained tobacco use, indicating nicotinic acetylcholine receptors may also be a potential target.[61,62] Presynaptic striatal nicotinic acetylcholine receptors play a role in the regulation of dopamine release, and animal studies indicate that these receptors are decreased with nigrostriatal damage. Studies in MPTP-lesioned monkeys have indicated that nicotine may attenuate LIDs without worsening parkinsonian symptoms.[62] Human studies using nicotine, delivered by intravenous administration, transdermal patch, or gum have produced mixed results and in some cases, intolerable CNS side-effects. Selective nicotinic agonists that target nigrostriatal α4β2 and α6β2 nicotinic subtype receptors may provide an improved benefit to risk profile, but human studies are currently unavailable.

Modulating neurotransmitter activity is a promising approach, but these compounds have shown varying degrees of success in animal models and in preclinical studies. Some have not been effective or have had limited anti-dyskinetic benefit in proof-of-concept studies in humans (*e.g.*, tesofensine, perampanel, istradefylline, safinamide, nicotine patch or gum, nabilone, idazoxan or sarizotan).[10,46] Promising compounds under development (Figure 5.1) are discussed below.

5.4.4.1 Fipamezole (α2 Adrenergic Receptor Antagonist)

The activation of α2 adrenergic receptors on presynaptic noradrenergic terminals and postsynaptic GABA-ergic medium-sized spiny neurons may facilitate activation of direct striatal pathways in the basal ganglia and thus make the development of LIDs more likely.[63] In early studies, fipamezole—an antagonist of these receptors—reduced the severity of LIDs by 23% and 31% at 60 and 90 mg, respectively ($p<0.05$), without affecting the anti-parkinsonian response to levodopa.[64] A four-week double-blind, randomized, placebo-controlled multiple dose-escalating (30 to 90 mg) study (FJORD study, $n=179$) was conducted in the USA (115 subjects, 26 centers) and India (64 subjects, 7 centers) with peak-effect dyskinesias that were at least of moderate severity and present for $\geq 25\%$ of waking hours over the prior week. All subjects were on stable levodopa dosing regimens during the study.[58,65] Dyskinesias were evaluated when patients became "on" with levodopa and then again at 0.5 and 1 hours afterwards. Patients maintained self-assessment diaries to assess dyskinesias. Although the overall data showed no improvement in LID scale, there was a regional difference in the outcome results, presumably due to heterogeneity in the study populations. In the USA subjects, fipamezole at a dose of 90 mg (but not 30 or 60 mg) reduced LIDs (improvement compared to placebo, -1.9 ± 0.9; 95% CI 0.0, -3.8; $p=0.047$); but in the Indian subjects the study end-points for efficacy were not met. Dose responsiveness in relation to the change from baseline to day 28 average LID scale score was also demonstrated in the USA subjects ($p=0.014$). The most commonly reported adverse events were nausea, dysgeusia and oral hypoesthesias. Further large-scale studies are warranted to confirm the clinical findings and significance of the USA cohort results.

5.4.4.2 Preladenant (Adenosine A_{2A} Receptor Antagonist)

Preladenant (SCH-420814) was studied in a 12-week, randomized double-blind placebo-controlled dose-ranging study of 253 patients with moderate-to-severe PD who had motor fluctuations and dyskinesias on stable dopaminergic therapy.[66] Patients maintained self-assessment diaries to record dyskinesias and their severity. After 12 weeks, patients in both the 5- and 10-mg twice-daily dosing groups experienced an adjusted mean difference in awake "on" time of 1.4 hours per day ($p=0.024$) and 1.3 hours per day ($p=0.049$), respectively, compared to 0.2 hours per day in the placebo group, without overall significant worsening of dyskinesias. The most common adverse events were somnolence, nausea, constipation, and insomnia. The incidence of dyskinesia in the combined preladenant group *vs.* placebo was 9% *vs.* 13%. These data are encouraging and indicate the need for further study of the long-term effects of preladenant in minimizing the risk for dyskinesias.

5.4.4.3 Levetiracetam

Levetiracetam, an anti-seizure drug, has been evaluated for its impact on LIDs with mixed results.[29–31] Its exact mechanism of action is unknown. The efficacy and safety of levitiracetam was assessed in two randomized, double-blind placebo-controlled studies.[30,63] The first was a parallel-group 11-week pilot study in 32 PD patients with moderate-to-severe LIDs on stable dopaminergic therapy. Placebo or levetiracetam was administered twice daily (titrated from 250 to 2000 mg per day) as an add-on therapy.[30] Subjects underwent evaluation by the UPDRS and modified AIMS to assess dyskinesias. The mean changes of the modified AIMS from baseline were −1.5 (−26%) for levetiracetam ($p=0.33$) and +0.9 (+13%) for placebo ($p=0.59$) without significant differences between groups. Mean changes of the UPDRS item 32/33 sum score from baseline showed significant improvement of dyskinesia in the levetiracetam group [− 1.0 (−20%); $p=0.012$], but not in the placebo group [− 0.4 (−8%); $p=0.31$].

The second study included a cross-over period, but due to a period effect, data after the cross-over point were excluded from analysis.[67] PD participants with LIDs ($n=38$) received levetiracetam 500 mg per day, were assessed, titrated to 1000 mg per day and re-assessed, before and after cross-over. The "on" time with LIDs time decreased by 37 min (95% CI 0.59, 7.15; $p=0.02$) at 500 mg per day, and by 75 min (95% CI 3.3, 12.4; $p=0.002$) at 1000 mg per day. The "on" time without LIDs increased by 46 min (95% CI −1.55, −0.03; $p=0.04$) at 500 mg per day and 55 min (95% CI −10.39, −1.14; $p=0.018$) at 1000 mg per day. UPDRS 32 showed a decreased dyskinesia duration mean change 0.35 (95% CI 0.09, 0.5; $p=0.009$) at 1000 mg per day. Clinical global impression (CGI) showed LIDs decreased by 0.7 (95% CI 0.21, 1.18; $p=0.006$) at 1000 mg per day. "Off" time did not increase.

Levitiracetam was well-tolerated in both studies at doses up to 2000 mg per day, although CNS depressant effects sometimes limited upward titration. These two studies indicate that levitiracetam may have a mild but significant anti-dyskinetic effect, but further large-scale controlled studies are needed.

5.4.4.4 Pardoprunox

Pardoprunox, a partial agonist at D_2 and D_3 receptors and a full agonist at $5HT_1$ receptors, also binds with lower affinity to D_4, α1 adrenergic, and $5HT_7$ receptors.[68,69] It may have a lower tendency than other dopaminergic therapies to cause dyskinesias or neuropsychiatric side-effects. Its safety and tolerability were assessed in a randomized, controlled study of patients with advanced PD using a slow titration schedule to achieve the highest tolerated dose (0.3 to 42 mg per day) over seven weeks.[69] A significant increase in "on" time without troublesome dyskinesias was observed after weeks 2 ($p=0.045$), 5 ($p=0.031$), and 7 ($p=0.031$), but not at all study timepoints. A low incidence of dyskinesias was reported spontaneously as an adverse event (3.9%; 2 of 51 patients) in the pardoprunox group. Further studies are needed to evaluate the potential of minimizing the risk for dyskinesias with pardoprunox.

5.4.4.5 IPX066 (Rytary™)

IPX066 (Rytary™) is an oral extended-release formulation of carbidopa–levodopa that is composed of beads that dissolve at varying rates and thus provide a smoother and more prolonged intraintestinal absorption time.[70]

Compared to immediate-release carbidopa–levodopa, IPX066 demonstrated a 30% higher C_{max}, but 87% higher levodopa exposure (measured by levodopa AUC) with a multiple-dosing regimen (after 8 days of dosing and a mean of 3.5 IPX066 doses per day compared to 5.4 for immediate-release carbidopa–levodopa). Thus, this formulation allows for more sustained levodopa levels with less variability and potentially a decreased frequency of dosing with fewer motor fluctuations (and LIDs). An early-stage, cross-over comparison study of IPX066 and standard immediate-release carbidopa was conducted in 27 patients with advanced PD. Treatment with IPX066 resulted in a greater increase in "on" time without troublesome dyskinesia (+1.3 hours) compared to the immediate-release carbidopa–levodopa group (+0.43 hours, $p = 0.002$); however, there was no difference between the two groups in "on" time with troublesome dyskinesias. Phase 3 studies in a larger sample population have been completed in both early and advanced stages of PD and are being reviewed by the FDA, but as of writing, these studies are only available in abbreviated (poster) format.

5.5 Conclusion

Although carbidopa–levodopa remains the most efficacious treatment for managing the disabling motor symptoms of PD, the emergence of levodopa-induced dyskinesias of varying severity inevitably occurs with prolonged therapy. Certain oral dopaminergic therapies (*e.g.*, dopamine agonists) delay the need for initiating levodopa therapy, and thereby delay the onset of the dyskinesias. Alternate delivery methods utilizing parenteral therapies to enable continuous dopaminergic therapy are being developed. Recent research continues to identify rational pharmacologic targets; however treatments based on these targets are in the early stages of development. Therefore, there remains an unmet need for anti-dyskinetic therapies or for those that prevent dyskinesias from occurring altogether.

References

1. J. E. Ahlskog and M. D. Muenter, *Mov. Disord.*, 2001, **16**, 448.
2. R. Pahwa, S. A. Factor, K. E. Lyons, W. G. Ondo, G. Gronseth, H. Bronte-Stewart, M. Hallett, J. Miyasaki, J. Stevens, W. J. Weiner and Quality Standards Subcommittee of the American Academy of Neurology, *Neurology*, 2006, **66**, 983.
3. G. C. Cotzias, P. S. Papavasiliou and R. Gellene, *New Engl. J. Med*, 1969, **280**, 337.
4. J. G. Nutt, K. A. Chung and N. H. Holford, *Neurology*, 2010, **74**, 1191.
5. J. Jankovic, *Mov. Disord.*, 2005, **20**(11), S11.
6. R. A. Hauser, M. P. McDermott and S. Messing, *Arch. Neurol.*, 2006, **63**, 1756.
7. V. Kostic, S. Przedborski, E. Flaster and N. Sternic, *Neurology*, 1991, **41**, 202.

8. L. Mazzella, M. D. Yahr, L. Marinelli, N. Huang, E. Moshier and A. Di Rocco, *Parkinsonism Relat. Disord.*, 2005, **11**, 151.
9. S. Ku and G. A. Glass, *Mov. Disord.*, 2010, **25**, 1177.
10. M. D. Gottwald and M. J. Aminoff, *Ann. Neurol.*, 2011, **69**, 919.
11. P. Jenner, *Nat. Rev. Neurosci.*, 2008, **9**, 665.
12. P. Calabresi, M. Di Filippo, V. Ghiglieri and B. Picconi, *Mov. Disord.*, 2008, **23**(3), S570.
13. A. Nadjar, C. R. Gerfen and E. Bezard, *Prog. Neurobiol.*, 2009, **87**, 1.
14. M. A. Cenci, *Parkinsonism Relat. Disord.*, 2007, **13**(3), S263.
15. C. W. Olanow, J. A. Obeso and F. Stocchi, *Nat. Clin. Pract. Neurol.*, 2006, **2**, 382.
16. A. Antonini, *Parkinsonism Relat. Disord.*, 2007, **13**, S24.
17. P. Calbresi, M. DiFilippo, V. Ghiglieri, N. Tambasco and B Picconi, *Lancet Neurol.*, 2010, **9**, 1106.
18. C. R. Gerfen, S. Miyachi, R. Paletzki and P. Brown, *J. Neurosci.*, 2002, **22**, 5042.
19. F. Gardoni, B. Picconi, V. Ghiglieri, F. Polli, V. Bagetta, G. Bernardi, F. Cattabeni, M. Di Luca and P. Calabresi, *J. Neurosci.*, 2006, **26**, 2914.
20. B. Picconi, D. Centonze, K. Hakansson, G. Bernardi, P. Greengard, G. Fisone, M. A. Cenci and P. Calabresi, *Nat. Neurosci.*, 2003, **6**, 501.
21. C. Colosimo, P. Martinez-Martin, G. Fabbrini, R. A. Hauser, M. Merello, J. Miyasaki, W. Poewe, C. Sampaio, O. Rascol, G. T. Stebbins, A. Schrag and C. G. Goetz, *Mov. Disord.*, 2010, **25**, 1131.
22. R. Constantinescu, M. Romer, M. P. McDermott, C. Kamp, K. Kieburtz and the CALM-PD Investigators of the Parkinson Study Group, *Mov. Disord.*, 2007, **22**, 1317.
23. O. Rascol, D. J. Brooks, A. D. Korczyn, P. P. De Deyn, C. E. Clarke, A. E. Lang, M. Abdalla and the 056 Study Group, *Mov. Disord.*, 2006, **21**, 1844.
24. N. J. Crosby, K. H. Deane and C. E. Clarke, *Cochrane Database Syst. Rev.*, 2003, **2**, CD003467.
25. L. V. Metman, P. Del Dotto, P. van den Munchkof, J. Fang, M. M. Mouradian and T. N. Chase, *Neurology*, 1998, **50**, 1323.
26. E. Wolf, K. Seppi, R. Katzenschlager, G. Hochschorner, G. Ransmayr, P. Schwingenschuh, E. Ott, I. Kloiber, D. Haubenberger, E. Auff and W. Poewe, *Mov. Disord.*, 2010, **25**, 1357.
27. E. Luginger, G. K. Wenning, S. Bosch and W. Poewe, *Mov. Disord.*, 2000, **15**, 873.
28. B. J. Snow, L. Macdonald, D. McAuley and W. Wallis, *Clin. Neuropharmacol.*, 2000, **23**, 82.
29. T. A. Zesiewicz, K. L. Sullivan, J. L. Maldonado, W. O. Tatum and R. A. Hauser, *Mov. Disord.*, 2005, **20**, 1205.
30. M. Wolz, M. Lohle, K. Strecker, U. Schwanebeck, C. Schneider, H. Reichmann, X. Grahlert, J. Schwarz and A. Storch, *J. Neural. Transm.*, 2010, **117**, 1279.
31. K. E. Lyons and R. Pahwa, *Clin. Neuropharmacol.*, 2006, **29**, 148.

32. M. Contin, P. Martinelli, F. Albani, C. Scaglione, P. Avoni, G. Rizzo and A. Baruzzi, *Clin. Neuropharmacol.*, 2007, **30**, 122.
33. D. Deleu, Y. Hanssens and M. G. Northway, *Drugs Aging*, 2004, **21**, 687.
34. D. Devos and the French Duodopa Study Group, *Mov. Disord.*, 2009, **24**, 993.
35. Parkinson Study Group Investigators, *Arch. Neurol.*, 2004, **61**, 1044.
36. Parkinson Study Group CALM Cohort Investigators, *Arch. Neurol.*, 2009, **66**, 563.
37. Parkinson Study Group, *JAMA, J. Am. Med. Assoc.*, 2000, **284**, 1931.
38. W. Poewe, O. Rascol, P. Barone, R. A. Hauser, Y. Mizuno, M. Haaksma, L. Salin, N. Juhel, A. H. Schapira and the Extended Release Pramipexole Study Group, *Neurology*, 2011, **77**, 759.
39. O. Rascol, D. J. Brooks, A. D. Korczyn, P. P. De Deyn, C. E. Clarke and A. E. Lang, *New Eng. J. Med.*, 2000, **342**, 1484.
40. R. A. Hauser, O. Rascol, A. D. Korczyn, A. Jon Stoessl, R. L. Watts, W. Poewe, P. P. De Deyn and A. E. Lang, *Mov. Disord.*, 2007, **22**, 2409.
41. R. L. Watts, K. E. Lyons, R. Pahwa, K. Sethi, M. Stern, R. A. Hauser, W. Olanow, A. M. Gray, B. Adams and N. L. Earl, *Mov. Disord.*, 2010, **25**, 858.
42. R. Pahwa, M. A. Stacy, S. A. Factor, K. E. Lyons, B. P. Hersh, L. W. Elmer, D. D. Truong, N. L. Earl and the EASE-PD Study Investigators, *Neurology*, 2007, **68**, 1108.
43. R. O. Stoessel and F Kietburtz, *Ann. Neurol.*, 2010, **68**, 3.
44. F. Stocchi, O. Rascol, K. Kieburtz, W. Poewe, J. Jankovic, E. Tolosa, P. Barone, A. E. Lang and C. W. Olanow, *Ann. Neurol.*, 2010, **68**, 18.
45. B. Elahi, N. Phielipp and R. Chen, *Can. J. Neurol. Sci.*, 2012, **39**, 465.
46. G. Fabbrini, J. M. Brotchie, F. Grandas, M. Nomoto and C. G. Goetz, *Mov. Disord.*, 2007, **22**, 1379.
47. S. Miyamoto, G. E. Duncan, C. E. Marx and J. A. Lieberman, *Mol. Psychiatry*, 2005, **10**, 79.
48. F. Durif, B. Debilly, M. Galitzky, D. Morand, F. Viallet, M. Borg, S. Thobois, E. Broussolle and O. Rascol, *Neurology*, 2004, **62**, 381.
49. *Apokyn®* *(apomorphine hydrochloride)*, Prescription Solutions, Morristown, USA, 2006.
50. C. M. Stibe, A. J. Lees, P. A. Kempster and G. M. Stern, *Lancet*, 1988, **1**, 403.
51. P. Kanovsky, D. Kubova, M. Bares, H. Hortova, H. Streitova, I. Rektor and V. Znojil, *Mov. Disord.*, 2002, **17**, 188.
52. A. Colzi, K. Turner and A. J. Lees, *J. Neurol. Neurosurg. Psychiatry*, 1998, **64**, 573.
53. A. J. Manson, K. Turner and A. J. Lees, *Mov. Disord.*, 2002, **17**, 1235.
54. P. J. Garcia Ruiz, A. Sesar Ignacio, B. Ares Pensado, A. Castro Garcia, F. Alonso Frech, M. Alvarez Lopez, J. Arbelo Gonzalez, J. Baiges Octavio, J. A. Burguera Hernandez, M. Calopa Garriga, D. Campos Blanco, B. Castano Garcia, M. Carballo Cordero, J. Chacon Pena, A. Espino Ibanez, A. Gorospe Onisalde, S. Gimenez-Roldan, P. Granes Ibanez,

J. Hernandez Vara, R. Ibanez Alonso, F. J. Jimenez Jimenez, J. Krupinski, J. Kulisevsky Bojarsky, I. Legarda Ramirez, E. Lezcano Garcia, J. C. Martinez-Castrillo, D. Mateo Gonzalez, F. Miquel Rodriguez, P. Mir, E. Munoz Fargas, J. Obeso Inchausti, J. Olivares Romero, J. Olive Plana, P. Otermin Vallejo, B. Pascual Sedano, V. Perez de Colosia Rama, I. Perez Lopez-Fraile, A. Planas Comes, V. Puente Periz, M. C. Rodriguez Oroz, D. Sevillano Garcia, P. Solis Perez, J. Suarez Munoz, J. Vaamonde Gamo, C. Valero Merino, F. Valldeoriola Serra, J. M. Velazquez Perez, R. Yanez Bana and I. Zamarbide Capdepon, *Mov. Disord.*, 2008, **23**, 1130.
55. D. Nyholm, *Exp. Rev. Neurother.*, 2006, **6**, 1403.
56. M. C. Kurth, J. W. Tetrud, C. M. Tanner, I. Irwin, G. T. Stebbins, C. G. Goetz and J. W. Langston, *Neurology*, 1993, **43**, 1698.
57. D. Nilsson, D. Nyholm and S. M. Aquilonius, *Acta Neurol. Scand.*, 2001, **104**, 343.
58. K. Eggert, C. Schrader, M. Hahn, M. Stamelou, A. Russmann, R. Dengler, W. Oertel and P. Odin, *Clin. Neuropharmacol.*, 2008, **31**, 151.
59. A. Munoz, Q. Li, F. Gardoni, E. Marcello, C. Qin, T. Carlsson, D. Kirik, M. Di Luca, A. Bjorklund, E. Bezard and M. Carta, *Brain*, 2008, **131**, 3380.
60. R. A. Hauser, M. Cantillon, E. Pourcher, F. Micheli, V. Mok, M. Onofrj, S. Huyck and K. Wolski, *Lancet Neurol.*, 2011, **10**, 221.
61. M. Quik, L. Z. Huang, P. Neeraja, T. Bordia, C Campos and X. A. Perez, *Biochem. Pharmacol.*, 2009, **78**, 677.
62. M. Quik, K. O'Leary and C. M. Tanner, *Mov. Disord.*, 2008, **12**, 1641.
63. J. M. Savola, M. Hill, M. Engstrom, H. Merivuori, S. Wurster, S. G. McGuire, S. H. Fox, A. R. Crossman and J. M. Brotchie, *Mov. Disord.*, 2003, **18**, 872.
64. P. A. LeWitt, L. R. Kingery and J. Savola, *Mov. Disord.*, 2009, **24**, S435.
65. P. A. Lewitt, R. A. Hauser, M. Lu, A. P. Nicholas, W. Weiner, N. Coppard, M. Leinonen and J. M. Savola, *Neurology*, 2012, **79**, 163.
66. R. A. Hauser, M. Cantillon, E. Pourcher, F. Micheli, V. Mok, M. Onofrj, S. Hyk and K. Wolsk, *The Lancet Neruology*, 2011, **10**, 221.
67. P. Stathis, S. Konitsiotis, G. Tagaris and D. Peterson, *Mov. Disord*, 2011, **26**, 264.
68. J. Bronzova, C. Sampaio, R. A. Hauser, A. E. Lang, O. Rascol, A. Theeuwes, S. V. van de Witte, G. van Scharrenburg and G. Bruegel, *Mov. Disord.*, 2010, **25**, 738.
69. R. A. Hauser, J. Bronzova, C. Sampaio, A. E. Lang, O. Rascol, A. Theeuwes, S. V. van de Witte and the Pardoprunox Study Group, *Eur. Neurol.*, 2009, **62**, 40.
70. R. A. Hauser, A. L. Ellenbogen, L. V. Metman, A. Hsu, M. J. O'Connell, N. B. Modi, H. Yao and S. H. Kell, *Mov. Disord.*, 2011, **26**, 2246.

CHAPTER 6

D_3 Receptor Agonists and Antagonists as Anti-Parkinsonian Therapeutic Agents

MARK JOHNSON AND ALOKE DUTTA*

Department of Pharmaceutical Sciences, Eugene Applebaum College of Pharmacy & Health Sciences, Wayne State University, Detroit, USA
*Email: adutta@wayne.edu

6.1 Introduction: The D_3 Receptor

Dopamine (DA) neurons and their associated receptors have long been known to be implicated in the pathogenesis of Parkinson's disease (PD). The DA receptors, phylogenetically classified as members of the biogenic amine receptors and part of the 'rhodopsin-like' subfamily, belong to the super-family of membrane-bound proteins, termed "G-protein coupled receptors".[1] Until 1990, the DA receptor population of the brain and periphery was believed to consist solely of the D_1 and D_2 subtypes.[2–4] Cloning of these two receptors led to the discovery of several additional, low-abundance DA receptors, including D_3, D_4 and D_5.[5–7] The DA receptor subtypes, D_1–D_5, were further classified on the basis of their inherent biochemical and pharmacological properties. The D_1-like receptors, including D_1 and D_5, were found to be stimulatory in nature, thereby activating the second messenger enzyme, adenylyl cyclase, to produce cyclic adenosine monophosphate (cAMP). In contrast, D_2-like receptors,

RSC Drug Discovery Series No. 34
Emerging Drugs and Targets for Parkinson's Disease
Edited by Ana Martinez and Carmen Gil

Published by the Royal Society of Chemistry, www.rsc.org

including D_2, D_3 and D_4, negatively couple to adenylyl cyclase and decrease the production of cAMP. The cloning of the D_3 receptor, initially undertaken by Sokoloff and colleagues, using cDNA from the rat and probes derived from the D_2 receptor sequence, became of particular interest due to new hypotheses that proposed the D_3 receptor as a therapeutic target for neuropsychiatric disorders and PD.[8] D_3-initiated signaling cascades have been found to include: stimulation or inhibition of adenylyl cyclase, increased extracellular acidification and alterations in Ca^{2+} and K^+ currents, each of which are able to be blocked by pertussis toxin, suggesting coupling to a G_i isoform.[9–15] Hydrophobicity analysis indicates the D_3 receptor structure to constitute seven transmembrane domains and to be characterized by a relatively long third intracellular loop and short caboxy termini.[16] Computer models demonstrate that the seven transmembrane domains of the D_3 receptor conform to idealized α-helices.[17]

6.1.1 D_3 Localization and Distribution in the Brain

A variety of approaches has been used to identify the location of D_3 receptors including molecular, pharmacological and immunological methods. Due to the lack of truly selective pharmacological tools or extensive validation of those in current use, much of what is known about the distribution of D_3 receptors is based on localization of receptor mRNA. *In situ* hybridization studies in rat brain reveal that D_3 mRNA is expressed preferentially in the limbic brain regions.[16] High levels of D_3 mRNA are observed in the islands of Calleja, *nucleus accumbens* and olfactory tubercle.[8,18–21] Low levels of D_3 mRNA are found in the cerebral cortex, caudate–putamen, ventral pallidum, *substantia nigra pars reticula*, ventral tegmental area and cerebellar cortex.[18,19] The distribution of D_3 mRNA and associated receptors was found to be similar in rat and human brain, although receptor localization in human brain is somewhat less restricted, with moderate amounts of D_3 receptors found in the basal ganglia and cortical regions.[16,22–24]

In the *nucleus accumbens*, D_3 receptor mRNA colocalizes with that of TrkB, a high-affinity, membrane-bound receptor, for which the brain-derived neurotrophic factor (BDNF) is a ligand.[25,26] In wild-type $BDNF^{+/+}$ mice, D_3 receptor binding and mRNA increases significantly in the *nucleus accumbens* from postnatal days 9–14 to 17–23, whereas in homozygous $BDNF^{-/-}$ mice, D_3 receptor binding and mRNA are low early in development and do not increase at later stages, implicating BDNF for normal D_3 receptor expression.[25] Similar to other D_2-like receptors, D_3 receptors are located both pre and postsynaptically. Neurochemical studies suggest a role for D_3 receptors as synthesis- and/or release-modulating autoreceptors.[16] Decreased extracellular DA, induced by low doses of PD-128907, measured by brain microdialysis, was found to be attenuated in D_3 receptor-deficient mice, suggesting that a subset of D_3 receptors control DA release.[27]

6.1.1.1 Significance of D_3 Receptors in Parkinson's Disease

Although it has been assumed that D_2 receptor stimulation is necessary for the relief of motor dysfunction in PD, DA agonists used in the treatment of PD have, in many cases, as high or higher affinity for D_3 receptors. It is thought that meso-limbic D_3 receptors may play a role in relief of PD symptoms, as the limbic *striatum* is known to be involved in aspects of movement, such as goal-directed behaviors and locomotor activity.[28] Locomotor stimulation is observed in 6-hydroxydopamine (6-OHDA)-lesioned rats at doses of D_3-preferring agonists that are inhibitory in normosensitive rats, suggesting that D_3-preferring agonists may be a viable option for the anti-Parkinsonian treatment of DA-depleted animals.[29] A postmortem study of PD cases revealed that the number of D_3 receptors was reduced by 45% in the basal ganglia, while the number of D_2 receptors was elevated in similar regions by 15–25%.[30] Furthermore, similar studies have shown that non-responders to anti-Parkinsonian medication correlate with a lower D_3 receptor number (−48%). In contrast, PD cases that remain responsive to anti-Parkinsonian drugs correlate with elevated levels (+25%) of D_3 receptors.[31] Evidence suggests that a decrease in D_3 receptor numbers in the early stages of PD is followed by an elevation of D_3 receptor numbers after initiation of levodopa (L-DOPA) therapy, the gold-standard treatment option in clinical PD cases. However, disease progression likely results in further meso-limbic DA neuron loss and other receptors along with a permanent reduction in D_3 receptor number in critical striatal regions. In advanced stages of the disease, anti-Parkinsonian drugs are no longer able to effectively treat many of the predominant symptoms of PD and the patient is judged to be unresponsive.[28]

6.2 D_3 Receptor-Selective Ligands

The individual and characteristic expression patterns of the DA receptor subtypes in the CNS suggest that they mediate different actions of DA and, therefore, may exert different therapeutic effects.[32] The unique distribution of D_3 receptors in the limbic and nigrostriatal brain regions make it a promising target to treat neurological diseases, including PD. The D_3 receptor is known to contribute to the control of movement, emotion and cognition.[23,33,34] The D_3 receptor, while playing a role in the control of motor activity under normal conditions, may have an expanded and more significant role in the Parkinsonian brain, in which its expression and distribution patterns have been found to be markedly altered. The development of truly selective pharmacological tools to probe the pathophysiological significance of the D_3 receptor in PD have been uniquely challenging due to the sequence similarity between D_2-like receptors, specifically, D_2 and D_3 receptors. Sequence alignment studies reveal the D_2 and D_3 receptors to have a 50% overall sequence identity, along with a 63% sequence similarity. Moreover, when considering only transmembrane domains, a sequence identity of 79% and a sequence similarity of 90% are reported, when comparing D_2 and D_3 receptors.[35,36] However, homology modeling of D_2-like receptors has revealed differences in relative orientation and transmembrane packing, which has contributed to recent progress in the design and development of D_3-selective agents.[37–39]

6.2.1 D_3 Receptor-Selective Agonists

Following the identification of the D_3 receptor in 1990, several DA receptor agonists have been evaluated for their D_3 receptor affinity and selectivity.[40] The endogenous ligand, DA, binds to the high-affinity state of the human D_3 receptor at low nanomolar (K_i, $D_3 = 3.9$ nM) affinity, while exhibiting low selectivity ($D_2/D_3 = 0.4$) over D_2 receptors.[41] Rigidization of the aminoethyl side-chain of DA has yielded the 2-aminotetralins, S-5-OH- and R-7-OH-DPAT (Figure 6.1). Studies have indicated moderate D_3 selectivity for both S-5-OH- and R-7-OH-DPAT ($D_2/D_3 = 26$ and 60, respectively).[42] Bioisosteric mimicry of the catechol moiety of DA has been used in an effort to obtain improved bioavailability, along with selectivity for D_3 receptors. Introduction of the catechol bioisostere, 2-aminothiazole, led to the identification of pramipexole, which maintains a high D_3 affinity (0.5–8.5 nM) and varying selectivity ($D_2/D_3 = 8$–193), depending upon assay conditions.[43–45] Mitogenesis assays indicate the intrinsic activity of pramipexole to be 100% at D_2 and 80% at D_3 receptors.[45] Pramipexole is a well known anti-Parkinsonian drug (Figure 6.1). Ropinirole binds with moderate affinity to D_3 (69 nM), while only weakly to D_2 (1380 nM) receptors. Mitogenesis studies indicate preferential, full agonist activity of ropinirole at D_3 (EC_{50}, $D_3 = 1.2$ nM, $D_2/D_3 = 8.3$) receptors.[1]

Currently, a number of DA agonists are being used in the clinic to treat PD. Some of the traditional ergoline agonists are bromocriptine, lisuride and apomorphine.[46] More recently, the non-ergoline drugs, pramipexole and ropinirole, have been added to the list of available DA agonists (Figure 6.1).[47,48] The non-ergot class of DA agonists was developed to address some of the undesirable side-effects exhibited by the ergot derivatives. Notably, pramipexole and ropinirole have exhibited preferential affinity for D_3 receptors.

Recently, a 'hybrid approach' to drug design has been utilized by connecting 2-aminoteralins or bioisosteric equivalents to arylpiperazine fragments, yielding D_3-selective agonists. Hybrid derivative, 1, shows a high affinity for D_3 (4.1 nM) and a 510-fold selectivity over D_2 receptors, while also acting as a full agonist at both D_2 (90%) and D_3 (86%) receptors in mitogenesis assays.[49,50] Pramipexole derivative, D-264 (Figure 6.1), exhibits a pronounced affinity for D_3 (0.92 nM) and appreciable selectivity over D_2 ($D_2/D_3 = 253$) receptors.[51] In functional studies, D-264 displays potent ($EC_{50} = 1.51$ nM) agonist activity at D_3 receptors, although selectivity (22-fold) over D_2 receptors is diminished.[52] Furthermore, replacement of the biphenyl group of D-264 with an indole-2-carboxamide moiety yielded D-440 (Figure 6.1). D-440 maintains a high D_3 affinity (1.84 nM), along with a >2-fold increase in selectivity ($D_2/D_3 = 583.2$) over D_2 receptors, compared to D-264.[52] Functional studies indicate D-440 to be a highly potent ($EC_{50} = 0.26$ nM, $D_2/D_3 = 438$) and

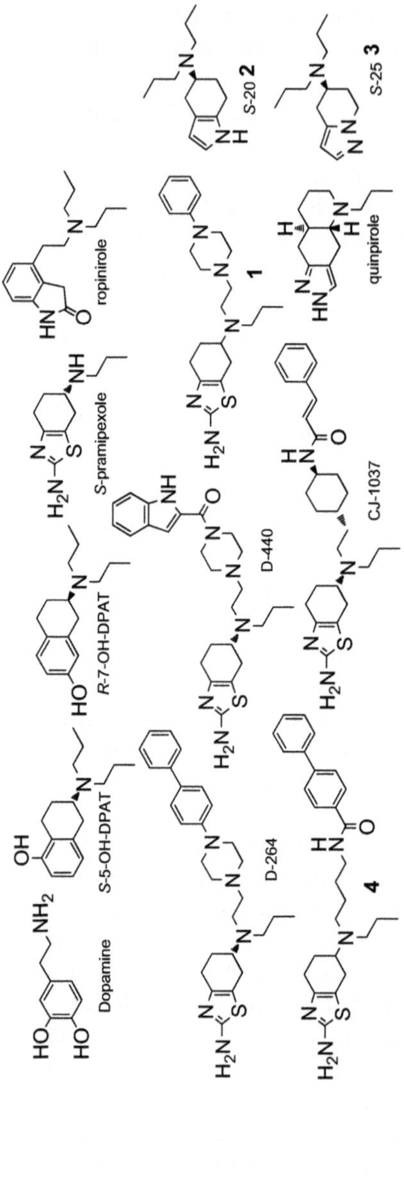

Figure 6.1 D$_3$ receptor-selective agonists.

selective, D_3 receptor agonist.[52] Recent synthetic efforts have produced CJ-1037, a pramipexole-based, cinnamide-derivative, which incorporates a *trans*-cyclohexyl group in the alkyl spacer and has led to much improved D_3 selectivity (800- and >30 000-fold over D_2 and D_1-like receptors, respectively). *In vivo* functional studies indicate that CJ-1037 exhibits partial agonist properties at D_3 and no activity at D_2 receptors, as assessed by yawning-induction in rats.[53]

Quinpirole contains a bioisosteric pyrazole ring system, binding with a high affinity (0.96 nM) and selectivity ($D_2/D_3 = 133$) for D_3 receptors.[1] Additionally, quinpirole displays full agonism at both D_2 and D_3 receptors in mitogenesis assays.[11] In a novel series of analogues, the pyrrole moiety was investigated as a catechol bioisostere. The 1-aza derivative, *S*-2, displays moderate potency (38 nM) and high selectivity ($D_2/D_3 = 316$) for D_3 over other DA receptor subtypes.[54] The aminotetrahydropyrazolo derivative, *S*-3, exhibits a high D_3 affinity (K_i, $D_3 = 4.0$ nM, $D_2/D_3 = 45$) and selectivity over D_2 receptors, along with potent (EC$_{50} = 3.4$ nM) intrinsic activity (82%) in stimulating mitogenesis at D_3 receptors.[55] The pramipexole derivative, compound **4**, displays agonistic properties, along with excellent affinity (0.50 nM) at D_3 receptors and a 400-fold selectivity over D_2 receptors.[56]

6.2.2 D_3 Receptor-Selective Antagonists

One strategy used to produce preferential D_3 antagonists has been based upon modification or replacement of the hydroxyl function within 5-OH- or 7-OH-DPAT (Figure 6.2).[1] Conversion of 7-OH-DPAT into a tricyclic ether produced S14297, which binds with high affinity to human D_3 receptors (13 nM) and displays a 23-fold selectivity over D_2 receptors.[57] S14297 exhibits excellent selectivity over all other DA receptor subtypes (>1 µM) and is regarded as a functional D_3 antagonist, inhibiting a 7-OH-DPAT-mediated decrease in DA levels in the *nucleus accumbens*.[57,58] Incorporation of various *p*-substituted phenyl moieties onto the 6-position of the aminotetralin scaffold allowed the discovery of the sulfonyl derivative, GR218231, which shows a high affinity for D_3 receptors (1.3 nM) and exceptional selectivity ($D_2/D_3 = 380$) over D_2 and other DA receptor subtypes.[1,59] In addition, GR218231 acts as a selective, human D_3 antagonist in neurochemical experiments.[60] Iodine substitution, replacing the methoxy group of GR218231, yielded compound **5**, a ligand with excellent D_3 affinity (0.16 nM) and 1300-fold selectivity over D_2 receptors.[59] Efforts at conformational restriction of 5-OH- and 7-OH-DPAT, in which the *N* side-chain is elongated and modified, produced S33084. S33084 binds with a high affinity (0.32 nM) and selectivity ($D_2/D_3 = 100$) to D_3 receptors, along with a 200-fold selectivity over 40 other binding sites.[60,61] Furthermore, S33084 potently inhibits stimulation of [^{35}S]guanosine 5′-O-[gamma-thio]triphosphate ([^{35}S]GTPγS) binding to Gα, mediated by human D_3 receptors, demonstrating antagonistic properties.[60] Due to its selective binding and functional profile, S33084 serves as an important tool to study the pathophysiological significance of the D_3 receptor.[1] Ring contraction of the aminotetralin scaffold produced a series of aminoindan derivatives. The 5,6-dimethoxy derivative, U99194A, displays moderate D_3 affinity (31 nM) and 32-fold selectivity over D_2 receptors.[62] Moreover, U99194A is devoid of agonist activity in mitogenesis and extracellular acidification assays and antagonizes the stimulating effects of DA.[63] Extension of the aminoindan nucleus with the 2-aminothiazole moiety produced GMC1111, which binds preferentially to D_3 receptors (K_i, $D_3 = 1.4$ nM, $D_2/D_3 = 19$), while exhibiting antagonistic properties, as assessed by mitogenesis assay.[64] Recently, GMC1111 was investigated in 6-OHDA-lesioned marmosets, where it reduced both Parkinson symptoms and LID.[65]

Strategies to avoid rapid clearance of the N-propyl-2-aminotetralin class of D_3-selective antagonists, via depropylation, led to the identification of tetrahydroisoquinolines as a lead structure (Figure 6.2).[66] Modification of the N-arylcarbamide moiety yielded a 3-indolyl derivative, compound **6**, which displayed a high affinity for D_3 (4.0 nM) and selectivity (160-fold) over D_2 receptors, along with potent (EC_{50}, $D_3 = 1.3$ nM, $D_2 = 250$ nM) and preferential antagonism at D_3 receptors. Moreover, compound **6** displayed good brain penetrance, low blood clearance (27 mL min^{-1} kg^{-1}) and a long plasma half-life (2.5 h).[66] The cinnamide derivative, ST 198, is devoid of intrinsic activity and displays a high D_3 affinity (12 nM), while having a selectivity of 65-fold over D_2 receptors and >400-fold over the remaining DA receptor subtypes.[67,68] Rigidization of the butyl spacer, along with introduction of a 6-cyano substituent on the isoquinoline nucleus and optimization of the arylcarbamide moiety led to the identification of SB 277011, a potent (10 nM) and selective (100-fold vs. D_2 receptors, >100-fold for over 60 other receptors and ion channels) D_3 receptor antagonist ($EC_{50} = 4.0$ nM) with high brain penetrance (brain : blood = 3.6 : 1) and good oral bioavailabilty (43%) in rats.[69] Ring contraction, along with cinnamide substitution in the arylcarbamide region of SB 277011, produced compound **7**. Analog **7** was found to have a high D_3 (5.0 nM) affinity and selectivity ($D_2/D_3 = 100$) over D_2 receptors, along with potent antagonism ($EC_{50} = 5$ nM) at D_3 receptors. Furthermore, compound **7** exhibits high oral bioavailability (77%), low plasma clearance (14 mL min^{-1} kg^{-1}) and a long half-life (5.2 h).[70] In the 2-methoxybenzamide class of D_3 antagonists, the naphthamide analog, S-nafadotride, displays a high D_3 (0.3 nM) affinity and 10-fold selectivity over D_2 receptors. Additionally, in mitogenesis experiments, nafadotride antagonizes quinpirole-induced mitogenesis.[71] Extension of the benzamide moiety, along with addition of an arylpiperazine side-chain, produced the 2,3-dichloro-phenylpiperazine derivative, NGB 2849. NGB 2849 displays a high affinity (0.9 nM) for D_3 receptors and selectivity ($D_2/D_3 = 290$) over other DA receptor subtypes. Moreover, NGB 2849 displays potent ($EC_{50} = 6.8$ nM) antagonistic properties at D_3 receptors in mitogenesis assays.[72] The closely related naphthamide derivative, BP 897 (Figure 6.2), a high-affinity (0.92 nM) D_3 receptor ligand, displays a 66-fold preference over D_2 receptors.[73] Studies aimed at determining the intrinsic activity of BP 897 at

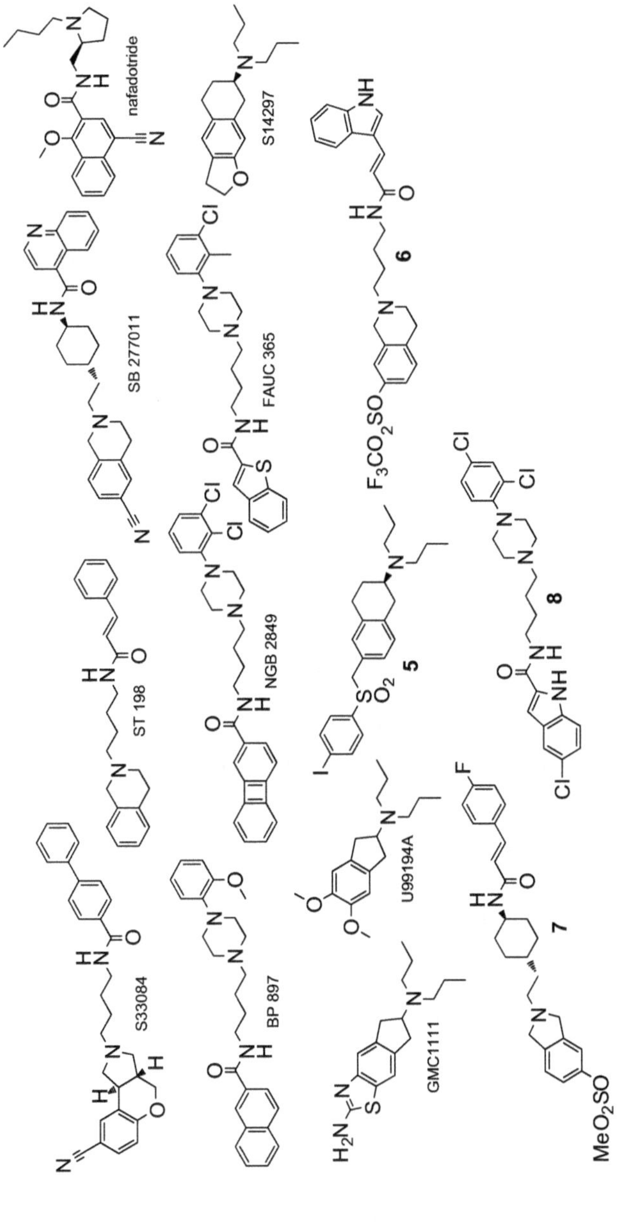

Figure 6.2 D_3 receptor-selective antagonists and partial agonists.

D_3 receptors indicate mixed, partial agonist/antagonist properties. BP 897 displays potent ($EC_{50} = 1$ nM) partial agonist (59% relative to quinpirole) activity at human D_3 receptors, assessed as a decrease in forskolin-induced cAMP synthesis.[73] On the other hand, BP 897 dose-dependently ($IC_{50} = 3.1$ nM) antagonizes the [^{35}S]GTPγS stimulating effect of DA.[74] Bioisosteric replacement of the naphthamide moiety of BP 897 with benzo[b]thiophene, produced FAUC 365, which displays antagonistic properties with subnanomolar affinity (0.50 nM) for D_3 receptors and remarkable selectivity ($D_2/D_3 = 7200$) over related receptors.[75] Further bioisosteric modification of the arylcarboxamide moiety of BP 897 led to the 5-chloro-indole-2-carboxamide derivative, compound **8**. Compound **8**, discovered to be a partial agonist in [^{35}S]GTPγS assays, displays subnanomolar D_3 affinity (0.38 nM) and excellent selectivity over D_2 ($D_2/D_3 > 26\,000$) and related receptors (D_1, $5HT_{1A}$, $\alpha_1 > 21\,000$-fold).[76]

6.3 Role of D_3 Receptors in Levodopa-Induced Dyskinesias

PD diagnosis typically occurs after substantial (70–80%) loss of dopaminergic neurons in the nigrostriatal pathway. Therefore, since deficits of >50% in the nigrostriatal pathway can be accommodated without the expression of PD symptoms, there must be compensatory mechanisms used by surviving dopaminergic neurons and other striatal cells to alleviate the effects of the progressive loss of dopaminergic innervations. Compensatory mechanisms might include changes in postsynaptic DA receptor density and/or sensitivity of specific DA receptor subtypes.[77] The circuits that connect the *striatum* to the output nuclei comprise the "direct" and "indirect" pathways. Neurons of the direct pathway express D_1 receptors and project to the *globus pallidus pars interna* and *substantia nigra pars reticula*. On the other hand, indirect pathway neurons express D_2 receptors and influence the output structures through a series of connections that involve the *globus pallidus par externa* and subthalamic nucleus.[78] Activation of the direct nigrostriatal pathway disinhibits thalamocortical neurons and aids motor activity, whereas activation of the indirect pallidostriatal pathway enhances inhibition on thalamocortical neurons and reduces motor activity.[79] These output pathways exert a dual action on the pallido–thalamo–cortical circuit, the balanced efforts of each contributing to co-ordinated movements in normal subjects.

Evidence suggests that both PD symptoms and LIDs arise from an imbalance between these two pathways.[78] The onset of LIDs is correlated with both over-reduction of the firing frequency of the internal segment *globus pallidus* (ISGP) and an abnormal firing pattern, which result in reduced inhibition of motor thalamic nuclei, and thus, over-activation of cortical motor areas.[80]

L-DOPA, the biosynthetic precursor to DA, is used therapeutically in PD patients to increase DA levels in the CNS.[81] Once administered, L-DOPA crosses the blood–brain barrier, where it is taken up into surviving DA neurons and is converted to DA. L-DOPA treatment initially reduces motor symptoms, but chronic use ultimately induces debilitating and pharmacoresistant involuntary movements.[82] Furthermore, the pharmacokinetic profile of L-DOPA varies significantly, causing abrupt transitions between "on" (active) and "off" (inactive) phases.[83] Recent studies have aimed to understand the pathophysiological changes that accompany LIDs. In 6-OHDA-lesioned rats, the motor-stimulating effect of L-DOPA, measured as contraversive rotations, becomes enhanced upon repeated intermittent administration. The intensity of response to an L-DOPA challenge increased three- to four-fold following chronic L-DOPA administration and progressively declined upon withdrawal of the treatment regime. The timecourse of the change in rotational behavior paralleled that of D_3 receptor protein appearance and disappearance in the caudate–putamen. Furthermore, SKF 38393, a selective D_1 receptor partial agonist, was found to produce similar, yet less extensive, D_3 receptor induction compared to L-DOPA. Moreover, SCH 23390, a selective D_1 receptor antagonist, completely prevented L-DOPA-mediated D_3 receptor induction (Figure 6.3).[84] In monkeys rendered Parkinsonian with 1-methyl-4-phenyl-1,2,3,6-tetrahydropyridine (MPTP), expression of D_3 receptors was found to decrease significantly (68%) in the caudate nucleus. MPTP-intoxicated monkeys receiving chronic doses of L-DOPA for several months resulted in LID symptoms in five of nine cases. Interestingly, D_3 receptor expression was found to be higher in the putamen (+154%) of MPTP-intoxicated monkeys with LID compared to non-dyskinetic and normal monkeys. Furthermore, D_3 receptor density in the putamen correlated with the incidence and severity of LIDs. PD-like symptoms and LIDs were associated with down- and up-regulation of D_3 receptor expression. Therefore, normalization of D_3 receptors in the putamen and ISGP of dyskinetic animals may account for the significant improvements in LIDs that have been noted with D_3 receptor-selective ligands.[68]

Interestingly, recent studies suggest that decreases in D_3 receptor expression in the 6-OHDA-lesioned rat model, MPTP-treated primate model and in PD patients, depends on deprivation of the anterogradely-transported factor, BDNF.[85] Furthermore, BDNF deprivation reduced D_3 expression selectively, as homologous D_1 and D_2 receptor expression was either unchanged, or only marginally decreased by 6-OHDA lesion.[25,84] During L-DOPA treatment of 6-OHDA-lesioned rats, infusion of a BDNF antagonist into the denervated *striatum* impaired both induction of D_3 mRNA and protein expression.[25] Moreover, the BDNF antagonist dose-dependently inhibited behavioral sensitization to L-DOPA, suggesting that behavioral sensitization is triggered by BDNF.[84] Changes in D_3 receptor expression, localization and function during chronic L-DOPA treatment, leading to both motor benefits and motor abnormalities, further implicate an essential role for D_3 receptors in the pathology of the PD brain and as a therapeutic target.

6.3.1 D$_3$ Receptor Modulation in the Treatment of Levodopa-Induced Dyskinesias

D$_3$-Selective ligands have been investigated for their potential in reducing LIDs. Recently, BP 897, a D$_3$ receptor-selective mixed agonist/antagonist, was studied *in vivo* to evaluate its ability to reduce LIDs. BP 897 was administered to dyskinetic, MPTP-intoxicated monkeys in combination with L-DOPA. Interestingly, BP 897 reduced LIDs by 66%, while having almost no effect on the motor recovery provided by L-DOPA. D$_3$ receptor-selective antagonists have also shown promise in their ability to reduce LID in PD animal models. The D$_3$ receptor-selective antagonist, ST 198 (K_i D$_3$ = 12 nM, D$_2$ = 780 nM, D$_2$/D$_3$ = 65), has been shown to significantly attenuate LIDs when given to dyskinetic monkeys in conjunction with L-DOPA, although this effect is accompanied by a re-appearance of PD-like symptoms and locomotor deficits. Studies have consistently implicated a role for D$_3$ receptors in both the generation of dyskinesia and the therapeutic effect of L-DOPA. The apparent deterioration in PD-like symptoms suggests an antagonism of L-DOPA, by ST 198, at D$_3$ receptors and possibly at D$_2$ receptors.[68] This indication contrasts with previous studies, in which neither D$_3$ receptor blockade, nor invalidation, impaired locomotor activity in normal rodents.[60,86,87] Thus, DA denervation, followed by treatment with L-DOPA may cause a re-distribution of the functions of D$_2$ and D$_3$ receptors.[68] A recent study has demonstrated that the D$_3$ receptor-selective antagonist, PG01037, is able to attenuate L-DOPA-dependent, abnormal involuntary movements (AIMs) in the unilaterally lesioned rat.[88] Furthermore, simultaneous administration of L-DOPA and PG01042, a D$_3$ receptor-selective agonist, caused attenuation of AIMs in a dose-dependent manner. Moreover, neither PG01037 nor PG01042 showed any deleterious side-effects on the motor co-ordination or agility of L-DOPA-treated animals.[81]

D$_3$-Selective ligands with varying intrinsic activity have been correlated in terms of their ability to attenuate L-DOPA-dependent AIMs in the 6-OHDA-lesioned rat model of PD. WC 10, intrinsically classified as a D$_3$ weak partial agonist/antagonist, attenuated AIMs (IC$_{50}$ = 6.6 mg kg^{-1}) when administered simultaneously with L-DOPA–benserazide (both 8 mg kg^{-1}), although some aspects of motor activity and co-ordination were negatively affected. Additionally, WC 10 dramatically and rapidly reduced AIMs at their most intense level when administered following L-DOPA–benserazide treatment. WC 44 (Figure 6.3), classified as a D$_3$ agonist and D$_2$ partial agonist, attenuated AIM scores (IC$_{50}$ = 5.5 mg kg^{-1}) dose-dependently, while locomotor activity was unaffected at low doses (3 mg kg^{-1}). Furthermore, the prolongation of involuntary movements, as seen with WC 10, was not observed in the temporal plot of WC 44. D$_3$ receptor-selective partial agonists, WC 21 and WC 26, each significantly (80 and 90% for WC 21 and WC 26, respectively) and potently reduced AIM scores (IC$_{50}$ = 3.8 and 5.8 mg kg^{-1}, respectively). The locomotor side-effects observed at high doses (10 mg kg^{-1}) of D$_3$-selective agonist, WC 44, may be attributed to coblockade of D$_2$ receptors. These

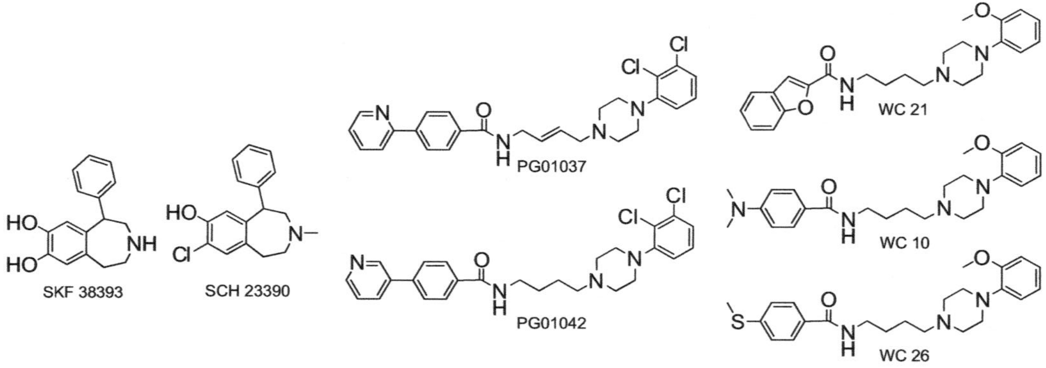

Figure 6.3 D$_3$ receptor-selective antagonists and partial agonists.

findings suggest that designing D_3 receptor-selective agonists with higher D_2/D_3 affinity ratios may be a feasible strategy to develop pharmacotherapeutic agents to treat LIDs.[77] In a similar study, the D_3 receptor-selective antagonist, nafadotride, was evaluated in 6-OHDA-lesioned, L-DOPA-induced, dyskinetic rats to examine its ability to reduce dyskinesias. Nafadotride (2 mg kg^{-1}) reduced all four distinct types of AIMs produced by L-DOPA. Additionally, nafadotride (Figure 6.2) did not reduce the anti-Parkinsonian effect of L-DOPA.

DA agonists directly activate postsynaptic DA receptors, bypassing the presynaptic synthesis of DA and the degenerating nigrostriatal system, avoiding the potential neurotoxicity and side-effects caused by L-DOPA.[78] A recent study compared the effects of L-DOPA and the D_3 receptor-preferring agonist, pramipexole, in early PD, indicating significantly fewer motor fluctuations and less development of dyskinesias in the pramipexole-treated group after two years. Therefore, it is hypothesized that D_3 receptor stimulation will not cause behavioral sensitization and/or dyskinesias in the absence of L-DOPA therapy.[28] Recent studies indicate that D_1 receptor expression levels increase significantly at the plasma membrane in animal models of LID.[89,90] Furthermore, when dyskinetic animals are co-administered L-DOPA and D_3 antagonist, ST 198, known to improve dyskinesia severity in rodent and non-human primate models of PD, membrane-bound D_1 receptor density is restored to normal levels.[90] Thus, D_3-mediated reduction in D_1 receptor expression in the *striatum* is proposed as a possible mechanism for the alleviation of dyskinesias provided by D_3-selective ligands. Current knowledge of LID pathophysiology suggests that effective anti-dyskinetic therapy should target the D_3/D_1 receptor signaling cascade, as targeting D_2 receptor signaling seems to produce anti-therapeutic effects deleterious to patients.[79]

6.4 Neuroprotective Action of D_3 Receptor-Preferring Agonists

An interesting development in the use of DA agonists for the treatment of PD is that some have proven to be neuroprotective.[28] DA agonists may exert their neuroprotective effects *via* direct scavenging of free radicals, up-regulating the activity of radical scavenging enzymes, stabilizing mitochondrial membrane potentials or by inhibiting apoptosis.[91–93] Clinical studies demonstrate that the D_3 receptor-preferring agonists, pramipexole and ropinirole, reduce PD disease progression *via* PET-imaging of nigrostriatal terminals.[94] Pramipexole (1 mg kg^{-1}) exhibits potent protection of DA neurons against MPTP-induced loss in mice.[95,96] Neuroprotection against MPTP was measured as a reduction in both striatal DA depletion and the loss of tyrosine hydroxylase (TH)-positive striatal neurons.[95,97,98] Pramipexole has recently been examined for its effect on MPTP-induced neurotoxicity in aged mice, as aged humans are predominantly affected by PD. Pramipexole significantly reduced the impact of MPTP on dopamine transporter (DAT)-labeling of DA fibers and TH-positive neurons of the *substantia nigra*.[31] In this case, neuroprotection offered by pramipexole was thought to originate in the induction of protective factors, such as the anti-apoptotic protein, Bcl-2.[99]

In primary, mesencephalic cultures, pramipexole (10 μM) was found to significantly protect cells from 1-methyl-4-phenylpyridinium (MPP$^+$)-induced neurotoxicity, with protection paralleling the time-dependent increment of glial-cell-derived neurotrophic factor (GDNF) and BDNF levels. Moreover, blockade of GDNF and BDNF with neutralizing antibodies, and D$_3$ receptors with D$_3$ receptor-preferring antagonists, negated the neuroprotection. Blockade of the neuroprotective effect of pramipexole, utilizing a D$_3$ receptor-preferring antagonist, indicates that the neurotrophic, and, in turn, neuroprotective action of pramipexole is D$_3$ receptor-mediated.[100]

6.4.1 D$_3$ Receptor-Independent Neuroprotection

Evidence from preclinical studies indicates that L-DOPA not only induces motor fluctuations, but also may be toxic to DA neurons.[101–103] Evidence of increased lipid peroxidation, free iron, decreased glutathione levels, altered expression of anti-oxidant enzymes and impaired mitochondrial function in the postmortem PD brain has led to the oxidant stress hypothesis. The oxidant stress hypothesis theorizes that a partial loss of nigral DA neurons leads to compensatory increases in DA synthesis, and therefore, to increases in oxidative metabolism of DA. L-DOPA metabolism may exacerbate oxidative stress and overwhelm natural anti-oxidant defenses, such as glutathione, catalase and superoxide dismutase, thereby furthering DA cell loss.[104] Although several studies have differing views regarding DA neurotoxicity, there has been increasing interest in developing drugs that reduce the biochemical abnormalities that accompany chronic L-DOPA therapy, and thus, the course of PD.[93,105,106] These hypotheses predict that DA agonists, which bypass oxidative DA metabolism, should be neuroprotective. Specifically, D$_3$ receptor-preferring agonists should decrease DA synthesis and release by D$_3$ autoreceptor activation, which would decrease DA metabolism and the overall ROS load. In rostral, mesencephalic cells, pramipexole and other D$_3$ receptor-preferring agonists have shown the ability to not only reverse the progressive loss of cells in culture over time, but also to increase cell proliferation. In contrast, D$_2$ agonists (U95666, bromocriptine and pergolide), along with D$_2$ and D$_3$ antagonists, did not alter the survival of tyrosine hydroxylase immunoreactive (THir) cells. L-DOPA neurotoxicity (TD$_{50}$ = 10–20 μM) is thought to occur in a receptor-independent fashion, as the D- and L-isomer of DOPA produce equivalent, dose-dependent reductions in THir cell count.[104,107] Studies indicate that the neuroprotective action of pramipexole against DA, hydrogen peroxide and 6-OHDA result, in part, from anti-oxidant mechanisms.[91,101] Furthermore, the inactive, R-(+) enantiomer of pramipexole was found to be as potent as the D$_3$ receptor-active, S-(−) enantiomer in inhibiting mitochondrial permeability transition, along with reducing caspase activity and apoptosis, suggesting that the neuroprotective mechanism of pramipexole may be receptor-independent.[108] However, the evidence for a neuroprotective benefit of pramipexole in the clinic is not clear at this point.

6.4.1.1 D_3 Receptor-Dependent Neuroprotection and Neurorestoration

Interestingly, studies also demonstrate the neuroprotective action of pramipexole in mesencephalic culture to be, in part, receptor-dependent.[104] Pramipexole has been shown to dose-dependently ($ED_{50} = 500$ pM) attenuate L-DOPA-induced, THir cell loss.[109] Furthermore, the neuroprotective effect of pramipexole was increased by co-incubation with D_3 receptor-preferring agonists (7-OH-DPAT and PD 128907), while D_3 receptor-preferring antagonists dose-dependently inhibited neuroprotection.[110] Expression of the anti-apoptotic protein Bcl-2 and its homologue, Bcl-X_1, increases in response to oxidative stress, inhibiting cell death.[111] Interestingly, treatment of primary mesencephalic cultures with pramipexole increased the expression of Bcl-X_1 and reduced the neurotoxicity of L-DOPA.[104] Evidence suggests that *via* increased Bcl-X_1 expression, pramipexole is able to stabilize the mitochondrial transition pore.[112] Additionally, pramipexole attenuates tumor necrosis factor-α (TNFα)-induced THir cell loss in mesencephalic cultures.[104] Interestingly, recent findings have indicated increased levels of TNFα and TNFα receptors in peripheral leukocytes, cerebral spinal fluid and the *substantia nigra* of PD patients.[113–116] Therefore, pramipexole may be able to block neurodegenerative actions of pro-inflammatory cytokines, in addition to its functions as a D_3 receptor-preferring agonist and anti-oxidant.[104]

An expanding area of neuroprotection research is that of the endogenous production of neurotrophic factors.[117] Striatal neurotrophic activity functions as a regulatory feedback mechanism designed to maintain the cytoarchitectural integrity of the nigrostriatal pathway.[118,119] BDNF, GDNF and fibroblast-derived growth factor (FGF) have been shown to protect DA neurons exposed to neurotoxins, such as 6-OHDA and MPP^+.[120,121] The mechanism by which neurotrophic factors exert their protective effect is proposed to involve alterations in pro-apoptotic signal transduction.[122] The limitation of neurotrophic factors as potential therapeutic candidates is their lack of CNS penetration. Elevation of endogenous neurotrophic factors by drugs that can penetrate the CNS and induce their natural production is an alternative approach to increase neurotrophic activity in the PD brain.[100] The recently discovered D_3 receptor-preferring agonist, D-264, has shown promise as a neuroprotective therapy in two *in vivo* PD animal models. In MPTP- and lactacystin-treated mice, in which degeneration of dopaminergic pathways is known to occur, pretreatment with D-264 has been shown to: (1) dose-dependently increase the number of TH-positive neurons; (2) significantly attenuate lactacystin-induced inhibition of proteasome activity; (3) inhibit pro-inflammatory, microglia activation; (4) dose-dependently reduce the activation of astrocytes; and (5) increase BDNF and GDNF levels. Moreover, the D_3 receptor-selective antagonist, U99194, significantly altered the neuroprotective effects of D-264, indicating a role for D_3 receptor activation in its neuroprotection.[123] Mesencephalic cells have been shown to constitutively produce a soluble trophic factor, termed the "DA autotrophic factor", which enhances the survival and growth of DA neurons.[124,125] Conditioned media transfer experiments have been performed to elucidate the trophic effect of several D_3 receptor-preferring agonists. Media of donor cultures exposed to either pramipexole, 7-OH-DPAT or PD-128907 was found to dose-dependently increase THir cell counts in recipient cultures.[110] Molecular analysis of the conditioned medium identified a 35 kDa protein that, when isolated and co-incubated with mesencephalic cultures, increased THir cell counts, confirming its neurotrophic effect.[126] Additionally, it is hypothesized that increased DA autotrophic factor may up-regulate $Bcl-X_1$ expression and lead to reduced apoptosis.[104]

D_3-Preferring agonists have also been explored for their ability to induce neurogenesis.[127–129] The existence of neural stem cells in the CNS of adult mammals has been demonstrated convincingly over several years.[130] Evidence suggests that the rate of adult neurogenesis can be modulated and the differentiation of precursor cells into neurons can be increased; findings that are relevant to strategies that aim to restore dopaminergic pathways in PD.[130,131] Interestingly, D_3 receptor-preferring agonists were found to increase subventricular zone (SVZ) cell numbers *in vivo*, *via* recruitment of D_3 receptors, an effect that reflects enhanced mitogenesis rather than decreased apoptosis. Following intrasubventricular (ICV) infusion of 7-OH-DPAT, cell proliferation doubled in the SVZ and rostral migratory stream, whereas infusion of D_1 receptor agonist, SKF 82958, was found to be ineffective. Co-infusion of 7-OH-DPAT along with the D_3 receptor-selective antagonist, SB 277011A, blocked the increase of bromodeoxyuridine (BrdU)-positive cells, consistent with a role for D_3 receptors in mediating cell-proliferative actions of D_3 agonists.[127] D_3 receptors are found on dopaminergic neurons of the *substantia nigra*, which, in the adult, contains progenitor cells with neurogenic potential.[132–134] Recent studies have shown that a two-week infusion of 7-OH-DPAT into the third ventricle causes a six-fold elevation of BrdU-labeled cells in the ventricular lining and *substantia nigra pars compacta*. The BrdU-positive cells expressed a neuronal phenotype and were colabeled with TH, indicating dopaminergic characteristics.[128] Furthermore, following ICV infusion of 7-OH-DPAT to 6-OHDA-lesioned rats, BrdU-positive cells increased two-fold in the *substantia nigra*, ipsilateral to the lesion. Moreover, the newly synthesized cells developed a mature, neuronal phenotype and their increase correlated with a time-dependent elevation in striatal innervations. Finally, these changes were paralleled by an improvement in motor function on the lesioned side. These findings suggest that D_3 receptor-preferring agonists could provide a disease-modifying effect by partially restoring the nigrostriatal pathway in the PD brain and further implicate the role of D_3 receptor activation in neuroprotective and neurorestorative therapy for PD.[129]

6.5 Conclusion

The D_3 receptor has gained much attention as a potential therapeutic target for the treatment of PD. The localization of D_3 receptors in the limbic and nigrostriatal brain regions make it a promising target to treat PD, as it is known to contribute to the control of movement, emotion and cognition. Although D_2 receptor stimulation is thought to be necessary for the relief of motor symptoms, the majority of DA agonists used in the treatment of PD have equal or higher affinity for the D_3 receptor. The development of selective pharmacological tools to investigate the role of D_3 receptors in PD has been challenging due to the sequence similarity between D_2 and D_3 receptors. However, D_3 receptor ligands with varying intrinsic activity and selectivity over D_2 and related receptor subtypes have been developed. L-DOPA, the endogenous precursor to DA, is used therapeutically to increase DA levels in the CNS of PD patients. L-DOPA treatment, while initially reducing motor symptoms, ultimately causes involuntary movements, termed L-DOPA-induced dyskinesia (LIDs). Interestingly, the D_3 receptor has been shown to play a role not only in the development of LIDs, but also in their treatment. D_3 receptor-selective ligands of varying intrinsic activity have shown the ability to reduce LIDs in PD animal models. Furthermore, D_3 receptor-preferring agonists have shown promise as a neuroprotective PD therapy. These molecules have shown the ability to protect DA neurons from neurotoxic insults and may act *via* receptor-independent and receptor-dependent mechanisms. Therefore, due to accumulating evidence suggesting the D_3 receptor as a therapeutic target in PD, D_3 receptor-selective ligands may act as disease-modifying therapy, treating both the symptoms and progression of the disease.

Acknowledgements

This work was supported by National Institute of Neurological Disorders and Stroke/National Institute of Health (NS047198, AKD).

References

1. F. Boeckler and P. Gmeiner, *Pharmacol. Ther.*, 2006, **112**, 281.
2. P. Seeman and D. Grigoriadis, *Neurochem. Int.*, 1987, **10**, 1.
3. L. Vallar and J. Meldolesi, *Trends Pharmacol. Sci.*, 1989, **10**, 74.
4. B. Levant in *CNS Neurotransmitters and Neuromodulators: Dopamine*, ed. T. W. Stone, CRC Press, Boca Raton, 1996, p. 77.
5. J. R. Bunzow, H. H. M. Van Tol, D. K. Granoly, P. Albert, J. Salon, M. Christie, C. A. Machida, K. A. Neve and O. Civelli, *Nature*, 1988, **336**, 783.
6. F. J. Monsma, L. C. Mahan, L. D. McVittie, C. R. Gerfen and D. R. Sibley, *Proc. Natl. Acad. Sci. U. S. A.*, 1990, **87**, 6723.

7. Q.-Y. Zhou, D. K. Grandy, L. Thambi, J. A. Kushner, H. H. M. Van Tol, R. Cone, D. Pribnow, J. Salon, J. R. Bunzow and O. Civelli, *Nature*, 1990, **347**, 76.
8. P. Sokoloff, B. Giros, M. P. Martres, M. L. Bouthenet and J. C. Schwartz, *Nature*, 1990, **347**, 146.
9. C. L. Chio, M. E. Lajiness and R. M. Huff, *Mol. Pharmacol.*, 1994, **45**, 51.
10. B. A. Cox, M. P. Rosser, M. R. Kozlowski, K. M. Duwe, R. L. Neve and K. A. Neve, *Synapse*, 1995, **21**, 1.
11. C. Pilon, D. Lévesque, V. Dimitriadou, N. Griffon, M. P. Martres, J. C. Schwartz and P. Sokoloff, *Eur. J. Pharmacol.*, 1994, **268**, 129.
12. M. N. Potenza, G. F. Graminski, C. Schmauss and M. R. Lerner, *J. Neurosci.*, 1994, **14**, 1463.
13. G. R. Seabrook, J. A. Kemp, S. B. Freedman, S. Patel, H. A. Sinclair and G. McAllister, *Br. J. Pharmacol.*, 1994, **111**, 391.
14. L. X. Liu, L. H. Burgess, A. M. Gonzalez, D. R. Sibley and L. A. Chiodo, *Synapse*, 1996, **24**, 156.
15. P. Werner, N. Hussy, G. Buell, M. A. Jones and R. A. North, *Mol. Pharmacol.*, 1996, **49**, 656.
16. B. Levant, *Pharmacol. Rev.*, 1997, **49**, 231.
17. C. D. Livingstone, *Biochem. J.*, 1992, **287**, 277.
18. M. L. Bouthenet, E. Souil, M. P. Martres, P. Sokoloff, B. Giros and J. C. Schwartz, *Brain Res.*, 1991, **564**, 203.
19. G. Mengod, M. T. Villaro, G. B. Landwehrmeyer, M. I. Martinez-Mir, H. B. Niznik, R. K. Sunahara, P. Seeman, B. F. O'Dowd, A. Probst and J. M. Palacios, *Neurochem. Int.*, 1992, **20**, 33.
20. B. Landwehrmeyer, G. Mengod and J. M. Palacios, *Eur. J. Neurosci.*, 1993, **5**, 145.
21. J. Diaz, C. Pilon, B. Le Foll, C. Gros, A. Triller and J. C. Schwartz, *Neuroscience*, 1995, **65**, 731.
22. L. Herroelen, J.-P. De Backer, N. Wilczak, A. Flamez, G. Vauquelin and J. De Keyser, *Brain Res.*, 1994, **648**, 222.
23. A. Murray, H. L. Ryoo, E. V. Gurevich and J. N. Joyce, *Proc. Natl. Acad. Sci. U. S. A.*, 1994, **91**, 11271.
24. R. A. Lahti, R. C. Roberts and C. A. Tamminga, *NeuroReport*, 1995, **6**, 2505.
25. O. Guillin, J. Diaz, P. Carroll, N. Griffon, J. C. Schwartz and P. Sokoloff, *Nature*, 2001, **411**, 86.
26. P. Sokoloff, J. Diaz, B. Le Foll, O. Guillin, L. Leriche, E. Bezard and C. Gross, *CNS Neurol. Disord. : Drug Targets*, 2006, **5**, 25.
27. A. Zapata and T. S. Shippenberg, *Neuropharmacology*, 2001, **41**, 351.
28. J. N. Joyce and H. Thomas, *Pharmacol. Ther.*, 2001, **90**, 231.
29. M. Van den Buuse and A. C. Lambrechts, *Eur. J. Pharmacol.*, 1993, **243**, 169.
30. H. L. Ryoo, D. Pierrotti and J. N. Joyce, *Mov. Disord.*, 1998, **13**, 788.
31. J. N. Joyce, H. Ryoo, E. C. Gurecivh, C. Adler and T. Beach, *Parkinsonism Relat. Disord.*, 2001, **7**, 225.

32. J. N. Joyce and J. H. Meador-Woodruff, *Neuropsychopharmacology*, 1997, **16**, 375.
33. B. Landwehrmeyer, G. Mengod and J. M. Palacios, *Brain Res. Mol. Brain Res.*, 1993, **18**, 187.
34. P. Seeman, A. Wilson, P. Gmeiner and S. Kapur, *Synapse*, 2006, **60**, 205.
35. F. Boeckler and H. Lanig, *J. Med. Chem.*, 2005, **48**, 694.
36. G. H. Gonnet, M. A. Cohen and S. A. Benner, *Science*, 1992, **256**, 1443.
37. K. Palczewski, T. Kumasaka, T. Hori, C. A. Behnke, H. Motoshima, B. A. Fox, I. LeTrong, D. C. Teller, T. Okada, R. E. Stenkamp, M. Yamamoto and M. Miyano, *Science*, 2000, **289**, 739.
38. D. C. Teller, C. A. Okada, K. Behnke, R. E. Palczewski and R. E. Stenkamp, *Biochemistry*, 2001, **40**, 7761.
39. T. Okada, Y. Fujiyoshi, M. Silow, J. Navarro, E. M. Landau and Y. Shichida, *Proc. Natl. Acad. Sci. U. S. A.*, 2002, **99**, 5982.
40. P. Sokoloff, B. Giros, M. P. Martres, M. L. Bouthenet and J. C. Schwartz, *Nature*, 1990, **347**, 146.
41. P. Sokoloff, M. Andrieux, R. Besancon, C. Pilon, M. P. Martres, B. Giros and J. C. Schwartz, *Eur. J. Pharmacol.*, 1992, **225**, 331.
42. L. A. van Vliet, P. G. Tepper, D. Dijkstra, G. Damsma, H. Wikström, T. A. Pugsley, H. C. Akunne, T. G. Heffner, S. A. Glase and L. D. Wise, *J. Med. Chem.*, 1996, **39**, 4233.
43. J. Mierau, F. J. Schneider, H. A. Ensinger, C. L. Chio, M. E. Lajiness and R. M. Huff, *Eur. J. Pharmacol.*, 1995, **290**, 29.
44. F. Sautel, N. Griffon, D. Lévesque, C. Pilon, J. C. Schwartz and P. Sokoloff, *NeuroReport*, 1995, **6**, 329.
45. S. Perachon, J. C. Schwartz and P. Sokoloff, *Eur. J. Pharmacol.*, 1999, **366**, 293.
46. A. J. Lees and G. M. Stern, *J. Neurol., Neurosurg. Psychiatry*, 1981, **44**, 1020.
47. U. K. Rinne, F. Bracco, C. Chouza, E. Dupont, O. Gershanik, J. F. Marti Masso, J. L. Montastruc and C. D. Marsden, *Drugs*, 1998, **55**(1), 23.
48. O. Rascol, D. J. Brooks, E. R. Brunt, A. D. Korczyn, W. H. Poewe and F. Stocchi, *Mov. Disord.*, 1998, **13**, 39.
49. A. K. Dutta, M. E. Reith and X. S. Fei, *Bioorg. Med. Chem. Lett.*, 2002, **12**, 619.
50. A. K. Dutta, S. K. Venkataraman, X. S. Fei, R. Kolhatkar, S. Zhang and M. Reith, *Bioorg. Med. Chem.*, 2004, **12**, 4361.
51. S. Biswas, F. Fernando, S. Zhang, J. Zhen, M. Reith and A. K. Dutta, *J. Med. Chem.*, 2008, **51**, 3005.
52. M. Johnson, T. Antonio, M. E. Reith and A. K. Dutta, *J. Med. Chem.*, 2012, **55**, 5826.
53. J. Chen, G. T. Collins, J. Zhang, C. Y. Yang, B. Levant, J. Woods and S. Wang, *J. Med. Chem.*, 2008, **51**, 5905.
54. M. Bergauer, H. Hubner and P. Gmeiner, *Tetrahedron*, 2004, **60**, 1197.
55. J. Elsner, F. W. Boeckler, F. W. Heinemann, H. Hubner and P. Gmeiner, *J. Med. Chem.*, 2005, **48**, 5771.

56. K. Y. Avenell, I. Boyfield, M. S. Hadley, C. N. Johnson, D. J. Nash, G. J. Riley and G. Stemp, *Bioorg. Med. Chem. Lett.*, 1999, **9**, 2715.
57. M. J. Millan, J. L. Peglion, J. Vian, J. M. Rivet, M. Brocco, A. Gobert, A. Newman-Tancredi, C. Dacquet, K. Bervoets and S. Girardon, *Eur. J. Pharmacol.*, 1994, **260**, R3.
58. J. M. Rivet, V. Audinot, A. Gobert, J. L. Peglion and M. J. Millan, *Eur. J. Pharmacol.*, 1994, **265**, 175.
59. P. J. Murray, L. A. Harrison, M. R. Johnson, G. M. Robertson, D. I. C. Scopes, M. Stokes, S. Wadman, J. W. E. F. Whitehead, A. G. Hayes, G. J. Kilpatrick, C. Large, C. M. Stubbs and M. P. Turpin, *Bioorg. Med. Chem. Lett*, 1996, **6**, 403.
60. M. J. Millan, A. Gobert, A. Newman-Tancredi, F. Lejeune, D. Cussac, J. M. Rivet, V. Audinot, T. Dubuffet and G. Lavielle, *J. Pharmacol. Exp. Ther.*, 2000, **293**, 1048.
61. T. Dubuffet, A. Newman-Tancredi, D. Cussac, V. Audinot, A. Loutz and M. J. Millan, *Bioorg. Med. Chem. Lett.*, 1999, **9**, 2059.
62. S. R. Haadsma-Svensson, *J. Med. Chem.*, 2001, **44**, 4716.
63. S. R. Haadsma-Svensson and K. A. Svensson, *CNS Drug Rev.*, 1998, **4**, 42.
64. L. A. van Vliet, N. Rodenhuis, H. Wikström, T. A. Pugsley, K. A. Serpa, L. T. Meltzer, T. G. Meffner, L. D. Wise, M. E. Lajiness, R. M. Huff, K. Scensson, G. R. Haenen and A. Bast, *J. Med. Chem.*, 2000, **43**, 3549.
65. R. Klintenberg, J. Arts, M. Jongsma, H. Wikström, L. Gunne and P. E. Andren, *Eur. J. Pharmacol.*, 2003, **459**, 231.
66. N. E. Austin, K. Y. Avenell and I. Boyfield, *Bioorg. Med. Chem. Lett.*, 1999, **9**, 179.
67. U. R. Mach, A. E. Hackling, S. Perachon, S. Ferry, C. G. Wermuth, J. C. Schwartz, P. Sokoloff and H. Stark, *ChemBioChem*, 2004, **5**, 508.
68. E. Bezard, C. Gross and J. M. Brotchie, *Nat. Med.*, 2003, **9**, 762.
69. G. Stemp, T. Ashmeade, C. L. Branch, M. S. Hadley, A. J. Hunter, C. N. Johnson, D. J. Nash, K. M. Thewlis, A. K. Vong, N. E. Austin, P. Jeffrey, K. Y. Avenell, I. Boyfield, J. J. Hagan, D. N. Middlemiss, A. Reavill, G. J. Riley, C. Routledge and M. Wood, *J. Med. Chem.*, 2000, **43**, 1878.
70. N. E. Austin, K. Y. Avenell and I. Boyfield, *Bioorg. Med. Chem. Lett.*, 2001, **11**, 685.
71. F. Saute, N. Griffon, P. Sokoloff, J. C. Schwartz, C. Launay, P. Simon, J. Costentin, A. Schoenfelder, F. Garrido and A. Mann, *J. Pharmacol. Exp. Ther.*, 1995, **275**, 1239.
72. J. Yuan, X. Chen, R. Brodbeck, R. Primus, J. Braun, J. W. Wasley and A. Thurkauf, *Bioorg. Med. Chem. Lett.*, 1998, **8**, 2715.
73. M. Pilla, S. Perachon, F. Sautel, F. Garrido, A. Mann, C. G. Wermuth, J. C. Schwartz, B. J. Everitt and P. Sokoloff, *Nature*, 1999, **400**, 371.
74. K. Wicke and J. Garcia-Ladona, *Eur. J. Pharmacol.*, 2001, **424**, 85.
75. L. Bettinetti, K. Schlotter, H. Hubner and P. Gmeiner, *J. Med. Chem.*, 2002, **45**, 4594.

76. G. Campiani, S. Butini, F. Trotta, C. Fattorusso, B. Catalanotti, F. Aiello, S. Gemma, V. Nacci, E. Novellino, J. A. Stark, A. Cagnotto, E. Fumagalli, F. Carnovali, L. Cervo and T. Mennini, *J. Med. Chem.*, 2003, **46**, 3822.
77. R. Kumar, L. R. Riddle, S. A. Griffin, W. Chu, S. Vangveravong, J. Neisewander, R. H. Mach and R. R. Luedtke, *Neuropharmacology*, 2009, **56**, 956.
78. C. Monville, E. M. Torres and S. B. Dunnett, *Brain Res. Bull.*, 2005, **68**, 16.
79. A. Berthet and E. Bezard, *Parkinsonism Relat. Disord.*, 2009, **15S**, S8.
80. T. Boraud, E. Bezard, B. Bioulac and C. Gross, *Brain*, 2001, **124**, 546.
81. L. R. Riddle, R. Kumar, S. A. Griffin, P. Grunt, A. H. Newman and R. R. Luedtke, *Neuropharmacology*, 2011, **60**, 284.
82. E. Bezard, J. M. Brotchie and C. E. Gross, *Nat. Rev. Neurosci.*, 2001, **2**, 577.
83. J. N. Joyce and M. J. Millan, *Curr. Opin. Pharmacol.*, 2007, **7**, 100.
84. R. Bordet, S. Ridray, S. Carboni, J. Diaz, P. Sokoloff and J. C. Schwartz, *Proc. Natl. Acad. Sci. U. S. A.*, 1997, **94**, 3363.
85. D. Lévesque, M. P. Martres, J. Diaz, N. Griffon, C. H. Lammers, P. Sokoloff and J. C. Schwartz, *Proc. Natl. Acad. Sci. U. S. A.*, 1995, **92**, 1719.
86. C. Reavill, S. G. Taylor, M. D. Wood, T. Ashmeade, N. E. Austin, K. Y. Avenell, I. Boyfield, C. L. Branch, J. Cilia, M. C. Coldwell, M. S. Hadley, A. J. Hunter, P. Jeffrey, F. Jewitt, C. N. Johnson, D. N. Jones, A. D. Medhurst, D. N. Middlemiss, D. J. Nash, G. J. Riley, C. Routledge, G. Stemp, K. M. Thewlis, B. Trail, A. K. Vong and J. J. Hagan, *J. Pharmacol. Exp. Ther.*, 2000, **294**, 1154.
87. D. Accili, C. S. Fishburn, J. Drago, H. Steiner, J. E. Lachnowicz, B. H. Park, E. B. Gauda, E. J. Lee, M. H. Cool, D. R. Sibley, C. R. Gerfen, H. Westphal and S. Fuchs, *Proc. Natl. Acad. Sci. U. S. A.*, 1996, **93**, 1945.
88. R. Kumar, L. Riddle, S. A. Griffin, P. Grundt, A. H. Newman and R. R. Luedtke, *Neuropharmacology*, 2009, **56**, 944.
89. C. Guigoni, E. Doudnikoff, Q. Li, B. Bloch and E. Bezard, *Neurobiol. Disord.*, 2007, **26**, 452.
90. A. Berthet, G. Porras, E. Doudnikoff, H. Stark, M. Cador, E. Bezard and B. Bloch, *J. Neurosci.*, 2009, **29**, 4829.
91. L. Zou, J. Jankovic, D. B. Rowe, W. Xie, S. H. Appel and W. Le, *Life Sci.*, 1999, **64**, 1275.
92. W. D. Le and J. Jankovic, *Drugs Aging*, 2001, **18**, 389.
93. A. H. Schapira, D. P. Mikhailidis and J. Davar, *J. Am. Med. Assoc.*, 2004, **291**, 358.
94. C. E. Clarke and M. M. Guttman, *Lancet*, 2002, **360**, 1767.
95. Y. Kitamura, Y. Kohno, M. Nakazawa and Y. Nomura, *Jpn. J. Pharmacol.*, 1997, **74**, 51.
96. L. Zou, J. Xu, J. Jankovic, Y. He, S. H. Appel and W. Le, *Neurosci. Lett.*, 2000, **281**, 167.

97. K. W. Lange, W. D. Rausch, W. Gsell, M. Naumann, E. Oestreicher and P. Riederer, *Transm. Suppl.*, 1994, **43**, 183.
98. D. Muralikrishnan and K. P. Mohanakumar, *FASEB J.*, 1998, **12**, 905.
99. K. Takata, Y. Kitamura, J. Kakimura, Y. Kohno and T. Taniguchi, *Brain Res.*, 2000, **872**, 236.
100. F. Du, R. Li, Y. Huang, X. Li and W. Le, *Eur. J. Neurosci.*, 2005, **22**, 2422.
101. W. D. Le, J. Jankovic, W. Xie and S. H. Appel, *J. Neural Transm.*, 2000, **107**, 1165.
102. P. Barone, *Neurology*, 2003, **61**, S12.
103. R. Tintner and J. Jankovic, *Expert Opin. Invest. Drugs*, 2003, **12**, 1803.
104. P. M. Carvey, S. O. McGuire and Z. D. Ling, *Parkinsonism Relat. Disord.*, 2001, **7**, 213.
105. R. M. Kostrzewa, P. Nowak, J. P. Kostrzewa, R. A. Kostrzewa and R. Brus, *Amino Acids*, 2005, **28**, 157.
106. E. Fernandez-Espejo, *Mol. Neurobiol.*, 2004, **29**, 15.
107. Z. D. Ling, S. C. Pieri and P. M. Carvey, *Clin. Neuropharmacol.*, 1996, **19**, 360.
108. N. A. Abramova, D. S. Cassarino, S. M. Khan, T. W. Painter and J. P. Bennett, *J. Neurosci. Res.*, 2002, **67**, 494.
109. P. M. Carvey, S. Pieri and Z. D. Ling, *J. Neural Transm.*, 1997, **104**, 209.
110. Z. D. Ling, H. C. Robie, C. W. Tong and P. M. Carvey, *J. Pharmacol. Exp. Ther.*, 1999, **289**, 202.
111. D. E. Bredesen, *Ann. Neurol.*, 1995, **38**, 839.
112. J. P. Bennett and M. F. Piercey, *J. Neurol. Sci.*, 1999, **163**, 25.
113. G. P. Boka, P. Anglade, D. Wallach, F. Javoy-Agid, Y. Agid and E. C. Hirsch, *Neurosci. Lett.*, 1994, **172**, 151.
114. P. Bongioanni, G. A. Fontana and T. Pantaleo, *Neurodegeneration*, 1996, **5**, 351.
115. B. Brugg, P. P. Michel, Y. Agid and M. Ruberg, *J. Neurochem.*, 1996, **66**, 733.
116. S. Hunot, B. Brugg, D. Ricard, P. P. Michel, M. P. Muriel, M. Ruberg, A. B. Faucheuz, Y. Agid and E. C. Hirsch, *Proc. Natl. Acad. Sci. U. S. A.*, 1997, **93**, 7531.
117. T. J. Collier and C. E. Sortwell, *Drugs Aging*, 1999, **4**, 261.
118. P. M. Carvey, L. R. Ptak, D. Lin, E. S. Lo, C. M. Buhrfiend, G. E. Drucker and J. Z. Fields, *Biochem. Behav.*, 1993, **46**, 195.
119. P. M. Carvey, L. R. Ptak, S. T. Nath, D. K. Sierens, E. J. Mufson, C. G. Goetz and H. L. Klawans, *Exp. Neurol.*, 1993, **120**, 149.
120. K. Krobert, I. Lopez-Colberg and L. A. Cunningham, *Exp. Neurol.*, 1997, **145**, 511.
121. P. B. Kirschner, B. G. Jenkins, J. B. Schulz, S. P. Finkelstein, R. T. Matthews, B. R. Rosen and M. F. Beal, *Brain Res.*, 1996, **713**, 178.
122. L. Yang, R. T. Matthews, J. B. Schulz, T. Klickgether, A. W. Liao, J. C. Martinou, J. B. Penney, B. T. Hyman and M. F. Beal, *J. Neurosci.*, 1998, **18**, 8145.

123. C. Li, S. Biswas, X. Li, A. K. Dutta and W. Le, *J. Neurosci. Res.*, 2010, **88**, 2513.
124. R. Dal Toso, O. Giorgi, C. Soranzo, G. Kirschner, G. Ferrari, M. Favaron, D. Benvegnu, D. Presti, S. Vincini and G. Toffano, *J. Neurosci.*, 1988, **8**, 733.
125. A. Prochiantz, M. C. Daguet, U. DiPorzio, A. Herbert and J. Glowinski, *Prog. Brain Res.*, 1983, **58**, 365.
126. Z. D. Ling, C. W. Tong and P. M. Carvey, *Brain Res.*, 1998, **791**, 137.
127. J. M. Van Kampen, T. Hagg and H. A. Robertson, *Eur. J. Neurosci.*, 2004, **19**, 2377.
128. J. M. Van Kampen and H. A. Robertson, *Neuroscience*, 2005, **136**, 381.
129. J. M. Van Kampen and C. B. Eckman, *J. Neurosci.*, 2006, **26**, 7272.
130. P. Ferretti, *Curr. Neurovasc. Res.*, 2004, **1**, 215.
131. E. Gould and C. G. Gross, *J. Neurosci.*, 2002, **22**, 619.
132. J. Diaz, C. Pilon, B. Foll, C. Gros, A. Triller, J. C. Schwartz and P. Sokoloff, *J. Neurosci.*, 2000, **20**, 8677.
133. D. C. Lie, G. Dziewczapolski, A. R. Willhoite, B. K. Kaspar, C. W. Shults and F. H. Gage, *J. Neurosci.*, 2002, **22**, 6639.
134. M. Zhao, S. Momma, K. Delfani, M. Carlen, R. M. Cassidy, C. B. Johansson, H. Brismar, O. Shupliakov, J. Frisen and A. M. Janson, *Proc. Natl. Acad. Sci. U. S. A.*, 2003, **100**, 7925.

CHAPTER 7
Protein Phosphatases in Parkinson's Disease

PETR HENEBERG

Third Faculty of Medicine, Charles University in Prague, Prague, Czech Republic
Email: petr.heneberg@lf3.cuni.cz

7.1 Introduction

Several pieces of evidence suggest the involvement of a wide spectrum of tyrosine, serine/threonine and lipid phosphatases in Parkinson's disease (PD). Here we review the currently available data. However, experimental evidence is still very limited and the genome-wide sequencing or proteomic analyses of dorsal motor nuclei and of the olfactory bulbs and nuclei may identify more phosphatases involved in the pathophysiology of PD.

7.2 Protein Tyrosine Phosphatases

The involvement of protein tyrosine phosphatases in the pathophysiology of PD is well illustrated by the effect of the general protein tyrosine phosphatase inhibitor, peroxovanadium (potassium bisperoxo(1,10-phenanthroline)-oxovanadate(V) also known as bpV(phen) and pervanadate). When applied at 3–10 µM concentration for two weeks close to the *substantia nigra* starting immediately after a unilateral moderate injury by injection of 6-hydroxydopamine into the rat mid-brain, 75% of pervanadate-treated neurons survived compared to only 45% in vehicle-infused rats, while degeneration of the dopaminergic projections to the *neostriatum* was also reduced.[1] The effects of pervanadate are known to synergize with those of brain-derived neurotrophic factor.[2] However, our understanding of the role of particular protein tyrosine phosphatases in the onset and progression of PD is limited. Below is summarized our current knowledge on this subject, highlighting the signaling events with potential for pharmacotherapeutic targeting.

7.2.1 Receptor Protein Tyrosine Phosphatase β/ζ

Receptor protein tyrosine phosphatase β/ζ (RPTPβ/ζ, PTPRB, PTPRZ) plays an important role in the neural system since it mediates the protection of oligodendrocytes from demyelination, provides protection from muscular sclerosis (autoimmune encephalomyelitis), and serves as a target for *Helicobacter pylori* VacA protein (of its m1 family).[3,4]

Several pieces of evidence suggest involvement of RPTPβ/ζ in PD. RPTPβ/ζ participates in downstream signaling of pleiotrophin and midkine, the key factors for survival of the injured dopaminergic neurons *in vitro* and *in vivo* which are known to be up-regulated in the *substantia nigra* of subjects with PD. These two growth factors bind RPTPβ/ζ and cause its dimerization and inactivation.[5–7] Inactivated RPTPβ/ζ allows increased phosphorylation of its various substrates including β-catenin, β-adducin, p190 RhoGAP, Magi 1, Fyn kinase, ALK kinase, TrkA and GIT1/Cat-1.[8,9] Of note is that ALK kinase also serves as a direct substrate of pleiotrophin, but ALK dephosphorylation by RPTPβ/ζ does not affect this alternative pleiotrophin signaling pathway.[10] In experimental animals, the RPTPβ/ζ expression is up-regulated in the *striatum* (but not in the mesencephalon) of lesioned rats and further increased by L-3,4-dihydroxyphenylalanine treatment.[11]

RPTPβ/ζ activity can be inhibited by 4-trifluoromethylsulfonylbenzyl 4-trifluoromethylsulonylphenyl ether. This trifluoromethyl sulfonyl compound inhibits RPTPβ/ζ with IC_{50} 3.5 μM, being marginally (two-fold) selective over RPTPε, almost 10-fold selective over RPTPσ, and by more than one order of magnitude selective over SHP-2, PTP1B, PTP-MEG2 and RPTPμ.[12]

Several thiophosphotyrosylated peptides serve also as selective RPTPβ/ζ inhibitors. Among them, the phospholipase-Cγ-derived TAEPDpYGALYE peptide has K_i 3 μM (K_M 1 μM) against RPTPβ/ζ, being selective over LAR-D1 and CD45 by two orders of magnitude. Compared to the full-length substrate, the affinity of RPTPβ/ζ for this peptide was three-fold lower. Similar substrate-derived peptides were successfully developed from insulin receptor and Src, however their affinities for RPTPβ/ζ were substantially lower.[13]

RPTPβ/ζ migrates occasionally to the lipid rafts (*e.g.*, upon binding of the VacA toxin), the migration can be inhibited by a common cholesterol-sequestering drug, methyl-β-cyclodextrin.[14] Treatment with *O*-glycosidase or formation of a double-point mutant T748A, T749A in the QTTQP motif of RPTPβ/ζ leads to a decrease in VacA binding. The VacA–RPTPβ/ζ binding motif should be of pharmacotherapeutic interest since it promises high selectivity for RPTPβ/ζ and RPTPα.[4]

Receptor isoforms of RPTPβ/ζ can be cleaved by metalloproteinases and presenilin/γ-secretase, causing the release of the intracellular part of RPTPβ/ζ into the cytosol.[15]

7.2.2 Protein Tyrosine Phosphatase PTP-PEST

PTP-PEST (PTPN12) is largely known for its ability to regulate cell migration and immune cell signaling. However, PTP-PEST was also shown to bind N-terminal SH3 domains of the CAS family proteins, including NEDD9.[16] NEDD9 promotes the outgrowth of neurites and of neural progenitor populations, and naturally occurring polymorphisms in the NEDD9 promoter region have been linked to susceptibility to both Alzheimer's and Parkinson's diseases.[17,18] The link between PD and NEDD9 (and subsequently PTP-PEST) is still questionable since the most recent study on this topic failed to confirm the predicted association.[19] Despite the fact that direct linkage requires further corroboration, the NEDD9 involvement in reversal of neurodegenerative states is clear-cut, and NEDD9 mRNA is also known to be substantially down-regulated in response to TREM2 loss of function in dementia.[18,20]

Selective inhibition of PTP-PEST can be reached when using silencing RNAs (such as those reported by Chellaiah *et al.*[21]) or anti-sense N-terminal cDNA.[22] PTP-PEST activity can be inhibited by caspase-3, which cleaves PTP-PEST specifically at the DSPD motif, and thus restricts association of PTP-PEST with its interaction partners.[23] Auranofin and auranofin-derived gold(I) N-heterocyclic carbene complexes inhibit PTP-PEST with IC_{50} values of 9 and 7 μM, respectively. They are selective over PTPs from other families (*e.g.*, CD45 or TC-PTP), but are only marginally selective over other members of the PTP-PEST family (IC_{50} values against PEP are 200 and 11 μM, respectively).[24]

Importantly, the effects of PTP-PEST on NEDD9 signaling are affected by several other phosphatases. Among them is SHP-2, which binds NEDD9 at focal adhesion, inhibits extracellular matrix-induced NEDD9 phosphorylation, and thus restricts NEDD9-dependent cell migration.[25] Also serine/threonine phosphatase PP2A binds and dephosphorylates NEDD1, which leads to the de-adhesion-induced loss of apparent M_r of NEDD9 from 115 to 105 kDa.[26] Among the confirmed targets of PP2A on NEDD1 is Ser369.[27]

7.2.3 Striatum-Enriched Protein Tyrosine Phosphatase

Striatum-enriched protein tyrosine phosphatase (STEP, PTPN5) is preferentially expressed in the neurons of the central nervous system, where it regulates dopaminergic and glutaminergic neurotransmission.[3] Among the major STEP targets are ERK1/2 Tyr204/187, stress-activated protein kinase p38 Tyr182, the Src family tyrosine kinase Fyn Tyr420, Pyk2 Tyr402, N-methy-D-aspartate receptors (NMDARs; subunit GluN2B Tyr1472), and α-amino-3-hydroxy-5-methyl-4-isoxazolepropionic acid receptors (AMPARs; subunit GluA2). The dephosphorylation by STEP leads to inactivation of ERK, p38, Fyn and Pyk2, while the dephosphorylation of NMDARs and

AMPARs leads to their internalization and endocytosis. Dephosphorylation of NMDARs can be also indirect, involving inhibition of the Fyn-mediated phosphorylation of GluN2B Tyr1472.[28] Importantly, STEP is involved in striatal neurodegeneration, which is associated with several diseases including the L-3,4-dihydroxyphenylalanine-unresponsive Parkinsonism subtype of multiple-system atrophy.[29] Typically, this disease state is characterized by excessive accumulation and prolonged retention of dopamine in the nigrostriatal synapse, while in healthy individuals, the released dopamine is rapidly cleared and recycled into presynaptic neurons.[30]

Involvement of STEP in PD onset and progression stems from its ability to rapidly inactivate phosphorylated ERK kinase.[31] STEP itself is phosphorylated on Ser residues *via* the D_1 receptor and the cAMP-dependent PKA pathway. STEP Ser221 phosphorylation sterically inhibits the binding to phosphorylated ERK, and thus prevents its dephosphorylation, prolongs its activity and blocks its nuclear translocation. In striatal neurons treated for 15 min with dopamine or the D_1 receptor agonist (SKF R-38393) STEP is hyperphosphorylated and overexpressed when compared to its pretreatment levels. SKF R-38393-induced STEP phosphorylation is blocked by pretreatment with SCH 23390 (a D_1 receptor antagonist) or KT5720 (a PKA inhibitor, analog of staurosporine), but not upon pretreatment with sodium meta-bisulfite (an antioxidant), mPKCi (a PKC inhibitor), wortmannin (a PI3K inhibitor) or genistein (a PTK inhibitor).[30–32] Also use of numerous inhibitors of the ERK pathway blocks the activation of STEP translation in neuronal tissue. These include, *e.g.*, SL327 (an ERK/MAPK inhibitor), LY294002 (a PI3K inhibitor), and rapamycin (an mTOR inhibitor).[33] Src inhibition was reported to mimic some of the effects mediated by STEP phosphatase.[34]

Importantly, STEP dephosphorylation (and thus activation) is mediated by calcineurin, the calcium-dependent serine/threonine phosphatase already implicated in the pathology of PD (*cf.* Section 3.3).[32,35] Treatment with FK506 (a calcineurin inhibitor) leads to STEP inhibition both *in vitro* and *in vivo* (in mice at $1 \, mg \, kg^{-1}$ *i.p.*).[35] Physiologically, the calcineurin-dependent pathway starts usually with NMDARs or AMPARs stimulation, which initiates calcium influx and activation of calcineurin/PP2B, resulting in DARPP-32 inactivation (Thr34 dephosphorylation), preventing DARPP-32 from blocking the PP1 action on STEP Ser221.[28] The calcineurin/PP2B-mediated DARPP-32 Thr34 dephosphorylation can be reversed *via* the cAMP-mediated D_1 receptor/PKA pathway (the same pathway that can lead to direct STEP Ser221 phosphorylation) since PKA also phosphorylates DARPP-32 Thr34 and thus activates it, leading to the inhibition of PP1 activity and inhibition of STEP Ser221 dephosphorylation by PP1.[28] A detailed review of the DARPP-32 contribution to striatal signaling was published by Walaas *et al.*[36]

Full-length STEP (also called $STEP_{61}$ because of its M_r) can be inactivated by calpain-mediated cleavage between Ser224 and Leu225 in its KIM domain. Although the PTP domain of truncated STEP ($STEP_{33}$) is intact, the truncation leads to the inability of $STEP_{33}$ to interact with its substrates. Proteolytic cleavage of STEP was shown to affect, *e.g.*, p38 phosphorylation and cell death signaling pathways.[28,37] Extrasynaptic GluN2B-containing NMDARs are known to mediate STEP cleavage as inferred from the experiments involving the GluN2B antagonist ifenprodil.[37,38] The cleavage can be experimentally blocked by the peptide spanning the cleavage site, resulting in neuroprotective effects against, *e.g.*, glutamate excitotoxicity and oxygen-glucose deprivation.[28,37]

Dimerization of STEP was reported. The dimerization is mediated by Cys65 and Cys76 within the hydrophobic region of the N-terminus of the full-length STEP, and thus is absent in shorter forms of STEP (STEP$_{46}$, STEP$_{33}$). STEP dimerization decreases its enzymatic activity and is dependent on the redox conditions of the cellular microenvironment.[39] Although STEP$_{46}$ does not undergo the above dimerization, H_2O_2-mediated oligomerization of STEP$_{46}$ was reported,[39] although the mechanism is unknown and it may be an artifact of very high H_2O_2 concentrations. As with many other PTPs,[40] reactive Cys in the STEP PTP domain can be oxidized by reactive oxygen species (ROS), which thus can substantially decrease its activity.[39] Since several methods were recently developed to target specifically the differentially oxidized PTP active sites,[41,42] the modulation of STEP activity *via* oxidation-state-specific antibodies or redox probes is promising the efficient modulation of STEP activity in the near future.

STEP was also recently identified as a locally translated protein. STEP mRNA and protein are found in punctae along dendrites and near postsynaptic densities in hippocampal cultures. Its translocation is up-regulated within synaptoneurosomes following metabotropic glutamate receptor activation *via* *R,S*-3,5-dihydroxyphenylglycine.[28,43] The importance of STEP translocation and its applicability in pharmacotherapy remains to be solved, however. Generally the mRNAs important for the plastic changes are transported along dendrites, where they reside in a suppressed state until the appropriate synaptic stimulus permits their rapid translation at distinct synaptic sites to induce or maintain synaptic plasticity.

7.2.4 Src Homology 2 Domain-Containing Phosphatase 2

Although the Src homology 2 domain-containing phosphatase 2 (SHP-2, PTPN11) is under-reported in relation to PD, at least two pieces of evidence suggest its potentially very important role. First of all, SHP-2 is capable of activating ERK signaling, which is downstream of activated D_1 receptors in the *striatum*.[44] The importance of this pathway in the pathogenesis of PD is obvious from the clinical data involving the regimens based on dopamine replacement and from the beneficial effects of the use of dopamine receptor agonists in subjects with degenerated nigrostriatal dopaminergic neurons.[45] The pharmacotherapeutic potential of the modification of the ERK pathway is severely limited by its ubiquitous expression and involvement in numerous crucial signaling events in multiple tissues.[46]

The second piece of evidence of potential involvement of SHP-2 in the onset and progression of PD is indirect, although highly relevant. It stems from the recently emerging evidence of the important role of SHP-2 in mitochondria.[47,48] Since severe mitochondrial dysfunction is associated with PD, testing the role of mitochondrial SHP-2 in PD deserves considerable attention. A substantial number of the neuroprotective agents tested to be used against PD are already pro-mitochondrial drugs.[49]

The currently available experimental drugs targeting SHP-2 specifically are very limited, and none is clinically approved. Silencing RNA was used against SHP-2 to uncouple D_1 receptors from the ERK pathway.[50] Most of the experiments in other disease models reported very efficient knock-downs resulting from the targeting of SHP-2 nt 199-219.[3] Efficient SHP-2 inhibition can be reached using *bis*(4-trifluoromethylsulfonamidophenyl)-1,4-diisopropylbenzene sold commercially by Calbiochem as the inhibitor PTPi IV. This inhibitor works as a phosphate mimetic, with IC_{50} 1.8 µM against SHP-2.[3] The cell-permeable inhibitor 8-hydroxy-7-(6-sulfonaphthalen-2-yl)diazenyl-quinoline-5-sulfonic acid with IC_{50} 0.3 µM is also distributed by the same company under the commercial name NSC-87877. Ohio State University filled a patent (WO 2003/093498) on sulfanilic acid azochromotrop, which is claimed to be selective over other PTPs. Among the clinically approved drugs, stibogluconate inhibits SHP-2, but SHP-2 serves as only one of its numerous targets (*e.g.*, selectivity for SHP-2 is lower by one order of magnitude when compared to SHP-1).[3]

Were SHP-2 activity to be targeted for the purpose of PD treatment, the active compound would have to be directed towards the specific cell types, since SHP-2 also regulates numerous growth factor receptors and its aberrantly expressed transcripts are also involved in cancer onset and progression.

7.2.5 Phosphatase and Tensin Homolog Deleted on Chromosome 10

Some familial forms of PD are associated with the phosphatase and tensin homolog deleted on chromosome 10 (PTEN)-induced putative kinase-1 (PINK1, PARK6). PINK1 mutations are the second most common cause of recessive and early-onset Parkinsonism; 1 to 8% mutation frequencies have been reported in most of the early-onset or familial cohorts.[51] PTEN is a tyrosine and lipid phosphatase, while PINK1 serves as a Ser/Thr kinase with a high degree of homology for the Ca^{2+}/calmodulin kinase family that localizes to mitochondria. In the *Drosophila* model, PINK1 is thought to act within the same signaling pathway as PARKIN and is thought to affect mitophagy.[52–55] PARKIN translocation to mitochondria relies on PINK1 expression. PINK1 spans the outer mitochondrial membrane with its kinase domain facing the cytosol, and is thought to directly phosphorylate PARKIN Thr175 (but not Thr217). Other than PARKIN, PINK1 is known to phosphorylate only a few proteins, namely TRAP1 and the serine protease HtrA2/Omi.[55] PINK1 can be cleaved, which is thought to protect the cells against various stressors. Only uncleaved PINK1 binds Beclin 1, and stimulates autophagy.[56] The cleaved PINK1 fragment is thought to exhibit a short half-life, and is preferentially degraded by a proteasomal-like activity, probably by PARL protease.[55] Inhibition of mitochondrial fusion mediated by α-synuclein is thought to be alleviated by PINK1, PARKIN and DJ-1 overexpression.[57] Clinically reported PINK1 heterozygous mutations exert a gene dosage effect, suggesting that haploinsufficiency of PINK1 is the most likely mechanism that increases the PINK1-dependent susceptibility to dopaminergic cellular loss.[51] Among the upstream regulators of PINK1 is MARK2/Par-1, which phosphorylates PINK1 at Thr313 (the residue frequently mutated in PD, the mutation of which leads to abnormal mitochondrial distribution in neurons). Thr313 phosphorylation of cleaved PINK1 enhances mitochondrial anterograde transport and increases the fraction of stationary mitochondria, while the Thr313 phosphorylation of full-length PINK1 enhances mitochondrial retrograde transport.[58] PINK1 can also be autophosphorylated at Ser228 and Ser402 following the decrease of mitochondrial potential. Some of the PD-associated mutations prevent the above effect and inhibit PARKIN recruitment onto depolarized mitochondria (experimentally proven using dephosphorylation-mimicking Ser–Ala mutations and phosphorylation-mimicking Ser–Asp mutations).[59]

Loss of PINK1 (or expression of its kinase-dead mutant) leads to the dynamin-related protein 1 (Drp1) dephosphorylation in a calcineurin-dependent manner. Calcineurin inhibitor FK506 blocks Drp1 dephosphorylation and prevents loss of mitochondrial integrity in PINK1-deficient cells, but does not fully rescue the mitochondrial membrane potential.[60]

In zebrafish, morpholino oligonucleotide PINK1 knock-down displays moderately decreased numbers of central dopaminergic neurons, and alterations of mitochondrial functions caused in part by increased caspase-3 activity, increased GSK3β activity and increased ROS levels. Application of anti-oxidants (100 μM reduced glutathione, or 100 μM N-acetylcysteine) was shown to normalize the ROS levels and to improve the associated phenotype. Inhibition of elevated GSK-3β activity by 50 mM LiCl or by 10 μM SB216763 (3-(2,4-dichlorophenyl)-4-(1-methyl-1H-indol-3-yl)1H-pyrrole-2,5-dione) partially reverted the changes caused by PINK1 knock-down.[61] Another possible link to the ROS-mediated signaling was recently published by Chien et al.,[62] who found that PINK1 G309D overexpression increases apoptosis of SH-SY5Y neuronal cells following H_2O_2 or 1-methyl-4-phenylpyridinium treatment in the heme oxygenase-1-dependent and TRAP1-dependent manners. The H_2O_2-induced heme oxygenase-1 induction was Akt- and ERK-dependent, and phosphorylation of these two proteins (but not of p38) was inhibited in cells expressing PINK G309D or in those with down-regulated TRAP1.[62]

In the fruit fly as well as in the mouse, PINK1 deficiency or mutation affects the function of the mitochondrial respiratory chain Complex I, resulting in mitochondrial depolarization and increased sensitivity to apoptotic stress. In the fruit fly neurons, PINK1 deficiency alters synaptic function since the reserve pool of synaptic vesicles does not undergo mobilization during rapid stimulation. PINK1 deficiency can be rescued by adding ATP to the synapse.[63]

Besides PINK1, the already above-mentioned DJ-1 protein also binds directly to the PTEN phosphatase. The DJ-1 binding leads to PTEN inhibition. The DJ-1 activity depends on the redox status of its Cys106, and the redox status of DJ-1 Cys106 is thought to regulate PTEN activity, leading to cell proliferation.[64] However, the PTEN–DJ-1 interactions in PD are only poorly understood.

Interestingly, the role of PTEN in PD was mostly studied in relation to PINK1 aberrant signaling. However, from a more general point of view, PTEN is a very well known regulator of the major cell growth and survival signaling pathway, namely the PI3K/Akt pathway, where it antagonizes the function of PI3K by dephosphorylating PIP3, a key product of PI3K responsible for the activation of downstream targets including Akt. Involvement of PTEN in Akt signaling should be of high interest in PD as well since the Akt- and ERK-mediated pathways are already among the current targets of pharmacotherapeutic modulation of PD. Recently, Akundi et al. reported that PINK1-deficient embryonic fibroblasts and primary cortical neurons display reduced Akt phosphorylation in response to IGF-1 or insulin.[65] In IGF-1-treated PINK1-deficient cells, phosphorylation of GSK-3β, phosphorylation of ribosomal protein S6 (suggesting decreased mTOR activity), and nuclear exclusion of FoxO1 were also decreased. PINK1 deletion also led to the abrogation of IGF-1-mediated protection against staurosporine-induced metabolic dysfunction.[65]

Note also that PTEN can undergo S-nitrosylation by low NO concentrations at Cys83, which inhibits PTEN enzymatic activity and consequently stimulates the downstream Akt pathway. The S-nitrosylation of PTEN was reported in ischemic mouse brain in the core and penumbra regions (at sites with low NO concentrations), while S-nitrosylated Akt was found only in the ischemic core (at the site with high NO concentration). S-Nitrosylation of PTEN is expected to contribute to cell survival by means of enhanced Akt signaling and have neuroprotective effects (in contrast to S-nitrosylated Akt).[66] Besides ROS-mediated regulation of PTEN activity, NOS clearly appear as physiological regulators of PTEN signaling.

7.2.6 Dual-Specificity Protein Phosphatase 1

Dual-specificity protein phosphatase 1 (DUSP1, MKP-1) dephosphorylates ERK2 on both Thr183 and Tyr185. In the striatum of 6-hydroxydopamine-lesioned rats, DUSP1 is inhibited in response to D_2 receptor stimulation. However, in PC12 cells (rat pheochromocytoma cell line), the D_2 and D_3 receptor agonist quinpirole did not lead to any changes in protein tyrosine phosphatase activity, whereas ERK was activated by D_2 receptor stimulation. The quinpiperole-elicited inhibition of tyrosine phosphatases in lesioned *striata* can be inhibited by the D_2 receptor antagonist spiperone, while the D_1 receptor agonist 2,3,4,5-tetrahydro-7,8-dihydroxy-1-phenyl-1*H*-3-benzazepine (SKF 38393) does not affect the tyrosine phosphatase signaling. Thus, experimental modulation of D_2 receptor signaling suggests that 6-hydroxydopamine-induced striatal denervation leads to abnormal coupling between D_2 receptors and the MAPK/DUSP pathway. Unilateral inhibition of total protein tyrosine phosphatases by sodium vanadate leads to the contralateral rotation of the exposed animals in response to D_2 receptor stimulation.[67] Substantial changes in DUSP1 activity can clearly explain at least part of the observed ERK hyperphosphorylation in 6-hydroxydopamine-lesioned rats. Other phosphatases (involved in the regulatory cascade counteracting the MAPK pathway) may be also involved and are worth examining in the near future.[68]

Several inhibitors of DUSP1 are known. Among them NU-126, which inhibits DUSP1 at IC_{50} $28.8 \pm 2.9\,\mu M$, but has poor selectivity over a range of other phosphatases including VHR, MKP3, Cdc25B and PTP1B.[69] Another DUSP1 inhibitor is NSC 95397. NSC 95397 inhibits DUSP1 activity at IC_{50} $13\,\mu M$, however it is not selective over DUSP6 and Cdc25.[70] Sanguinarine inhibits DUSP1 at IC_{50} $17.3 \pm 1.2\,\mu M$, however it is not selective over DUSP14.[71] Chelerythrine inhibits DUSP1 at IC_{50} $16.2\,\mu M$, however it is not selective over DUSP14,[71] and PSI2106 inhibits DUSP1 at IC_{50} $8.0\,\mu M$ but is not selective over DUSP6.[72] Expression of DUSP1 is enhanced by glucocorticoids.[73] Gene silencing of DUSP1 was reported by Moncho-Amor et al.[74] DUSP1 redox regulation may also be considered to be of pharmacotherapeutic interest since the oxidative stress present in PD is thought to be a causative agent of cell death and simultaneously it is thought to be linked to the DUSPs linked to the MAPK pathway as documented by Levinthal and DeFranco.[75] Results presented by Huo et al. suggest the possible existence of a regulatory circuit between ROS and DUSPs since these authors found dexamethasone to inhibit the NOX-dependent ROS production via suppression of DUSP1-dependent MAPK pathways in activated microglia.[76] Oxidative stress emerging in response to depletion of the anti-oxidant glutathione can be efficiently elicited by administration of high glutamate concentrations.[75] Targeted redox regulation (i.e., using bidentate inhibitors which allow precise selectivity and which also retain the activity of their parental compounds) is of high pharmacotherapeutic interest not only in PD treatment, but in any diseases involving aberrant activation of virtually any tyrosine and dual-specificity phosphatases.

7.3 Protein Serine/Threonine Phosphatases

In contrast to protein tyrosine phosphatases, there are several advantages of the targeting protein serine/threonine phosphatases involved in PD.[77] Among them is their limited functional redundancy when compared to the tyrosine phosphatases (and also to the kinases). PD subjects may benefit from their activation, which can be achieved more effectively than quantitative inhibition. Finally, there are already several protein serine/threonine phosphatase activators, which seem to be very promising and which further highlight the importance of this enzyme family in drug discovery.

7.3.1 Protein Phosphatase 1

Protein phosphatase 1 (PP1) is a major, highly conserved Ser/Thr phosphatase regulating cell cycle, metabolism, neuronal signaling and transcription, and has been extensively reviewed elsewhere.[78] Lifting the fog on PP1 signaling is complicated by the existence of a very diverse set of PP1 substrates. These may include neurotransmitter receptors (the AMPA receptor GluR1 subunit, the NMDA receptor NR1 subunit, *etc.*), ion channels and pumps (N/P-type Ca^{2+} channels, Na^+ channel, Na^+, K^+-ATPase, *etc.*), and transcriptions factors (CREB, c-Fos, ΔFosB, *etc.*). For the nearly complete list of PP1α interacting proteins in human brain the reader is referred to the recent report by Esteves *et al.*[79] Of special interest are dual substrates for PP1 and PKA (GluR1 Ser845, NR1 Ser897), where the ability of PKA to activate PP1 serves as a positive feedback mechanism.[80] The catalytic subunit of PP1 (PP1c) is regulated by a plethora of interacting proteins (some considered to be subunits), most of them binding through the K/R-K/R-V/I-X-F/V docking sequence. The subunits may have either activating or inhibitory roles, and those discussed in this chapter have already been shown to play a role in PD.

PPP1R15A (GADD34; protein phosphatase 1 regulatory subunit 15A) recruits PP1 to dephosphorylate the translation initiation factor eIF2α phosphatase. Its overexpression is capable of restoring the vital translation rates of global protein synthesis by phosphorylated eIF2α during prion disease, rescuing synaptic deficits and neuronal loss, and thereby significantly increasing the survival and facilitating the recovery of cells from stress. The effect can be mimicked by the reduction of levels of prion protein (*e.g.*, by siRNA), which affects the same downstream target – phosphorylated eIF2α. eIF2α dephosphorylation can be inhibited by salubrinal, which not only increases the total levels of phosphorylated eIF2α, but also exacerbates neurotoxicity and reduces survival in the experimental prion-diseased mouse model.[81,82] Activation of PPP1R15A by eIF2α serves as a negative feedback loop since it terminates UPR signaling and allows translational recovery. Activation of the unfolded protein response and/or increased phosphorylation of eIF2α are characteristically observed in subjects with Parkinson's, Alzheimer's and prion diseases.[82] The TGF-β signaling pathway is down-regulated by PPP1R15A since it promotes dephosphorylation of TGFB1 by PP1. The activity of PPP1R15A can be regulated *via* Lyn kinase-mediated tyrosine phosphorylation.[83,84]

Also PPP1R1B (DARPP-32; the protein phosphatase 1 regulatory subunit 1B) is known to be involved in the regulation of PP1 signaling in PD. PPP1R1B acts as a substrate of cAMP-dependent PKA kinase enriched in dopamine-innervated brain areas.[85] Once phosphorylated at Thr34, PPP1R1B serves as one of the three related PKA-regulated inhibitors of PP1 phosphatase activity, being conserved across the *subphylum vertebrata*.[85–87] In a rat model of PD induced by the intrastriatal injection of 6-hydroxydopamine, PPP1R1B signaling in the *striatum* of hemi-Parkinsonian rats can be regulated by inhibitors of nNOS (*e.g.*, 7-nitroindazole). Mild dorsolateral striatal 6-hydroxydopamine lesions induce nNOS up-regulation and an increase in Thr34-phosphorylated

PPP1R1B. The application of nNOS inhibitors reduces the 6-hydroxydopamine-induced dopaminergic damage in the dorsolateral *striatum* and *substantia nigra*, reverses the changes in PPP1R1B Thr34 phosphorylation levels, and increases the total PPP1R1B levels, suggesting that nNOS inhibitors may have a beneficial effect in restoring the neurotransmitter imbalance of the basal ganglia circuitry in Parkinsonism.[88] The phosphorylation of PPP1R1B Thr34 is mediated by PKA and cGMP-dependent protein kinases, while the dephosphorylation is mediated by calcineurin (PP2B) and by PP2A.[85] Calcineurin-mediated dephosphorylation of PPP1R1B Thr34 can me initiated, e.g., by the activation of NMDA or AMPA receptors, which increase intracellular Ca^{2+},[89,90] or by the activation of D_2 receptors *via* PLC and subsequent changes in intracellular Ca^{2+} in striatal neurons.[91] The regulatory events mediated by PPP1R1B phosphorylated at Thr34 require the simultaneous presence of dephosphorylated Ser97. Ser97 is responsible for cytonuclear shuttling of PPP1R1B, its dephosphorylation allows the presence of PPP1R1B in the nucleus (where it inhibits PP1).[85] Extensive information on PPP1R1B regulation *via* (de)phosphorylation of its regulatory threonines and serines was provided by Nishi *et al.*[80] The mechanism of PP1c inhibition by PPP1R1B is poorly understood and displays mixed competitive and uncompetitive kinetics;[92,93] the interaction between PP1c and PPP1R1B is probably bidentate, involving distant KKIQF and PpTP motifs.[85]

The third case, where PP1 activity or expression was linked directly to PD, is the model of chronic striatal dopamine depletion.[94] Loss of nigrostriatal dopamine inputs in PD or in its animal models results in morphological alterations in striatal medium spiny neurons (>90% of total striatal neurons), and in the impairment of multiple forms of cortico–striatal synaptic plasticity. These effects can be reversed by dopamine replacement with levodopa, which prevents hyperphosphorylation of CaMKIIα Thr286. Interestingly, the CaMKIIα Thr286 hyperphosphorylation lasts for up to 18 months following dopamine depletion, while the hyperphosphorylation of downstream CaMKII target GluR1 Ser831 is detected after 9–18 months, but not 3–6 weeks, after dopamine depletion.[95] One of the mechanisms allowing CaMKIIα Thr286 hyperphosphorylation is thought to be mediated *via* inhibition of regulatory phosphatases, namely PP1 and its subunit PPP1R1B discussed above. However, direct inhibition of PP1 activity has not been demonstrated so far. Instead, regulation of PP1 localization is known, being mediated by spinophilin and neurabin, and being critical for normal synaptic plasticity and dendritic spine morphology.[94] Chronic dopamine depletion enhances the association of the PP1γ isoform with spinophilin, which selectively decreases the PP1γ activity, and probably allows hyperphosphorylation of CaMKIIα Thr286 and several other striatal proteins. Neurabin fragment (amino acids 146–493) at 1 μM concentration is capable of inhibiting ∼90% of PP1γ activity, while the activity of PP1β remains unaffected.[94]

Wang *et al.*[96] published an anecdotal report on the involvement of PPP1R12A (MYPT) in PD. PPP1R12A is a nearly ubiquitously expressed PP1 regulatory subunit, with the highest expression in muscle cells. However,

Wang et al.[96] found it to be down-regulated in the cerebrospinal fluid of human subjects with PD. The consequences of this finding are largely unclear, a major function of PPP1R12A is the regulation of the catalytic subunit of PP1 and the mediation of its binding to myosin. As a part of the PP1 catalytic complex, it is involved in PLK1 dephosphorylation, and is capable of HIF1AN-dependent HIF1A inhibition.[97,98]

Regulation of PP1 signaling also plays a role in the amantadine-mediated beneficial effects on subjects with PD.[99] In a model of controlled cortical impact injury in rats, chronic hypophosphorylation of PPP1R1B Ser34 and an increase in PP1 activity were reported (while the phosphorylation of PPP1R1B Thr75 was unchanged). Amantadine given daily for two weeks was shown to restore the PPP1R1B Ser34 phosphorylation back to normal values and to reduce PP1 activity. Besides that, it was also reported to cause a decrease of PPP1R1B Thr75 phosphorylation, which is consistent with its proposed activity as a partial NMDA receptor antagonist and partial dopamine agonist.[100]

Inhibitors of PP1 were discussed in detail by Sheppeck et al.[101] Among the activators of PP1 is, e.g., the microcystin analogue reported by Tappan and Chamberlin,[102] which acts as a PP1 direct agonist and phospho-DARP antagonist at IC_{50} 2.0 μM. In addition, the C_6 ceramide can be utilized for PP1 activation (see Section 7.3.2). So far, the PP1 activators have not been tested on PD models.

7.3.2 Protein Phosphatase 2A

Protein phosphatase 2A (PP2A) is the most frequently reported phosphatase in studies on PD. However, similarly to PP1, this molecule retains very diverse functions through the human body, which hampers its targeting in the course of disease onset and progression. PP2A is known for its modulation of phosphorylase B casein kinase 2, mitogen-stimulated S6 kinase and MAP-2 kinase. It dephosphorylates, e.g., SV40 large T antigen, p53, and RAF1. PP2A consists of a heterodimeric core enzyme composed of a PPP2CA catalytic subunit and a PPP2R1A constant regulatory subunit (PR65) which mediates associations with three families of regulatory subunits and other cellular and viral proteins. Among the interesting features of PP2A is its ability to be reversibly methyl esterified on Leu309, which may play a role in holoenzyme assembly. Tyrosine or threonine phosphorylation of PP2A results in its inactivation. PP2A probably reactivates itself by autodephosphorylation.

One of the typical signs of PD is the aggregation of α-synuclein hyperphosphorylated on Ser129 (Lewy bodies), the chaperone-like protein known to be mutated or triplicated in rare forms of PD. Soluble α-synuclein stimulates the activity of PP2A, while its phosphorylation on Ser129 attenuates its stimulatory effects. Aggregated α-synuclein displays half of the PP2A inhibitory activity when compared to the soluble version.[103] Interestingly, PP2A is known to dephosphorylate several proteins interacting with α-synuclein, and serving as critical regulators of brain function. Among them are the dopamine regulatory enzyme tyrosine hydroxylase, ERK1/2 and the microtubule-associated protein tau. Tyrosine hydroxylase becomes hyperphosphorylated in mouse brains displaying high levels of aggregated α-synuclein.[104] Tau becomes hyperphosphorylated in brain regions with low PP2A activity,[105] or expression.[106–109] PP2A also directly dephosphorylates the α-synuclein Ser129. This reaction was shown to be significantly enhanced by carboxyl methylation of the PP2A catalytic subunit. Experimental animals raised on a diet supplemented with eicosanoyl-5-hydroxytryptamide (an agent, which enhances PP2A methylation) dramatically reduce α-synuclein Ser129 phosphorylation and α-synuclein aggregation, leading to enhanced neuronal activity, increased dendritic arborizations, reduced astroglial and microglial activation, and improved motor performance. The active compound, eicosanoyl-5-hydroxytryptamide, is known to be present in coffee, which may partly explain the repeatedly reported link between the coffee consumption and reduced risk of PD onset.[110–113] The levels of eicosanoyl-5-hydroxytryptamide and of the closely related N-alkanoyl-5-hydroxytryptamides in coffee usually reach $\sim 0.5 \,\mathrm{mg\,g}^{-1}$.[114]

PP2A participates in the manganese-stimulated PKCδ/PP2A signaling pathway leading to the negative regulation of tyrosine hydroxylase and the apoptotic cell death of dopaminergic neurons. Manganese exposure results in manganism, a neurological disorder similar to PD. The mechanism of action involves manganese-induced caspase-3-dependent cleavage of PKCδ. Responses to acute and chronic manganese exposure differ – exposure to 3–10 μM manganese for three hours results in an increase of tyrosine hydroxylase activity and Ser40 phosphorylation, while exposure to 0.1–1 μM manganese for 24 h results in the opposite effect since it stimulates a dose-dependent decrease in tyrosine hydroxylase activity. The chronic (but not the acute) response can be inhibited by rottlerin (a PKCδ inhibitor); both PKCδ and PP2A activities are increased following chronic manganese exposure and respond well to the rottlerin treatment.[115]

Methamphetamine (0.5 to 3.0 mM for 24 h) was shown to decrease PP2A activity (but not PP2A expression) in the dopaminergic cell line MN9D. Methamphetamine exposure was also associated with changes in phospho-ERK1/2 and phospho-Akt, and with the up-regulation of the pro-survival protein Bcl-2. Low concentrations of methamphetamine are thought to result in a decreased vulnerability of affected cells to the subsequent oxidative stress occurring, e.g., in PD.[116]

Based on studies of gene transfer animal models, overexpression of PARKIN was shown to increase the protein levels of PP2A subunits B and C by ~60%. Treatment of PARKIN-overexpressing cells with okadaic acid (a PP2A inhibitor) abolished the predicted downstream effects, namely the PP2A-induced decrease of α-synuclein. The PARKIN–PP2A–α-synuclein signaling was abolished when overexpressing PARKIN T240R mutant. Of note is that coimmunoprecipitation experiments did not reveal any direct interaction between PARKIN and PP2A.[117]

Use of the PP2A inhibitors okadaic acid, microcystin LR or fostriecin was shown to lead to perikaryal hyperphosphorylation of neurofilament proteins and may induce neuronal apoptosis. The observed effect can be reversed by the peptidyl-prolyl isomerase Pin1 *in vitro*. Since the aberrant hyperphosphorylation of neurofilament proteins within the cell bodies is characteristic for neurodegenerative disorders including PD (under normal conditions, they are phosphorylated in the axonal compartment), Pin1 is being considered as a pharmacotherapeutic target to modulate PP2A-mediated signaling in neurodegenerative disorders.[118]

The action of fostriecin (a PP2A inhibitor) and C_6 ceramide (a PP2A activator) was shown to affect the action of ketone bodies on hippocampal long-term potentiation in rats independently of the ROS levels altered by ketone bodies and independently of externally added H_2O_2 or rotenone. These data suggest that oxidative impairment of hippocampal long-term potentiation is associated with PP2A activation and can be prevented by PP2A inhibition mediated by ketone bodies.[119]

Chen *et al.* linked PP2A to the endoplasmic reticulum stress involved in PD onset and progression.[120] The endoplasmic reticulum stress (in response to, *e.g.*, 6-hydroxydopamine treatment) induces GSK3β dephosphorylation on Ser9 mediated by PP2A. At 1 nM, okadaic acid was shown to block the above dephosphorylation reaction.[120]

PP2A activity can be targeted using a plethora of well-established inhibitors. Among them are the above-mentioned okadaic acid (IC_{50} 1 nM; selective over PP1 which has IC_{50} 10–15 nM), cantharidin (IC_{50} 200 nM; selective over PP1 which has IC_{50} 1100 nM), fostriecin (IC_{50} 1.5 nM; partially selective over PP4 which has IC_{50} 3 nM, and over PP1 which has IC_{50} 131 000 nM), microcystin-LR (IC_{50} 0.04 nM; selective over PP1 which has IC_{50} 1.7 nM), nodularin (IC_{50} 0.026 nM; selective over PP1 which has IC_{50} 1.8 nM, and over PP2B which has IC_{50} 1800 nM), and calyculin A (IC_{50} 0.5–1.0 nM; only marginally selective over PP1 which has IC_{50} 2 nM).[68] The effects of chronic microcystin-LR expression on the brain proteome were recently analyzed by Wang *et al.*[121] The effects of okadaic acid and calyculin A on mesencephalic neurons were reported by Zeevalk *et al.*[122] PP2A inhibitors were discussed in detail by Sheppeck *et al.*[101]

Activation of Bα containing PP2A was reached by inhibiting PP2A demethylation mediated by eicosanoyl-5-hydroxytryptamide (EHT). EHT administration to α-synuclein transgenic mice reduced α-synuclein Ser129 phosphorylation, its subsequent aggregation, and ameliorated the associated neuropathology and behavioral aberrations. EHT action on PP2A is mediated by its antagonistic effect on PME, reported to occur at IC_{50} 3.9 μM.[113] In addition, ceramides were shown to inhibit PP2A demethylation,[123] to block binding of the endogenous PP2A inhibitor I2,[124] and thus to allosterically activate PP2A. C_2 ceramide is known to work as an antagonist of PME and a direct agonist of PP2A, potentiating 350% of the basal PP2A activity at 20 μM,[123,125] while C_6 ceramide is known to affect PP2A as an I2 antagonist (but is also a PP1 direct agonist), increasing the basal PP2A activity to 250% at 15 μM.[126] The ceramide analog glucosylceramide, which accumulates in Gaucher disease mouse models with mutations in the PD-associated glucocerebrosidase gene, is an inhibitor of PP2A and leads to α-synuclein aggregation.[77,127] Note also the enormous number of PP2A modulatory compounds tested in other disease models and safely administered to humans. These include (but are not limited to) fingolimod, memantine, tocopherol (vitamin E), and sodium selenite.[77]

7.3.3 Protein Phosphatase 3

The α-subunit of protein phosphatase 3 (PP3, formerly PP2B or calcineurin A) is a calcium-dependent, calmodulin-stimulated protein phosphatase known for dephosphorylation of DNM1L, HSPB1 and SSH1. In PD, PP3 is known to be regulated by Pin1, which enhances the formation of α-synuclein inclusions. It is thought that the mechanism involves the transcription factor NFAT. Dephosphorylation of NFAT by PP3 prevents binding of Pin1 and alters the phosphorylation-dependent shuttling of NFAT between the cytoplasm and nucleus.[128,129] The mechanism functions as a regulatory circuit since overexpression of Pin1 prevents PP3 from dephosphorylation of NFAT. However, the detailed mechanism of action is unclear since Pin1 is also thought to promote Ser/Thr dephosphorylation, and some authors speculate that Pin1 may act directly on PP3 similarly to cyclophilin and FKBP prolyl isomerases.[129]

Two decades ago it was hypothesized that PP3 may play an important role in the pathogenesis of PD based on the use of the immunosuppressive agents cyclosporine A and FK-506, which were thought to elicit a dopamine-like effect upon dopaminoceptive neurons in the striatum – leading to increased DARPP-32 phosphorylation and PP1 inhibition.[130] However, since then, the field has not shifted forward significantly and the exact role of PP3 in PD remains to be elucidated. Calcineurin is now known as an important mediator of N-methyl-D-aspartate receptor-dependent spine loss in hippocampal neurons, and L-type Ca^{2+} channel-dependent activation of MEF2. Application of calcineurin inhibitors (1 μM ascomycin and 4 μM cyclosporine A) leads to the attenuation of depolarization-induced spine loss in pallidostriatal medium spiny neurons in cortico–striatal cocultures during high (35 mM) potassium treatment.[131,132]

Numerous PP3 inhibitors are known, some even in clinical use. Note that most of them are used to alter signaling in immune or malignant cells. However, when properly targeted, their use in PD is not ruled out. Among the known inhibitors are pimecrolimus and tacrolimus (be aware of their possible tumorigenic side-effects), cyclosporine, and many others (*cf.* the detailed reviews elsewhere).[133–136]

7.3.4 PH Domain Leucine-Rich Repeat Protein Phosphatase 1

PH domain leucine-rich repeat protein phosphatase 1 (PHLPP1, PLEKHE1, SCOP) was found to be a downstream target of ubiquitin C-terminal hydroxylase (UCH-L1), the mutations of which (S18Y) are thought to be protective for PD. PHLPP1 mediates dephosphorylation of Akt1 Ser473,

PRKCB isoform β-II Ser660 and PRKCA Ser657, suggesting it plays a role in survival and apoptosis. Overexpression of UCH-L1 down-regulates PHLPP1 and thus boosts Akt signaling.[137]

7.4 Future Views

Identification and understanding of the PD-associated signaling pathways are valuable for providing insight into the pathogenic mechanism of this neurodegenerative disorder. Among the mechanisms of interest are autophagy, detoxification, redox regulation, *etc*. Of special interest should be the utilization of our broad and still increasing knowledge on cancer-associated pathways. PD and cancer are two pathologic processes resulting from excessive signaling by one of the two sets of opposing forces: those driving cell death and those promoting cell survival. While PD results from the excessive death of dopaminergic neurons, tumorigenesis is driven by excessive cell survival of the target issue. The underlying mechanisms can be generalized, since the death/survival signals are driven by the same set of signaling pathways in almost any cell type.[138] Epidemiological studies have already shown a decreased cancer incidence in PD subjects,[139] and most of the genes associated with familial forms of PD are already implicated in tumorigenesis (α-synuclein, PARK1, PTEN, UCH-L1, *etc*.).[138]

Acknowledgments

The study was supported in part by the UNCE project 204015 and by the PRVOUK project P31/2012 from Charles University in Prague, by the Czech Science Foundation project P301/12/1686, and by the Internal Grant Agency of the Ministry of Health of the Czech Republic project NT13663-3/2012.

References

1. P. Yang, A. Dankowski and T. Hagg, *Eur. J. Neurosci.*, 2007, **25**, 1332.
2. X. Lu, D. Maysinger and T. Hagg, *Exp. Neurol.*, 2002, **178**, 259.
3. P. Heneberg, *Curr. Med. Chem.*, 2009, **16**, 706.
4. P. Heneberg, *Curr. Med. Chem.*, 2012, **19**, 1530.
5. M. Fukada, H. Kawachi, A. Fujikawa and M. Noda, *Methods*, 2005, **35**, 54.
6. K. Meng, A. Rodriguez-Peña, T. Dimitrov, W. Chen, M. Yamin, M. Noda and T. F. Deuel, *Proc. Natl. Acad. Sci. U. S. A.*, 2000, **97**, 2603.
7. N. Sakaguchi, H. Muramatsu, K. Ichihara-Tanaka, N. Maeda, M. Noda, T. Yamamoto, M. Michikawa, S. Ikematsu, S. Sakuma and T. Muramatsu, *Neurosci. Res.*, 2003, **45**, 219.

8. E. Gramage, Y. B. Martín and G. Herradon, *Pharmacol. Biochem. Behav.*, 2012, **101**, 387.
9. E. Gramage, Y. B. Martín, P. Ramanah, C. Pérez-García and G. Herradón, *Neuroscience*, 2011, **190**, 307.
10. P. Perez-Pinera, W. Zhang, Y. Chang, J. A. Vega and T. F. Deuel, *J. Biol. Chem.*, 2007, **282**, 28683.
11. J. E. Ferrario, A. E. Rojas-Mayorquín, M. Saldaña-Ortega, C. Salum, M. Z. Gomes, S. Hunot and R. Raisman-Vozari, *J. Neurochem.*, 2008, **107**, 443.
12. P. Huang, J. Ramphal, J. Wei, C. Liang, B. Jallal, G. McMahon and C. Tang, *Bioorg. Med. Chem.*, 2003, **11**, 1835.
13. H. Cho, R. Krishnaraj, M. Itoh, E. Kitas, W. Bannwarth, H. Saito and C. T. Walsh, *Protein Sci.*, 1993, **2**, 977.
14. M. Nakayama, J. Hisatsune, E. Yamasaki, Y. Nishi, A. Wada, H. Kurazono, J. Sap, K. Yahiro, J. Moss and T. Hirayama, *Infect. Immunol.*, 2006, **74**, 6571.
15. J. P. Chow, A. Fujikawa, H. Shimizu, R. Suzuki and M. Noda, *J. Biol. Chem.*, 2008, **283**, 30879.
16. J. F. Côté, A. Charest, J. Wagner and M. L. Tremblay, *Biochemistry*, 1998, **37**, 13128.
17. Y. Li, A. Grupe, C. Rowland, P. Holmans, R. Segurado, R. Abraham, L. Jones, J. Catanese, D. Ross, K. Mayo, M. Martinez, P. Hollingworth, A. Goate, N. J. Cairns, B. A. Racette, J. S. Perlmutter, M. C. O'Donovan, J. C. Morris, C. Brayne, D. C. Rubinsztein, S. Lovestone, L. J. Thal, M. J. Owen and J. Williams, *Hum. Mol. Genet.*, 2008, **17**, 759.
18. N. Tikhmyanova, J. L. Little and E. A. Golemis, *Cell. Mol. Life Sci.*, 2010, **67**, 1025.
19. J. Chapuis, F. Moisan, G. Mellick, A. Elbaz, P. Silburn, F. Pasquier, D. Hannequin, C. Lendon, D. Campion, P. Amouyel and J. C. Lambert, *Hum. Mol. Genet.*, 2008, **17**, 2863.
20. E. Chouery, V. Delague, A. Bergougnoux, S. Koussa, J. L. Serre and A. Megarbane, *Hum. Mutat.*, 2008, **29**, E194.
21. M. A. Chellaiah, D. Kuppuswamy, L. Lasky and S. Linder, *J. Biol. Chem.*, 2007, **282**, 10104.
22. D. Davidson and A. Veillette, *EMBO J.*, 2001, **20**, 3414.
23. M. Halle, Y. C. Liu, S. Hardy, J. F. Theberge, C. Blanchetot, A. Bourdeau, T. C. Meng and M. L. Tremblay, *Mol. Cell Biol.*, 2007, **27**, 1172.
24. D. Krishnamurthy, M. R. Karver, E. Fiorillo, V. Orru, S. M. Stanford, N. Bottini and A. M. Barrios, *J. Med. Chem.*, 2008, **51**, 4790.
25. K. Yo, S. Iwata, Y. Hashizume, S. Kondo, S. Nomura, O. Hosono, H. Kawasaki, H. Tanaka, N. H. Dang and C. Morimoto, *Biochem. Biophys. Res. Commun.*, 2009, **382**, 210.
26. M. Zheng and P. J. McKeown-Longo, *J. Cell Sci.*, 2006, **119**, 96.
27. V. Hivert, J. Pierre and J. Raingeaud, *Biochem. Pharmacol.*, 2009, **78**, 1017.

28. S. M. Goebel-Goody, M. Baum, C. D. Paspalas, S. M. Fernandez, N. C. Carty, P. Kurup and P. J. Lombroso, *Pharmacol. Rev.*, 2012, **64**, 65.
29. I. Ghorayeb, P. O. Fernagut, I. Aubert, E. Bezard, W. Poewe, G. K. Wenning and F. Tison, *Mov. Disord.*, 2000, **15**, 531.
30. J. Chen, M. Rusnak, P. J. Lombroso and A. Sidhu, *Eur. J. Neurosci.*, 2009, **29**, 287.
31. S. Paul, G. L. Snyder, H. Yokakura, M. R. Picciotto, A. C. Nairn and P. J. Lombroso, *J. Neurosci.*, 2000, **20**, 5630.
32. S. Paul, A. C. Nairn, P. Wang and P. J. Lombroso, *Nat. Neurosci.*, 2003, **6**, 34.
33. Y. Hu, Y. Zhang, D. V. Venkitaramani and P. J. Lombroso, *J. Neurochem.*, 2007, **103**, 531.
34. K. A. Pelkey, R. Askalan, S. Paul, L. V. Kalia, T. H. Nguyen, G. M. Pitcher, M. W. Salter and P. J. Lombroso, *Neuron*, 2002, **34**, 127.
35. Y. S. Choi, S. L. Lin, B. Lee, P. Kurup, H. Y. Cho, J. R. Naegele and P. J. Lombroso, *J. Neurosci.*, 2007, **27**, 2999.
36. S. I. Walaas, H. C. Hemmings, P. Greengard and A. C. Nairn, *Front. Neuroanat.*, 2011, **5**, 50.
37. J. Xu, P. Kurup, Y. Zhang, S. M. Goebel-Goody, P. H. Wu, A. H. Hawasli, M. L. Baum, J. A. Bibb and P. J. Lombroso, *J. Neurosci.*, 2009, **29**, 9330.
38. R. Poddar, I. Deb, S. Mukherjee and S. Paul, *J. Neurochem.*, 2010, **115**, 1350.
39. I. Deb, R. Poddar and S. Paul, *J. Neurochem.*, 2011, **116**, 1097.
40. P. Heneberg, L. Dráberová, M. Bambousková, P. Pompach and P. Dráber, *J. Biol. Chem.*, 2010, **285**, 12787.
41. A. Haque, J. N. Andersen, A. Salmeen, D. Barford and N. K. Tonks, *Cell*, 2011, **147**, 185.
42. B. Boivin, M. Yang and N. K. Tonks, *Sci. Signal.*, 2010, **3**, I2.
43. Y. Zhang, D. V. Venkitaramani, C. M. Gladding, Y. Zhang, P. Kurup, E. Molnar, G.L. Collingridge and P. J. Lombroso, *J. Neurosci.*, 2008, **28**, 10561.
44. J. M. Cunnick, S. Meng, Y. Ren, C. Desponts, H. G. Wang, J. Y. Djeu and J. Wu, *J. Biol. Chem.*, 2002, **277**, 9498.
45. M. Stacy and A. Galbreath, *Clin. Neuropharmacol.*, 2008, **31**, 51.
46. S. Yoon and R. Seger, *Growth Factors*, 2006, **24**, 21.
47. A. Arachiche, O. Augereau, M. Decossas, C. Pertuiset, E. Gontier, T. Letellier and J. Dachary-Prigent, *J. Biol. Chem.*, 2008, **283**, 24406.
48. M. Salvi, A. Stringaro, A. M. Brunati, E. Agostinelli, G. Arancia, G. Clari and A. Toninello, *Cell. Mol. Life Sci.*, 2004, **61**, 2393.
49. A. H. V. Schapira, *J. Neurol. Neurosurg. Psychiatry*, 2005, **76**, 1472.
50. C. Fiorentini, C. Mattanza, G. Collo, P. Savoia, P. Spano and C. Missale, *J. Neurochem.*, 2011, **117**, 253.
51. E. K. Tan, F. S. Refai, M. Siddique, K. Yap, P. Ho, S. Fook-Chong and Y. Zhao, *Hum. Mutat.*, 2009, **30**, 1551.

52. I. E. Clark, M. W. Dodson, C. Jiang, J. H. Cao, J. R. Huh, J. H. Seol, S. J. Yoo, B. A. Hay and M. Guo, *Nature*, 2006, **441**, 1162.
53. J. Park, S. B. Lee, S. Lee, Y. Kim, S. Song, S. Kim, E. Bae, J. Kim, M. Shong, J. M. Kim and J. Chung, *Nature*, 2006, **441**, 1157.
54. A. C. Poole, R. E. Thomas, L. A. Andrews, H. M. McBride, A. J. Whitworth and L. J. Pallanck, *Proc. Natl. Acad. Sci. U. S. A.*, 2008, **105**, 1638.
55. C. Vives-Bauza and S. Przedborski, *Trends Mol. Med.*, 2011, **17**, 158.
56. S. Michiorri, V. Gelmetti, E. Giarda, F. Lombardi, F. Romano, R. Marongiu, S. Nerini-Molteni, P. Sale, R. Vago, G. Arena, L. Torosantucci, L. Cassina, M. A. Russo, B. Dallapiccola, E. M. Valente and G. Casari, *Cell Death Differ.*, 2010, **17**, 962.
57. F. Kamp, N. Exner, A. K. Lutz, N. Wender, J. Hegermann, B. Brunner, B. Nuscher, T. Bartels, A. Giese, K. Beyer, S. Eimer, K. F. Winklhofer and C. Haass, *EMBO J.*, 2010, **29**, 3571.
58. D. Matenia, C. Hempp, T. Timm, A. Eikhof and E. M. Mandelkow, *J. Biol. Chem.*, 2012, **287**, 8174.
59. K. Okatsu, T. Oka, M. Iguchi, K. Imamura, H. Kosako, N. Tani, M. Kimura, E. Go, F. Koyano, M. Funayama, K. Shiba-Fukushima, S. Sato, H. Shimizu, Y. Fukunaga, H. Taniguchi, M. Komatsu, N. Hattori, K. Mihara, K. Tanaka and N. Matsuda, *Nat. Commun.*, 2012, **3**, 1016.
60. A. Sandebring, K. J. Thomas, A. Beilina, M. van der Brug, M. M. Cleland, R. Ahmad, D. W. Miller, I. Zambrano, R. F. Cowburn, H. Behbahani, A. Cedazo-Mínguez and M. R. Cookson, *PLoS ONE*, 2009, **4**, e5701.
61. O. Anichtchik, H. Diekmann, A. Fleming, A. Roach, P. Goldsmith and D. C. Rubinsztein, *J. Neurosci.*, 2008, **28**, 8199.
62. W.-L. Chien, T.-R. Lee, S.-Y. Hung, K.-H. Kang, M.-J. Lee and W.-M. Fu, *J. Neurochem.*, 2011, **117**, 643.
63. V. A. Morais, P. Verstreken, A. Roethig, J. Smet, A. Snellinx, M. Vanbrabant, D. Haddad, C. Frezza, W. Mandemakers, D. Vogt-Weisenhorn, R. Van Coster, W. Wurst, L. Scorrano and B. De Strooper, *EMBO Mol. Med.*, 2009, **1**, 99.
64. Y.-C. Kim, H. Kitaura, T. Taira, S. M. M. Iguchi-Ariga and H. Ariga, *Int. J. Oncol.*, 2009, **35**, 1331.
65. R. S. Akundi, L. Zhi and H. Büeler, *Neurobiol. Disord.*, 2012, **45**, 469.
66. N. Numajiri, K. Takasawa, T. Nishiya, H. Tanaka, K. Ohno, W. Hayakawa, M. Asada, H. Matsuda, K. Azumi, H. Kamata, T. Nakamura, H. Kara, M. Minami, S. A. Lipton and T. Uehara, *Proc. Natl. Acad. Sci. U. S. A.*, 2011, **108**, 10349.
67. X. Zhen, C. Torres, G. Cai and E. Friedman, *Mol. Pharmacol.*, 2002, **62**, 1356.
68. N. Jailkhani, V. K. Chaudhri and K. V. Rao, *Anticancer Agents Med. Chem.*, 2011, **11**, 64.

69. J. S. Lazo, R. Nunes, J. J. Skoko, P. E. Queiroz de Oliveira, A. Vogt and P. Wipf, *Bioorg. Med. Chem.*, 2006, **14**, 5643.
70. A. Vogt, P. R. McDonald, A. Tamewitz, R. P. Sikorski, P. Wipf, J. J. Skoko and J. S. Lazo, *Mol. Cancer Ther.*, 2008, **7**, 330.
71. A. Vogt, A. Tamewitz, J. Skoko, R. P. Sikorski, K. A. Giuliano and J. S. Lazo, *J. Biol. Chem.*, 2005, **280**, 19078.
72. J. S. Lazo, J. J. Skoko, S. Werner, B. Mitasev, A. Bakan, F. Koizumi, A. Yellow-Duke, I. Bahar and K. M. Brummond, *J. Pharmacol. Exp. Ther.*, 2007, **322**, 940.
73. C. Nunes-Xavier, C. Romá-Mateo, P. Ríos, C. Tárrega, R. Cejudo-Marín, L. Tabernero and R. Pulido, *Anticancer Agents Med. Chem.*, 2011, **11**, 109.
74. V. Moncho-Amor, I. Ibañez de Cáceres, E. Bandres, B. Martínez-Poveda, J.L. Orgaz, I. Sánchez-Pírez, S. Zazo, A. Rovira, J. Albanell, B. Jiménez, F. Roio, C. Belda-Iniesta, J. García-Foncillas and R. Perona, *Oncogene*, 2011, **30**, 668.
75. D. J. Levinthal and D. B. DeFranco, *J. Biol. Chem.*, 2005, **280**, 5875.
76. Y. Huo, P. Rangarajan, E.-A. Ling and S. T. Dheen, *BMC Neurosci.*, 2011, **12**, 49.
77. S. P. Braithwaite, M. Voronkov, J. B. Stock and M. M. Mouradian, *Neurochem. Int.*, 2012, **6**, 899.
78. M. Bollen, W. Peti, M. J. Ragusa and M. Beullens, *Trends Biochem. Sci.*, 2010, **35**, 450.
79. S. L. Esteves, S. C. Domingues, O. A. da Cruz e Silva, M. Fardilha and E. F. da Cruz e Silva, *OMICS*, 2012, **16**, 3.
80. A. Nishi, M. Kuroiwa and T. Shuto, *Front. Neuroanat.*, 2011, **5**, 43.
81. M. Boyce, K. F. Bryant, C. Jousse, K. Long, H. P. Harding, D. Scheuner, R. J. Kaufman, D. Ma, D. M. Coen, D. Ron and J. Yuan, *Science*, 2005, **307**, 935.
82. J. A. Moreno, H. Radford, D. Peretti, J. Steinert, N. Verity, M. G. Martin, M. Halliday, J. Morgan, D. Dinsdale, C. A. Ortori, D. A. Barrett, P. Tsaytler, A. Bertolotti, A. E. Willis, M. Bushell and G. R. Mallucci, *Nature*, 2012, **485**, 507.
83. A. V. Grishin, O: Azhipa, I. Semenov and S. J. Corey, *Proc. Natl. Acad. Sci. U. S. A.*, 2001, **98**, 10172.
84. H. Molina, D. M. Horn, N. Tang, S. Mathivanan and A. Pandey, *Proc. Natl. Acad. Sci. U. S. A.*, 2007, **104**, 2199.
85. M. Yger and J.-A. Girault, *Front. Behav. Neurosci.*, 2011, **5**, 56.
86. P. Svenningsson, A. Nishi, G. Fisone, J. A. Girault, A. C. Nairn and P. Greengard, *Ann. Rev. Pharmacol. Toxicol.*, 2004, **44**, 269.
87. A. Nishi, Y. Watanabe, H. Higashi, M. Tanaka, A. C. Nairn and P. Greengard, *Proc. Natl. Acad. Sci. U. S. A.*, 2005, **102**, 1199.
88. J. E. Yuste, M. B. Echeverry, F. Ros-Bernal, A. Gomez, C. M. Ros, C. M. Campuzano, E. Fernandez-Villalba and M. T. Herrero, *Neuropharmacology*, 2012, **63**, 1258.

89. S. Halpain, J.-A. Girault and P. Greengard, *Nature*, 1990, **343**, 369.
90. A. Nishi, G. L. Snyder, A. C. Nairn and P. Greengard, *J. Neurochem.*, 1999, **72**, 2015.
91. S. Hernandez-Lopez, T. Tkatch, E. Perez-Garci, E. Gallaraga, J. Bargas, H. Hamm and D. J. Surmeier, *J. Neurosci.*, 2000, **20**, 8987.
92. H. C. Hemmings, P. Greengard, H. Y. L. Tung and P. Cohen, *Nature*, 1984, **310**, 503.
93. H. C. Hemmings, A. C. Nairn, J. I. Elliott and P. Greengard, *J. Biol. Chem.*, 1990, **265**, 20369.
94. A. M. Brown, A. J. Baucum, M. A. Bass and R. J. Colbran, *J. Biol. Chem.*, 2008, **283**, 14286.
95. A. M. Brown, A. Y. Deutch and R. J. Colbran, *Eur. J. Neurosci.*, 2005, **22**, 247.
96. E. S. Wang, Y. Sun, J. G. Guo, X. Gao, J. W. Hu, L. Zhou, J. Hu and C. C. Jiang, *Acta Neurol. Scand.*, 2010, **122**, 350.
97. A. Zagorska, M. Deak, D. G. Campbell, S. Banerjee, M. Hirano, S. Aizawa, A. R. Prescott and S. R. Alessi, *Sci. Signal.*, 2010, **3**, RA25.
98. S. Yamashiro, Y. Yamakita, G. Totsukawa, H. Goto, K. Kaibuchi, M. Ito, D. J. Hartshorne and F. Matsumura, *Dev. Cell*, 2008, **14**, 787.
99. M. Brenner, A. Haass, P. Jacobi and K. Schimrigk, *J. Neurol.*, 1989, **236**, 153.
100. J. W. Bales, H. Q. Yan, X. Ma, Y. Li, R. Samarasinghe and C. E. Dixon, *Exp. Neurol.*, 2011, **229**, 300.
101. J. E. Sheppeck, C. M. Gauss and A. R. Chamberlin, *Bioorg. Med. Chem.*, 1997, **5**, 1739.
102. E. Tappan and A. R. Chamberlin, *Chem. Biol.*, 2008, **15**, 167.
103. J. Wu, H. Lou, T. N. M. Alerte, E. K. Stachowski, J. Chen, A. B. Singleton, R. L. Hamilton and R. G. Perez, *Neuroscience*, 2012, **207**, 288.
104. T. N. M. Alerte, A. A. Akinfolarin, E. E. Friedrich, S. A. Mader, C.-S. Hong and R. G. Perez, *Neurosci. Lett.*, 2008, **435**, 24.
105. C. X. Gong, T. J. Singh, I. Grundke-Iqbal and K. Iqbal, *J. Neurochem.*, 1993, **61**, 921.
106. S. Kins, A. Crameri, D. R. Evans, B. A. Hemmings, R. M. Nitsch and J. Götz, *J. Biol. Chem.*, 2001, **276**, 38193.
107. E. Sontag, A. Luangpirom, C. Hladik, I. Mudrak, E. Ogris, S. Speciale and C. L. White, *J. Neuropathol. Exp. Neurol.*, 2004, **63**, 287.
108. A. Schild, L. M. Ittner and J. Götz, *Biochem. Biophys. Res. Commun.*, 2006, **343**, 1171.
109. N. Deters, L. M. Ittner and J. Götz, *Biochem. Biophys. Res. Commun.*, 2009, **379**, 400.
110. W. Hellenbrand, H. Boeing, B. P. Robra, A. Seidler, P. Vieregge, P. Nischan, J. Joerg, W. H. Oertel, E. Schneider and G. Ulm, *Neurology*, 1996, **47**, 644.

111. P. A. Fall, M. Fredrikson, O. Axelson and A. K. Granérus, *Mov. Disord.*, 1999, **14**, 28.
112. M. D. Benedetti, J. H. Bower, D. M. Maraganore, S. K. McDonnell, B. J. Peterson, J. E. Ahlskog, D. J. Schaid and W. A. Rocca, *Neurology*, 2000, **55**, 1350.
113. K.-W. Lee, W. Chen, E. Junn, J.-Y. Im, H. Grosso, P. K. Sonsalla, X. Feng, N. Ray, J. R. Fernandex, Y. Chao, E. Masliah, M. Voronkov, S. P. Braithwaite, J. B. Stock and M. M. Mouradian, *J. Neurosci.*, 2011, **31**, 6963.
114. R. Lang and T. Hofmann, *Eur. Food Res. Technol.*, 2005, **220**, 638.
115. D. Zhang, A. Kanthasamy, V. Anantharam and A. Kanthasamy, *Toxicol. Appl. Pharmacol.*, 2011, **254**, 65.
116. A. El Ayadi and M. J. Zigmond, *PLoS ONE*, 2011, **6**, e24722.
117. P. J. Khandelwal, S. B. Dumanis, L. R. Feng, K. Maguire-Zeiss, G. W. Rebeck, H. A. Lashuel and C. E. H. Moussa, *Mol. Neurodegener.*, 2010, **5**, 47.
118. P. Rudrabhatla, W. Albers and H. C. Pant, *J. Neurosci.*, 2009, **29**, 14869.
119. M. Maalouf and J. M. Rho, *J. Neurosci. Res.*, 2008, **86**, 3322.
120. G. Chen, K. A. Bower, C. Ma, S. Fang, C. J. Thiele and J. Luo, *FASEB J.*, 2004, **18**, 1162.
121. M. Wang, D. Wang, L. Lin and H. Hong, *Chemosphere*, 2010, **81**, 716.
122. G. D. Zeevalk, L. P. Bernard, L. Manzino and P. K. Sonsalla, *J. Pharmacol. Exp. Ther.*, 2001, **298**, 925.
123. C. L. Chen, C. F. Lin, C. W. Chiang, M. S. Jan and Y. S. Lin, *Mol. Pharmacol.*, 2006, **70**, 510.
124. A. Mukhopadhyay, S. A. Saddoughi, P. Song, I. Sultan, S. Ponnusamy, C. E. Senkal, C. F. Snook, H. K. Arnold, R. C. Sears, Y. A. Hannun and B. Ogretmen, *FASEB J.*, 2009, **23**, 751.
125. R. T. Dobrowsky, C. Kamibayashi, M. C. Mumby and Y. A. Hannun, *J. Biol. Chem.*, 1993, **268**, 15523.
126. C. E. Chalfant, Z. Szulc, P. Roddy, A. Bielawska and Y. A. Hannun, *J. Lipid Res.*, 2004, **45**, 496.
127. Y. H. Xu, Y. Sun, H. Ran, B. Quinn, D. Witte and G. A. Grabowski, *Mol. Genet. Metab.*, 2011, **102**, 436.
128. W. Liu, H. D. Youn, X. Z. Zhou, K. P. Lu and J. O. Liu, *FEBS Lett.*, 2001, **496**, 105.
129. P. Rudrabhatla and H. C. Pant, *J. Alzheimer's Dis.*, 2010, **19**, 389.
130. S. Wera and J. Neyts, *Med. Hypotheses*, 1994, **43**, 132.
131. S. W. Flavell, C. W. Cowan, T. K. Kim, P. L. Greer, Y. Lin, S. Paradis, E. C. Griffith, L. S. Hu, C. Chen and M. E. Greenberg, *Science*, 2006, **311**, 1008.
132. X. Tian, L. Kai, P. E. Hockberger, D. L. Wokosin and D. J. Surmeier, *Mol. Cell. Neurosci.*, 2010, **44**, 94.
133. L. M. Margassery, J. Kennedy, F. O'Gara, A. D. Dobson and J. P. Morrissey, *J. Microbiol. Meth.*, 2012, **88**, 63.

134. C. Sommerer, S. Meuer, M. Zeier and T. Giese, *Clin. Chim. Acta*, 2012, **413**, 1379.
135. M. Naganuma, T. Fujii and M. Watanabe, *J. Gastroenterol.*, 2011, **46**, 129.
136. K. A. Al Johani, A. M. Hegarty, S. R. Porter and S. Fedele, *J. Am. Acad. Dermatol.*, 2009, **61**, 829.
137. S. Hussain, O. Foreman, S. L. Perkins, T. E. Witzig, R. R. Miles, J. van Deursen and P. J. Galardy, *Leukemia*, 2010, **24**, 1641.
138. R. H. Kim and T. W. Mak, *Br. J. Cancer*, 2006, **94**, 620.
139. A. B. West, V. L. Dawson and T. M. Dawson, *Trends Neurosci.*, 2005, **28**, 348.

The α-Synuclein Hypothesis

CHAPTER 8
Synuclein and Parkinson's Disease: An Update

KURT A. JELLINGER

Institute of Clinical Neurobiology, Vienna, Austria
Email: kurt.jellinger@univie.ac.at

8.1 Introduction: α-Synuclein and Disease

α-Synuclein is a protein that is highly abundant in the nervous system. Although its physiological functions and its role in neurodegeneration are not fully understood, it is implicated in the pathogenesis of a number of neurodegenerative disorders summarized as α-synucleinopathies, which include Parkinson's disease (PD), dementia with Lewy bodies (DLB), multiple-system atrophy (MSA), and other disorders.[1] PD, one of the most frequent neurodegenerative disorders, is a progressive multi-organ disease with variegated neurological and non-motor symptoms. It features degeneration of the dopaminergic nigrostriatal system, responsible for the core motor deficits, and multi-focal involvement of the central, peripheral and autonomic nervous system, and other organs, with widespread occurrence of presynaptic, intracytoplasmic, axonal, and dendritic depositions of fibrillary hyperphosphorylated α-synuclein that forms amyloid-like inclusions in selected neuronal populations.[2] Abnormal aggregates of α-synuclein occur in three major types of inclusions: (1) as intracellular and intraneuritic deposits (Lewy bodies and Lewy neurites in PD and DLB); (2) glial cytoplasmic inclusions (GCI) predominantly affecting oligodendroglia in MSA;[3] and (3) in giant axonal swellings (spheroids) in these and other rare diseases. These inclusions are widely accepted as diagnostic morphological hallmarks of α-synucleinopathies,[4] although α-synuclein aggregates also affect both astroglia and microglia.[5] The solubility of α-synuclein varies with the course of disease,[6] between disorders, and with the neuronal populations affected. The clinical phenotypes of PD are related to diffuse progression of Lewy pathology,[7,8] suggested to result from a prion-like seeding of α-synuclein.[9,10,10a] Differentiation of cognitive disorders that are related to different pathologies (PD-dementia/PDD, pure DLB, and DLB with Alzheimer-like pathology or the Lewy body variant of AD) may be difficult. The etiology of synucleinopathies is complex, with interaction of genetic and environmental risk factors,[11,12] although familial components may indicate genetic factors. The role of aging as an essential risk factor for developing PD[13] is characterized by reduced cellular stress defenses and compensatory capacity following injury, explained through the view that a loss of α-synuclein function is ultimately linked to neurodegeneration in PD. The recognition of the heterogeneity within synucleinopathies is important for the classification of their phenotypes,[14] as a basis for further therapy options. The complex mechanisms underlying synaptic derangement in PD and other α-synucleinopathies was recently reviewed.[15]

8.2 The Synuclein Family

α-Synuclein is the only member of a family of natively unfolded proteins that has been linked to disease.[16] It includes α-synuclein, a 14 kDa neuronal protein encoded by the 6-exon SNCA (PARK1) gene (OMIM 163890) coded on chromosome 4q.21, β-synuclein and γ-synuclein, encoded by other distinct genes (chromosome 5 and 10, respectively), that share significant sequences at the amino acid level. The synucleins are small (127–140 amino acids), natively soluble unfolded proteins, which are highly charged and have low hydropathy. Human α-synuclein was originally described as the precursor protein for the non-amyloid component (NACP) in Alzheimer's disease (AD) amyloid plaques. SNCA has two paralogous genes named SNCB (OMIM 602569) and SNCG (OMIM 602998), with which it shares an N-terminal domain, while β-synuclein lacks many amino acid residues in the NAC region. γ-Synuclein, initially described as "breast cancer associated protein 1", is smaller than α-synuclein and β-synuclein proteins due to a shorter C-terminal region, but contains much of the NAC region.[17] The SNCA gene is highly conserved across species with only a few amino acids differing between the human and rodent sequences.[17] The most prominent feature of α-synuclein is the hydrophobic NAC domain, lacking in the other synucleins, which seems to be important to form the aggregates or fibrillary structures present in Lewy body disease.

8.3 The Biochemistry of α-Synuclein

8.3.1 Structure

α-Synuclein is a 140 amino acid protein with three functional domains: an α-helical N-terminal domain, a central hydrophobic non-amyloidogenic component (NAC) domain, critical for aggregation, and an acidic unstructured

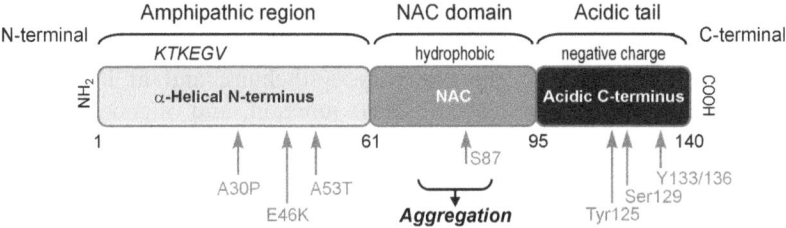

Figure 8.1 Schematic representation of the structure, protein domains and phosphorylation sites of human α-synuclein.

C-terminal, intrinsically disordered domain that is critical for the chaperone-like activity of α-synuclein, and inhibits β-sheet (protofibril and fibril) formation. The N-terminus contains six α-helical repeats of 11 residues with variations of the conserved central sequence KTKEGV, which is thought to be important in anchoring α-synuclein.[18] The isolated NAC domain forms amyloid structures, and small deletions within it can reduce the propensity of α-synuclein to aggregate. The amino terminal helices are important in α-synuclein's interaction with membranes, and both the intrinsic disorder of soluble monomers and the amino terminal helices are important for its biological interactions and functions.[19] The C-terminal domain appears to suppress α-synuclein aggregation and mediates its activity in promoting proliferation of dopaminergic cells. C-Terminal truncations have been associated with aggregates in human brain and in animal models.[20] α-Synuclein contains several phosphorylation sites for protein kinases (Figure 8.1).

The predominant physiological species of α-synuclein is a helically folded tetramer that resists aggregation,[21] while others have described soluble α-synuclein as a disordered free monomer,[22] complemented by a small fraction of helical trimers and tetramers.[23] α-Synuclein is potentially prone to mis-fold and has a strong tendency to self-aggregate *in vivo*, resulting in toxicity. It is an intrinsically disordered protein but a very dynamic molecule that can adopt different conformational states. Helical membrane-bound α-synuclein forms a partially folded stage as the key intermediate in aggregation, and provides the seeds responsible for deposition of the disordered free cytosolic form.[24] Folding and mis-folding occur on membranes;[25] mis-folded isoforms may accumulate as free α-synuclein in the cell. Its secondary structure determined by its environment and its conformation depend upon whether it is cytoplasmic or membrane-bound. Nuclear magnetic resonance studies of α-synuclein fibrils showed that the core extends with a repeated structural motif, thus disagreeing with the previously proposed fold.[26] Rapid exchange of α-synuclein between bound and unbound states provide mechanisms to ensure that stable cellular structures remain dynamic and susceptible to regulation. α-Synuclein may become structured upon interaction with other proteins and/or biological membranes.[22]

8.3.2 Localization and Regulation

In normal brain, α-synuclein is predominantly expressed in CNS neurons in the neocortex, hippocampus, striatum, thalamus, and cerebellum, where it is localized in the cytosole and at presynaptic terminals. From there it is transported to the neuronal perikarya by axonal flow. It also occurs in olfactory receptor neurons, the olfactory epithelium, and at low levels in skeletal muscle, cells of the neuromuscular junction, and oligodendrocytes.[17] α-Synuclein expression occurs in human and rat brain somata, dendrites, and glia that are susceptible to cellular α-synuclein aggregation.[5] It is localized outside of the nervous system in multiple organs,[27] which suggests that its function is not exclusive to the brain and related diseases, but is also associated with non-neurological disorders. Endogenous α-synuclein is normally distributed in both cytosolic and membrane-bound forms, contradicting the assertion that it is exclusively a cytoplasmic protein. It is restricted to axon terminals, which led to the acceptance that it was a cytoplasmic, presynaptic protein.[28]

Levels of α-synuclein are regulated by a balance of synthesis, degradation, and secretion. The ubiquitin–proteasomal system (UPS) and the autophagy/lysosome pathway (ALP) are the two major control systems postmitotic neurons use to maintain intracellular proteostasis.[29] While the UPS degrades most short-lived soluble proteins in the cytoplasm, nucleus and endoplasmic reticulum,[30,31] the ALP refers to the degradation of intracellular components, protein and organelles in lysosomes.[32]

Proteasomal dysfunction results in the accumulation of SUMOylated α-synuclein; contributing to inclusion formation, while deficits in SUMOlysation may contribute to α-synuclein aggregation.[33] Membrane-bound α-helical α-synuclein can aggregate spontaneously, but does not require membranes to form (proto)fibrils, and membranous binding may prevent self-aggregation. Partially folded α-synuclein is thought to be a precursor in the process of fibril formation, although α-synuclein appears to take on a wide range of folded and/or soluble oligomeric forms depending on its environment. Monomeric α-synuclein readily forms aggregates/fibrils *in vitro* when compared to its propensity to form tetramers *in vitro*, which suggests the fibrillogenic and tetrameric forms are generated through different pathways.[21] Soluble folding intermediates may be essential for its aggregation by a cascade comprising initially soluble oligomers, then insoluble oligomers, and finally fibrils present in inclusions/Lewy bodies (LBs) and GCIs.[16] Iron may promote α-synuclein aggregation either directly or *via* increased levels of oxidative stress, while metal-catalyzed oxidation of α-synuclein inhibits filament formation with increased formation of β-sheet-rich oligomers or protofibrils.[34] α-Synuclein expression is regulated by iron mainly at the translational level, supporting the role of iron in the translational control of α-synuclein.[35] Elevated α-synuclein expression and/or alteration of its aggregation properties cause the re-distribution of the protein within the cell, enhancing endoplasmic reticulum–mitochondrial interactions.[36]

α-Synuclein is subject to a number of post-translational modifications,[37,38] the most important, phosphorylation at Ser129 by kinases, is linked to disease progression, although other modifications, (ubiquitinylation and truncation) also have functional consequences.[20] The critical role of truncated α-synuclein for its progressive aggregation was shown recently.[39] Tyrosine phosphorylation of α-synuclein has been less well studied, and its functional consequences are unclear. Unlike tau, whose phosphorylation is critical for its binding to microtubules and regulating axonal transport, enhanced phosphatase activity reduced the phosphorylation and aggregation of α-synuclein. The ubiquitin-independent proteasome pathway or a ubiquitin-dependent pathway after dephosphorylation have both been implicated.[40] Pre- and post-translational changes of α-synuclein could be responsible for different pathways leading to its aggregation.[41] However, the observation of aggregated α-synuclein by and of itself does not prove that aggregation is important; all available data prove that this deposition of α-synuclein occurs, but not that it is causal.[42]

The mechanisms for α-synuclein degradation remain unclear. Either monomeric forms can be degraded by the proteasome, or only a small portion of soluble cell-derived intermediates as oligomers, not including monomeric α-synuclein, are targeted to the 26S proteasome for degradation.[16] Two separate lysosomal pathways – chaperone-mediated autophagy (CMA) and macro-autophagy or a more general lysosomal function – may be the initiating factors in α-synuclein degradation.[43] Wild-type (wt) α-synuclein but not mutant forms are degraded by CMA, whereas all forms are degraded by macroautophagy.[44] Inhibition of CMA leads to increased aggregation of high-molecular-weight and detergent-insoluble α-synuclein species, while enhanced CMA-dependent degradation occurs under conditions of stress induced by an excess of α-synuclein. Reduced expression of substances regulating CMA in PD brains, supports the notion that its dysfunction is implicated in PD pathogenesis.[45] The interplay between different α-synuclein degradation pathways, based on *in vivo* studies, provided a model for the role of the UPS and the different subtypes of ALP in the evolution and progression of α-synuclein pathology.[32]

8.3.3 Physiological Functions

The physiological function of α-synuclein is incompletely understood. This multi-functional protein is implied in many cellular processes that co-ordinate nuclear and synaptic events, neuronal plasticity, cytoskeletal function, vesicle fusion and recycling, synaptic integrity and transmission, neuronal differentiation and regeneration.[46,47] It colocalizes with synaptophysin, which may support synaptic vesicle association, while others found no effect of overexpressed α-synuclein on synaptic efficacy. α-Synuclein has functions on signal transduction, gene expression, axonal transport of synaptic vesicles, microtubule and membrane stability, mitochondrial function, and anchoring cytoskeletal compartments to plasma membranes, while it is not required for neuronal development. α-Synuclein is an important component of the mitochondrial function/Ca^{2+} homeostasis that is essential to neuron function and survival.[48] It has a fundamental role in the control of presynaptic terminal size and synaptic structure, acting as a molecular chaperone *via* the carboxy-terminal region, is important for folding and re-folding of synaptic proteins, protein phosphorylation, and as a negative regulator for dopamine synthesis by affecting the activity of its key enzyme, tyrosin hydroxylase.[48] Overexpression, mutant forms, and loss of α-synuclein will likely have a detrimental effect on synaptic function and neurotransmission, in particular dopamine handling and biosynthesis.[48] α-Synuclein has an anti-oxidant function, and participates in regulating neurotransmitter homeostasis, in dopaminergic neurons and there is an age-related increase of α-synuclein in neuromelanin-containing neurons in the substantia nigra (SN) that are selectively vulnerable in PD.[49]

Mounting evidence suggests a protective role of α-synuclein at the synapse, where it has a non-classical chaperone activity, modulating the SNARE (soluble N-ethylmaleimide-sensitive factor attachment protein receptor) complex assembly.[50] SNARE dysfunction is suggested to be the initial trigger of mislocation and accumulation of α-synuclein, and is probably an important pathomechanism of α-synucleinopathies.[51] On the other hand, α-synuclein promotes SNARE-complex assembly *in vivo* and *in vitro*.[51a] α-Synuclein interacts with "heat shock protein 40" (Hsp 40) and HSP 70, presynaptic molecular chaperones, which contributes to maintaining the integrity of synaptic nerve terminals, and might protect against synaptic degeneration.[18,52] Many other functions and posttranslational modifications have been attributed to α-synuclein, including participation in oxidative stress production, ubiquitination, nitration, glycation, phosphorylation, *etc.*[47] Glycation of α-synuclein has important pathological implications, since it leads to the formation of advanced glycation products (AGEs), which generate large amounts of reactive oxygen species (ROS) and *vice versa* leading to cell death.[53] α-Synuclein re-distributes neuromelanin lipid in the *substantia nigra* in PD, associates with many proteins, and regulates the activity of several enzymes, *e.g.*, mitogen-activating protein kinases (MAPKs), increases cellular ferriductase activity and iron(II) levels in dopaminergic cells leading to their selective loss in PD,[53] while it prevents cytochrome *c* release and apoptosis through inhibition of the MAPK signaling pathway,[54] and negatively regulates complex I activity.

8.3.4 Genetics

The functional role of α-synuclein in the pathogenesis of PD is evidenced by the causal relationship between genetic mutations and disease, and gene expression profiling of nigral dopaminergic neurons gave insight into PD pathology.[55–60] There are 18 PARK loci currently described, with six confirmed pathogenic PD genes, *i.e.*, SNCA (PARK 1,4), PRKN (PARK 2,3), UCHL1 (PARK 5) PINK1 (PARK 6,8), DJ1 (PARK 7), ATP13A2 (PARK 9) associated with a typical Parkinsonian phenotype in the majority of affected gene carriers.[55,61] The neuropathology of genetic PD has been summarized recently.[62,63]

Currently 16 PARK loci have been identified with autosomal dominant genes, such as SNCA (PARK1/4), linked to chromosome 4q21, LRRK2 (leucine-rich repeat kinase 2), dardarin/PARK8, chromosome 12q12, PARK 17 (VPS35/vacuolar protein sorting 35), eIF-4G1 (eukaryotic translation initiation factor 4G1), and gene triplications have been found to cause autosomal dominant forms of PD that are clinically comparable to sporadic PD (sporadic PD), but with variable pathology.[63,63a] PARK2 (PARKIN), linked to chromosome 6q25.2-q27, PINK1/PARK6 (phosphatase and tensin homologue/PTEN-induced kinase 1, DJ1/PARK7 (a multi-functional redox-sensitive protein leading to suppression of apoptosis), both linked to chromosome 1p36, PARKIN-encoding ubiquitin carboxy-terminal hydrolase (UCHL1), linked to chromosome 6q25.2-27, PARK9 (ATP13A2/ATPase type 13A2) (as Parkinson-dystonia/Kufor–Rakeb syndrome), PARK14 (PLAG6/phospholipase A2 group IV), and PARK15 (FBXO7/F-box protein 7) are autosomal recessive juvenile onset, most of these cases showing no LBs.[63] SNCA, DJ1, ATP13A2, and PRKN also contribute to sporadic PD.[55] Newly discovered genes associated with autosomal-dominant PD include GBA (glucosidase-β-acid), eIFAG1 (eukaryotic translation factor 4 γ-protein 1), presenting with an α-synuclein Parkinsonian phenotype.[61,63]

The other PARK loci/genes are either provisional or associated with atypical Parkinsonism, while carriers of non-PARK loci/genes frequently present with a mixed phenotype.[61] The LRRK2 and PRKN mutations are the most prevalent.[58] The SNCA and LRRK2 mutations are pathologically characterized by α-synuclein deposition, albeit in some LRRK2 cases, tau and TDP-43 were identified.[61] SNCA genetic variants predict faster motor symptom progression in PD.[63,64] There might be indirect interactions between α-synuclein and LRRK2.[20] Widespread expression of LRRK2 in human brain, particularly in brainstem, suggests its association with early-stage pathology in PD,[65] but its exact biological function remains unclear. DJ-1 acts in parallel to the PINK1/PARKIN pathway to control mitochondrial function and autophagy.[66] The DJ1-C57 mouse model showing early onset unilateral loss of dopaminergic nigral neurons progressing to bilateral degeneration of the nigrostriatal axis provides a tool to study the preclinical aspects of neurodegeneration. Mutations of PARK2, PINK1, DJ-1, and LRRK2 are more frequent in patients with early onset PD, but are lower than previously estimated.[67] DJ-1 acts as a chaperone for α-synuclein.[60] Recent evidence pointed to a link between 14-3-3 proteins and PD in relation to both α-synuclein and LRRK2; some isoforms have been found in LBs in human PD and animal models.[68] LRRK 2 is phosphorylated at Ser910 and Ser935, which is decreased in most LRRK2 pathological mutants, and 14-3-3 binding is impaired accordingly. 14-3-3 may act as a regulator of LRRK2-mediated cellular functions, such as autophagy, neurite growth or cytoskeletal dynamics. Common variation within the LRRK2 gene has been shown to be a risk factor for PD.[69]

The PARKIN, PINK1 and DJ1 mutations impact mitochondrial function and/or oxidative stress responses, they are causative for mitophagy acting in the removal of damaged mitochondrial organelles, with PARKIN acting downstream of and modulated (phosphorylated and activated) by PINK1, but the mechanism by which they confer neuroprotection is not clear.[70] PARKIN and PINK1 also regulate mitochondrial mobility in axons, but whether they promote mitophagy remains unclear. PARKIN-targeted mitochondria are accumulated in the somatodendritic regions where mature lysosomes are predominantly located. Differential location to the inner and outer membrane may regulate

PINK1 function.[70a] Increasing evidence indicates that several pathogenic PD genes have a role in regulating protein stability, such as proteasome (α-synuclein), protein stabilizing and degredating and endoplasmic reticulum stress response (PINK1, PARKIN, DJ-1), autophagy (UPS35, LRRK2) and interact with cellular systems for homeostasis, such as mitochondrial maintenance (PINK1, DJ1, PARKIN, FBXO7), synaptic homeostasis (α-synuclein), ALP (α-synuclein, PARKIN, PINK 1), axonal transport, vesicle traffic, and neurite outgrowth (LRRK2), metabolic dysfunction (PARKIN), and UPS (α-synuclein, PARKIN, DJ1, UCHL1).[7,71,72]

Three single point mutations in α-synuclein are associated with early-onset PD: A53T, A30P, and E46K. Ala53Thr (A53T), identified in a large Italian family (Contursi) and in Greek kindreds shows both α-synuclein and tau pathology.[73] The A53T mutation was also found in DLBD, while the relevance of DJ-1 mutation for DLB is not known. Ala30Pro (A30P), in a German kindred, shows similarities to PD but with more severe pathology.[74] E46K or Glu46Lys reported in a Spanish family with autosomal dominant Parkinsonism, and dementia is associated with widespread Lewy pathology, referred to as "DLB".[75] These mutations have different effects on the amyloidogenicity, vesicle-binding and pore-like activity of α-synuclein, supporting the role of oligomer pore formation in the pathogenesis of PD.[76] Sequence variants in eukaryotic translation initiation factor 4-γ (eIF4G1) in two American families have been associated with DLB.[77] The effects of A53T and A30P mutants on the fragmentation, conformation, and association of α-synuclein in the presence of the 20S proteasome suggest that 20S mediated truncation of α-synuclein may play a role in both familial and sporadic PD.[78] Typical PD with Lewy pathology was recently described in a patient with heterozygous (R275W) mutation in the PARK2 gene.[79]

Multiplications and point mutations of the α-synuclein locus cause familial PD.[11] Genome-wide association studies (GWAS) confer an increased risk of polymorphic gene variants for late-onset sporadic PD, evaluating tyrosin hydroxylase genes and rare loci suggest heterogeneity of these diseases.[63,80,81,81a]

A meta-analysis revealed 10 gene sets with previously unknown association with PD pinpoint defects in mitochondrial electron transport and glucose utilization, that occur early in disease pathogenesis, while genes controlling bioenergetics are under-expressed.[82] Only the α-synuclein species with increased aggregating propensities, human wt and A30P, triggered degeneration of nigral neurons, and wt α-synuclein showed the highest degree of toxicity after long-term overexpression, suggesting that this promotes progressive neuronal degeneration.[41] Expansion of Rep1, a polymorphic mixed-dinucleotide repeat in the SNAC promoter region, is associated with elevated risk of sporadic PD.[83] Variants of all three members of the synuclein family affect the risk of developing DLB, and detection of a gene for familial DLB in 2q35.q36 emphasized the genetic heterogeneity of DLB.[84] α-Synucleinopathy associated with G51D SNCA mutations showing neuronal α-synuclein and oligodendroglial inclusions may represent a link between PD and MSA.[84a] The tau gene MAPT H1 is associated with reduced AD pathology.[85]

Mutations of the GBA gene at chromosome 1q21, which encodes glucocerebrosidase (GCase), suggest a link between PD and DLB with Gaucher disease through their overexpression promoting α-synuclein accumulation.[86] GBA mutations are the most frequent genetic risk factor, particularly for familial PD,[87] and cerebrosidase is present in α-synuclein inclusions on LB disorders.[88] GBA mutations, altering the endoplasmic reticulum and lysosomes,[89] may overwhelm the ability of UPS to remove accumulated α-synuclein and GBA-PD is associated with altered membrane metabolism that may promote α-synuclein aggregation and neurotoxicity,[90] while overexpression of α-synuclein inhibits the intracellular trafficking and lysosomal activity of wt GBA.[91] Membrane-bound α-synuclein interacts with GCase and inhibits enzyme activity.[91a] GBA mutation may not only increase the risk of both PD and DLB, and present diverse patterns of Lewy pathology, but also influence the natural history of PD and increase the risk of LBD with and without AD pathology.[92] Recent studies revealed a significant decrease of GCase activity in PD brains with GBA mutations, most pronounced in the SN, leading to mitochondrial dysfunction and decreased macroautophagy flux.[91] Loss of GCAase activity did not immediately raise SNCA concentrations, but first lead to neuronal ubiquitinopathy and axonal spheroids.[86] wt glucocerebrosidase protein expression changes *in vitro* could contribute to the GCase deficiency observed in sporadic PD brains.[91] The pattern of dopamine loss in patients with both PD and Gaucher disease was similar to sporadic PD, but decreased resting activity showed a pattern characteristic of diffuse Lewy body disease.[88]

8.4 α-Synuclein and Neurodegeneration

PD belongs to a group of neurodegenerative disorders caused by protein mis-folding, summarized as "protein mis-folding/conforming disease" or "proteinopathies".[1] Proteins implicated in neurodegeneration can be neither re-folded to their normal configuration nor degraded by proteasomes, which leads to their abnormal turn-over, elevated concentration, aggregation, and accumulation of insoluble deposits.[29] Protein folding and re-folding are both mediated by a network of molecules called chaperones and cochaperones, which, associated with the UPS and ALP pathways, remove irreversibly misfolded proteins. Dysfunction of those systems resulting from environmental toxins, mutation in α-synuclein and PARKIN as well as macroautophagic pathway failure because of oxidative stress are essential in the pathogenesis of PD.[93] Molecular chaperones have a central role in maintaining protein homeostasis in order to prevent or modulate neurodegeneration, and by diminishing α-synuclein neurotoxicity play a neuroprotective role.[94] Whether α-synuclein concentrations are increased in PD brains is unclear. While sensitive assays in combination with *in vivo* dialysis demonstrated relatively high concentrations of α-synuclein in brain interstitial fluids, normal levels in the cytosolic fraction and no correlations with nigral Lewy body intensity have been found.[95] No widespread extranigral α-synuclein accumulation in PD, as suggested by immunohistochemical reports,[96] has been confirmed by neurochemical methods demonstrating only mildly increased high-molecular-weight α-synuclein in the putamen.[95,97] This suggests that α-synuclein pathology revealed by immunohistochemistry might not be caused by its accumulation but rather by conformational changes. A recently described monoclonal antibody (5GA) distinguishes pathological from physiological forms of α-synuclein.[98] Expression of phosphorylated α-synuclein is increased in PD, DLB and AD with Lewy pathology. PD shows a significant increase in soluble and lipid-associated phosphorylated α-synuclein over the disease course, with progressive decrease of soluble α-synuclein in DLB and PD,[6] but no corresponding decrease in α-synuclein mRNA levels,[99] while both nitrated and phosphorylated α-synuclein are significantly elevated in many experimental models.[100]

Increases in phosphorylated α-synuclein may promote neurotoxicity, oligomer formation, trafficking block, alterations in axonal transport motor proteins,[101] produce a specific defect in synaptic vesicle recycling that precedes detectable neuropathology,[102] and reduce the ability to regulate tyrosin hydroxylase.[103] The re-distribution of α-synuclein from the synapses, where it performs crucial neuronal functions, to the cell body either due to maladaptive developmental plasticity-like signals an age-related impairment of axonal transport, may represent a primary mechanism of loss-of-function in aging and disease.[48] Despite speculation as to how aberrant protein activity might lead to neurodegeneration, it is not certain that α-synuclein aggregation is the primary cause or an epiphenomenon in the pathogenic process of PD. Studies of cellular and animal models have not only confirmed that mitochondrial dysfunctions, abnormal protein aggregation, oxidative stress, impaired bioenergetics, prion-like behavior of misfolded proteins, and dysregulation of calcium homeostasis are the main contributors to PD.[2,12,104] Recent studies have shown the deleterious effects of α-synuclein on newly generated neurons, in particular on their dendritic outgrowth and spine development, thus having a negative impact on adult neurogenesis and neuronal malfunction.[105] There is evidence for the role of extracellular α-synuclein in modulating glial functions[5] and neuroinflammation[16] explaining its neurotoxic effect as a pathogenic factor in PD.[106] However, better characterization of the features that make selective neuronal populations vulnerable in PD,[107,108] and the role of inflammation are clearly needed.

8.4.1 α-Synuclein Neurotoxicity and the Oligomer Toxicity Hypothesis

The relation of α-synuclein behavior to toxicity is governed by several conditions. Expression levels are critical for toxicity, and some variants may have different biophysical properties compared to the phosphorylated wt protein. The ability of α-synuclein to undergo self-assembly into a multitude of multi-meric forms adds further complexity in understanding its function and toxicity.[24] These facts raise some caveats about comparison of the properties of α-synuclein and its concentration-dependent behavior.

The tendency of A30P to accumulate as oligomers instead of mature fibrils suggested that α-synuclein may have a similar toxic mechanism to intermediates of other proteins, such as β-amyloid, tau protein, prions and polyglutamine peptides.[109] The "toxic oligomer hypothesis"[110] gained support *via* model systems with increased neurotoxicity by overexpression of α-synuclein that exhibited increased propensity to form oligomeric, prefibrillar structures and decreased formation of fibrillar aggregates. α-Synuclein mutants are particularly prone to formation of such oligomers, being the consequence of abnormal membrane interactions, alterations in vesicle traffic, involvement of mitochondria, or lysosomal membranes.[23,24,28] This may be a result of the toxic action of substances produced during the early phases of protein mis-folding.[111] Accumulation of mis-folded α-synuclein in the endoplasmic reticulum is the main event leading to induction of the endoplasmic reticulum stress-related unfolded protein response, induced by oligomeric species of α-synuclein and activated in nigral dopaminergic neurons in human PD and experimental models.[112] Therefore, endoplasmic reticulum stress response is a key cellular function that is disrupted leading to neuronal cell death in familial and sporadic PD. The role of unfolded protein stress in the endoplasmic reticulum and mitochondria in neurodegeneration has been reviewed recently.[113] Which particular α-synuclein species are toxic has been debated. Some favored the fully fibrillar or the intermediate soluble oligomeric species, but recent evidence indicates that early oligomeric forms and not the final protein aggregates are responsible for its toxicity.[114] Small intermediates termed "soluble oligomers" in the aggregation process might lead to synaptic dysfunction and neuronal death,[115] whereas large, insoluble deposits might function as a reservoir of bioactive oligomers.[116] The polymerization of α-synuclein from unstructured monomer to mature amyloid fibrils is a multi-step process that proceeds through the formation of altered-sized oligomers and polymers. However, the molecular mechanisms through which α-synuclein abnormally aggregates and contributes to neurodegeneration in these disorders remains unknown.[117] Spectrometric data suggest that polypeptide sequences may contribute to the formation of oligomeric aggregates with distinct biochemical properties.[118]

Spheroidal oligomers contain a significant amount of α-helical structure, which decreases in protofilaments, while the β-sheet structure content of α-synuclein increases from spheroidal oligomers, through protofibrils, to fibrils. Loss of dopaminergic nigral cells in animals with α-synuclein variants that form oligomers (E57K, E35K) showed that these are toxic *in vivo* and might disrupt membranes.[114] Dopamine and its metabolites inhibit the conversion of protofibrils to fibrils and may promote protofibril accumulation. This could partly explain the selective vulnerability of dopamine neurons to α-synuclein-mediated toxicity.[119] Since α-synuclein oligomers function as inhibitors of dopamine release, which provides a clue, at the molecular level, to their neurotoxicity.[119a] Intervention in the early part of the aggregation pathway by prevention of oligomer formation or increased clearance may be neuroprotective.[120]

Oligomeric species have been isolated from cells, from human and mouse brain. Accumulation of toxic α-synuclein oligomers within endoplasmic reticulum occurs in α-synucleinopathies *in vivo*.[121] While small-sized oligomers are not resistant to proteinase K digestion, the generation of soluble oligomers and aggregates consisting of fibrillar α-synuclein resistant to proteinase K digestion is required for the induction of neuronal degeneration. Fibrillar and protofibrillar α-synuclein variants also cause divergent axonal lesions, exemplifying that they induce neurotoxicity by various means.[41] Damage of dopamine synaptic vesicles in nigral neurons by pathological interaction with α-synuclein induces a vicious circle of dysregulated cytosolic dopamine and further damage to targeting dopamine neurons.[122] Oligomeric forms of α-synuclein were shown to increase synaptic transmission and to impair long-term potentiation, mediated by NMDA (*N*-methyl-D-aspartate) receptor activation.[123] The UPS renders mutated or damaged proteins less toxic than their soluble forms, which suggests that the ubiquitinated proteins in LBs may be a manifestation of a cytoprotective response designed to stabilize and detoxify otherwise toxic proteins, and to eliminate damaged cellular components and to delay neurodegeneration. A novel hypothesis on the pathogenesis of PD suggests that functional α-synuclein is critical to cell survival and that its reduction, whether through aggregation or reduced expression, may lead to neurodegeneration in PD.[48] Although the direct mechanism whereby α-synuclein potentiates neuronal survival is yet to be elucidated, a growing body of evidence demonstrates that in certain conditions α-synuclein serves to promote neuronal survival.

Recent studies of an animal model of PD indicate that attachment of CL1 peptide to the C-terminal of α-synuclein enhanced its toxicity to dopaminergic neurons in an age-dependent manner and induced the formation of Lewy-body-like α-synuclein aggregation in the SN.[124] These results suggest that enhanced aggregate formation of α-synuclein may not be protective, contrary to some previous suggestions.[24] Intermediate-stage prefibrillary oligomers and protofibrillary forms can induce damage to neurons both *in vitro* and *in vivo*, presumably by forming pores in membranes.[114] However, further studies suggest that α-synuclein oligomers do not necessarily form pore-like structures but that local structural rearrangements of the protein lead to insertion of specific regions into the hydrophobic core of the lipid bilayer, thereby disrupting the lipid packing of cellular membranes.[125] With normal aging, the protein is concentrated in the cell body and when coupled with post-translational modifications (*i.e.*, phosphorylation, nitration, truncation, *etc.*), or enhanced expression due to genetic multiplication or promoter polymorphisms that increase the stability of oligomers and decrease the stability of tetramers, the likelihood of fibril formation increases. Once insoluble fibrils appear, the reaction will be further pushed in favor of the protection of more insoluble inclusions, the sequestration of α-synuclein into these inclusions will bring the concentration of soluble α-synuclein at the synapse below that which is required for proper cellular functions. Neuron dysfunction leads to neuronal death and the clinical and neuropathological manifestation of PD and other synucleinopathies.

In conclusion, formation of toxic oligomers, loss of functional soluble protein, and dysregulation of crucial cellular processes may share equal burden in the formation of events that either sensitize the neuron to other insults, or sets in motion a cascade of events that itself culminates in cell death.[48]

8.4.2 Mitochondrial Involvement in Parkinson's Disease

Mitochondrial alterations are important in the multi-factorial pathogenic process of PD.[12,55,126–129] Mitochondria are involved in many critical pathways, like ATP (adenosine triphosphate) generation, regulation of the electron transport chain, calcium homeostasis, mitochondrial dynamics, microtubule-dependent cellular traffic, the autophagy/lysosomal pathway, and programmed cell death. Mitochondrial dysfunction triggering increased free tubulin destabilizes the microtubule network and promotes α-synuclein oligomerization.[130] Mis-folded α-synuclein accumulates within both the inner and outer mitochondria membrane and can induce dysfunction and fragmentation of mitochondria, causing energy depletion, which is relevant for neuronal viability.[131] α-Synuclein may impact transcriptional regulation *via* numerous promoters, which have effects on mitochondrial function.[132] Specific interaction of α-synuclein and cyto-oxygenase (COX), the key enzyme of the mitochondrial respiratory chain, suggests that α-synuclein aggregation contributes to mitochondrial dysfunction. α-Synuclein controls mitochondrial Ca^{2+} homeostasis by modulating the interaction between mitochondria and the endoplasmatic reticulum.[36] Overexpression of α-synuclein impairs mitochondrial Complex I function, resulting in its reduced activity, increased free radical production, and mitophagy, decreasing respiration and increasing free radical production, or Complex IV activity. It may bind to mitochondrial membranes leading to fragmentation, loss of transmembrane potential and neuronal death.[131] Endogenous mitochondrial neuronal uncoupling proteins play a vital role in protecting neurons against various pathogenic stresses associated with PD.[33,133] The mitochondrial chaperone protein TRAP1 mitigates α-synuclein toxicity.[133] Dysfunction in both the mitochondrial and autophagic pathways and oxidative stress are important.[134] Mitochondrial metabolic control of microtubule dynamics impairs the autophagic pathway,[135] and disorders of mitochondrial axonal transport may impair synaptic transmission.[136] While α-synuclein impairs the normal dynamics of mitochondria in animal models of PD,[137] others do not support a direct association between mitochondrial respiratory chain dysfunction and α-synuclein pathology.[138] Mitochondrial dysfunction has been seen in a chronic PD-relevant model using rotenone, representing very early changes in mitochondrial dynamics that may precede neuronal death. α-Synuclein disrupts mitogen-activated protein kinase (MAPK)-controlled stress signaling in yeast and human cells, which results in inefficient cell protection responses and cell death, supporting a unique mechanism of PD pathology.[139]

Structural changes of mitochondria occur with even low overexpression of α-synuclein and in the virtual absence of structural defects in other intracellular organelles.[131] A reduction in α-synuclein results in the disruption of cellular processes such as mitochondrial function and autophagy.[36] Consequently, α-synuclein-related pathology and aggregation represents a *de facto* loss-of-function mechanism of toxicity.[48] Which α-synuclein species cause these effects is not clear. Its effects on mitochondria could be related to interactions with (pre)synaptic vesicles, the lysosomal membrane or the endoplasmic reticulum–Golgi apparatus.[16] They may induce release of oxidative species, which in turn leads to secondary induction of α-synuclein oligomerization, and aggregation, thus, creating a vicious cycle. Reduction of cerebral mitochondrial metabolism occurs early in PD, but whether mitochondrial dysfunction is a primary or secondary event, or part of a multi-factorial pathogenic process remains to be elucidated.[140] Much evidence suggests the involvement of α-synuclein in mitochondrial dysfunction, oxidative damage, protein misfolding, abnormal autophagy, and respiratory chain deficits in the pathogenesis of PD,[141] but to what extent dysregulated mitochondrial function contributes to sporadic PD remains to be elucidated.

8.4.3 Lysosomal Dysfunction and Autophagy

Overexpression of α-synuclein impairs macroautophagy, a main route for clearance of aggregate-prone intracytoplasmic proteins, whereas its depletion enhances this pathway. Increases in macroautophagy lead to a decrease in α-synuclein load and impairment of neuronal function, while its inhibition protects against toxic effects. Aberrant α-synuclein can bind to the membranes of lysosomes, inhibiting CMA,[142] lysosomal function, and the proteasome. Dysfunction of both the UPS and autophagy in PD has been established by both animal and cell culture models.[32] Deficiency of ATP13A2 (PARK9) causing Kufor–Rakeb syndrome (early-onset Parkinsonism, pyramidal degeneration and dementia) leads to lysosomal dysfunction, α-synuclein accumulation and neurotoxicity[143] *via* mitochondrial dysregulation.[144] In human post-mortem tissue, UPS components and activities as well as constituents of ALP are reduced in the *substantia nigra* of PD patients compared to controls.[32]

A close relationship between mitochondrial function and autophagy/mitophagy, which is crucial for degradation of surplus or damaged mitochondria, is necessary to orchestrate numerous metabolic pathways in the cell.[134] SNARE has a key role both in the genesis of autophagosomes and in autophagosome-lysosomal function.[134a] Defects in one of these elements could impair the other, resulting in increased risk for disease.[145] Increases in macroautophagy lead to decreased α-synuclein load and improvements of neuronal function, whereas its inhibition protects against toxic effects. Aberrant α-synuclein affects the lysosome, its degradation becomes diminished, and further lysosomal damage occurs, suggesting that dysfunctional autophagy is one of the failing cellular mechanisms involved in the pathogenesis of PD.[146] Despite increased α-synuclein accumulation with A53T mutations; it is not increased above that observed in sporadic PD; suggesting that mutated α-synuclein protein is not removed by macroautophagy.[147] The pathways inducing autophagic cell death or programmed necrosis, which is relevant to PD, include the death receptors, PCL2 family, caspases, calpain, edk5, p53, autophagy, mitophagy, and mitochondrial fragmentation. These have been evidenced by postmortem PD studies and various genetic models.[148] The complex relationship between autophagy and oxidative stress in PD has been reviewed.[149]

8.4.4 Oxidative and Nitrative Injuries

In PD, many compromised anti-oxidant systems are suggested to underlie cellular vulnerability to progressive oxidative stress, which generates excessive reactive oxygen species (ROS) or free radicals in the SN with subsequent cell damage. Overexpression of human wt or mutant α-synuclein elevates the aggregation of intracellular ROS, and increases the cytotoxicity of dopamine oxidative products. Nitration of α-synuclein, signifying the presence of reactive nitrogen species, is a major signature of PD.[150]

Increase of iron in the SN with a shift in the iron ratio Fe(II):Fe(III) of 2:1 compared to 1:2 in controls can promote dopamine synthesis, with increased generation of reactive metabolites. α-Synuclein increases cellular ferrireductase activity and Fe(II) levels in dopaminergic neurons leading to their selective loss.[53] This suggests that iron and α-synuclein act in concert for disease propagation. Peroxynitrite, formed by reduced superoxide dismutase (SOD), induces the aggregation of α-synuclein *in situ*, and nitrated α-synuclein is found in the core of LBs.[150] Formation of α-synuclein protofibrils is stimulated by translational modifications that occur under conditions of oxidative stress, while its aggregation is inhibited by anti-oxidants and proteins with chaperone activity.[47] Advanced glycation products (AGEs) promote deposition of abnormal proteins, which in turn sustain local oxidative stress and inflammatory response leading to neurodegeneration.[53,151] These findings indicate a multi-component process in its pathogenesis and cell death pathways.[152]

8.4.5 α-Synuclein and Neuroinflammation

α-Synuclein can trigger inflammation and activation of microglia, which, by releasing toxic factors or by phagocytosing cells, and degrading α-synuclein more avidly than neurons or astrocytes form a self-perpetuating cycle for neurodegeneration.[153] Overexpression of mutant α-synuclein modulates microglia cells releasing pro-inflammatory cytokines, nitric oxide, elevated levels of arachidonic acid metabolizing enzymes, oxidative stress, and excessive levels of reactive oxygen species triggering more inflammation. This supports the notion that α-synuclein has an important role in the initiation and maintenance of inflammation through activation of a pro-inflammatory response in microglia.[154] This may occur after neuronal death, but it is also possible that α-synuclein is released *via* exocytosis or that cleaved portions are presented *via* anti-gene presentation, which could lead to a vicious cycle of inflammatory response, release of α-synuclein and further inflammation.

In PD, nigral cell degeneration is associated with astroglial reaction and HMC class II positive microglia that may be both inducing factors or sequelae of neuronal death.[155] Although a specific receptor for α-synuclein binding to microglia is still unknown, these cells can take up extracellular α-synuclein,[154] which in turn triggers the release of immune modulatory mediators. Neuroinflammation in PD is a double-edged sword that is protective in the early stage of neurodegeneration but becomes detrimental with disease progresssion.[156] How neuroinflammation may contribute to the prion-like behavior of α-synuclein was reviewed recently.[157]

8.5 α-Synuclein Interaction with Other Proteins

Overlap between synucleinopathies and other protein opathies suggests interactions of pathological proteins engaging common downstream pathways.[158] The co-occurrence of α-synuclein, tau and other proteins in various neurodegenerative disorders[159] highlights the interface between these mis-folded proteins, which may be co-aggregated in the same brain, in the same region or even in the same cell in human brain and transgenic mice. Accumulated α-synuclein (promoted by oxidative stress) has a stimulatory effect on tau phosphorylation by glycogen-synthase kinase-3β(GSK-3β), while Hsp70 may suppress α-synuclein-mediated tau phosphorylation in initial disease stages.[160] Interaction with tubulin suggests that α-synuclein could be a microtubule-associated protein similar to tau. PD-associated risk factors, *e.g.*, environmental toxins and α-synuclein mutation, may promote tau phosphorylation, causing microtubule instability, which leads to neuronal loss in the PD brain. The independent and joint effects of the SCA and MAPT (tau) genes in PD have been described,[161] and the MAPT H1 haplotype may be a risk factor for PD.[85] The α-synuclein and tau genotypes interact to influence the progression of PD, which is more prominent in the early disease stages.[162]

Whereas α-synuclein can spontaneously polymerize into amyloidogenic fibrils *in vitro*, tau polymerization requires an inducing agent, *e.g.*, α-synuclein seeds.[163] Oxidatively modified α-synuclein is degraded by the proteasome and plays a pro-aggregation role for tau, and is an *in vivo* regulator for tau phosphorylation at Ser262 leading to deposition of both proteins,[164,165] and further promotes the recruitment of tau to protein inclusions in oligodendroglial cells. The E46K modification of α-synuclein may induce both Lewy-like and tau pathologies,[166] while tau enhances α-synuclein aggregation and toxicity and disrupts inclusion formation in cellular models.[167]

Increased accumulation of phospho-tau in the *striata* of PD patients and in the A53T transgenic mouse model,[168] related to increased activity of GSK-3β, is stimulated by α-synuclein that associates with the actin cytoskeleton, and by GSK-3β.[160,164] Tauopathy in PD *striata* is restricted to dopaminergic neurons, whereas degeneration of the frontal cortex, associated with increased α-synuclein deposits because of reduced proteasomal activity, is not associated with tauopathy.[97] Thus, tauopathy in PD has a restricted pattern of distribution,[169] which differs from its generalized affect in AD.

There are strong interactions between α-synuclein, tau and β-amyloid, particularly in their oligomeric forms, which might synergistically promote their mutual aggregation and *vice versa*.[170,171] Cross-seeding between dissimilar proteins that share β-sheet structures has been described, *e.g.*, for tau and α-synuclein.[172] Tau phosphorylation was found in synapse-enriched fractions of the frontal cortex in PD and AD [173] and in brainstem of α-synuclein mice. Other links are suggested by the colocalization of both proteins in neurofibrillary tangles and LBs, especially in neuronal populations vulnerable for both aggregates.[174] Co-occurrence of abnormal depositions of tau, α-synuclein and TDP-43 in AD, DLB and other neurodegenerative diseases highlight the interface between these disorders.[175]

Others have suggested that β-amyloid is more likely to promote the deposition of α-synuclein than tau, indicating synergistic actions of β-amyloid and α-synuclein.[176] β-Amyloid and α-synuclein act as seeds and affect each other's aggregation, which may explain molecular interactions between AD and LBD pathologies.[177] β-Amyloid peptides enhanced α-synuclein accumulation and neuronal deficits in a transgenic mouse model, while α-synuclein-induced synapse damage is enhanced by β-amyloid-42,[178] and LB formation may be triggered, at least in part, by AD pathology. Protein clearance mechanisms of α-synuclein and β-amyloid in LB disorders have been reviewed recently.[179]

PD and AD could be linked by progressive accumulation of phospho-tau, GSK-3β, and α-synuclein, and activation of caspase and caspase-cleft delta-tau may represent a common way to abnormal intracellular accumulation of both α-synuclein and tau, promoted by β-amyloid deposition, thus unifying the pathology of AD and LB disease. This suggests a complex continuum characterized by variable amounts of pathological proteins generated by the same stimulus probably depending on genetic and environmental factors. Despite documented colocalization of α-synuclein and tau in LBs,[174] the basic mechanisms (differences in proteasomal and GSK-3β activities, oxidative stress in the presence of α-synuclein, *etc.*), of the synergistic effects of α-synuclein, phospho-tau, β-amyloid, and other proteins, suggesting a dualism or triad of amyloidogenic neurodegeneration, remain to be elucidated.

8.6 α-Synuclein Spread and Disease Propagation

Mounting evidence implicates that cell–cell transmission of disease-specific proteins and their propagation underlies disease propagation in neurodegenerative disorders by a transneuronal spread through neural networks,[180,181] and also to other organs. The concept that α-synuclein lesions ramify within the CNS by a seeding-like process is supported by the fact that fetal dopamine transplants in the *striatum* in PD patients surviving more than five years may develop α-synuclein-positive LBs. These data imply a host-to-graft propagation, and a neuron-to-neuron transmission or trans-synaptic spread of mis-folded α-synuclein appears important for the propagation of PD, and that LBs in transplanted dopaminergic neurons may develop similarly to those in the host SN.[10] Neural grafts placed into transgenic mice expressing human α-synuclein take up the human protein and form α-synuclein-positive aggregates,[182] and intracerebral inoculation of synthetic α-synuclein initiates a rapidly progressive neurodegenerative synucleinopathy in mice.[180] Fibrillary α-synuclein internalized by primary neurons is transported in axons in both directions, but not by trans-synaptic transfer,[183] and can be transported along peripheral nerves from the gut to the brain.[20]

The effects of LBs in the grafted neurons are unclear. Oligomers of α-synuclein can recruit and aggregate endogenously, expressed by cortical neurons, and this effect increases with time and with concentration of applied oligomers.[184] Secreted α-synuclein can recruit endogenous forms in the recipient cells, act as a permissive template, and promote the mis-folding of small aggregates. Some of the α-synuclein uptake from the extracellular space may occur *via* endocytosis, although additional mechanisms such as *via* exosomes, might also contribute.[185] Secretion of α-synuclein and disease propagation have been reviewed.[186] Changes in the properties of serine-129-phosphorylated α-synuclein occur with progression of Lewy pathology.[183a]

α-Synuclein fibrillation starts *in vitro* with soluble oligomers forming a nucleus, but once the nucleus forms, aggregates form rapidly. Accumulation of pathologic α-synuclein leads to decrease of synaptic proteins, impairment of neuronal connectivity and neuron death.[180,187] Secreted forms of α-synuclein may be biologically important because of their potential for causing paracrine effects on adjacent cells, decreasing the viability of recipient neurons in a concentration-dependent form, and this effect is largely mediated by oligomeric species.[28,184] Extracellular α-synuclein may play a role in disease propagation[123] and could further trigger a neuroinflammatory response through microglial activation.[5,16] Extracellular monomeric α-synuclein can directly translocate across plasma membrane and enter neuronal cells without being degraded by the proteolytic system.[186b] Antibody-aided clearance of extracellular α-synuclein prevents cell-to-cell aggregate transmission.[186a]

Like prion protein PrP, α-synuclein is unstructured in aqueous buffers, while adopting an α-helical structure when membrane-bound, which can become β-sheet when present at high concentration or in mutant form. *In vivo* approaches in cell culture could not discriminate between a "prion-like" mechanism – host-derived translocated α-synuclein inducing its mis-folding generated in the graft – *versus* simple translocation of the protein from the host to the graft.[182] Prion infection promotes the accumulation of α-synuclein in aged human α-synuclein mice,[188] suggesting that α-synuclein pathology could be induced in cells and spread by a "prion-like" mechanism transmitting the conformationally altered α-synuclein.[10] There is also direct evidence that aggregated α-synuclein can be transmitted from affected nerve cell to healthy dopamine neurons, thereby triggering the neurodegenerative process.[9] Although the mechanism of seeding remains uncertain, there is evidence that prions can be conveyed between neurons by trans-synaptic transport. Thus, the propagation of α-synuclein lesions by cell-to-cell passage appears to be similar to that in other neurodegenerative disorders.[189–191] Early sites of Lewy pathology in PD are the olfactory bulb and enteric plexuses, lending support to the "dual hit hypothesis" suggesting that pathogenic α-synuclein may reach the brain *via* a consecutive network of projection neurons,[191] *e.g.*, from the olfactory system and lower brainstem to the cortex, which may be influenced by changes of conformation of α-synuclein in the olfactory system or gut.[190a] However, recent studies found no evidence to support concerns that proteins underlying PD transmit disease to humans.[190b]

8.7 Neuropathology of Lewy Body Disorders

8.7.1 Sporadic Parkinson's Disease

In sporadic PD, the essential neuropathology is neuronal loss not only in the dopaminergic SN but in many other parts of the central, peripheral, and autonomic nervous systems and other organs, associated with α-synuclein-positive Lewy pathology throughout these systems.[2,192] The recently improved criteria for the neuropathological diagnosis of PD require these two key features – neuronal loss in the SN *compacta*, and Lewy pathology.[107] Standardized methods for the assessment of these changes by use of a semi-quantitative grading system and immunohistological methods for the detection of Lewy pathology have been proposed.[14,107,193] Recent studies have confirmed the multi-organ distribution of α-synuclein and Lewy pathologies, with negative involvement of major parts of sensory components in the nervous system,[194] and the muscular-skeletal system and sciatic nerve.[8] Spinal cord lesions in sporadic PD were reviewed.[195]

8.7.1.1 Formation and Development of α-Synuclein/Lewy Pathology

Biochemical increase of phospho-α-synuclein precedes α-synuclein aggregation which is followed by formation of LBs and dystrophic neurites,[196,197] but does not necessarily correlate with LB staging.[198] Loosely packed α-synuclein filaments as premature "pale neurites" are initiated at axon collaterals and extend centripetally into proximal segments.[199] These changes and the formation of "pale bodies" precede the development of LBs. The early intra-axonal aggregation of α-synuclein could damage the parental neurons by interfering with axonal transport.[200] However, the presence or absence of immunostaining for α-synuclein cannot be interpreted as evidence that the cell suffers or is free of dysfunction, but also could reflect a successful response to proteolytic stress. Reduced thyrosin hydroxylase immunoreactivity in neurons may represent a cytoprotective mechanism, but it can also be preserved in neurons with α-synuclein accumulation.[2,7] Changes in the properties of Ser129 phosphorylated α-synuclein occur with progression of Lewy-type histopathology in human brains.[201]

LBs occur in two types. "Classic Lewy bodies" are spherical cytoplasmic intraneuronal inclusions, 8–30 μm in diameter with a hyaline eosinophilic core, concentric lamellar bands, and a narrow pale-stained halo, forming single or multiple inclusions. Ultra-structurally, they are non-membrane-bound, granulofilamentous structures composed of radially arranged, 7–20 nm intermediate filaments associated with electron-dense granular material and vesicular structures, the core showing densely packed filaments and dense granular material and the periphery with radially arranged 10 nm filaments. "Cortical Lewy bodies" are eosinophilic, rounded, angular, or reniform structures without halo, are poorly organized with a felt-like arrangement composed of 7–27 nm wide filaments, and are mostly devoid of a central core.[202] α-Synuclein adopts an altered three-dimensional structure and undergoes N-terminal ubiquitination but the mechanism of its aggregation serving as a nidus for LB formation *in vivo* has not yet been elucidated. Both classic and cortical LBs share immunochemical and biochemical characteristics, the major components being p-α-synuclein, ubiquitin, phosphorylated neurofilaments and many other substances.[2,7] Sequestration of choline acetyltransferase and tyrosin hydroxylase within LBs, suggest that they may disrupt cholinergic and catecholaminergic neurotransmitter production.[203] Demonstration of the autophagy adapter protein NBR1 in LBs[204] and of ubiquilin-1 (UBQLN1) in cytoplasmic and nuclear inclusions in several proteinopathies[205] suggests that it is involved in their formation.[204] α-Synuclein can be recovered from purified LBs from PD and DLB brain. Colocalization of α-synuclein, synphilin and PARKIN within LBs suggests that PARKIN plays a role in post-translational modification of α-synuclein, which results in changes in protein size, structure-enhancing fibrillation and formation of LBs.[2] Proteomic analysis of cortical LBs revealed 296 proteins related to multiple or unknown functions; in brainstem Lewy bodies 90 proteins were identified, whereas another study identified 1263 proteins in the SN.[196,206]

The formation of LBs runs through several phases. Classic LBs show initial intraneuronal appearance of dust-like α-synuclein particles related to neuromelanin or lipofuscin, homogenous deposition of α-synuclein and ubiquitin in the center, stepwise condensation to ubiquitinated filamentous inclusion, and final degradation to extraneuronal LBs after disappearance of the involved neuron.[196] Cortical LBs show diffuse α-synuclein and ubiquitin labeling, while subcortical LBs have a distinct, central ubiquitin domain with α-synuclein appearing in the periphery and ubiquitination being the later event. Initial accumulation of α-synuclein in the neuronal cytoplasm is followed by stepwise accumulation of dense filaments, spreading to dendrites, deformation of LBs, and final degradation by astroglial processes. LBs are accompanied by dystrophic neurites, which according to three-dimensional studies, may evolve into LBs, with ubiquitin at the core and neurofilaments at the outermost layer.[207] The role of LBs in promoting neurotoxicity or neuroprotection remains poorly understood.

8.7.1.2 *Relationship between α-Synuclein and Lewy Pathology*

Semi-quantitative assessment of LBs in a large autopsy series of PD, proposed a staging of the spread of Lewy pathology to designate a predictable sequence of lesions in the brain beginning in the lower brainstem and anterior olfactory nucleus with caudo–rostral progression to the neocortex.[96,194] The validity of this six-stage scheme, which corresponds roughly to the classification of LB disorders into three phenotypes – brainstem predominant, limbic/transitional, and diffuse neocortical – has been a matter of vigorous debate, since between 6.3 and 47% of all cases of autopsy-proven PD and 18% of incidental LB disease did not follow a caudo–rostral spread of Lewy pathology.[198,208–210] Whereas the old classification left 45–50% of individuals unclassified,[211] a recently proposed unifying system for LB diseases, distinguishing four stages, correlates α-synuclein pathology with nigrostriatal degeneration, reduction of striatal tyrosin hydroxylase concentration, nigral cell loss, and clinical scores.[208] This is supported by an increase of phosphorylated α-synuclein restricted to the olfactory bulb and brainstem in the early stages of Lewy pathology.[197] Despite several limitations, the brain pathology for most patients with typical PD can be predicted using Braak's scheme.[212]

Recent assessment of clinical markers for the starting of PD estimated a pro-dromal interval of about 4.5 years, while DLB appeared to have a slower progression.[213] The duration and severity of motor dysfunction in PD, the corresponding decrease of dopamine transporter (DAT) and vesicular monoamine transporter 2 (Vmat2) immunoreactivity in the *striatum* are inversely correlated with the total α-synuclein burden and neuronal loss in the SN,[214,215] but not with LB counts in the SN, which supports the concept of synaptic dysfunction and/or impairment of axonal transport.[216]

Recent studies suggested that nigral dopaminergic cell loss rather than α-synuclein, tau or amyloid-β pathology are associated with decreased striatal 123IH-CIT binding in the *striatum* in autopsy-confirmed subjects with AD, DLB, and PDD, although dysfunctional nigral neurons may have an additional affect on striatal uptake *in vivo*. Other recent studies showed that dopaminergic nigral neurons inhibit striatal output through release of GABA, whose transport is dependent on Vmat2.[217]

About 15% of SN neurons contain LBs and may survive for 7.5 years (2% loss per year after clinical diagnosis of PD).[218,219] Although α-synuclein aggregates may be cytotoxic, LBs are considered markers of ongoing neuronal damage, or might even be harmless end-products of sequestration of toxic molecules as a type of cell-protective mechanism,[2,196] rather than a cause of neuronal cell loss. However, the mechanism responsible for the regulation of the machinery that handles toxic waste by segregating it into aggregates is still poorly understood.

Overexpression of human α-synuclein in rat nigral neurons leading to deficient dopamine release preceding outright neuron loss *via* decreased presynaptic vesicle density suggests that lack of dopamine is due to axonal fiber loss.[220] This assumption was recently confirmed in a spontaneous autosomal recessive rat model.[221] Decline in axonal transport proteins (kinesin) preceding other nigral cell pathology and alterations in dopaminergic markers (tyrosin hydroxylase) in early stages of sporadic PD were recapitulated in a rat genetic PD model, supporting the hypothesis of dopaminergic neurodegeneration following a "dying-back" pattern involving axonal transport disruption.[101] Recent studies showed lower neuron densities in the *substantia nigra* before

Lewy deposition in the nigrostriatal system, suggesting that neurodegeneration and cerebellar dysfunction precedes Lewy pathology in the *substantia nigra*.[220a] Myenteric neuron loss despite α-synucleine aggregates in the gastrointestinal tract is not a feature of PD.[222]

This supports a dying-back mechanism in diseases with neuronal α-synuclein pathology in which dysfunction starts at the synapse and leads to axonal degeneration and α-synuclein accumulation in LBs and Lewy neurites.[122] These data and the recent demonstration of accumulation of small α-synuclein aggregates at presynaptic terminals in PD and DLB suggest that related synaptic dysfunction and axon degeneration, not nerve cell loss, may be the primary determinant of progression of the neurodegeneration in PD, and loss of neurons is an epiphenomenon after the loss of synapses, defining PD as a "synaptopathy".[223,224] Pre- and postsynaptic sites may represent adequate targets for early therapeutic intervention.[224]

8.7.2 Dementia with Lewy Bodies and Parkinson's Disease

DLB and PDD are considered to be parts of an α-synuclein-associated disease spectrum. DLB, a progressive disorder associated with some core clinical neuropsychiatric features, is considered to be the second most common neurodegenerative dementia syndrome in the elderly.[225] An arbitrary cut-off is used: if dementia develops first or within one year of PD diagnosis, then DLB is diagnosed, while if PD follows first or within one year of PD diagnosis, this suggests PDD. Distinction between both disorders may be difficult.[7,226] Their pathological hallmarks are α-synuclein/Lewy pathology or a variable mixture of LB and AD pathologies, which may interact synergistically.[227] Both cortical and subcortical α-synuclein lesions have been suggested to be predominant,[228] whereas for others, AD lesions were more important, the β-amyloid load being similar to that in AD.[229] Occipital and posterior parietotemporal hypometabolism is a distinguishing feature of DLB that is independent of amyloid pathology.[230] The severity and extent of α-synuclein are scored semiquantatively in specific brain regions.[225,231] Neurofibrillary tangles, being frequent in both DLB and PDD, are often restricted to limbic regions, while extensive tau pathology may be absent. Between 10 and 50% of PDD cases had enough AD lesions to attain the pathological diagnosis of definite AD, but this may also be related to higher Braak LB stages in the absence of significant AD pathology.[232] A recent clinicopathologic study identified three subgroups of PDD: (1) predominantly synucleinopathy (Braak LB stages 5–6; 38%); (2) synucleinopathy with β-amyloid deposition but minimal or no tau pathology (59%); and (3) synucleinopathy and AD with considerable neocortical tauopathy (Braak tau stages 5–6; 3%). Patients in group II had significantly shorter survival than those with pure synucleinopathy.[233] Reduced cortical cholinergic innervation in both DLB and PDD is similar and more severe than in AD, while synaptic loss in DLB is similar to that in AD.[233] Elevated ApoE ε4 frequency in DLB and PDD cases, in which the overall brain neuritic plaque burden was lower, indicate that ApoE ε4 may contribute to neurodegeneration through mechanisms unrelated to amyloid processing.[234]

Despite many similarities between DLB and PDD, there are some morphological differences, in particular, higher amyloid plaque load in the *striatum*, usually absent in non-demented PD,[235,236] and more severe α-synuclein load in hippocampal CA 2/3 areas.[2] DLB cases had more severe β-amyloid load than PDD, while others reported higher β-amyloid load in cortical and subcortical areas.[227] DLB is usually associated with only mild cerebrovascular lesions, showing an inverse relationship with the severity of Lewy pathology.[237]

In conclusion, both DLB and PDD show heterogenous pathology and neurochemistry, which depend on the different lesion patterns, supporting the hypothesis that these α-synuclein-related disorders and AD share a common underlying molecular pathogenesis.

8.8 Animal Models of Parkinson's Disease

Experimental models suggested to shed light on the neuropathobiology of PD come from essentially four sources: pharmacological, *e.g.*, reserpine, toxic *e.g.*, MPTP (the pyridine toxin 1-methyl-4-phenyl-1,2,3,6-tetrahydropyridine),[238] rotenone and paraquat,[239–241] transgenic,[59,242,243] and cellular.[244] During recent years a myriad of different models carrying mutations similar to those found in humans in *Drosophila melanogaster*,[245,246] *Caenorhabditis Elegans*,[247] rodents,[221] and mamalians,[248] have been developed to study the pathogenesis of this disease. Although some genetic models reproduced the key features of PD, they did not succeed in reproducing both the pathology and progressive degenerating process in human PD.[249–252] A recently developed BAC α-synuclein transgenic rat developing misfolding and modification of human α-synuclein and a behavioral phenotype recapitulating some changes in PD, associated with severe impact on dopaminergic nerve terminal degeneration, is concomitant with accumulating evidence correlating α-synuclein aggregation to synaptic pathologies in disease progression in PD.[252] The sole model of DLB reported is an α-synuclein transgenic mouse.[253]

Viral PD models comprising α-synuclein and LRRK-2-based overexpression, or mimicking PARKIN loss of function by overexpression of PARKIN substrates,[254] viral-vector-mediated α-synuclein lesion as a chronic and progressive model[119] and other recent genetic models may provide valuable insights into PD mechanisms in order to contribute to the development of therapeutic targets. The relevance of both pathogenic and etiologic models and their limits for new therapeutic studies have been reviewed recently.[48,249,255]

8.9 α-Synuclein as a Biomarker for Synucleinopathies

Although α-synuclein is the main constituent of LBs, its detection in the cerebrospinal fluid (CSF) and plasma of PD patients has yielded inconsistent results.[256,257] However, increased phosphorylated α-synuclein was detected in PD samples, which was not the case for oligo-α-synuclein or oligo-*p*-α-synuclein.[258]

Confirming reports about relatively low CSF α-synuclein concentrations in both PD and DLB,[259,260] recent studies showed significantly lower α-synuclein levels in PD, DLB.[261,262] CSF levels of α-synuclein oligomers were increased in PD patients,[263,264] while others were unable to detect them in PD.[265]

Recently, alterations in phosphorylated α-synuclein in the CSF of PD patients were detected.[266] However, CSF α-synuclein alone did not provide relevant information for PD diagnosis, showing low specificity, but a better performance was obtained with the total tau to α-synuclein ratio.[261] Different levels of CSF biomarkers have been reported in different phenotypes of PD,[267] while others did not see such biomarker changes.[268] Low CSF levels of L-DOPA and its metabolites (dihydroxyphenylacetic acid or DOPAC and dihydroxyphenylglycol or DHPG) indicate central dopaminergic and noradrenergic lesions in synucleinopathies.[269] The role of biomarkers in LB disorders was reviewed recently.[270,270a]

CSF α-synuclein is currently unsuitable for differentiating between PD and atypical Parkinsonism,[271] while the level of neurofilament light chain alone may allow it.[272] Postmortem CSF levels of oligomeric α-synuclein and phosphorylated α-synuclein (PS-129) did not distinguish PD and DLB from progressive supranuclear palsy (PSP) or control groups,[273] while PS-129 CSF concentration, when combined with total α-synuclein concentration, contributed to distinguishing PD from MSA and PSP.[139] These conflicting data imply the need for the development of novel techniques to specifically visualize pathological proteins in biological fluids, e.g., total/oligomeric α-synuclein and DJ-1,[274] and a combined approach may be the most promising way for the identification of (imaging and protein) biomarkers in LB disorders.[270,275] Serum concentrations of α-synuclein and its antibodies are not significantly affected in PD,[275a] while others detected reduced α-synuclein antibody levels in PD.[275b] The search for effective biomarkers for diagnosis and surveillance of PD continues. The usefulness of CSF endolysosomal enzyme levels, either alone or in combination with other markers, remains to be established in future studies.[276]

8.10 Conclusions and Outlook for the Future

α-Synuclein is a small, soluble neural protein with a predominantly presynatic location in the brain and in many other organs. Its physiological functions are not fully understood. Under pathological conditions, due to gene mutations and exogenous factors, or both. α-Synuclein undergoes post-translational changes and aggregation leading to the formation of insoluble intraneuronal and axonal deposits, LBs and Lewy neurites.[4] These histological hallmarks of PD and DLB are associated with loss of specific neuronal populations and destruction of specific neuronal networks causing multi-fold neurochemical changes and inducing a variety of clinical symptoms. These are sequelae of complicated molecular changes due to mitochondrial dysfunction, autophagy, oxidative and nitrative changes, disorders or calcium and iron homeostasis, neuroinflammation, and other deleterious factors leading to energy deficiency and cell death. The neurotoxicity of α-synuclein is suggested to be caused by soluble oligomers or intermediate proteins and not by insoluble aggregates. Whether LBs and other protein deposits are detrimental or protective is a matter of discussion. They may either be innocent bystanders or represent an end-stage of failed cytoprotective elimination of toxic proteins as a defense mechanism against processes underlying cell death. PD is considered the result of an encounter between one or several environmental triggers and one or more susceptibility alleles, the interaction of which produces a premotor syndrome followed by the clinical PD phenotype over a period of years or decades in the sense of a "complex disease".[277]

The demonstration of presynaptic deposition of α-synuclein in PD and DLB suggests they represent synaptopathies, and cell loss may be an epiphenomenon. Interaction of α-synuclein with other pathological proteins may explain the frequent overlap between various proteinopathies, *e.g.*, PD and DLB with AD. Prion-like interneuronal spreading of α-synuclein and other pathological proteins may represent a seminal mechanism of their propagation leading to disease progression.[10,172,174,175] While most of the available animal models do not exactly reproduce the molecular and morphological key features of human PD, new viral and genetic models may provide deeper insights into their pathogenesis. The demonstration that rotenone, an environmental toxin, promotes the release of α-synuclein by enteric neurons, which is taken up by presynaptic neurites and retrogradely transported to the soma where it accumulates, provided new insights into the initiation and progression of PD pathology.[278]

Increasing evidence suggests that α-synuclein is an interesting therapeutic target in PD,[279] but future clinical trials need more exact analysis of α-synuclein excreted into body fluids as possible biomarkers reflecting α-synuclein misfolding in the CNS, to enable a more accurate diagnosis as the basis of future therapies.[15,274] They may allow detection of α-synuclein pathology even before the onset or clinical signs and symptoms. Such surrogate markers of disease onset and progression would be important tools for clinical trials aiming to achieve disease modification. α-Synuclein synaptic pathology and other targets may be considered for the development of new pharmaceutical and gene-based strategies.[223,280] Early intervention into the aggregation process by development of ligands that can bind to mis-folded proteins by inhibition of its aggregation[120,281] and secretion,[186] modification of α-synuclein phosphorylation,[20,282] or truncation, are using enhanced degradation and proteolytic clearance of α-synuclein as potential therapeutic targets in PD.[32,277,283–288] Compounds that potentially dissolve vesicle-bound α-synuclein oligomers into monomers, leaving the lipid vesicles intact, may also provide a promising strategy for developing novel therapeutic drugs for PD.[289,290] Cholinesterase inhibitors, according to currently available evidence, have a positive impact on global assessment, cognitive function, behavioral disturbance, and activities of daily living rating scales, while the effect in DLB remains unclear.[291] Recent synthesis and *in vitro* evaluation of phenothiazine derivatives with high affinity and selectivity for aggregated α-synuclein may open further therapeutic aspects.[292] Intravenous immunoglobuline does not prevent α-synuclein aggregation, but may reduce its neurotoxicity through an unknown mechanism.[293] Other potential neuroprotective strategies for PD may be aimed at manipulating the mitochondrial pool,[294] at targeting both apoptotic and non-apoptotic pathways all of which may simultaneously occur in PD patients.[295] New synaptic and molecular targets for neuroprotection in PD,[296] trophic factor delivery,[297] and cell therapy for PD[298] were critically discussed. A new ABC α-synuclein transgenic rat model recapitulating both clinical and pathological features of PD will open up research for understanding the role of α-synuclein in dopaminergic synapse function and discovery of novel therapeutic methods. Recent studies in a transgenic mouse model that showed antibody-aided clearance of extracellular α-synuclein preventing cell-to-cell aggregate transmission, provides a basis for immunotherapy for PD/DLB.[298] The clinical implications of α-synuclein oligomers for PD and their role as potential treatment targets have been reviewed recently.[299,300,301] Immunotherapy with antibodies promoting removal of extracellular α-synuclein species may delay the progression of the disease.[186] Overcoming these problems to allow the development of novel candidate drugs to prevent or cure PD and other proteinopathies is one of the major challenges of modern neuroscience.

References

1. K. A. Jellinger in *Encyclopedia of Movement Disorders*, ed. K. Kompoliti and L. Verhagen Metman, Academic Press, Oxford, 2010, vol. 3, p. 203.
2. K. A. Jellinger, *Mov. Disord.*, 2012, **27**, 8.
3. K. A. Jellinger and P. L. Lantos, *Acta Neuropathol.*, 2010, **119**, 657.
4. M. Goedert, M. G. Spillantini, K. Del Tredici and H. Braak, *Nat. Rev. Neurol.*, 2013, **9**, 13.
5. L. Fellner, K. A. Jellinger, G. K. Wenning and N. Stefanova, *Acta Neuropathol.*, 2011, **121**, 675.
6. J. Zhou, M. Broe, Y. Huang, J. P. Anderson, W.-P. Gai, E. A. Milward, M. J. Porritt, D. W. Howells, A. J. Hughes, X. Wang and G. Halliday, *Acta Neuropathol.*, 2011, **121**, 695.
7. K. A. Jellinger, *Transl. Neurosci.*, 2012, **3**, 75.
8. T. G. Beach, C. H. Adler, L. I. Sue, L. Vedders, L. Lue, C. L. White III, H. Akiyama, J. N. Caviness, H. A. Shill, M. N. Sabbagh and D. G. Walker, *Acta Neuropathol.*, 2010, **119**, 689.
9. C. W. Olanow and K. McNaught, *Mov. Disord.*, 2011, **26**, 1056.
10. S. George, N. L. Rey, N. Reichenbach, J. A Steiner and P. Brundin, *Brain Pathol*, 2013, **23**, 350; (a) N. P. Visanji, P. L. Brooks, L.-N. Hazrati and A. E Lang, *Acta Neuropathol Comm.*, 2013, **1**, 2.
11. M. J. Devine, K. Gwinn, A. Singleton and J. Hardy, *Mov. Disord.*, 2011, **26**, 2160.
12. A. Pilsl and K. F. Winklhofer, *Acta Neuropathol.*, 2012, **123**, 173.
13. T. J. Collier, N. M. Kanaan and J. H. Kordower, *Nat. Rev. Neurosci.*, 2011, **12**, 359.

14. G. M. Halliday, J. L. Holton, T. Revesz and D. W. Dickson, *Acta Neuropathol.*, 2011, **122**, 187.
15. A. Bellucci, M. Zaltieri, L. Navarria, J. Grigoletto, C. Missale and P. Spano, *Brain Res.*, 2012, **1476**, 183.
16. K. Vekrellis, M. Xilouri, E. Emmanouilidou, H. J. Rideout and L. Stefanis, *Lancet Neurol.*, 2011, **10**, 1015.
17. A. Surguchov, *Int. Rev. Cell. Mol. Biol.*, 2008, **270**, 225.
18. A. Rekas, K. J. Ahn, J. Kim and J. A. Carver, *Proteins*, 2012, **80**, 1316.
19. O. Ullman, C. K. Fisher and C. M. Stultz, *J. Am. Chem. Soc.*, 2011, **133**, 19536.
20. S. P. Braithwaite, J. B. Stock and M. M. Mouradian, *Rev. Neurosci.*, 2012, **23**, 191.
21. T. Bartels, J. G. Choi and D. J. Selkoe, *Nature*, 2011, **477**, 107.
22. B. Fauvet, M. K. Mbefo, M. B. Fares, C. Desobry, S. Michael, M. T. Ardah, E. Tsika, P. Coune, M. Prudent, N. Lion, D. Eliezer, D. J. Moore, B. Schneider, P. Aebischer, O. M. El-Agnaf, E. Masliah and H. A. Lashuel, *J. Biol. Chem.*, 2012, **287**, 15345.
23. T. Gurry, O. Ullman, C. K. Fisher, I. Perovic, T. Pochapsky and C. M. Stultz, *J. Am. Chem. Soc.*, 2013, **135**, 3865.
24. L. Breydo, J. W. Wu and V. N. Uversky, *Biochim. Biophys. Acta*, 2012, **1822**, 261.
25. I. Dikiy and D. Eliezer, *Biochim. Biophys. Acta*, 2012, **1818**, 1013.
26. G. Comellas, L. R. Lemkau, A. J. Nieuwkoop, K. D. Kloepper, D. T. Ladror, R. Ebisu, W. S. Woods, A. S. Lipton, J. M. George and C. M. Rienstra, *J. Mol. Biol.*, 2011, **411**, 881.
27. R. Barbour, K. Kling, J. P. Anderson, K. Banducci, T. Cole, L. Diep, M. Fox, J. M. Goldstein, F. Soriano, P. Seubert and T. J. Chilcote, *Neurodegener. Dis.*, 2008, **5**, 55.
28. L. Stefanis, *Cold Spring Harb. Perspect. Med.*, 2012, **2**, a009399.
29. D. Ebrahimi-Fakhari, I. Cantuti-Castelvetri, Z. Fan, E. Rockenstein, E. Masliah, B. T. Hyman, P. J. McLean and V. K. Unni, *J. Neurosci.*, 2011, **31**, 14508.
30. M. H. Smith, H. L. Ploegh and J. S. Weissman, *Science*, 2011, **334**, 1086.
31. J. H. Claessen, L. Kundrat and H. L. Ploegh, *Trends Cell Biol.*, 2012, **22**, 22.
32. D. Ebrahimi-Fakhari, L. Wahlster and P. J. McLean, *Acta Neuropathol.*, 2012, **124**, 153.
33. P. Krumova, E. Meulmeester, M. Garrido, M. Tirard, H. H. Hsiao, G. Bossis, H. Urlaub, M. Zweckstetter, S. Kugler, F. Melchior, M. Bahr and J. H. Weishaupt, *J. Cell Biol.*, 2011, **194**, 49.
34. J. Levin, T. Högen, A. S. Hillmer, B. Bader, F. Schmidt, F. Kamp, H. A. Kretzschmar, K. Bötzel and A. Giese, *J. Parkinson's Dis.*, 2011, **1**, 205.
35. F. Febbraro, M. Giorgi, S. Caldarola, F. Loreni and M. Romero-Ramos, *Neuroreport*, 2012, **23**, 576.

36. T. Cali, D. Ottolini, A. Negro and M. Brini, *J. Biol. Chem.*, 2012, **287**, 17914.
37. A. Oueslati, M. Fournier and H. A. Lashuel, *Prog. Brain Res.*, 2010, **183**, 115.
38. K. E. Paleologou and O. M. El-Agnaf, *Subcell Biochem*, 2012, **65**, 109.
39. K. Prasad, T. G. Beach, J. Hedreen and E. K. Richfield, *Brain Pathol.*, 2012, **22**, 811.
40. Y. Machiya, S. Hara, S. Arawaka, S. Fukushima, H. Sato, M. Sakamoto, S. Koyama and T. Kato, *J. Biol. Chem.*, 2010, **285**, 40732.
41. G. Taschenberger, M. Garrido, Y. Tereshchenko, M. Bahr, M. Zweckstetter and S. Kugler, *Acta Neuropathol.*, 2012, **123**, 671.
42. M. R. Cookson, *Mol. Neurodegener.*, 2009, **4**, 9.
43. M. Xilouri, O. R. Brekk and L. Stefanis, *Mol. Neurobiol.*, 2013, **47**, 537.
44. M. A. Lynch-Day, K. Mao, K. Wang, M. Zhao and D. J. Klionsky, *Cold Spring Harb. Perspect. Med.*, 2012, **2**, a009357.
45. S. K. Mak, A. L. McCormack, A. B. Manning-Bog, A. M. Cuervo and D. A. Di Monte, *J. Biol. Chem.*, 2010, **285**, 13621.
46. G. Esposito, F. Ana Clara and P. Verstreken, *Dev. Neurobiol.*, 2012, **72**, 134.
47. J. C. Rochet, B. A. Hay and M. Guo, *Prog. Mol. Biol. Transl. Sci.*, 2012, **107**, 125.
48. N. M. Kanaan and F. P. Manfredsson, *J. Parkinson's Dis.*, 2012, **2**, 249.
49. S. Anwar, O. Peters, S. Millership, N. Ninkina, N. Doig, N. Connor-Robson, S. Threlfell, G. Kooner, R. M. Deacon, D. M. Bannerman, J. P. Bolam, S. S. Chandra, S. J. Cragg, R. Wade-Martins and V. L. Buchman, *J. Neurosci.*, 2011, **31**, 7264.
50. S. K. Mak, A. L. McCormack, J. W. Langston, J. H. Kordower and D. A. Di Monte, *Exp. Neurol.*, 2009, **220**, 359.
51. C. E. Chua and B. L. Tang, *Brain Res. Rev.*, 2011, **67**, 268; (a) J. Burre, M. Sharma, T. Tsetsenis, V. Buchman, M. R. Etherton and T. C. Sudhof, *Science*, 2010, **329**, 1663.
52. S. N. Witt, *Mol. Neurobiol.*, 2013, **47**, 552.
53. E. Guerrero, P. Vasudevaraju, M. L. Hegde, G. B. Britton and K. S. Rao, *Mol. Neurobiol.*, 2013, **47**, 525.
54. R. E. Musgrove, A. E. King and T. C. Dickson, *Neurotox. Res.*, 2012.
55. K. R. Kumar, K. Lohmann and C. Klein, *Curr. Opin. Neurol.*, 2012, **25**, 466.
56. A. Ferree, M. Guillily, H. Li, K. Smith, A. Takashima, R. Squillace, M. Weigele, J. J. Collins and B. Wolozin, *Neurodegener. Dis.*, 2012, **10**, 238.
57. T. Gasser, J. Hardy and Y. Mizuno, *Mov. Disord.*, 2011, **26**, 1042.
58. S. Saiki, S. Sato and N. Hattori, *J. Neurol. Neurosurg. Psychiatry*, 2012, **83**, 430.
59. O. Corti, S. Lesage and A. Brice, *Physiol. Rev.*, 2011, **91**, 1161.
60. I. Martin, V. L. Dawson and T. M. Dawson, *Annu. Rev. Genomics. Hum. Genet.*, 2011, **12**, 301.

61. S. Fujioka and Z. K. Wszolek, *Neurodegener. Dis.*, 2012, **10**, 257.
62. M. Poulopoulos, O. A. Levy and R. N. Alcalay, *Mov. Disord.*, 2012, **27**, 831.
63. H. Houlden and A. B. Singleton, *Acta Neuropathol.*, 2012, **124**, 325; (a) K. M. Doherty, L. Silveira-Moriyama, L. Parkkinen, D. G. Healy, M. Farrell, N. E. Mencacci, Z. Ahmed, F. M. Brett, J. Hardy, N. Quinn, T. J. Counihan, T. Lynch, Z. V. Fox, T. Revesz, A. J. Lees and J. L. Holton, *JAMA Neurol.*, 2013, **70**, 571.
64. B. Ritz, S. L. Rhodes, Y. Bordelon and J. Bronstein, *PLoS One*, 2012, **7**, e36199.
65. A. Puschmann, E. Englund, O. A. Ross, C. Vilarino-Guell, S. J. Lincoln, J. M. Kachergus, S. A. Cobb, A. L. Tornqvist, S. Rehncrona, H. Widner, Z. K. Wszolek, M. J. Farrer and C. Nilsson, *Parkinsonism Relat. Disord.*, 2012, **18**, 332.
66. K. J. Thomas, M. K. McCoy, J. Blackinton, A. Beilina, M. van der Brug, A. Sandebring, D. Miller, D. Maric, A. Cedazo-Minguez and M. R. Cookson, *Hum. Mol. Genet.*, 2011, **20**, 40.
67. L. L. Kilarski, J. P. Pearson, V. Newsway, E. Majounie, M. D. Knipe, A. Misbahuddin, P. F. Chinnery, D. J. Burn, C. E. Clarke, M. H. Marion, A. J. Lewthwaite, D. J. Nicholl, N. W. Wood, K. E. Morrison, C. H. Williams-Gray, J. R. Evans, S. J. Sawcer, R. A. Barker, M. M. Wickremaratchi, Y. Ben-Shlomo, N. M. Williams and H. R. Morris, *Mov. Disord.*, 2012, **27**, 1522.
68. A. Kurz, C. May, O. Schmidt, T. Muller, C. Stephan, H. E. Meyer, S. Gispert, G. Auburger and K. Marcus, *J. Neural. Transm.*, 2012, **119**, 297.
69. I. F. Mata, H. Checkoway, C. M. Hutter, A. Samii, J. W. Roberts, H. M. Kim, P. Agarwal, V. Alvarez, R. Ribacoba, P. Pastor, O. Lorenzo-Betancor, J. Infante, M. Sierra, P. Gomez-Garre, P. Mir, B. Ritz, S. L. Rhodes, A. Colcher, V. Van Deerlin, K. A. Chung, J. F. Quinn, D. Yearout, E. Martinez, F. M. Farin, J. Y. Wan, K. L. Edwards and C. P. Zabetian, *Mov. Disord.*, 2012, **27**, 1823.
70. X. Wang, D. Winter, G. Ashrafi, J. Schlehe, Y. L. Wong, D. Selkoe, S. Rice, J. Steen, M. J. LaVoie and T. L. Schwarz, *Cell*, 2011, **147**, 893; (a) J. F. Trempe and E. A. Fon, *Front Neurol.*, 2013, **4**, 38; (b) S. M. Jin, M. Lazarou, C. Wang, L. A. Kane, D. P. Narendra and R. J. Youle, *J. Cell. Biol.*, 2010, **191**, 933.
71. N. P. Reynolds, A. Soragni, M. Rabe, D. Verdes, E. Liverani, S. Handschin, R. Riek and S. Seeger, *J. Am. Chem. Soc.*, 2011, **133**, 19366.
72. G. Piccoli, S. B. Condliffe, M. Bauer, F. Giesert, K. Boldt, S. De Astis, A. Meixner, H. Sarioglu, D. M. Vogt-Weisenhorn, W. Wurst, C. J. Gloeckner, M. Matteoli, C. Sala and M. Ueffing, *J. Neurosci.*, 2011, **31**, 2225.
73. K. Markopoulou, D. W. Dickson, R. D. McComb, Z. K. Wszolek, L. Katechalidou, L. Avery, M. S. Stansbury and B. A. Chase, *Acta Neuropathol.*, 2008, **116**, 25.

74. K. Seidel, L. Schols, S. Nuber, E. Petrasch-Parwez, K. Gierga, Z. Wszolek, D. Dickson, W. P. Gai, A. Bornemann, O. Riess, A. Rami, W. F. Den Dunnen, T. Deller, U. Rub and R. Kruger, *Ann. Neurol.*, 2010, **67**, 684.
75. J. J. Zarranz, J. Alegre, J. C. Gomez-Esteban, E. Lezcano, R. Ros, I. Ampuero, L. Vidal, J. Hoenicka, O. Rodriguez, B. Atares, V. Llorens, E. Gomez Tortosa, T. del Ser, D. G. Munoz and J. G. de Yebenes, *Ann. Neurol.*, 2004, **55**, 164.
76. I. F. Tsigelny, Y. Sharikov, W. Wrasidlo, T. Gonzalez, P. A. Desplats, L. Crews, B. Spencer and E. Masliah, *FEBS J.*, 2012, **279**, 1000.
77. S. Fujioka, C. Sundal, A. J. Strongosky, M. C. Castanedes, R. Rademakers, O. A. Ross, C. Vilarino-Guell, M. J. Farrer, Z. K. Wszolek and D. W. Dickson, *Acta Neuropathol.*, 2013, **125**, 425.
78. K. A. Lewis, A. Yaeger, G. N. DeMartino and P. J. Thomas, *J. Bioenerg. Biomembr.*, 2010, **42**, 85.
79. C. Ruffmann, M. Zini, S. Goldwurm, M. Bramerio, S. Spinello, D. Rusconi, M. Gambacorta, F. Tagliavini, G. Pezzoli and G. Giaccone, *Acta Neuropathol.*, 2012, **123**, 901.
80. M. F. Keller, M. Saad, J. Bras, F. Bettella, N. Nicolaou, J. Simon-Sanchez, F. Mittag, F. Buchel, M. Sharma, J. R. Gibbs, C. Schulte, V. Moskvina, A. Durr, P. Holmans, L. L. Kilarski, R. Guerreiro, D. G. Hernandez, A. Brice, P. Ylikotila, H. Stefansson, K. Majamaa, H. R. Morris, N. Williams, T. Gasser, P. Heutink, N. W. Wood, J. Hardy, M. Martinez, A. B. Singleton and M. A. Nalls, *Hum. Mol. Genet.*, 2012, **21**, 4996.
81. N. Pankratz, G. W. Beecham, A. L. DeStefano, T. M. Dawson, K. F. Doheny, S. A. Factor, T. H. Hamza, A. Y. Hung, B. T. Hyman, A. J. Ivinson, D. Krainc, J. C. Latourelle, L. N. Clark, K. Marder, E. R. Martin, R. Mayeux, O. A. Ross, C. R. Scherzer, D. K. Simon, C. Tanner, J. M. Vance, Z. K. Wszolek, C. P. Zabetian, R. H. Myers, H. Payami, W. K. Scott and T. Foroud, *Ann. Neurol.*, 2012, **71**, 370; (a) P. Holmans, V. Moskvina, L. Jones, M. Sharma, A. Vedernikov, F. Buchel, M. Sadd, J. M. Bras, F. Bettella, N. Nicolaou, J. Simon-Sanchez, F. Mittag, J. R. Gibbs, C. Schulte, A. Durr, R. Guerreiro, D. Hernandez, A. Brice, H. Stefansson, K. Majamaa, T. Gasser, P. Heutink, N. W. Wood, M. Martinez, A. B. Singleton, M. A. Nalls, J. Hardy, H. R. Morris and N. M. Williams, *Hum. Mol. Genet.*, 2013, **22**, 1039.
82. B. Zheng, Z. Liao, J. J. Locascio, K. A. Lesniak, S. S. Roderick, M. L. Watt, A. C. Eklund, Y. Zhang-James, P. D. Kim, M. A. Hauser, E. Grunblatt, L. B. Moran, S. A. Mandel, P. Riederer, R. M. Miller, H. J. Federoff, U. Wullner, S. Papapetropoulos, M. B. Youdim, I. Cantuti-Castelvetri, A. B. Young, J. M. Vance, R. L. Davis, J. C. Hedreen, C. H. Adler, T. G. Beach, M. B. Graeber, F. A. Middleton, J. C. Rochet and C. R. Scherzer, *Sci. Transl. Med.*, 2010, **2**, 52ra73.
83. S. M. Goldman, F. Kamel, G. W. Ross, S. A. Jewell, G. S. Bhudhikanok, D. Umbach, C. Marras, R. A. Hauser, J. Jankovic, S. A. Factor,

S. Bressman, K. E. Lyons, C. Meng, M. Korell, D. F. Roucoux, J. A. Hoppin, D. P. Sandler, J. W. Langston and C. M. Tanner, *Ann. Neurol.*, 2012, **71**, 40.
84. B. Meeus, A. Verstraeten, D. Crosiers, S. Engelborghs, M. Van den Broeck, M. Mattheijssens, K. Peeters, E. Corsmit, E. Elinck, B. Pickut, R. Vandenberghe, P. Cras, P. P. De Deyn, C. Van Broeckhoven and J. Theuns, *Neurobiol. Aging*, 2012, **33**, 629e5.
85. C. Wider, O. A. Ross, K. Nishioka, M. G. Heckman, C. Vilarino-Guell, B. Jasinska-Myga, N. Erketin-Taner, R. Rademakers, N. R. Graff-Radford, D. C. Mash, S. Papapetropoulos, R. Duara, H. Uchikado, Z. K. Wszolek, M. J. Farrer and D. W. Dickson, *J. Neurol. Neurosurg. Psychiatry*, 2012, **83**, 424.
86. V. Cullen, S. P. Sardi, J. Ng, Y. H. Xu, Y. Sun, J. J. Tomlinson, P. Kolodziej, I. Kahn, P. Saftig, J. Woulfe, J. C. Rochet, M. A. Glicksman, S. H. Cheng, G. A. Grabowski, L. S. Shihabuddin and M. G. Schlossmacher, *Ann. Neurol.*, 2011, **69**, 940.
87. S. Lesage, M. Anheim, C. Condroyer, P. Pollak, F. Durif, C. Dupuits, F. Viallet, E. Lohmann, J. C. Corvol, A. Honore, S. Rivaud, M. Vidailhet, A. Durr and A. Brice, *Hum. Mol. Genet.*, 2011, **20**, 202.
88. O. Goker-Alpan, J. C. Masdeu, P. D. Kohn, A. Ianni, G. Lopez, C. Groden, M. C. Chapman, B. Cropp, D. P. Eisenberg, E. D. Maniwang, J. Davis, E. Wiggs, E. Sidransky and K. F. Berman, *Brain*, 2012, **135**, 2440.
89. M. Kurzawa-Akanbi, P. S. Hanson, P. G. Blain, D. J. Lett, I. G. McKeith, P. F. Chinnery and C. M. Morris, *J. Neurochem.*, 2012, **123**, 298.
90. K. Brockmann, R. Hilker, U. Pilatus, S. Baudrexel, K. Srulijes, J. Magerkurth, A. K. Hauser, C. Schulte, I. Csoti, C. D. Merten, T. Gasser, D. Berg and E. Hattingen, *Neurology*, 2012, **79**, 213.
91. M. E. Gegg, D. Burke, S. J. R. Heales, J. M. Cooper, J. Hardy, N. W. Wood and A. H. V. Schapira, *Ann. Neurol.*, 2012, **72**, 455; (a) T. L. Yap, A. Velayati, E. Sidransky and J. C. Lee, *Mol. Genet. Metab.*, 2013, **108**, 56.
92. N. Seto-Salvia, J. Pagonabarraga, H. Houlden, B. Pascual-Sedano, O. Dols-Icardo, A. Tucci, C. Paisan-Ruiz, A. Campolongo, S. Anton-Aguirre, I. Martin, L. Munoz, E. Bufill, L. Vilageliu, D. Grinberg, M. Cozar, R. Blesa, A. Lleo, J. Hardy, J. Kulisevsky and J. Clarimon, *Mov. Disord.*, 2012, **27**, 393; (a) S. E. Winder-Rhodes, J. R. Evans, M. Ban, S. L. Mason, C. H. Williams-Gray, T. Foltynie, R. Duran, N. E. Mencacci, S. J. Sawcer and R. A. Barker, *Brain*, 2013, **136**, 392.
93. C. Cook, C. Stetler and L. Petrucelli, *Cold Spring Harb. Perspect. Med.*, 2012, **2**, a009423.
94. M. S. Gorbatyuk, A. Shabashvili, W. Chen, C. Meyers, L. F. Sullivan, M. Salganik, J. H. Lin, A. S. Lewin, N. Muzyczka and O. S. Gorbatyuk, *Mol. Ther.*, 2012, **20**, 1327.

95. J. Tong, H. Wong, M. Guttman, L. C. Ang, L. S. Forno, M. Shimadzu, A. H. Rajput, M. D. Muenter, S. J. Kish, O. Hornykiewicz and Y. Furukawa, *Brain*, 2010, **133**, 172.
96. H. Braak, K. Del Tredici, U. Rub, R. A. de Vos, E. N. Jansen Steur and E. Braak, *Neurobiol. Aging*, 2003, **24**, 197.
97. J. Wills, J. Jones, T. Haggerty, V. Duka, J. N. Joyce and A. Sidhu, *Exp. Neurol.*, 2010, **225**, 210.
98. G. G. Kovacs, U. Wagner, B. Dumont, M. Pikkarainen, A. A. Osman, N. Streichenberger, I. Leisser, J. Verchère, T. Baron, I. Alafuzoff, H. Budka, A. Perret-Liaudet and I. Lachmann, *Acta Neuropathol.*, 2012, **124**, 37.
99. J. G. Quinn, D. T. Coulson, S. Brockbank, N. Beyer, R. Ravid, J. Hellemans, G. B. Irvine and J. A. Johnston, *Brain Res.*, 2012, **1459**, 71.
100. A. Ulusoy and D. A. Di Monte, *Mol. Neurobiol.*, 2013, **47**, 484; (a) A. Gold, Z. T. Turkalp and D. G. Munoz, *Mov Disord*, 2013, **28**, 237.
101. Y. Chu, G. A. Morfini, L. B. Langhamer, Y. He, S. T. Brady and J. H. Kordower, *Brain*, 2012, **135**, 2058.
102. V. M. Nemani, W. Lu, V. Berge, K. Nakamura, B. Onoa, M. K. Lee, F. A. Chaudhry, R. A. Nicoll and R. H. Edwards, *Neuron*, 2010, **65**, 66.
103. H. Lou, S. E. Montoya, T. N. Alerte, J. Wang, J. Wu, X. Peng, C. S. Hong, E. E. Friedrich, S. A. Mader, C. J. Pedersen, B. S. Marcus, A. L. McCormack, D. A. Di Monte, S. C. Daubner and R. G. Perez, *J. Biol. Chem.*, 2010, **285**, 17648.
104. E. C. Hirsch, P. Jenner and S. Przedborski, *Mov. Disord.*, 2013, **28**, 24.
105. B. Winner, M. Regensburger, S. Schreglmann, L. Boyer, I. Prots, E. Rockenstein, M. Mante, C. Zhao, J. Winkler, E. Masliah and F. H. Gage, *J. Neurosci.*, 2012, **32**, 16906.
106. C. Pacheco, L. G. Aguayo and C. Opazo, *Front. Physiol.*, 2012, **3**, 297.
107. T. Yasuda, Y. Nakata and H. Mochizuki, *Mol. Neurobiol.*, 2013, **47**, 466.
108. D. Béraud, H. A. Hathaway, J. Trecki, S. Chasovskikh, D. A. Johnson, J. A. Johnson, H. J. Federoff, M. Shimoji, T. R. Mhyre and K. A. Maguire-Zeiss, *J. Neuroimmune Pharmacol.*, 2013, **8**, 94.
109. R. A. Moore, L. M. Taubner and S. A. Priola, *Curr. Opin. Struct. Biol.*, 2009, **19**, 14.
110. D. M. Walsh and D. J. Selkoe, *Protein Pept. Lett.*, 2004, **11**, 213.
111. J. J. Hoozemans, E. S. van Haastert, D. A. Nijholt, A. J. Rozemuller and W. Scheper, *Neurodegener. Dis.*, 2012, **10**, 212.
112. E. Colla, P. Coune, Y. Liu, O. Pletnikova, J. C. Troncoso, T. Iwatsubo, B. L. Schneider and M. K. Lee, *J. Neurosci.*, 2012, **32**, 3306.
113. S. Bernales, M. M. Soto and E. McCullagh, *Front Aging Neurosci.*, 2012, **4**, 5.
114. B. Winner, R. Jappelli, S. K. Maji, P. A. Desplats, L. Boyer, S. Aigner, C. Hetzer, T. Loher, M. Vilar, S. Campioni, C. Tzitzilonis, A. Soragni, S. Jessberger, H. Mira, A. Consiglio, E. Pham, E. Masliah, F. H. Gage and R. Riek, *Proc. Natl. Acad. Sci. U. S. A.*, 2011, **108**, 4194.
115. D. R. Brown, *IUBMB Life*, 2010, **62**, 334.

116. B. S. Gadad, G. B. Britton and K. S. Rao, *J. Alzheimer's Dis.*, 2011, **24**(2), 223.
117. H. A. Lashuel, C. R. Overk, A. Oueslati and E. Masliah, *Nat. Rev. Neurosci.*, 2013, **14**, 38.
118. C. Vlad, M. I. Iurascu, S. Slamnoiu, B. Hengerer and M. Przybylski, *Methods Mol. Biol.*, 2012, **896**, 399.
119. A. Ulusoy, T. Bjorklund, K. Buck and D. Kirik, *Neurobiol. Dis.*, 2012, **47**, 367; (a) B. K. Choi, M. G. Choi, J. Y. Kim, Y. Yang, Y. Lai, D. H. Kweon, N. K. Lee and Y. K. Shin, *Proc. Natl. Acad. Sci. U. S. A.*, 2013, **110**, 4087.
120. Z. Sultana, K. E. Paleologou, K. M. Al-Mansoori, M. T. Ardah, N. Singh, S. Usmani, H. Jiao, F. L. Martin, M. M. Bharath, S. Vali and O. M. El-Agnaf, *Neuroscience*, 2011, **199**, 303.
121. E. Colla, P. H. Jensen, O. Pletnikova, J. C. Troncoso, C. Glabe and M. K. Lee, *J. Neurosci.*, 2012, **32**, 3301.
122. M. Lundblad, M. Decressac, B. Mattsson and A. Bjorklund, *Proc. Natl. Acad. Sci. U. S. A.*, 2012, **109**, 3213.
123. Z. S. Martin, V. Neugebauer, K. T. Dineley, R. Kayed, W. Zhang, L. C. Reese and G. Taglialatela, *J. Neurochem.*, 2012, **120**, 440.
124. O. W. Wan and K. K. Chung, *PLoS One*, 2012, **7**, e38545.
125. M. T. Stockl, N. Zijlstra and V. Subramaniam, *Mol. Neurobiol.*, 2013, **47**, 613.
126. P. W. Ho, J. W. Ho, H. F. Liu, D. H. So, Z. H. Tse, K. H. Chan, D. B. Ramsden and S. L. Ho, *Transl. Neurodegener.*, 2012, **1**, 3.
127. P. Coskun, J. Wyrembak, S. E. Schriner, H. W. Chen, C. Marciniack, F. Laferla and D. C. Wallace, *Biochim. Biophys. Acta*, 2012, **1820**, 553.
128. M. Karbowski and A. Neutzner, *Acta Neuropathol.*, 2012, **123**, 157.
129. M. K. McCoy and M. R. Cookson, *Antioxid. Redox Signaling*, 2012, **16**, 869.
130. A. R. Esteves, D. M. Arduino, D. F. Silva, C. R. Oliveira and S. M. Cardoso, *Parkinson's Dis.*, 2011, **2011**, 693761.
131. K. Nakamura, V. M. Nemani, F. Azarbal, G. Skibinski, J. M. Levy, K. Egami, L. Munishkina, J. Zhang, B. Gardner, J. Wakabayashi, H. Sesaki, Y. Cheng, S. Finkbeiner, R. L. Nussbaum, E. Masliah and R. H. Edwards, *J. Biol. Chem.*, 2011, **286**, 20710.
132. A. Siddiqui, S. J. Chinta, J. K. Mallajosyula, S. Rajagopolan, I. Hanson, A. Rane, S. Melov and J. K. Andersen, *Free Radical Biol. Med.*, 2012, **53**, 993.
133. E. K. Butler, A. Voigt, A. K. Lutz, J. P. Toegel, E. Gerhardt, P. Karsten, B. Falkenburger, A. Reinartz, K. F. Winklhofer and J. B. Schulz, *PLoS Genet.*, 2012, **8**, e1002488.
134. J. Lee, S. Giordano and J. Zhang, *Biochem. J.*, 2012, **441**, 523; (a) K. Moreau, M. Renna, D. C. Rubinsztein, *Trends. Biochem. Sci.*, 2013, **38**, 57.
135. S. M. Cardoso, A. R. Esteves and D. M. Arduino, *Neurodegener. Dis.*, 2012, **10**, 38.

136. Q. Cai, M. L. Davis and Z. H. Sheng, *Neurosci. Res.*, 2011, **70**, 9.
137. W. Xie and K. K. Chung, *J. Neurochem.*, 2012, **122**, 404.
138. A. K. Reeve, T. K. Park, E. Jaros, G. R. Campbell, N. Z. Lax, P. D. Hepplewhite, K. J. Krishnan, J. L. Elson, C. M. Morris, I. G. McKeith and D. M. Turnbull, *Arch. Neurol.*, 2012, **69**, 385.
139. S. Wang, B. Xu, L. C. Liou, Q. Ren, S. Huang, Y. Luo, Z. Zhang and S. N. Witt, *Proc. Natl. Acad. Sci. U. S. A.*, 2012, **109**, 16119.
140. H. Bueler, *Exp. Neurol.*, 2009, **218**, 235.
141. S. Mullin and A. Schapira, *Mol. Neurobiol.*, 2013, **47**, 587.
142. A. R. Winslow and D. C. Rubinsztein, *Autophagy*, 2011, **7**, 429.
143. M. Usenovic, E. Tresse, J. R. Mazzulli, J. P. Taylor and D. Krainc, *J. Neurosci.*, 2012, **32**, 4240.
144. A. M. Gusdon, J. Zhu, B. Van Houten and C. T. Chu, *Neurobiol. Dis.*, 2012, **45**, 962.
145. K. Okamoto and N. Kondo-Okamoto, *Biochim. Biophys. Acta*, 2012, **1820**, 595.
146. L. G. Friedman, M. L. Lachenmayer, J. Wang, L. He, S. M. Poulose, M. Komatsu, G. R. Holstein and Z. Yue, *J. Neurosci.*, 2012, **32**, 7585.
147. Y. Huang, F. Chegini, G. Chua, K. Murphy, W. Gai and G. M. Halliday, *Transl. Neurodegener.*, 2012, **1**, 2.
148. K. Venderova and D. S. Park, *Cold Spring Harb. Perspect. Med.*, 2012, 2.
149. E. Janda, C. Isidoro, C. Carresi and V. Mollace, *Mol. Neurobiol.*, 2012, **46**, 639.
150. C. Cook, L. Petrucelli in *Parkinson's Disease, Second Edition*, ed. R. F. Pfeiffer, Z. K. Wszolek and M. Ebadi, CRC Press, Boca Raton, FL, 2012, p. 559.
151. J. Li, D. Liu, L. Sun, Y. Lu and Z. Zhang, *J. Neurol. Sci.*, 2012, **317**, 1.
152. M. Ramalingam and S. J. Kim, *J. Neural. Transm.*, 2012, **119**, 891.
153. H. M. Gao, F. Zhang, H. Zhou, W. Kam, B. Wilson and J. S. Hong, *Environ. Health Perspect.*, 2011, **119**, 807.
154. L. Alvarez-Erviti, Y. Couch, J. Richardson, J. M. Cooper and M. J. Wood, *Neurosci. Res.*, 2011, **69**, 337.
155. E. C. Hirsch and S. Hunot, *Lancet Neurol.*, 2009, **8**, 382.
156. K. Sekiyama, S. Sugama, M. Fujita, A. Sekigawa, Y. Takamatsu, M. Waragai, T. Takenouchi and M. Hashimoto, *Parkinson's Dis.*, 2012, **2012**, 271732.
157. C. M. Lema Tome, T. Tyson, N. L. Rey, S. Grathwohl, M. Britschgi and P. Brundin, *Mol. Neurobiol.*, 2013, **47**, 561.
158. K. A. Jellinger, *J. Cell. Mol. Med.*, 2012, **16**, 1166.
159. L. Reiniger, A. Lukic, J. Linehan, P. Rudge, J. Collinge, S. Mead and S. Brandner, *Acta Neuropathol.*, 2011, **121**, 5.
160. F. Kawakami, M. Suzuki, N. Shimada, G. Kagiya, E. Ohta, K. Tamura, H. Maruyama and T. Ichikawa, *FEBS J.*, 2011, **278**, 4895.
161. A. Elbaz, O. A. Ross, J. P. Ioannidis, A. I. Soto-Ortolaza, F. Moisan, J. Aasly, G. Annesi, M. Bozi, L. Brighina, M. C. Chartier-Harlin, A. Destee, C. Ferrarese, A. Ferraris, J. M. Gibson, S. Gispert,

G. M. Hadjigeorgiou, B. Jasinska-Myga, C. Klein, R. Kruger, J. C. Lambert, K. Lohmann, S. van de Loo, M. A. Loriot, T. Lynch, G. D. Mellick, E. Mutez, C. Nilsson, G. Opala, A. Puschmann, A. Quattrone, M. Sharma, P. A. Silburn, L. Stefanis, R. J. Uitti, E. M. Valente, C. Vilarino-Guell, K. Wirdefeldt, Z. K. Wszolek, G. Xiromerisiou, D. M. Maraganore and M. J. Farrer, *Ann. Neurol.*, 2011, **69**, 778.
162. Y. Huang, D. B. Rowe and G. M. Halliday, *J. Parkinson's Dis.*, 2011, **1**, 271.
163. E. A. Waxman and B. I. Giasson, *J. Neurosci.*, 2011, **31**, 7604.
164. T. Haggerty, J. Credle, O. Rodriguez, J. Wills, A. W. Oaks, E. Masliah and A. Sidhu, *Eur J. Neurosci.*, 2011, **33**, 1598.
165. H. Y. Qureshi and H. K. Paudel, *J. Biol. Chem.*, 2011, **286**, 5055.
166. K. L. Emmer, E. A. Waxman, J. P. Covy and B. I. Giasson, *J. Biol. Chem.*, 2011, **286**, 35104.
167. N. Badiola, R. M. de Oliveira, F. Herrera, C. Guardia-Laguarta, S. A. Goncalves, M. Pera, M. Suarez-Calvet, J. Clarimon, T. F. Outeiro and A. Lleo, *PLoS One*, 2011, **6**, e26609.
168. J. Wills, J. Credle, T. Haggerty, J. H. Lee, A. W. Oaks and A. Sidhu, *PLoS One*, 2011, **6**, e17953.
169. T. Kaul, J. Credle, T. Haggerty, A. W. Oaks, E. Masliah and A. Sidhu, *BMC Neurosci.*, 2011, **12**, 79.
170. V. N. Uversky, *J. Neurochem.*, 2007, **103**, 17.
171. P. Lei, S. Ayton, D. I. Finkelstein, P. A. Adlard, C. L. Masters and A. I. Bush, *Int. J. Biochem. Cell Biol.*, 2010, **42**, 1775.
172. P. T. Kotzbauer, B. I. Giasson, A. V. Kravitz, L. I. Golbe, M. H. Mark, J. Q. Trojanowski and V. M. Lee, *Exp. Neurol.*, 2004, **187**, 279.
173. G. Muntane, E. Dalfo, A. Martinez and I. Ferrer, *Neuroscience*, 2008, **152**, 913.
174. H. Fujishiro, Y. Tsuboi, W. L. Lin, H. Uchikado and D. W. Dickson, *Acta Neuropathol.*, 2008, **116**, 17.
175. T. Arai, I. R. Mackenzie, M. Hasegawa, T. Nonoka, K. Niizato, K. Tsuchiya, S. Iritani, M. Onaya and H. Akiyama, *Acta Neuropathol.*, 2009, **117**, 125.
176. S. E. Marsh and M. Blurton-Jones, *Alzheimer's Res. Ther.*, 2012, **4**, 11.
177. K. Ono, R. Takahashi, T. Ikeda and M. Yamada, *J. Neurochem.*, 2012, **122**, 883.
178. C. Bate, S. Gentleman and A. Williams, *Mol. Neurodegener.*, 2010, **5**, 55.
179. C. J. Dunning, J. F. Reyes, J. A. Steiner and P. Brundin, *Prog. Neurobiol.*, 2012, **97**, 205.
180. K. C. Luk, V. M. Kehm, B. Zhang, P. O'Brien, J. Q. Trojanowski and V. M. Lee, *J. Exp. Med.*, 2012, **209**, 975.
181. A. Raj, A. Kuceyeski and M. Weiner, *Neuron*, 2012, **73**, 1204.
182. K. M. Danzer, L. R. Kranich, W. P. Ruf, O. Cagsal-Getkin, A. R. Winslow, L. Zhu, C. R. Vanderburg and P. J. McLean, *Mol. Neurodegener.*, 2012, **7**, 42.

183. E. C. Freundt, N. Maynard, E. K. Clancy, S. Roy, L. Bousset, Y. Sourigues, M. Covert, R. Melki, K. Kirkegaard and M. Brahic, *Ann. Neurol.*, 2012, **72**, 517; (a) D. G. Walker, L. F. Lue, C. H. Adler, H. A. Shill, J. N. Caviness, M. N. Sabbagh, H. Akiyama, G. E. Serrano, L. I. Sue and T. G. Beach, *Exp. Neurol.*, 2013, **240**, 190.
184. I. Russo, L. Bubacco and E. Greggio, *Am. J. Neurodegener. Dis.*, 2012, **1**, 217.
185. C. Hansen, E. Angot, A. L. Bergstrom, J. A. Steiner, L. Pieri, G. Paul, T. F. Outeiro, R. Melki, P. Kallunki, K. Fog, J. Y. Li and P. Brundin, *J. Clin. Invest.*, 2011, **121**, 715.
186. O. Marques and T. F. Outeiro, *Cell Death Dis.*, 2012, **3**, e350; (a) E. J. Bae, H. J. Lee, E. Rockenstein, D. H. Ho, E. B. Park, N. Y. Yang, P. Desplats, E. Masliah and S. J. Lee, *J. Neurosci.*, 2012, **32**, 13454.; (b) F. Cheng, G. Vivacqua and S. Yu, *J. Chem. Neuroanat.*, 2011, **42**, 242.
187. L. A. Volpicelli-Daley, K. C. Luk, T. P. Patel, S. A. Tanik, D. M. Riddle, A. Stieber, D. F. Meaney, J. Q. Trojanowski and V. M. Lee, *Neuron*, 2011, **72**, 57.
188. E. Masliah, E. Rockenstein, C. Inglis, A. Adame, C. Bett, M. Lucero and C. J. Sigurdson, *Prion*, 2012, **6**, 184.
189. L. Liu, V. Drouet, J. W. Wu, M. P. Witter, S. A. Small, C. Clelland and K. Duff, *PLoS One*, 2012, **7**, e31302.
190. C. Hansen and J. Y. Li, *Trends Mol. Med.*, 2012, **18**, 248; (a) M. Masuda-Suzukake, T. Nonaka, M. Hosokawa, T. Oikawa, T. Arai, H. Akiyama, D. M. Mann and M. Hasegawa, *Brain*, 2013, **136**, 1128; (b) D. J. Irwin, J. Y. Abrams, L. B. Schonberger, E. W. Leschek, J. L. Mills, V. M. Lee and J. Q. Trojanowski, *JAMA Neurol*, 2013, **70**, 462.
191. M. Polymenidou and D. W. Cleveland, *J. Exp. Med.*, 2012, **209**, 889.
192. D. W. Dickson, *Cold Spring Harb. Perspect. Med.*, 2012, **2**, a009258.
193. G. Halliday, A. Lees and M. Stern, *Mov. Disord.*, 2011, **26**, 1015.
194. H. Braak and K. Del Tredici, *Adv. Anat. Embryol. Cell Biol.*, 2009, **201**, 1.
195. K. Del Tredici and H. Braak, *Acta Neuropathol.*, 2012, **124**, 643.
196. K. Wakabayashi, K. Tanji, S. Odagiri, Y. Miki, F. Mori and H. Takahashi, *Mol. Neurobiol.*, 2013, **47**, 495.
197. L.-F. Lue, D. G. Walker, C. H. Adler, H. Shill, H. Tran, H. Akiyama, L. I. Sue, J. Caviness, M. N. Sabbagh and T. G. Beach, *Brain Pathol.*, 2012, **22**, 745.
198. K. A. Jellinger, *Biochim. Biophys. Acta*, 2009, **1792**, 730.
199. T. Kanazawa, E. Adachi, S. Orimo, A. Nakamura, H. Mizusawa and T. Uchihara, *Brain Pathol.*, 2012, **22**, 67.
200. H. C. Cheng, C. M. Ulane and R. E. Burke, *Ann. Neurol.*, 2010, **67**, 715.
201. D. G. Walker, L. F. Lue, C. H. Adler, H. A. Shill, J. N. Caviness, M. N. Sabbagh, H. Akiyama, G. E. Serrano, L. I. Sue and T. G. Beach, *Exp. Neurol.*, 2013, **240**, 190.
202. P. G. Ince, B. Clark, J. L. Holton, T. Revesz, S. B. Wharton in *Greenfield's Neuropathology*, ed. S. Love, D. N. Louis and D. W. Ellison, Hodder Arnold, London, 8th edn, 2008, p. 889.

203. B. N. Dugger, M. E. Murray, B. F. Boeve, J. E. Parisi, E. E. Benarroch, T. J. Ferman and D. W. Dickson, *Neuropathol. Appl. Neurobiol.*, 2012, **38**, 142.
204. S. Odagiri, K. Tanji, F. Mori, A. Kakita, H. Takahashi and K. Wakabayashi, *Acta Neuropathol.*, 2012, **124**, 173.
205. F. Mori, K. Tanji, S. Odagiri, Y. Toyoshima, M. Yoshida, T. Ikeda, H. Sasaki, A. Kakita, H. Takahashi and K. Wakabayashi, *Acta Neuropathol.*, 2012, **124**, 149.
206. K. D. van Dijk, H. W. Berendse, B. Drukarch, S. A. Fratantoni, T. V. Pham, S. R. Piersma, E. Huisman, J. J. Breve, H. J. Groenewegen, C. R. Jimenez and W. D. van de Berg, *Brain Pathol.*, 2012, **22**, 485.
207. T. Kanazawa, T. Uchihara, A. Takahashi, A. Nakamura, S. Orimo and H. Mizusawa, *Brain Pathol.*, 2008, **18**, 415.
208. T. G. Beach, C. H. Adler, L. Lue, L. I. Sue, J. Bachalakuri, J. Henry-Watson, J. Sasse, S. Boyer, S. Shirohi, R. Brooks, J. Eschbacher, C. L. White, 3rd, H. Akiyama, J. Caviness, H. A. Shill, D. J. Connor, M. N. Sabbagh and D. G. Walker, *Acta Neuropathol.*, 2009, **117**, 613.
209. M. E. Kalaitzakis, M. B. Graeber, S. M. Gentleman and R. K. Pearce, *Acta Neuropathol.*, 2008, **116**, 125.
210. L. Parkkinen, T. Pirttila and I. Alafuzoff, *Acta Neuropathol.*, 2008, **115**, 399.
211. J. Zaccai, C. Brayne, I. McKeith, F. Matthews and P. G. Ince, *Neurology*, 2008, **70**, 1042.
212. G. Halliday, H. McCann and C. Shepherd, *Expert Rev. Neurother.*, 2012, **12**, 673.
213. P. A. Agarwal and A. J. Stoessl, *Mov Disord*, 2013, **28**, 71.
214. H. Bernheimer, W. Birkmayer, O. Hornykiewicz, K. Jellinger and F. Seitelberger, *J. Neurol. Sci.*, 1973, **20**, 415.
215. A. DelleDonne, K. J. Klos, H. Fujishiro, Z. Ahmed, J. E. Parisi, K. A. Josephs, R. Frigerio, M. Burnett, Z. K. Wszolek, R. J. Uitti, J. E. Ahlskog and D. W. Dickson, *Arch. Neurol.*, 2008, **65**, 1074.
216. G. G. Kovacs, I. J. Milenkovic, M. Preusser and H. Budka, *Mov. Disord.*, 2008, **23**, 1608.
217. N. X. Tritsch, J. B. Ding and B. L. Sabatini, *Nature*, 2012, **490**, 262.
218. S. Greffard, M. Verny, A. M. Bonnet, D. Seilhean, J. J. Hauw and C. Duyckaerts, *Neurobiol. Aging*, 2010, **31**, 99.
219. L. Parkkinen, S. S. O'Sullivan, C. Collins, A. Petrie, J. L. Holton, T. Revesz and A. J. Lees, *J. Parkinson's Dis.*, 2011, **1**, 277.
220. M. N. Gaugler, O. Genc, W. Bobela, S. Mohanna, M. T. Ardah, O. M. El-Agnaf, M. Cantoni, J.-C. Bensadoun, R. Schneggenburger, G. W. Knott, P. Aebischer and B. L. Schneider, *Acta Neuropathol.*, 2012, **123**, 653; (a) J. M. Milber, J. V. Noorigian, J. F. Morley, H. Petrovitch, L. White, G. W. Ross and J. E. Duda, *Neurology*, 2012, **79**, 2307.
221. G. Stoica, G. Lungu, N. L. Bjorklund, G. Taglialatela, X. Zhang, V. Chiu, H. H. Hill, J. O. Schenk and I. Murray, *J. Neurochem.*, 2012, **122**, 812.

222. D. M. Annerino, S. Arshad, G. M. Taylor, C. H. Adler, T. G. Beach and J. G. Greene, *Acta Neuropathol.*, 2012, **124**, 665.
223. A. Bellucci, L. Navarria, M. Zaltieri, C. Missale and P. Spano, *Brain Res.*, 2012, **1432**, 95.
224. B. Picconi, G. Piccoli and P. Calabresi, *Adv. Exp. Med. Biol.*, 2012, **970**, 553.
225. I. G. McKeith, D. W. Dickson, J. Lowe, M. Emre, J. T. O'Brien, H. Feldman, J. Cummings, J. E. Duda, C. Lippa, E. K. Perry, D. Aarsland, H. Arai, C. G. Ballard, B. Boeve, D. J. Burn, D. Costa, T. Del Ser, B. Dubois, D. Galasko, S. Gauthier, C. G. Goetz, E. Gomez-Tortosa, G. Halliday, L. A. Hansen, J. Hardy, T. Iwatsubo, R. N. Kalaria, D. Kaufer, R. A. Kenny, A. Korczyn, K. Kosaka, V. M. Lee, A. Lees, I. Litvan, E. Londos, O. L. Lopez, S. Minoshima, Y. Mizuno, J. A. Molina, E. B. Mukaetova-Ladinska, F. Pasquier, R. H. Perry, J. B. Schulz, J. Q. Trojanowski and M. Yamada, *Neurology*, 2005, **65**, 1863.
226. D. Aarsland, E. Londos and C. Ballard, *Int. Psychogeriatr.*, 2009, **21**, 216.
227. Y. Compta, L. Parkkinen, S. S. O'Sullivan, J. Vandrovcova, J. L. Holton, C. Collins, T. Lashley, C. Kallis, D. R. Williams, R. de Silva, A. J. Lees and T. Revesz, *Brain*, 2011, **134**, 1493.
228. M. L. Kraybill, E. B. Larson, D. W. Tsuang, L. Teri, W. C. McCormick, J. D. Bowen, W. A. Kukull, J. B. Leverenz and M. M. Cherrier, *Neurology*, 2005, **64**, 2069.
229. B. N. Dugger, G. E. Serrano, L. I. Sue, D. G. Walker, C. H. Adler, H. A. Shill, M. N. Sabbagh, J. N. Caviness, J. Hidalgo, M. Saxon-LaBelle, G. Chiarolanza, M. Mariner, J. Henry-Watson and T. G. Beach, *J. Parkinson's Dis.*, 2012, **2**, 57.
230. K. Kantarci, V. J. Lowe, B. F. Boeve, S. D. Weigand, M. L. Senjem, S. A. Przybelski, D. W. Dickson, J. E. Parisi, D. S. Knopman, G. E. Smith, T. J. Ferman, R. C. Petersen and C. R. Jack Jr., *Neurobiol. Aging*, 2012, **33**, 2091.
231. I. Alafuzoff, P. G. Ince, T. Arzberger, S. Al-Sarraj, J. Bell, I. Bodi, N. Bogdanovic, O. Bugiani, I. Ferrer, E. Gelpi, S. Gentleman, G. Giaccone, J. W. Ironside, N. Kavantzas, A. King, P. Korkolopoulou, G. G. Kovacs, D. Meyronet, C. Monoranu, P. Parchi, L. Parkkinen, E. Patsouris, W. Roggendorf, A. Rozemuller, C. Stadelmann-Nessler, N. Streichenberger, D. R. Thal and H. Kretzschmar, *Acta Neuropathol.*, 2009, **117**, 635.
232. H. Braak, U. Rub, E. N. Jansen Steur, K. Del Tredici and R. A. de Vos, *Neurology*, 2005, **64**, 1404.
233. P. T. Kotzbauer, N. J. Cairns, M. C. Campbell, A. W. Willis, B. A. Racette, S. D. Tabbal and J. S. Perlmutter, *Arch. Neurol.*, 2012, **69**, 1326.
234. D. Tsuang, J. B. Leverenz, O. L. Lopez, R. L. Hamilton, D. A. Bennett, J. A. Schneider, A. S. Buchman, E. B. Larson, P. K. Crane, J. A. Kaye,

P. Kramer, R. Woltjer, J. Q. Trojanowski, D. Weintraub, A. S. Chen-Plotkin, D. J. Irwin, J. Rick, G. D. Schellenberg, G. S. Watson, W. Kukull, P. T. Nelson, G. A. Jicha, J. H. Neltner, D. Galasko, E. Masliah, J. F. Quinn, K. A. Chung, D. Yearout, I. F. Mata, J. Y. Wan, K. L. Edwards, T. J. Montine and C. P. Zabetian, *JAMA Neurol.*, 2013, **70**, 223.
235. G. M. Halliday, Y. J. Song and A. J. Harding, *J. Neural. Transm.*, 2011, **118**, 713.
236. K. A. Jellinger and J. Attems, *Acta Neuropathol.*, 2006, **112**, 253.
237. E. Ghebremedhin, A. Rosenberger, U. Rub, M. Vuksic, T. Berhe, H. Bickeboller, R. A. de Vos, D. R. Thal and T. Deller, *J. Neuropathol. Exp. Neurol.*, 2010, **69**, 442.
238. J. Bove and C. Perier, *Neuroscience*, 2012, **211**, 51.
239. J. R. Cannon and J. T. Greenamyre, *Prog. Brain Res.*, 2010, **184**, 17.
240. C. M. Tanner, F. Kamel, G. W. Ross, J. A. Hoppin, S. M. Goldman, M. Korell, C. Marras, G. S. Bhudhikanok, M. Kasten, A. R. Chade, K. Comyns, M. B. Richards, C. Meng, B. Priestley, H. H. Fernandez, F. Cambi, D. M. Umbach, A. Blair, D. P. Sandler and J. W. Langston, *Environ. Health Perspect.*, 2011, **119**, 866.
241. R. Nistico, B. Mehdawy, S. Piccirilli and N. Mercuri, *Int. J. Immunopathol. Pharmacol.*, 2011, **24**, 313.
242. T. M. Dawson, H. S. Ko and V. L. Dawson, *Neuron*, 2010, **66**, 646.
243. R. M. Welchko, X. T. Leveque and G. L. Dunbar, *Parkinson's Dis.*, 2012, **2012**, 128356.
244. T. Alberio, L. Lopiano and M. Fasano, *FEBS J.*, 2012, **279**, 1146.
245. H. Mizuno, N. Fujikake, K. Wada and Y. Nagai, *Parkinson's Dis.*, 2010, **2011**, 212706.
246. A. J. Whitworth, *Adv. Genet.*, 2011, **73**, 1.
247. C. Yao, R. El Khoury, W. Wang, T. A. Byrd, E. A. Pehek, C. Thacker, X. Zhu, M. A. Smith, A. L. Wilson-Delfosse and S. G. Chen, *Neurobiol. Dis.*, 2010, **40**, 73.
248. P. M. Antony, N. J. Diederich and R. Balling, *Mamm. Genome*, 2011, **22**, 401.
249. E. Bezard, Z. Yue, D. Kirik and M. G. Spillantini, *Mov. Disord.*, 2012, **28**, 61.
250. E. Bezard and S. Przedborski, *Mov. Disord.*, 2011, **26**, 993.
251. J. Blesa, S. Phani, V. Jackson-Lewis and S. Przedborski, *J. Biomed Biotechnol*, 2012, **2012**, 845618.
252. S. Nuber, F. Harmuth, Z. Kohl, A. Adame, M. Trejo, K. Schonig, F. Zimmermann, C. Bauer, N. Casadei, C. Giel, C. Calaminus, B. J. Pichler, P. H. Jensen, C. P. Muller, D. Amato, J. Kornhuber, P. Teismann, H. Yamakado, R. Takahashi, J. Winkler, E. Masliah and O. Riess, *Brain*, 2013, **136**, 412.
253. Y. Lim, V. M. Kehm, E. B. Lee, J. H. Soper, C. Li, J. Q. Trojanowski and V. M. Lee, *J. Neurosci.*, 2011, **31**, 10076.
254. K. Low and P. Aebischer, *Neurobiol. Dis.*, 2012, **48**, 189.
255. F. Blandini and M. T. Armentero, *FEBS J.*, 2012, **279**, 1156.

256. S. Jesse, P. Steinacker, S. Lehnert, F. Gillardon, B. Hengerer and M. Otto, *CNS Neurosci. Ther.*, 2009, **15**, 157.
257. B. Mollenhauer and C. Trenkwalder, *Mov. Disord.*, 2009, **24**, 1411.
258. P. G. Foulds, J. D. Mitchell, A. Parker, R. Turner, G. Green, P. Diggle, M. Hasegawa, M. Taylor, D. Mann and D. Allsop, *FASEB J.*, 2011, **25**, 4127.
259. B. Mollenhauer, V. Cullen, I. Kahn, B. Krastins, T. F. Outeiro, I. Pepivani, J. Ng, W. Schulz-Schaeffer, H. A. Kretzschmar, P. J. McLean, C. Trenkwalder, D. A. Sarracino, J. P. Vonsattel, J. J. Locascio, O. M. El-Agnaf and M. G. Schlossmacher, *Exp. Neurol.*, 2008, **213**, 315.
260. M. Eller and D. R. Williams, *Clin. Chem. Lab. Med.*, 2011, **49**, 403.
261. L. Parnetti, D. Chiasserini, G. Bellomo, D. Giannandrea, C. De Carlo, M. M. Qureshi, M. T. Ardah, S. Varghese, L. Bonanni, B. Borroni, N. Tambasco, P. Eusebi, A. Rossi, M. Onofrj, A. Padovani, P. Calabresi and O. El-Agnaf, *Mov. Disord.*, 2011, **26**, 1428.
262. B. Mollenhauer, E. Trautmann, P. Taylor, P. Manninger, F. Sixel-Doring, J. Ebentheuer, C. Trenkwalder and M. G. Schlossmacher, *Neurosci. Lett.*, 2013, **532**, 44.
263. M. R. Sierks, G. Chatterjee, C. McGraw, S. Kasturirangan, P. Schulz and S. Prasad, *Integr. Biol.*, 2011, **3**, 1188.
264. T. Tokuda, M. M. Qureshi, M. T. Ardah, S. Varghese, S. A. Shehab, T. Kasai, N. Ishigami, A. Tamaoka, M. Nakagawa and O. M. El-Agnaf, *Neurology*, 2010, **75**, 1766.
265. M. Bidinosti, D. R. Shimshek, B. Mollenhauer, D. Marcellin, T. Schweizer, G. P. Lotz, M. G. Schlossmacher and A. Weiss, *J. Biol. Chem.*, 2012, **287**, 33691.
266. Y. Wang, M. Shi, K. A. Chung, C. P. Zabetian, J. B. Leverenz, D. Berg, K. Srulijes, J. Q. Trojanowski, V. M. Lee, A. D. Siderowf, H. Hurtig, I. Litvan, M. C. Schiess, E. R. Peskind, M. Masuda, M. Hasegawa, X. Lin, C. Pan, D. Galasko, D. S. Goldstein, P. H. Jensen, H. Yang, K. C. Cain and J. Zhang, *Sci. Transl. Med.*, 2012, **4**, 121ra20.
267. H. Prikrylova Vranova, J. Mares, P. Hlustik, M. Nevrly, D. Stejskal, J. Zapletalova, R. Obereigneru and P. Kanovsky, *J. Neural. Transm.*, 2012, **119**, 353.
268. G. Alves, K. Bronnick, D. Aarsland, K. Blennow, H. Zetterberg, C. Ballard, M. W. Kurz, U. Andreasson, O. B. Tysnes, J. P. Larsen and E. Mulugeta, *J. Neurol. Neurosurg. Psychiatry*, 2010, **81**, 1080.
269. D. S. Goldstein, C. Holmes and Y. Sharabi, *Brain*, 2012, **135**, 1900.
270. L. Parnetti, A. Castrioto, D. Chiasserini, E. Persichetti, N. Tambasco, O. El-Agnaf and P. Calabresi, *Nat. Rev. Neurol.*, 2013, **9**, 131.
271. M. B. Aerts, R. A. Esselink, W. F. Abdo, B. R. Bloem and M. M. Verbeek, *Neurobiol. Aging*, 2012, **33**, 430 e1.
272. S. Hall, A. Ohrfelt, R. Constantinescu, U. Andreasson, Y. Surova, F. Bostrom, C. Nilsson, H. Widner, H. Decraemer, K. Nagga, L. Minthon, E. Londos, E. Vanmechelen, B. Holmberg, H. Zetterberg, K. Blennow and O. Hansson, *Arch. Neurol.*, 2012, **69**, 1445.

273. P. G. Foulds, O. Yokota, A. Thurston, Y. Davidson, Z. Ahmed, J. Holton, J. C. Thompson, H. Akiyama, T. Arai, M. Hasegawa, A. Gerhard, D. Allsop and D. M. Mann, *Neurobiol. Dis.*, 2012, **45**, 188.
274. B. Mollenhauer and J. Zhang, *Mov. Disord.*, 2012, **27**, 644; (a) S. Slaets, N. Le Bastard, J. Theuns, K. Sleegers, A. Verstraeten, E. De Leenheir, J. Luyckx, J. J. Martin, C. Van Broeckhoven and S. Engelborghs, *J. Alzheimers. Dis.*, 2013, **35**, 137.
275. K. Kasuga, M. Nishizawa and T. Ikeuchi, *Int. J. Alzheimer's Dis.*, 2012, **2012**, 437025; (a) L. M. Smith, M. C. Schiess, M. P. Coffey, A. C. Klaver and D. A. Loeffler, *PLoS One*, 2012, **7**, e52285; (b) D. Besong-Agbo, E. Wolf, F. Jessen, M. Oechsner, E. Hametner, W. Poewe, M. Reindl, W. H. Oertel, C. Noelker, M. Bacher and R. Dodel, *Neurology*, 2013, **80**, 169.
276. K. D. van Dijk, E. Persichetti, D. Chiasserini, P. Eusebi, T. Beccari, P. Calabresi, H. W. Berendse, L. Parnetti and W. D. van de Berg, *Mov. Disord.*, 2013, May 27, doi: 10.1002/mds.25495.
277. T. Kitada, J. J. Tomlinson, H. S. Ao, D. A. Grimes and M. G. Schlossmacher, *Curr. Treat. Options Neurol.*, 2012, **14**, 230.
278. F. Pan-Montojo, M. Schwarz, C. Winkler, M. Arnhold, G. A. O'Sullivan, A. Pal, J. Said, G. Marsico, J. M. Verbavatz, M. Rodrigo-Angulo, G. Gille, R. H. Funk and H. Reichmann, *Sci. Rep.*, 2012, **2**, 898.
279. P. Brundin and R. Olsson, *Expert Rev. Neurother.*, 2011, **11**, 917.
280. T. T. Rohn, *CNS Neurol.Disord. Drug Targets*, 2012, **11**, 174.
281. K. Vekrellis and L. Stefanis, *Expert Opin. Ther. Targets*, 2012, **16**, 421.
282. J. Wu, H. Lou, T. N. Alerte, E. K. Stachowski, J. Chen, A. B. Singleton, R. L. Hamilton and R. G. Perez, *Neuroscience*, 2012, **207**, 288.
283. K. S. Kim, Y. R. Choi, J. Y. Park, J. H. Lee, D. K. Kim, S. J. Lee, S. R. Paik, I. Jou and S. M. Park, *J Biol Chem*, 2012, **287**, 24862; (a) S. M. Park, K. S. Kim, Prion, 2013, 7, 121.
284. H. Harris and D. C. Rubinsztein, *Nat. Rev. Neurol.*, 2011, **8**, 108.
285. G. Marino, F. Madeo and G. Kroemer, *Curr. Opin. Cell. Biol.*, 2011, **23**, 198.
286. S. P. Sardi, P. Singh, S. H. Cheng, L. S. Shihabuddin and M. G. Schlossmacher, *Neurodegener. Dis.*, 2012, **10**, 195.
287. S. M. Park and K. S. Kim, *Prion*, 2012, 7.
288. D. C. Rubinsztein, P. Codogno and B. Levine, *Nat. Rev. Drug. Discov.*, 2012, **11**, 709.
289. L. V. Kalia, S. K. Kalia, P. J. McLean, A. M. Lozano and A. E. Lang, *Ann. Neurol.*, 2012, **73**, 155.
290. M. Rolinski, C. Fox, I. Maidment and R. McShane, *Cochrane Database Syst. Rev.*, 2012, 3, CD006504.
291. L. Yu, J. Cui, P. K. Padakanti, L. Engel, D. P. Bagchi, P. T. Kotzbauer and Z. Tu, *Bioorg. Med. Chem.*, 2012, **20**, 4625.
292. L. M. Smith, A. C. Klaver, M. P. Coffey, L. Dang and D. A. Loeffler, *Int. Immunopharmacol.*, 2012, **14**, 550.
293. A. H. Schapira, *Antioxid. Redox Signaling*, 2012, **16**, 965.
294. C. Perier, J. Bove and M. Vila, *Antioxid. Redox Signaling*, 2012, **16**, 883.

295. E. J. Bae, H. J. Lee, E. Rockenstein, D. H. Ho, E. B. Park, N. Y. Yang, P. Desplats, E. Masliah and S. J. Lee, *J. Neurosci.*, 2012, **32**, 13454.
296. P. Calabresi, M. Di Filippo, A. Gallina, Y. Wang, J. N. Stankowski, B. Picconi, V. L. Dawson and T. M. Dawson, *Mov. Disord.*, 2013, **28**, 51.
297. J. H. Kordower and A. Bjorklund, *Mov. Disord.*, 2013, **28**, 96.
298. A. Bjorklund and J. H. Kordower, *Mov. Disord.*, 2013, **28**, 110.
299. L. V. Kalia, S. K. Kalia, P. J. McLean, A. M. Lozano and A. E. Lang, *Ann. Neurol.*, 2013, **73**, 155.
300. H. A. Lashuel, C. R. Overk, A. Oueslati and E. Masliah, *Nat. Rev. Neurosci.*, 2013, **14**, 38.
301. J. Wagner, S. Ryazanov, A. Leonov, J. Levin, S. Shi, F. Schmidt, C. Prix, F. Pan-Montojo, U. Bertsch, G. Mitteregger-Kretzschmar, M. Geissen, M. Eiden, F. Leidel, T. Hirschberger, A. A. Deeg, J. J. Krauth, W. Zinth, P. Tavan, J. Pilger, M. Zweckstetter, T. Frank, M. Bahr, J. H. Weishaupt, M. Uhr, H. Urlaub, U. Teichmann, M. Samwer, K. Botzel, M. Groschup, H. Kretzschmar, C. Griesinger and A. Giese, *Acta. Neuropathol.*, 2013, **125**, 795.

Neuroprotective Therapies

CHAPTER 9
New Approaches to Neuroprotection in Parkinson's Disease

MARÍA ANGELES MENA,[a,b] JUAN PERUCHO,[a,b] JOSÉ LUIS LÓPEZ-SENDÓN[b,c] AND JUSTO GARCÍA DE YÉBENES*[b,c]

[a] Departamento de Investigación, Unidad de Neurofarmacología, Madrid, Spain; [b] Centro de Investigación Biomédica en red sobre Enfermedades Neurodegenerativas, Madrid, Spain; [c] Servicio de Neurología, Unidad de Enfermedades Neurodegenerativas, Madrid, Spain
*Email: jgyebenes@yahoo.com

9.1 Introduction

If one searches the literature published during the last 10 years for "Neuroprotection in Parkinson's disease" and enters these words without restriction, one obtains 926 papers published in this period of time. If the search is restricted to "human studies" the number of papers drops to 512, and if the restriction is increased to "clinical trials" the number is dramatically reduced to six. A detailed analysis of these studies reveals that only one study fulfills the requirements of a clinical trial[1] but it is only an open-label trial and, therefore, not a first class piece of scientific and clinical evidence.

This implies that with respect to neuroprotection in Parkinson's disease (PD), there has been a lot of speculation but very little data in recent years. This is not surprising, since efforts in previous years to modify the progression of PD with monoamine oxidase inhibitors (MAOIs), dopamine agonists, L-DOPA and other compounds[2–8] have fallen short of being convincing.

The lack of proof of neuroprotective effects in PD could be related to several issues. There is an important methodological problem in differentiating purely symptomatic improvement from disease-modifying effects on pure clinical end-points. Surrogate biomarkers of dopamine neuronal function, such as PET and SPECT, were considered some years ago as very promising tools to solve these problems, but these expectations were not fulfilled since the read-outs of these parameters were strongly modified by many drugs and by the placebo effect.[9,10]

In addition, in recent years there has been an increase in evidence favoring the view of PD not as a single molecular disorder with a unique cause and a single pathogenic mechanism, but rather as a conglomerate of diseases with some common clinical and pathologic features. In the majority of the university clinics that could be considered centers of excellence for PD, idiopathic PD accounts for 50–60% of patients with Parkinsonism. The rest of the patients suffer other diseases such as vascular Parkinsonism, drug-induced Parkinsonism, or secondary Parkinsonism, such as progressive supranuclear palsy, multiple-system atrophy and other rare diseases (Figure 9.1).

The patients with idiopathic PD can not be pooled as a single disease produced by a single cause and mediated by a single pathogenic mechanism. In the majority of these patients, the cause and the pathogenic mechanisms are unknown. In some cases, up to 15–25% according to the interests and type of practice of the PD clinics the disease appears to be present in several members of the same family, following a pattern of Mendelian inheritance. In the last 15 years, up to 16 genes and loci responsible for PD have been identified (Table 9.1) and several additional genetic risk factors have been identified.[11–15] The proteins coded by these genes have many different functions including regulation of neurotransmitter release, processing of abnormal proteins through the ubiquitin proteasome system or through autophagy and lysosomal

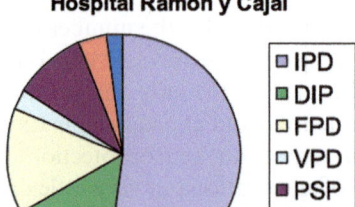

Figure 9.1 Classification of Parkinsonism at Hospital Ramón y Cajal, Madrid. IPD = idiopathic Parkinson's disease, DIP = drug-induced Parkinsonism, FPD = familial Parkinson's disease, VPD = vascular Parkinson's disease, PSP = progressive supranuclear palsy and MSA = multiple-system atrophy.

Table 9.1 Genes and loci responsible for PD, as related to Mendelian inheritance. D = dominant, R = recessive.

Locus	Gene	Inheritance and phenotype
PARK1, 4q21	SNCA A53P, A30T, E46K	D, typical, Lewy bodies
PARK2, 6q25	PARKIN	R, juvenile
PARK3, 2P13	DESCONOCIDO	D
PARK4, 4q21	SNCA	D, duplications and triplications
PARK5, 4p14	UCHL1	D, doubtful
PARK6, 1p36	PINK1	R, juvenile
PARK7, 1p36	DJ-1	R
PARK8, 12q12	LRRK2	D
PARK9, 1p36	ATP13A2	R, with dementia and spasticity
PARK10, 1p32	?	?
PARK11, 2q37	GIGYF2?	?
PARK12, Xq21	?	?
PARK13, 2p12	HTRA2	?
PARK14, 22q13	PLA2G6	R, with late-onset dystonia
PARK15, 22q12	FBXO7	R, with dementia and spasticity

elimination, mitochondrial function, kinase activity, regulation of phosphorylation of interacting proteins, and others.

If the causes of PD are so diverse, it will be difficult to find a single pharmaceutical compound able to prevent so many molecular abnormalities. Such a 'panacea' is not possible. It is more realistic to investigate different compounds acting through different neuroprotective mechanisms and to test these compounds in those patients whose PD is caused by such molecular abnormalities. For example, it is unlikely that an enhancer of the mitochondrial function could be neuroprotective in a group of patients with PD related to abnormal lysosomal function or to abnormal neurotransmitter release. Drugs should be tested, therefore, in patients whose disease is mediated by mechanisms related to the action of the putative neuroprotectants.

9.2 Pathogenic Mechanisms in Parkinson's Disease due to Genetic Defects

By definition, the cause of PD in idiopathic cases, which are the majority, is unknown. However, we can obtain valuable clues from the familial forms of the disease whose gene defects and functions of the mutant proteins are known. We should review briefly the best known of these defects in order to identify common mechanisms underlying PD without Mendelian inheritance.

9.2.1 α-Synuclein

α-Synuclein is so far the best-known player in the pathogenesis of PD. Three point mutations, A53P, A30T and E46K, produce autosomal dominant PD in families from the East Mediterranean, Germany and Spain, respectively.[16–18]

Familial PD has also been described in subjects with duplication or triplication of the α-synuclein gene and the severity of the disease is related to the dose of the protein.[2,19–21]

Additional data suggest the role of α-synuclein in idiopathic PD. α-Synuclein is a central constituent of the Lewy body, a spherical inclusion present in neurons in patients with PD and considered, in spite of its presence in other disease and its absence in some cases of genuine PD, as characteristic of this disease. α-Synuclein could be altered post-translationally by a number of processes including nitration in tyrosine residues, mediated by nitric oxide (NO), a toxic compound that increases with accumulation of free radicals,[22,23] or by phosphorylation at serine and threonine residues. In addition, the α-synuclein gene is the gene with strongest binding to PD in genome-wide analyses (GWAs) performed in thousands of patients worldwide.[24]

The role of α-synuclein in the pathogenesis of PD has been investigated in different *in silico* and *in vitro* models as well as in animal models of this disease from flies to mice. While α-synuclein is not essential for development and survival, since the suppression of this gene does not produce abnormalities, overexpression of the wild type as well as A30P and A53T α-synuclein in mutant Drosophila or mice causes a loss of dopamine neurons, motor deficits similar to those observed in patients with PD, expression of biochemical markers of cell aging and death and intraneuronal inclusions similar to Lewy bodies. E46K mutant α-synuclein is expected to be more toxic than the other α-synuclein mutants, since the change in the protein is more radical. The mutation substitutes a basic amino acid, lysine, for a dicarboxylic, glutamate. In addition, the substitution takes place in the middle of a highly conserved hexapeptide, KTKEGV, the third repeat of this sequence, critically important for the binding of α-synuclein to synaptic vesicles. In fact, patients with the E46K mutation develop widespread cortical pathology, as represented by the time to develop dementia from the time of the first symptom, and more widespread distribution of Lewy bodies.[18] *In vitro*, E46K mutant α-synuclein has higher affinity for artificial liposomes and greater tendency to auto aggregate than wild type, A30P or A53T α-synuclein.[25,26]

From the available information, it is difficult to pinpoint the exact mechanism that initiates the pathological changes that lead to nigrostriatal dopamine cell death in patients with PD related to point mutations of α-synuclein, increased doses of this protein or post-translational changes. The mechanism of toxicity is certainly related to a gain of function, since the pathology increases with the dose of α-synuclein and since dopamine neurons from α-synuclein knock-out animals are more resistant to dopamine neurotoxins.[27,28] α-Synuclein has many potential pathogenic mechanisms including the increase production of free radicals, since each molecule of dopamine metabolized to dihydroxyphenylacetic acid (DOPAC) produces a molecule of H_2O_2. The initial pathogenic mechanism of α-synuclein, therefore, is related to a higher production of free radicals, though this excess of free radicals contributes to other biochemical abnormalities such as mitochondrial impairment or increased aggregation of α-synuclein and other proteins actions, such as amyloidogenesis, impairment of mitochondrial function and others.[29] Its relatively selective effects on dopamine neurons suggest that the primary pathogenic event is related to dopamine function. *In vitro* studies have shown that α-synuclein plays a role in neurotransmitter release. Enhanced α-synuclein function impairs dopamine release from vesicles and increases the cytoplasmic, MAO-related turn-over of dopamine DOPAC. The activation of this neurotransmitter metabolic route increases H_2O_2 production and oxidative stress.

9.2.2 PARKIN

PARKIN mutations constitute probably the most frequent cause of hereditary Parkinsonism all over the world. The mutations of PARKIN were initially described in patients with PD from Japan, most often in young patients with a clinical phenotype that was called "autosomal recessive juvenile Parkinsonism" and was considered to be exclusive to that country. Further studies have shown that mutations of PARKIN are frequent in all investigated countries and that the 'juvenile' form is just a severe clinical phenotype linked to mutations which severely impair PARKIN function, such as single or multiple exon deletions, while the late-onset forms are mostly related to homozygotic or combined heterozygotic point mutations.[30,31] There are even patients with PARKIN-mutation-related PD with a single heterozygotic mutation. These patients have a late age at onset and a pattern of autosomal dominant inheritance.[32–34] There are also sporadic cases of PD with a single point mutation and one or more environmental risk factor for PD such as treatment with anti-psychotics or vital experiences with extraordinary stress (de Yébenes, personal communication).

PARKIN was described as an E3 ubiquitin ligase able to autoubiquitilate and ubiquitinate several substrates.[35–41] But that function does not explain the relatively selective involvement of dopamine neurons in PD related to PARKIN mutations since the expression of PARKIN is ubiquitous in brain and other tissues. Other functions of PARKIN which appear to be more relevant for the pathogenesis of PD are its role in neurotransmitter release, impaired mitochondria processing, tau metabolism and the formation of Lewy bodies.

PARKIN plays a role in neurotransmitter release. Studies *in vitro* from our own group, in mice with deletion of exon 3 of PARKIN have shown that the spontaneous release of dopamine and the vesicular release mediated by KCl were normal, but that amphetamine-induced release was reduced in these mice. In agreement with these data, motor activation induced by amphetamine in these animals was altered.[42] Spontaneous and vesicular release of dopamine is considered to be related to neural activity but amphetamine-induced liberation is considered to deal with the release of newly formed cytoplasmic dopamine, usually increased under conditions of stress. Impairment of cytoplasmic release would increase the cytoplasmic metabolism of dopamine, as in α-synuclein-related PD, and increase the production of free radicals. Susceptibility to disease in subjects with PARKIN mutations would be increased under stress conditions. In general, patients with PARKIN-related PD would be especially susceptible to increased dopamine metabolism mediated by stress.

The role of PARKIN in neurotransmitter release is not limited to dopamine but involves other neurotransmitters. We have shown that in the hippocampal slices of PARKIN-deficient mice, the amplitude of postsynaptic excitatory potentials is decreased, a finding that suggests abnormalities of long-term potentiation and memory consolidation, probably due to the impairment of glutamate release. These results are relevant for the cognitive deficits observed in the late stages of PD.

The function of PARKIN in mitochondria is related to its regulation of mitochondrial fusion/fission and mitoautophagy of impaired mitochondria; roles that are performed in co-operation with PINK[43–46] another protein whose mutations produce hereditary PD. The interaction between PARKIN and α-synuclein is suggested by the fact that PARKIN is an essential constituent of the typical inclusion found in the brain of patients with PD, the Lewy bodies, since patients with PD related to complete lack of PARKIN do not show Lewy bodies in their brain, while those with hemi-zygotic suppression of PARKIN do show these inclusions. The interaction of tau with PARKIN derives from the fact that PARKIN is essential for the stabilization of actin, which binds to tau in the microtubules. In the absence of PARKIN, tau becomes unstable in microtubules, floating in the cytoplasm, where it becomes hyperphosphorylated and translocates from its normal cellular compartment, the axon, to the nucleus and peri-nuclear area.[47,48] That interaction between PARKIN and tau provides the bases for the involvement of PARKIN in other neurological disease related to PD, the tauopathies.

9.2.3 Other Genes Involved in Hereditary Parkinsonism

The above-mentioned interaction between PARKIN and PINK-1, involves two proteins that work through the same pathways, PINK-1 upstream of PARKIN.[49,50] The gene that codes for PINK-1 protein is Park-6, whose mutations produce autosomal recessive PD. PINK-1 is a kinase that phosphorylates PARKIN and increases its function. Suppression of PINK-1 could be counteracted by PARKIN overexpression.[49,50] DJ-1 is a protein involved in oxidative stress.[51,52] The gene that codes for DJ-1 protein is Park-7, whose mutations produce autosomal recessive PD. DJ-1 is involved in the same pathways as PARKIN and PINK-1, since PINK-1 mutations can be compensated by overexpression of DJ-1.

LRRK-2 is the abbreviation for leucine-rich repeat kinase 2, a cacophonic name for the protein coded by gene Park-8, commonly used instead of the beautiful dardarin. Dardarin derives from the Basque onomatopoeia "dar-dar" (to shake), and it was initially used to describe a protein whose mutations were found in Basque families and were thought to be restricted to the Basque country.[18,53–55] We already know that although the R1441G mutation of Park-8 is more common in Basques, other mutations of this gene, such as the G2019S and others, are present in patients from all over the world.

The transmission of the disease caused by Park-8 mutations follows a pattern of autosomal dominant inheritance, although there are many subjects with these mutations and sporadic PD. Due to the pattern of incomplete penetrance, there are carriers of Park-8 mutations, members of families with other affected individuals, who remain free of symptoms beyond the age of diagnosis of the affected siblings. In accordance with these clinical inconsistencies in Park-8 mutation, the pathological features in the brain of affected individuals are quite inconsistent since some patients develop the full pathological spectrum of PD pathology, while others do not present Lewy bodies or relevant lesions in the nigrostriatal dopamine pathway. LRRK-2 protein is a kinase that interacts with other relevant proteins such as α-synuclein and tau. The pathogenic mechanism appears to be related to a gain of function and hyperphosphorylation of one or more proteins. Such an indirect effect, through the modification of other proteins, could explain the variability of the clinical phenotype and pathological features of Park-8-related PD.

The final genetic defects, which are linked to PD and we would like to discuss here, are related to lysosomal function, a mechanism quite different to those already discussed so far, and that could involve primary accumulation of abnormal proteins. Mutations of glucocerebrosidase (GCO) produce Gaucher's disease in infancy and PD in adulthood. There is no Mendelian pattern of inheritance of PD related to GCO mutations but the GCO locus is a high-risk locus for PD, in the same way that ApoE4 is a high risk for Alzheimer's disease. ATP13A2 is a phosphatase, also involved in lysosomal function, coded by the gene Park-9, whose mutations are associated with early-onset autosomal recessive PD.[56] Mutations of GCO and ATP13A2, therefore, produce PD through impairment of lysosomal function and accumulation of abnormal proteins.

9.3 Temporal Profile of Clinical Features and Pathological Changes Observed in Parkinson's Disease

9.3.1 The Temporal Spectrum on Clinical Findings

PD was initially described as a disease characterized by motor deficits, including akinesia, tremor and postural abnormalities as the key clinical features, to which rigidity was soon added. These features are very prominent and visible and therefore difficult to ignore by patients, carers and physicians, while other symptoms and signs that are more difficult to recognize, may be overlooked. Recent studies have shown, however, that patients with PD have many symptoms and signs, some of which are different from the motor deficits, and appear many years before them, and others which appear during the evolution of the characteristic clinical deficits. The problem with these symptoms and signs is that they are by no means specific to PD, since they appear in many subjects with normal neurological function and, therefore, are difficult to link with PD.

The first symptoms to appear in patients with PD are those called "personality traits", which are present for life and are consistent with a peculiar but not abnormal type of behavior which could be related to a difficulty in tolerating stress, and a tendency to avoid stress and stressful situations. Patients with PD, who develop motor symptoms in their presenility, are characterized by a type of behavior from adolescence with less violation of traffic rules, fewer changes of job, less tendency to divorce or to have sexual mates, less tendency toward alcoholism and drug addition, a strong respect for rules and ideas *etc.*, than their healthy siblings or population controls.

This type of behavior is probably the clinical correlate of their difficulty in releasing dopamine under situations of stress, which we mentioned as being characteristic of the first phase of the evolution of PARKIN-deficient mice. Dopamine is an essential neurotransmitter in the reward system, and inability to release dopamine blocks that system. Cytosolic dopamine is the newly synthesized neurotransmitter and it is released under conditions of stress. If the release of cytosolic dopamine is impaired, the patient is unable to respond to stress and the dopamine cell suffers an increased production of free radicals due to the above-mentioned MAO-mediated intracellular metabolism of dopamine, which produces excessive free radicals.

There are no systematic studies on the effect of stress in PD. Carlson and colleagues demonstrated that immobilization stress in rats produced enhancement of dopamine metabolism in brain.[57] Clinicians frequently present individual cases of PD starting after a very stressful experience. Jankovic and colleagues reported a couple of identical Jewish twins, both with PD, starting at 64 in one of the twins, who led a normal life, and in the other at 24, who was taken to a concentration camp during the Second World War.[58] We have seen a patient with a hemi-zygotic mutation of PARKIN, who should not have developed PD, starting with tremor that lasted his whole life, until death at 82. The symptoms started in 1936, when he was 13 and a junior member of a socialist trade union, and he was taken by an execution squad of fascists and underwent a faked shooting (de Yébenes, personal communication).

The second type of symptoms and signs occurring in PD is anosmia and autonomic dysfunction such as constipation, abnormal sweating and others. These symptoms are probably related to the initial pathological lesions observed in the brain of these patients, and also to lesions observed in the peripheral nervous system. These lesions are consistent with Lewy bodies and with α-synuclein immunoreactive lesions. As we shall discuss below, the first lesions appearing in the brain of patients with PD take place in the olfactory bulb and in the nucleus of the X cranial nerve, which could explain symptoms such anosmia and constipation. Other α-synuclein immunoreactive lesions have been observed in the peripheral nervous system. Minguez-Castellanos and colleagues[59] found Lewy bodies and α-synuclein immunoreactive lesions in a significant percentage of colonic and prostate gland biopsies of subjects without clinical features of PD who underwent these procedures to rule out cancer of these organs. Subsequent follow-up of these individuals showed that a large proportion of them developed PD in fewer than five years, suggesting that the peripheral nervous system is a place for early development of α-synuclein pathology.

It is difficult to pinpoint the mechanisms that render the olfactory bulb, the nucleus of the X cranial nerve and the peripheral autonomic nervous system exquisitely susceptible to α-synuclein immunoreactive lesions. It has been suggested that what these structures have in common is their relative poverty of a well-known neuroprotective agent, the glia (Lopez Mascaraque, personal communication). This is relevant because neurons in the olfactory bulb are in the process of neurogenesis during their entire life span and therefore their opportunity to unsheathe with astroglia is limited; and with regard to the neurons in the nucleus of the X cranial nerve and the peripheral autonomic nervous system, they have axons protected by unmyelinated fibbers which make them highly susceptible to damage.

The neuroprotective role of glia in PD has been suggested by many studies in experimental, both genetic and neurotoxic, models of this disease.[44,60–63] In some of these models, the presence of glia or culture medium obtained from cultured glia shift the neurotoxic effects of L-DOPA to neurotrophic.[64–68] This is particularly relevant for the pathogenesis of PD since we have already mentioned that in some of these models there is an abnormality of the release of cytoplasmic dopamine. The presence of glia is particularly relevant because it could switch this toxic neurotransmission abnormality into a harmless event.

The neuroprotective effects of glia are probably mediated by many compounds of both low (anti-oxidants) and high (neurotrophic factors and neuroprotective proteins) molecular weight.[44,61,67] Among the low-molecular-weight compounds, the most important is probably glutathione (GSH) gamma-L-Glutamyl-L-cysteinylglycine. GSH levels are critical for compensating the increased production of free radicals that takes place in mice with PARKIN deletions.[42,63,69–71] These animals present an up-regulation of GSH production during their youth and, at that time, they are able to maintain a number of nigrostriatal dopamine neurons that is close to normality. It is only during the last months of their life, when levels of GSH fall below normality, that a substantial drop-out of dopamine neurons is observed.[44,70]

The third type of symptoms that takes place in subjects with PD are those considered as the typical motor symptoms of this disease, mostly akinesia, rigidity, postural abnormalities and tremor at rest. Most of these symptoms are related to abnormal function and cell loss of the nigrostriatal dopamine neurons. It is estimated that the severity of nigrostriatal cell loss should be substantial before the symptoms appear, more than a 50% loss of neurons, since the remaining cells could compensate for the deficit by increasing their firing rate. This mechanism of compensation, however, is a deleterious vicious circle since the firing rate of these neurons is coupled with the metabolism of dopamine and an enhanced dopamine metabolism would further increase the production of free radicals that initiates the cell loss.

Nigrostriatal cell loss explains most but not all the motor symptoms observed in PD. The blockade of catecholamine synthesis with α-methyl-*p*-tyrosine produces akinesia and rigidity, but not tremor in experimental animals. The presence of tremor requires the involvement of other structures of the brain stem, including the raphe nuclei, rich in serotonin. Serotonin deficits also play an important role in the pathogenesis of depression, a common psychiatric disturbance present in patients with PD. The lesion of serotonin neurons is probably mediated by the same mechanisms that cause dopamine loss since serotonin is also metabolized *via* MAO through a free-radical-producing pathway.

The final group of symptoms that appears in PD is loss of the control of impulses, memory loss, perceptive and cognitive deficits and other symptoms with widespread involvement of the limbic system and the cerebral cortex. These findings imply extension of the abnormal dopamine function to the meso-limbic dopamine system and widespread abnormalities of tau protein and α-synuclein. These two proteins could be altered by a number of mechanisms. In mice lacking PARKIN, tau is unstable and detaches itself from actin in the microtubules, floating in the cytoplasm, where it becomes hyperphosphorylated and changes its cellular localization and function. It is unknown whether there are also abnormalities of tau phosphorylation as a consequence of mutations associated with kinases function gain, such as LRRK-2, or loss of function of phosphatases, as in the case of ATP13A2.

9.3.2 The Temporal Pattern of Pathological Changes

The neuropathological examination of the brains of patients, with and without clinical features of PD, has shown that α-synuclein-related pathology, including Lewy bodies and α-synuclein immunoreactive dystrophic neuritis, has a peculiar distribution and follows a special pattern of spreading which corresponds to the progression of the clinical symptoms.[72,73] According to that pattern, it has been proposed that from the pathological point of view PD could be subdivided in six pathological stages.

Stage I is characterized by the presence of α-synuclein pathology in the nucleus of the X cranial nerve and the olfactory bulb. These patients do not have classic Parkinsonian features but could have autonomic dysfunction and anosmia. Stage II is characterized by the spreading of the lesions to the locus coeruleus and nucleus raphe magnus. In addition, these patients could have depression. Stage III corresponds to the classical features of α-synuclein pathology in the nigrostriatal neurons and is associated with the characteristic motor features of PD. In Stage IV, the pathology spreads to the hippocampus and patients have memory disturbances. Stage V is associated with involvement of the medial frontal cortex and the limbic system, and from the clinical point of view is characterized by behavioral abnormalities and loss of impulse control. Stage VI is characterized by widespread extension of α-synuclein pathology to the cerebral cortex, and from the clinical point of view by dementia.

There have been criticisms of the pattern of spread of α-synuclein pathology postulated by Braak and his colleagues. One of the criticisms derives from the presence of Lewy bodies in the brains of patients without clinical features of PD. Up to 5% of the population aged over 65 present occasional Lewy bodies in the *substantia nigra*. It is unknown whether these subjects are just presymptomatic PD patients with lesions below the level of impaired neuronal dysfunction. On the other hand, these patients could be perfectly explained by our hypothesis regarding the role of glia in the initiation of the pathological changes that drive PD. Local mechanisms, including local ischemic insults, scars, *etc.*, could produce glial dysfunction leading to neuronal involvement.

Another criticism of the Braak schema is that it does not apply rigidly to all cases and there are occasional inconsistencies. This is not surprising since PD is not a disease with an unique pathogenesis but with multiple mechanisms of disease, and it is conceivable that the relative weight of the different mechanisms and their impact on different neuronal systems shows regional variability in the pattern and the severity of the lesions.

9.4 Neuroprotective Therapies: Towards Neuroprotection Based on Pathogenesis in Parkinson's Disease

In spite of all the work performed during the last quarter of a century, there is still no adequate neuroprotection for patients with PD. We believe that some of the reasons for this failure can be summarized as: (1) the putative neuroprotective treatments are initiated too late, when the damage to the nigrostriatal system is massive and irreversible; and (2) neuroprotective approaches are too naïve, as they are based on just one of the many pathogenic mechanisms of the disease.

(1) Neuroprotective treatments are initiated too late. It is estimated that at the time of diagnosis, the average neuronal loss in the nigrostriatal of patients with PD is over 50% of neurons. There are no estimations regarding neuronal loss in other brain regions or regarding the number or percentages of neurons with α-synuclein pathology or other histological lesions.

It is important to start neuroprotective treatments as soon as possible. Attempts have been made to anticipate the clinical diagnosis by using neuroimaging techniques in subjects with clinical features present in early Parkinsonism. For instance, radio nuclei studies and ecography of the *substantia nigra* have been applied to subjects with anosmia or constipation. The problem with this approach is that anosmia and constipation are fairly common in the general population, and neuroimaging studies are expensive and far from clear in their interpretation.

With the progress of neurogenetics a great number of genes linked to PD have been discovered and an increasing number of subjects have been identified who carry mutations producing PD or genotypes increasing the risk of it. These subjects could be closely followed up for the presence of clinical symptoms and a number of neuroprotective approaches are recommended including: avoiding drugs which induce Parkinsonism; avoiding stress as much as possible; and doing regular exercise.

The real problem occurs when the patient has no familial risk of PD. It is already known that the performance of simple movements, such as the frequency of blinking, the amplitude and speed of finger tapping between the index finger and the thumb, or the length of the stride, change with age in normal individuals and are different in normal subjects and in patients with PD. These tests are very easy to perform but they should be standardized and quantified. The mass application of these tests as a population screen, likewise the recording of blood pressure or heart rate by physicians and nurses, could provide a usable tool for the anticipation of diagnosis and neuroprotection in PD.

(2) Neuroprotective approaches are too simple as they are based on a single pathogenic mechanism of the disease. We have already seen that PD is a complex disease related to multiple pathogenetic mechanisms including: abnormalities in neurotransmitter function; increased free radical production; impaired mitochondrial function; excessive phosphorylation and defective phosphate activity; abnormal ubiquitination and impaired lysosomal functions; increased proteostasis, *etc*. We have already discussed that the pathological lesions in PD involve different neuronal systems according to the different stages of this disease. It would be, therefore, very naïve to think that a single drug acting on one specific pathway could stop the development of the disease.

According to the information described above, it is likely that neuroprotection could be achieved in PD only if it is started early, ideally before the presence of characteristic motor symptoms and signs, and only if the treatment targets the different pathogenic mechanisms. A summary of this complex neuroprotective strategy is presented in Table 9.2.

Reduction of stress is an easy mechanism of neuroprotection based on the involvement of cytoplasmic dopamine in the response to stress. Patients with a high risk for PD should develop strategies to avoid stress and to learn to relax when stress is unavoidable. General exercise is recommended at all times in patients with PD because of its neurotrophic effect due to its increased production of neurotrophic factors with effects on dopamine neurons, such as brain-derived neurotrophic factor (BDNF).[74] In addition, in patients with the already characteristic motor symptoms of PD, besides general exercise it is recommended that the patient should move most articulations in order to avoid immobilization-related joint degeneration. Special exercises are needed for voice maintenance, swallowing, breathing exercises, and others to keep the spine straight.

MAOIs are now used as neuroprotective agents in PD. Their value may only be limited to the first stages of the disease. GSH and other free radical

Table 9.2 Potential strategies for neuroprotection in PD.

Treatment	Mechanism of action	Purpose	Time of intervention
Reduce stress	Decreases cytoplasmic processing of dopamine	To reduce production of free radicals	Early stages
Exercise	Increases neurotrophic factors. Decreases risks of immobilization	To increase sprouting and survival of remaining neurons	Early stages
MAOIs	Decreases intracellular free radicals	To reduce production of free radicals	Early stages
GSH and free radical chelators	Decreases intracellular free radicals	To reduce the levels of free radicals	Early and late stages
Mitochondrial activity enhancers	Improve energy production	To rescue neurons from energy deprivation	Early and late stages
Inhibitors of phosphorylation	To reduce phosphorylation of proteins	To block proteostasis and abnormal deposition of proteins	Late stages
Potentiators of autophagy	To increase elimination of abnormal proteins	To block the accumulation of abnormal proteins	Late stages

scavengers would be useful in PD. GSH does not easily cross the blood–brain barrier. Its precursor, N-acetylcysteine has been occasionally used as a neuroprotective agent but never, to our knowledge, in a large clinical trial. The best option for the use of GSH and analogs appears to be using derivatives of GSH that cross the blood–brain barrier and could be converted in GSH in the nervous system. One such compound GSH-amide has been used in tardive dyskinesia with good results, but was never properly tested in PD.

Mitochondrial enhancers could be useful during the whole process of the disease. There are compounds that potentiate the activity of different mitochondrial respiratory chain complexes but the most frequently used ones are the potentiators of Complex 1. Inhibitors of kinases have been tested in other diseases such as Alzheimer's disease and Steele–Richardson syndrome. These compounds are also worth testing in PD. Potentiators of autophagy, such as trehalose, have been neuroprotective in animal models of PD and should also be tested in humans.

The methods used to test these putative neuroprotective agents are difficult. In some cases, such as in PD produced by a mutation of a single gene, where the pathogenic mechanism is simple and well known, it would be possible to perform a clinical trial with a single agent. In other cases, when the pathogenic mechanisms are more complex or unknown it would be necessary to treat the patients with a cocktail of substances with different molecular effects. In any case, only a detailed and thorough knowledge of the pathogenic mechanisms involved in each case would allow us efficient neuroprotective therapy in PD.

Acknowledgements

This work was supported in part by grants from the European Community and Comunidad de Madrid (CAM S2011/BMD-2308), Spanish Ministry of Health, FIS 2010/172, CIBERNED 2006/05/0059 and CIBERNED PI2010-6. The authors thank Mrs Claire Marsden for editorial help.

References

1. R. A. Hauser, M. F. Lew, H. Hurtig, W. G. Ondo, J. Wojcieszek and C. Fitzer-Attas, *Mov. Disord.*, 2009, **24**, 564.
2. S. Fahn, *Mov. Disord.*, 2008, **23**(3), S497.
3. S. Fahn, *Mov. Disord.*, 2010, **25**(1), S2.
4. C. G. Goetz, B. C. Tilley, S. R. Shaftman, G. T. Stebbins, S. Fahn, P. Martinez-Martin, W. Poewe, C. Sampaio, M. B. Stern, R. Dodel, B. Dubois, R. Holloway, J. Jankovic, J. Kulisevsky, A. E. Lang, A. Lees, S. Leurgans, P. A. LeWitt, D. Nyenhuis, C. W. Olanow, O. Rascol, A. Schrag, J. A. Teresi, J. J. van Hilten and N. LaPelle, *Mov. Disord.*, 2008, **23**, 2129.
5. C. Marras, A. E. Lang, S. W. Eberly, D. Oakes, S. Fahn, S. R. Schwid, C. Hyson and I. Shoulson, *Mov. Disord.*, 2009, **24**, 2370.
6. A. H. Schapira, *CNS Drugs*, 2011, **25**, 1061.
7. C. Singer, *Cleveland Clin. J. Med.*, 2012, **79**(2), S3.
8. R. Stowe, N. Ives, C. E. Clarke, K. Handley, A. Furmston, K. Deane, J. J. van Hilten, K. Wheatley and R. Gray, *Mov. Disord.*, 2011, **26**, 587.
9. L. J. Fuentes, P. J. Fernandez, G. Campoy, M. M. Antequera, J. Garcia-Sevilla and C. Antunez, *Dementia Geriatr. Cognit. Disord.*, 2010, **29**, 139.
10. S. Thobois, E. Broussolle and P. Remy, *Rev. Neurol.*, 2005, **161**, 385.
11. G. Charlesworth, S. Gandhi, J. M. Bras, R. A. Barker, D. J. Burn, P. F. Chinnery, S. M. Gentleman, R. Guerreiro, J. Hardy, J. L. Holton, A. Lees, K. Morrison, U. M. Sheerin, N. Williams, H. Morris, T. Revesz and N. W. Wood, *Neurobiol. Aging*, 2012, **33**, 838.
12. P. Pastor, *Neurology*, 2012, **79**, 619.
13. M. Sharma, J. P. Ioannidis, J. O. Aasly, G. Annesi, A. Brice, C. Van Broeckhoven, L. Bertram, M. Bozi, D. Crosiers, C. Clarke, M. Facheris, M. Farrer, G. Garraux, S. Gispert, G. Auburger, C. Vilarino-Guell, G. M. Hadjigeorgiou, A. A. Hicks, N. Hattori, B. Jeon, S. Lesage, C. M. Lill, J. J. Lin, T. Lynch, P. Lichtner, A. E. Lang, V. Mok, B. Jasinska-Myga, G. D. Mellick, K. E. Morrison, G. Opala, P. P. Pramstaller, I. Pichler, S. S. Park, A. Quattrone, E. Rogaeva, O. A. Ross, L. Stefanis, J. D. Stockton, W. Satake, P. A. Silburn, J. Theuns, E. K. Tan, T. Toda, H. Tomiyama, R. J. Uitti, K. Wirdefeldt, Z. Wszolek, G. Xiromerisiou, K. C. Yueh, Y. Zhao, T. Gasser, D. Maraganore and R. Kruger, *Neurology*, 2012, **79**, 659.

14. L. Soreq, Y. Ben-Shaul, Z. Israel, H. Bergman and H. Soreq, *Neurobiol. Dis.*, 2012, **45**, 1018.
15. Y. Wang and M. Yang, *Parkinsonism Relat. Disord.*, 2012, **18**, 102.
16. T. Gasser, *Ann. Neurol.*, 1998, **44**, S53.
17. M. H. Polymeropoulos, C. Lavedan, E. Leroy, S. E. Ide, A. Dehejia, A. Dutra, B. Pike, H. Root, J. Rubenstein, R. Boyer, E. S. Stenroos, S. Chandrasekharappa, A. Athanassiadou, T. Papapetropoulos, W. G. Johnson, A. M. Lazzarini, R. C. Duvoisin, G. Di Iorio, L. I. Golbe and R. L. Nussbaum, *Science*, 1997, **276**, 2045.
18. J. J. Zarranz, J. Alegre, J. C. Gomez-Esteban, E. Lezcano, R. Ros, I. Ampuero, L. Vidal, J. Hoenicka, O. Rodriguez, B. Atares, V. Llorens, E. Gomez Tortosa, T. del Ser, D. G. Munoz and J. G. de Yebenes, *Ann. Neurol.*, 2004, **55**, 164.
19. K. Gwinn, M. J. Devine, L. W. Jin, J. Johnson, T. Bird, M. Muenter, C. Waters, C. H. Adler, R. Caselli, H. Houlden, G. Lopez, A. Singleton and J. Hardy, *Mov. Disord.*, 2011, **26**, 2134.
20. J. H. Kim, B. L. Park, C. F. Pasaje, J. S. Bae, C. S. Park, B. J. Kim, M. Lee, W. H. Choi, T. M. Shin, I. G. Choi, J. Hwang, I. Koh, S. I. Woo and H. D. Shin, *J. Mol. Neurosci.*, 2012, **46**, 476.
21. E. Mutez, F. Lepretre, E. Le Rhun, L. Larvor, A. Duflot, V. Mouroux, J. P. Kerckaert, M. Figeac, K. Dujardin, A. Destee and M. C. Chartier-Harlin, *Hum. Mutat.*, 2011, **32**, E2079.
22. Y. Liu, M. Qiang, Y. Wei and R. He, *J. Mol. Cell Biol.*, 2011, **3**, 239.
23. D. K. Stone, T. Kiyota, R. L. Mosley and H. E. Gendelman, *Neurosci. Lett.*, 2012, **523**, 167.
24. C. M. Lill, J. T. Roehr, M. B. McQueen, F. K. Kavvoura, S. Bagade, B. M. Schjeide, L. M. Schjeide, E. Meissner, U. Zauft, N. C. Allen, T. Liu, M. Schilling, K. J. Anderson, G. Beecham, D. Berg, J. M. Biernacka, A. Brice, A. L. DeStefano, C. B. Do, N. Eriksson, S. A. Factor, M. J. Farrer, T. Foroud, T. Gasser, T. Hamza, J. A. Hardy, P. Heutink, E. M. Hill-Burns, C. Klein, J. C. Latourelle, D. M. Maraganore, E. R. Martin, M. Martinez, R. H. Myers, M. A. Nalls, N. Pankratz, H. Payami, W. Satake, W. K. Scott, M. Sharma, A. B. Singleton, K. Stefansson, T. Toda, J. Y. Tung, J. Vance, N. W. Wood, C. P. Zabetian, P. Young, R. E. Tanzi, M. J. Khoury, F. Zipp, H. Lehrach, J. P. Ioannidis and L. Bertram, *PLoS Genet.*, 2012, **8**, e1002548.
25. S. T. Chang and M. S. Chern, *Int. J. Clin. Pract. Suppl.*, 2005, **59**, 15.
26. N. Pandey, R. E. Schmidt and J. E. Galvin, *Exp. Neurol.*, 2006, **197**, 515.
27. R. E. Drolet, B. Behrouz, K. J. Lookingland and J. L. Goudreau, *Neurotoxicology*, 2004, **25**, 761.
28. C. E. Moussa, F. Mahmoodian, Y. Tomita and A. Sidhu, *Biochem. Biophys. Res. Commun.*, 2008, **365**, 833.
29. T. Yasuda and H. Mochizuki, *Apoptosis*, 2010, **15**, 1312.
30. S. Anders, B. Sack, A. Pohl, T. Munte, P. Pramstaller, C. Klein and F. Binkofski, *Brain*, 2012, **135**, 1128.

31. B. Koentjoro, J. S. Park, A. D. Ha and C. M. Sue, *Mov. Disord.*, 2012, **27**, 1299.
32. J. Hoenicka, L. Vidal, B. Morales, I. Ampuero, F. J. Jimenez-Jimenez, J. Berciano, T. del Ser, A. Jimenez, P. G. Ruiz and J. G. de Yebenes, *Arch. Neurol.*, 2002, **59**, 966.
33. T. Kitada, J. J. Tomlinson, H. S. Ao, D. A. Grimes and M. G. Schlossmacher, *Curr. Treat. Options Neurol.*, 2012, **14**, 230.
34. M. Sun, J. C. Latourelle, G. F. Wooten, M. F. Lew, C. Klein, H. A. Shill, L. I. Golbe, M. H. Mark, B. A. Racette, J. S. Perlmutter, A. Parsian, M. Guttman, G. Nicholson, G. Xu, J. B. Wilk, M. H. Saint-Hilaire, A. L. DeStefano, R. Prakash, S. Williamson, O. Suchowersky, N. Labelle, J. H. Growdon, C. Singer, R. L. Watts, S. Goldwurm, G. Pezzoli, K. B. Baker, P. P. Pramstaller, D. J. Burn, P. F. Chinnery, S. Sherman, P. Vieregge, I. Litvan, T. Gillis, M. E. MacDonald, R. H. Myers and J. F. Gusella, *Arch. Neurol.*, 2006, **63**, 826.
35. P. Choi, H. Snyder, L. Petrucelli, C. Theisler, M. Chong, Y. Zhang, K. Lim, K. K. Chung, K. Kehoe, L. D'Adamio, J. M. Lee, E. Cochran, R. Bowser, T. M. Dawson and B. Wolozin, *Brain Res. Mol. Brain Res.*, 2003, **117**, 179.
36. D. P. Huynh, D. T. Nguyen, J. B. Pulst-Korenberg, A. Brice and S. M. Pulst, *Exp. Neurol.*, 2007, **203**, 531.
37. T. Kitada, S. Asakawa, N. Hattori, H. Matsumine, Y. Yamamura, S. Minoshima, M. Yokochi, Y. Mizuno and N. Shimizu, *Nature*, 1998, **392**, 605.
38. H. Shimura, M. G. Schlossmacher, N. Hattori, M. P. Frosch, A. Trockenbacher, R. Schneider, Y. Mizuno, K. S. Kosik and D. J. Selkoe, *Science*, 2001, **293**, 263.
39. J. W. Um, D. S. Min, H. Rhim, J. Kim, S. R. Paik and K. C. Chung, *J. Biol. Chem.*, 2006, **281**, 3595.
40. Y. Yang, I. Nishimura, Y. Imai, R. Takahashi and B. Lu, *Neuron*, 2003, **37**, 911.
41. Y. Zhang, J. Gao, K. K. Chung, H. Huang, V. L. Dawson and T. M. Dawson, *Proc. Natl. Acad. Sci. U. S. A.*, 2000, **97**, 13354.
42. J. M. Itier, P. Ibanez, M. A. Mena, N. Abbas, C. Cohen-Salmon, G. A. Bohme, M. Laville, J. Pratt, O. Corti, L. Pradier, G. Ret, C. Joubert, M. Periquet, F. Araujo, J. Negroni, M. J. Casarejos, S. Canals, R. Solano, A. Serrano, E. Gallego, M. Sanchez, P. Denefle, J. Benavides, G. Tremp, T. A. Rooney, A. Brice and J. Garcia de Yebenes, *Hum. Mol. Genet.*, 2003, **12**, 2277.
43. H. Deng, M. W. Dodson, H. Huang and M. Guo, *Proc. Natl. Acad. Sci. U. S. A.*, 2008, **105**, 14503.
44. M. A. Mena and J. Garcia de Yebenes, *Neuroscientist*, 2008, **14**, 544.
45. K. Venderova, G. Kabbach, E. Abdel-Messih, Y. Zhang, R. J. Parks, Y. Imai, S. Gehrke, J. Ngsee, M. J. Lavoie, R. S. Slack, Y. Rao, Z. Zhang, B. Lu, M. E. Haque and D. S. Park, *Hum. Mol. Genet.*, 2009, **18**, 4390.

46. Y. Yang, Y. Ouyang, L. Yang, M. F. Beal, A. McQuibban, H. Vogel and B. Lu, *Proc. Natl. Acad. Sci. U. S. A.*, 2008, **105**, 7070.
47. J. Menendez, J. A. Rodriguez-Navarro, R. M. Solano, M. J. Casarejos, I. Rodal, R. Guerrero, M. P. Sanchez, J. Avila, M. A. Mena and J. G. de Yebenes, *Hum. Mol. Genet.*, 2006, **15**, 2045.
48. J. A. Rodriguez-Navarro, A. Gomez, I. Rodal, J. Perucho, A. Martinez, V. Furio, I. Ampuero, M. J. Casarejos, R. M. Solano, J. G. de Yebenes and M. A. Mena, *Hum. Mol. Genet.*, 2008, **17**, 3128.
49. I. E. Clark, M. W. Dodson, C. Jiang, J. H. Cao, J. R. Huh, J. H. Seol, S. J. Yoo, B. A. Hay and M. Guo, *Nature*, 2006, **441**, 1162.
50. J. Park, S. B. Lee, S. Lee, Y. Kim, S. Song, S. Kim, E. Bae, J. Kim, M. Shong, J. M. Kim and J. Chung, *Nature*, 2006, **441**, 1157.
51. C. P. Ramsey, E. Tsika, H. Ischiropoulos and B. I. Giasson, *Hum. Mol. Genet.*, 2010, **19**, 1425.
52. T. T. Pham, F. Giesert, A. Rothig, T. Floss, M. Kallnik, K. Weindl, S. M. Holter, U. Ahting, H. Prokisch, L. Becker, T. Klopstock, M. Hrabe de Angelis, K. Beyer, K. Gorner, P. J. Kahle, D. M. Vogt Weisenhorn and W. Wurst, *Genes Brain Behav.*, 2010, **9**, 305.
53. J. Alegre-Abarrategui, O. Ansorge, M. Esiri and R. Wade-Martins, *Neuropathol. Appl. Neurobiol.*, 2008, **34**, 272.
54. C. Paisan-Ruiz, S. Jain, E. W. Evans, W. P. Gilks, J. Simon, M. van der Brug, A. Lopez de Munain, S. Aparicio, A. M. Gil, N. Khan, J. Johnson, J. R. Martinez, D. Nicholl, I. M. Carrera, A. S. Pena, R. de Silva, A. Lees, J. F. Marti-Masso, J. Perez-Tur, N. W. Wood and A. B. Singleton, *Neuron*, 2004, **44**, 595.
55. A. Zimprich, B. Muller-Myhsok, M. Farrer, P. Leitner, M. Sharma, M. Hulihan, P. Lockhart, A. Strongosky, J. Kachergus, D. B. Calne, J. Stoessl, R. J. Uitti, R. F. Pfeiffer, C. Trenkwalder, N. Homann, E. Ott, K. Wenzel, F. Asmus, J. Hardy, Z. Wszolek and T. Gasser, *Am J. Hum. Genet.*, 2004, **74**, 11.
56. J. Hardy, *Neuron*, 2010, **68**, 201.
57. J. N. Carlson, L. W. Fitzgerald, R. W. Keller, Jr. and S. D. Glick, *Brain Res.*, 1991, **550**, 313.
58. J. J. Jankovic and E. Tolosa, *Parkinson's Disease and Movement Disorders*, Lippincott Williams & Wilkins, 2003.
59. A. Minguez-Castellanos, C. E. Chamorro, F. Escamilla-Sevilla, A. Ortega-Moreno, A. C. Rebollo, M. Gomez-Rio, A. Concha and D. G. Munoz, *Neurology*, 2007, **68**, 2012.
60. M. A. Mena, M. J. Casarejos and J. Garcia de Yebenes, *J. Neural Transm.*, 1999, **106**, 1105.
61. M. A. Mena, S. de Bernardo, M. J. Casarejos, S. Canals and E. Rodriguez-Martin, *Mol. Neurobiol.*, 2002, **25**, 245.
62. M. A. Mena, B. Pardo, C. L. Paino and J. G. De Yebenes, *Neuroreport*, 1993, **4**, 438.
63. R. M. Solano, M. J. Casarejos, J. Menendez-Cuervo, J. A. Rodriguez-Navarro, J. Garcia de Yebenes and M. A. Mena, *J. Neurosci.*, 2008, **28**, 598.

64. M. A. Mena, M. J. Casarejos, E. Rodríguez-Martín, R. Solano, J. Menendez and J. Garcia de Yebenes, in *Neurobiology of DOPA as a Neurotransmitter*, ed. Yoshimi Misu and Yoshio Goshima, CRC Press, 2006, pp. 303–319.
65. M. A. Mena, M. J. Casarejos, R. M. Solano and J. G. de Yebenes, *Curr. Top. Med. Chem.*, 2009, **9**, 880.
66. M. A. Mena, V. Davila, J. Bogaluvsky and D. Sulzer, *Mol. Pharmacol.*, 1998, **54**, 678.
67. M. A. Mena, V. Davila and D. Sulzer, *J. Neurochem.*, 1997, **69**, 1398.
68. E. Rodriguez-Martin, S. Canals, M. J. Casarejos, S. de Bernardo, A. Handler and M. A. Mena, *J. Neurochem.*, 2001, **78**, 535.
69. M. J. Casarejos, R. M. Solano, J. A. Rodriguez-Navarro, A. Gomez, J. Perucho, J. G. Castano, J. Garcia de Yebenes and M. A. Mena, *J. Neurochem.*, 2009, **110**, 1523.
70. J. A. Rodriguez-Navarro, R. M. Solano, M. J. Casarejos, A. Gomez, J. Perucho, J. G. de Yebenes and M. A. Mena, *J. Neurochem.*, 2008, **106**, 2143.
71. M. J. Casarejos, R. M. Solano, J. Menendez, J. A. Rodriguez-Navarro, C. Correa, J. Garcia de Yebenes and M. A. Mena, *J. Neurochem.*, 2005, **94**, 1005.
72. H. Braak, E. Ghebremedhin, U. Rub, H. Bratzke and K. Del Tredici, *Cell Tissue Res.*, 2004, **318**, 121.
73. H. Braak, C. M. Muller, U. Rub, H. Ackermann, H. Bratzke, R. A. de Vos and K. Del Tredici, *J. Neural Transm. Suppl.*, 2006, **70**, 89.
74. P. Rasmussen, P. Brassard, H. Adser, M. V. Pedersen, L. Leick, E. Hart, N. H. Secher, B. K. Pedersen and H. Pilegaard, *Exp. Physiol.*, 2009, **94**, 1062.

CHAPTER 10

Glutamate Receptor Modulators as Emergent Therapeutic Agents in the Treatment of Parkinson's Disease

SYLVAIN CÉLANIRE,*[a] BENJAMIN PERRY,[a] ROBERT LUTJENS,[a] SONIA POLI[a] AND IAN J. REYNOLDS*[b]

[a] Addex Therapeutics, Chemin des Aulx, Geneva, Switzerland; [b] Knopp Biosciences, Pittsburgh, USA
*Email: sylvain.celanire@addextherapeutics.com; ian.reynolds@tevapharm.com

10.1 Introduction

Parkinson's disease (PD) is clinically defined by the presence of a combination of slowness of movement (bradykinesia), poverty of movement (akinesia), muscle stiffness (rigidity) and tremor at rest. It is a neurodegenerative disorder characterized by a progressive loss of mid-brain dopaminergic (DA) neurons in the *substantia nigra pars compacta* (SNc) which input into the *striatum* and into different basal ganglia nuclei under normal conditions. This reduced or lack of dopaminergic input induces a major imbalance in the circuitry linking the basal ganglia to the thalamocortical motor circuit,[1] and is generally accepted to be the basis of the motor symptoms observed in PD. Specifically, it is believed that loss of striatal input induces an increased inhibition of the pallidostriatal synapse and disinhibition of the subthalamic nucleus (STN), where bursts of activity can be measured early in the course of dopamine depletion.[2,3] This hyperactivity may generate an excitotoxic insult and be the cause for continued loss of dopaminergic neurons in the SNc, participating in the progressive worsening of motor symptoms seen in PD patients. Preclinical studies showing that lesions of the STN reduce 6-hydroxy-dopamine (6-OHDA)-induced degeneration of SNc neurons confirm the validity of this hypothesis.[4] Therefore, counteracting the hyperactivity of the STN may prove a useful treatment of the motor symptoms of PD, while potentially also offering a way to slow disease progression.

In this chapter, we review the glutamatergic-based therapeutic arsenal as a novel emergent pharmacological class of agents in the symptomatic treatment of PD, the associated motor side-effects as well as the neuroprotective and disease-modification potential.

10.1.1 Glutamate Receptors: Nomenclature and Links with Parkinson's Disease

Glutamate plays a major role in transmitting signals throughout the whole body, especially in the brain where it serves as the main excitatory neurotransmitter as well as being the precursor of γ-aminobutyric acid (GABA) the main inhibitory neurotransmitter. Glutamate exerts its activity by binding to several glutamate receptors classified as ionotropic or metabotropic receptors (Figure 10.1). Ionotropic glutamate receptors (iGluRs) are ligand-gated

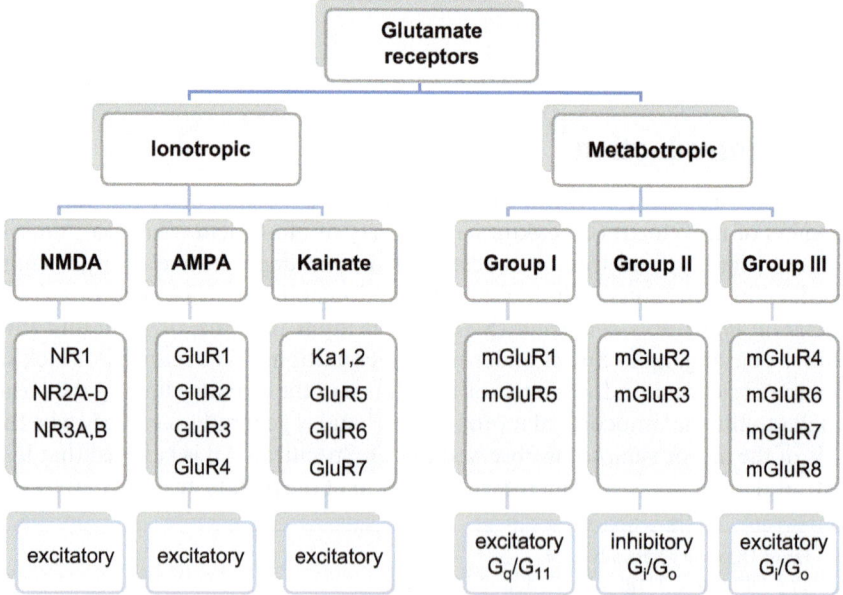

Figure 10.1 Ionotropic and metabotropic glutamate receptor classification and subtypes.

non-selective cation channels which allow the flow of K^+, Na^+ and sometimes Ca^{2+} in response to glutamate binding, classified as 2-amino-3-(5-methyl-3-oxo-1,2-oxazol-4-yl) propanoic acid (AMPA), kainate or N-methyl-D-aspartate (NMDA) receptors.[5] Metabotropic glutamate receptors (mGluRs) are 7-transmembrane, G-protein coupled receptors (GPCRs) of eight subtypes classified into three subgroups.[6] Contrary to the AMPA, NMDA receptors and mGluRs, very little is known of the target potential of kainate receptors to develop new therapies for PD, therefore this class will not be further discussed in this chapter.

10.1.1.1 Ionotropic Glutamate Receptors

Ionotropic glutamate receptors (iGluRs) present an interesting, if challenging, target in the context of PD, and have been the focus of substantial development effort.[7] NMDA receptors are located at key synapses within the basal ganglia circuitry, and electrophysiological studies indicate that they directly mediate excitation of neurons in the *striatum, globus pallidus* (GP), *substantia nigra pars reticulata* (SNr), STN and SNc.[8,9] Moreover, NMDA receptors are directly involved in excitotoxicity caused by high levels of glutamate;[10] excessive glutamatergic activity following the loss of dopaminergic neurons is thought to contribute to the progressive loss of neurons in the SNc, and therefore to the slow worsening of motor symptoms seen in PD. There is a rich pharmacology of both AMPA and NMDA receptors associated with orthosteric and allosteric modulation, which has identified a number of distinct mechanisms for modulating the activity of these receptors.[11] The activity of NMDA receptors is controlled by separate co-agonist sites that recognize glutamate and glycine, and occupation of both sites is necessary for receptor activation. Specific antagonists have been developed that target the glutamate and glycine sites, and there is an additional class of NMDA antagonist that binds to a site within the channel and are thus termed "channel blockers". NMDA receptors are comprised of several distinct subunits that include NR1, one or more of the NR2 (A–D) subunits, and possibly NR3 (Figure 10.1).[11] A series of NMDA antagonists that preferentially block NR2B-containing NMDA receptors has been developed, and these have been evaluated for anti-Parkinsonian properties. Key findings with these molecules will be described herein.

AMPA receptors, which, like NMDA receptors mediate fast excitatory neurotransmission at glutamatergic synapses, are highly expressed in the basal ganglia.[12,13] AMPA receptors are composed of different combinations of four subunits, GluR1–4, and are usually selective for monovalent cations, with some forms also showing permeability to calcium. We will describe several AMPA receptor antagonists and AMPA receptor potentiators showing potential for providing either anti-Parkinsonian efficacy or disease-modifying effects in preclinical models of PD.

10.1.1.2 Metabotropic Glutamate Receptors

Metabotropic glutamate receptors (mGluRs) are expressed pre- or postsynaptically and play a major role in the modulation of synaptic transmission.

There are eight subtypes of mGluRs, divided into three groups according to their sequence homology, ligand binding and G protein-coupling specificity (Figure 10.1).[6,14,15] Group I consists of mGluR1 and 5 receptors that are mainly postsynaptic, and couple to G_q to activate phospholipase C and increase intracellular calcium, leading to enhanced neuronal transmission. Group II includes mGluR2 and 3 receptors. Both are coupled to $G_{i/o}$ to inhibit adenylate cyclase, and are expressed mainly presynaptically on neurons where they inhibit neuronal transmission through regulation of neurotransmitter release while increasing production of neurotrophic factors when expressed on glial cells. Group III finally encompasses all other subtypes (mGluR4, 6, 7, and 8 receptors), that like Group II receptors, are coupled to $G_{i/o}$, and through presynaptic expression play an inhibitory role in neuronal transmission. mGluRs are differentially expressed throughout the basal ganglia,[16] and several lines of research investigating molecules specifically targeting different mGluR subtypes have generated interesting data in both preclinical and clinical studies that support approaches to tackle the motor symptoms of PD, as well as potentially offering new avenues for disease-modifying therapies. The focus here will be on molecules selectively inhibiting mGluR5 from Group I, or molecules activating mGluR2/3 of Group II or mGluR4 and R8 of Group III.

10.1.2 Allosteric Modulators *versus* Orthosteric Ligands

Allosteric modulators are an emerging class of orally available small molecule therapeutic agents that may offer a competitive advantage over classical drugs.[17] This potential stems from their ability to offer greater selectivity and better modulatory control at disease-mediating receptors. In contrast to orthosteric compounds that bind to the same site as the natural ligand of the receptor, allosteric modulators are non-competitive and bind to receptors at a different site. The activity of allosteric modulators is linked to the presence and activity of the natural ligand, to increase (positive allosteric modulator, PAM) or decrease (negative allosteric modulator, NAM) receptor activation in a dynamic fashion. By introducing an adjustment in gain for endogenous ligands the process of allosteric modulation allows an alternative to simply turning a receptor on or off. The orthosteric site is typically well conserved within the same family, as can be seen with the glutamate binding site of mGluRs for example. Targeting the allosteric sites of the receptor which have been less subject to conservation during evolution therefore offers new ways to identify highly selective molecules for the subtype of interest,[18] and the development of allosteric modulators is currently a topic of great interest.

10.1.3 Animal Models of Parkinson's Disease

The array of preclinical animal models of PD provides pharmacological tools for assessing the efficacy of novel therapeutic agents, not only by reversing Parkinsonian motor symptoms (akinesia, catalepsy, rigidity, bradykinesia, tremor, postural abnormalities) but also by addressing disease-modification potential in slowing the progressive neurodegeneration associated with the disease. The wide range of existing models and their relative strengths and weaknesses have been extensively reviewed by Duty and Jenner.[19] The models cited in the present review are briefly described hereafter. The pharmacological-based models include: (1) the haloperidol-induced catalepsy (HIC) model, in which novel therapeutics reverse the rigidity or catalepsy induced by a mixed D_1/D_2 receptor antagonist haloperidol; and (2) the reserpine-induced akinesia (RIA) model, in which reserpine, acting through vesicular monoamine transporter (Vmat), produces 85–95% dopamine (DA) depletion in the SNc and *striatum*, leading to profound akinesia (locomotor activity is measured in this case).

The two neurotoxin-based animal models are: (1) the 6-hydroxy dopamine (6-OHDA) model, producing a full unilateral lesion (>90%; measure of rotation or circling behavior, and forelimb akinesia) or a partial bilateral lesion (60–70%; measure of akinesia) in mice and rats; and (2) the 1-methyl-4-phenyl-1,2,3,6-tetrahydropyridine (MPTP)-treated non-human primate (NHP) model, which results in the degeneration of the *substantia nigra* by the toxic metabolite MPP^+, thereby inducing all known motor symptoms occurring in human PD.

Additionally, the MPTP-treated mouse model has been used to assess neuroprotective properties and disease-modification potential, but promising preclinical data in this model have not yet translated into drugs slowing disease progression in clinical trials.[20] Importantly, both 6-OHDA and MPTP-treated animal models have been used to assess abnormal involuntary movements (AIMs) known as dyskinesias, one of the major motor complications emerging upon repeated administration of L-DOPA. The management of these L-DOPA-induced dyskinesias (LIDs) is an ongoing challenge in PD treatment for which glutamate receptor modulators recently provided preclinical and clinical therapeutic validation.[21]

Finally, it has been shown that genes found to cause familial Parkinsonism are linked to mitochondrial dysfunction, and neurotoxins affecting mitochondria are used to produce degeneration of the nigrostriatal circuitry. The MitoPark mouse model is a recent genetic model for PD that replicates the slow and progressive development of the key symptoms seen in humans.[22,23]

10.2 Recent Progress of Ionotropic Glutamate Receptor Modulators in Parkinson's Disease

10.2.1 NMDA Receptor Blockers

The drug discovery efforts made in recent years around NMDA receptor modulators[24] in particular the NR2B subtype has been recently reviewed.[25] The main focus of this section will be on channel-blocking agents and on NR2B-selective antagonists. Although there have been reports that competitive glutamate site and NMDA receptor-associated glycine site antagonists have been effective in PD models, few of these compounds have advanced significantly in development.

10.2.1.1 Channel-Blocking NMDA Antagonists

Channel-blocking NMDA antagonists are effective in a variety of PD models that focus on restoring movement. MK801 (**1**, Figure 10.2) demonstrated both anti-Parkinsonian efficacy in the reversing of HIC in rodents and primates,[26,27] RIA in rodents as well as in animals depleted of dopamine following 6-OHDA or MPTP lesions.[28–30] However, the therapeutic window between improving locomotion and inducing overt CNS depression for MK801 (**1**) in primates is narrow.[31] Subsequent studies have explored the notion that less persistent channel blockade would be associated with fewer adverse effects.[32] Low-affinity channel blockers such as memantine (not shown) and amantadine (**2**) are effective in improving movement in animal models.[33,34] More novel channel blockers include neramexane (MRZ 2/579; **3**) which displays rapid channel kinetics and efficacy against PD symptoms in rodent models[35,36] but which also shows PCP-like discriminative stimulus activity.[37] Remacemide (**4**)[38,39] reverses motor deficits and potentiates L-DOPA responses in lesioned animals. In addition to improving movement, several channel blockers have been evaluated in models of LIDs. Several studies have shown that MK801 (**1**) improves dyskinesias in rodents and primates, although it has been suggested that anti-dyskinetic efficacy is only seen at doses that also reduce the anti-Parkinsonian effect of L-DOPA.[40–43] Interestingly the only marketed product currently available for patient use for dyskinesias is the low-affinity channel blocker amantadine, which is also effective in reducing AIMs in lesioned animals.[43,44]

It is clear that excitotoxicity contributes to neuronal loss in several *in vitro* and *in vivo* models of PD.[45–47] However, results have been somewhat mixed. Some studies have reported that MPTP-induced injury in mice is at least partially prevented by MK801 (**1**) while others have not found evidence for protection,[45,48–51] while protection against injury in primates appears to be more reliable.[52] A number of other NMDA blockers have neuroprotective properties in preclinical PD models, including amantadine (**2**), memantine and remacemide (**4**),[53,54] and some of these compounds have progressed into clinical studies for disease modification, such as remacemide (**4**).[54,55] However, none

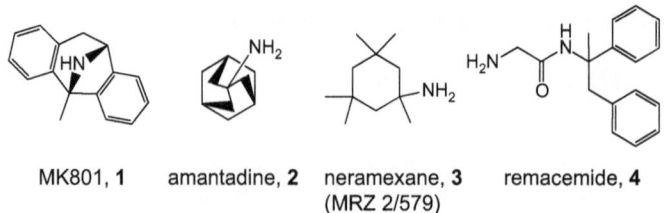

MK801, **1**   amantadine, **2**   neramexane, **3**   remacemide, **4**
(MRZ 2/579)

Figure 10.2 Selected non-competitive NMDA receptor blockers (**1–4**).

have yet demonstrated clinical efficacy, and it seems unlikely that effective neuroprotective doses of these compounds will be well tolerated.

10.2.1.2 NR2B-Selective Antagonists

The identification of subtypes of NMDA receptor allows for the possibility of engaging a subset of receptors and the potential for avoiding adverse effects. First-in-class ifenprodil (**5**, Figure 10.3) was initially identified as an atypical NMDA receptor antagonist based on [^{3}H]MK801 binding studies[56] and was subsequently shown to be an NR2B-preferring antagonist,[57] but with some off-target activity at α1-adrenoceptors. This enabled the development of a number of potent, brain penetrant and highly selective NR2B blockers (>100-fold

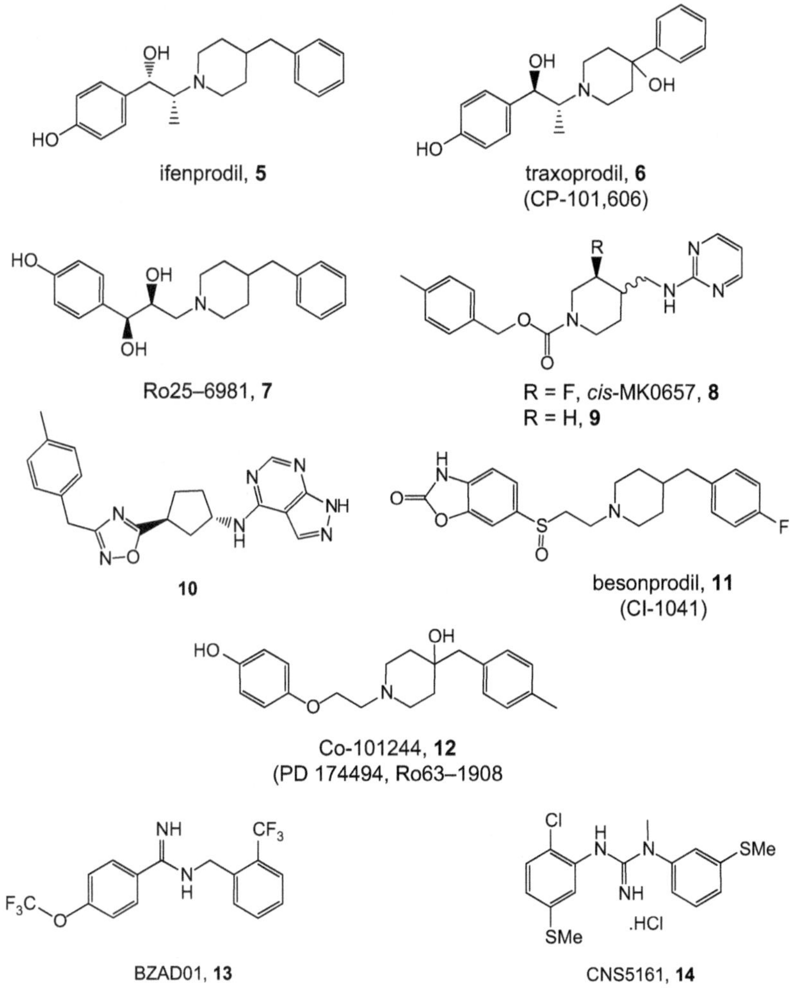

Figure 10.3 Selected NR2B receptor antagonists (**5**–**14**).

vs. NR2A receptors) which can be used as radioligands. Traxoprodil (CP101,606; **6**) has improved selectivity, although it has limited oral bioavailability.[58] Other potent, selective and orally bioavailable molecules include Ro25-6981 (**7**),[59] MK0657 (**8**) and some derivatives (**9** and **10**),[60,61] besonprodil (CI-1041, **11**),[62] Co-101244 (**12**)[63] and BZAD01 (**13**).[64]

Reversal of motor deficits in pharmacological or toxin-induced models appears to be a consistent property of NR2B antagonists. Early studies indicated that ifenprodil (**5**) reversed the locomotor deficit in reserpine-treated rats.[65] Traxoprodil (**6**) reversed HIC in rats, and partially reversed Parkinsonian symptoms in MPTP-lesioned monkeys.[66] Interestingly, traxoprodil (**6**) appeared to interact synergistically with L-DOPA in that study. Similarly, Ro25-6981 (**7**) reversed rotational behaviors in unilaterally 6-OHDA-lesioned rats and showed a trend towards symptom improvement in MPTP-lesioned marmosets.[67] More recently MK0657 (**8**), and compounds **9** and **10** demonstrated efficacy in reversing HIC in proportion to CNS exposures and NR2B occupancy.[60,61] Interestingly, NR2B antagonists appear to be additive or synergistic with L-DOPA in improving motor function. A number of studies have also shown that NR2B antagonists are effective in both reversing LIDs and preventing their onset. Blanchet and colleagues showed that Co-101244 (**12**) potentiated L-DOPA actions and reversed dyskinesias in MPTP-lesioned monkeys, while an NR2A-preferring competitive antagonist actually worsened symptoms.[68] Traxoprodil (**6**) has been reported to both improve and to worsen dyskinesias,[66,69] although both studies reported the enhancement of L-DOPA efficacy. CI-1041 (**11**) is also active against dyskinesias in monkeys.[62] While the precise mechanism underlying the effectiveness of NR2B antagonists against dyskinesias remains unclear, rodent and primate studies have suggested the development of a glutamate hypersensitivity and NR2B up-regulation following dopaminergic lesions.[70,71] Lastly, it is worth mentioning that BZAD01 (**13**) provided neuroprotection against 6-OHDA-induced lesions in rats.[72]

10.2.2 AMPA Receptor Modulators

With the major glutamatergic input into the *striatum* and subthalamic nuclei from the cortex, AMPA receptors in the *striatum* are well positioned to impact the functioning of motor control, and this has generated significant interest in the potential of AMPA receptor-directed therapeutics for the treatment of PD. Lagging behind the discovery of NMDA receptor ligands, both negative and positive AMPA receptor modulators have been developed, evaluated in preclinical PD models and have now advanced into clinical trials for PD (*vide infra*).

10.2.2.1 AMPA Receptor Antagonists

Early studies *in vivo* used the competitive AMPA antagonist NBQX (**15**; Figure 10.4) to investigate the role of AMPA receptors in motor deficits in lesioned rats and primates. These studies showed that NBQX (**15**) decreased rigidity and potentiated the effects of both L-DOPA and dopamine receptor agonists in their anti-Parkinsonian actions.[73–75] Direct administration of either NBQX (**15**) or the non-competitive antagonist GYKI-52466 (not shown) into the brain of 6-OHDA-lesioned rats improved akinesia,[76] although the monotherapy improvement does not appear to be a consistent feature of these compounds. Of the most characterized AMPA receptor NAMs, talampanel (LY-300164, **16**; Figure 10.4) and perampanel (E2007, **17**, recently approved for the treatment of epilepsy) have reached clinical testing in human (see Section 10.5.2). In preclinical studies, talampanel (**16**) enhanced the effects of L-DOPA in reversing the motor deficits in MPTP-lesioned monkeys.[77] Another key feature of advanced PD is the emergence of 'wearing off' episodes, whereby the duration of L-DOPA efficacy is diminished. Reduction in off time has emerged as a key clinical end point. Although animal models of wearing off are not well established, Marin and colleagues[78] were able to demonstrate that NBQX (**15**) increased the duration of L-DOPA action in 6-OHDA-lesioned rats.

The interaction between AMPA antagonists and L-DOPA is interesting in relation to the development and expression of dyskinesias. Although potentiating the effect of L-DOPA could be beneficial, the benefits would be limited if the potentiation was accompanied by exacerbation of LIDs. The precise mechanisms underlying the induction and expression of dyskinesias are still debated, however hyperactivity of striatal neurons containing GABA and enkephalin is correlated with the appearance of dyskinesias in rodent lesion models. Tezampanel (LY293558, **18**) was shown to diminish the 6-OHDA- and L-DOPA-induced increase in preproenkephalin mRNA in rats, suggesting that AMPA antagonists could delay the onset of dyskinesias.[79] In addition, the

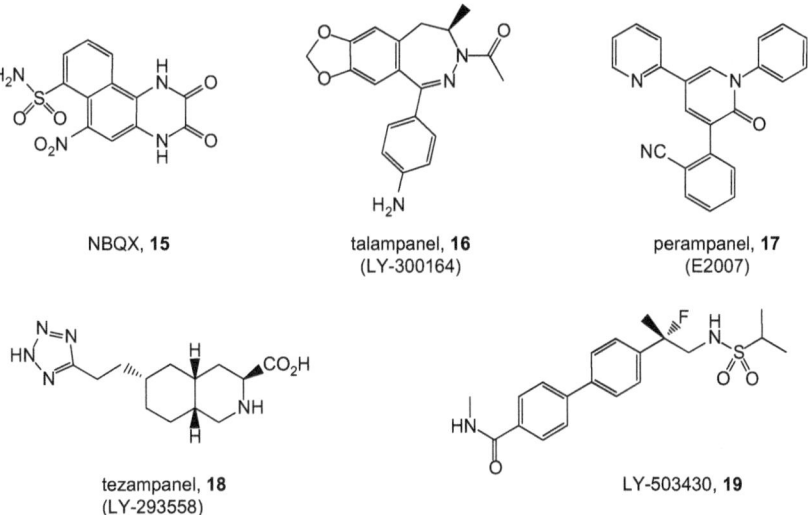

Figure 10.4 Selected AMPA receptor blockers (**15–18**) and a potentiator (**19**).

anti-epileptic drug topiramate (not shown), which blocks AMPA receptors as one of its mechanisms of action,[80] decreases established dyskinesias in MPTP-lesioned marmosets. Together with the evidence that dyskinesia induction is associated with increased expression of AMPA receptors in the *striatum* of MPTP-lesioned monkeys[81] some of which may be calcium permeable,[82] there is clearly a case to be made for the evaluation of AMPA receptor antagonists in the prevention or treatment of LIDs, either alone or in combination with NMDA receptor antagonists.[43]

10.2.2.2 AMPA Receptor Potentiators

Although glutamate-mediated neuronal injury (excitotoxicity) is predominantly mediated by NMDA receptor overactivation, it is clear that excessive activation of non-NMDA ionotropic glutamate receptors is also potentially toxic to neurons. This suggests that AMPA receptor antagonists might be effective as neuroprotective agents, and a limited number of studies have reported AMPA antagonist protection in PD models.[83] It is perhaps surprising then, that some compounds that increase AMPA receptor activity have also been reported to be neuroprotective. Under normal circumstances, AMPA receptors desensitize quite quickly, so that compounds which decrease desensitization increase AMPA receptor responses.[84] Initially developed as a cognitive enhancer, LY-503430 (**19**, Figure 10.4) substantially potentiates AMPA responses *in vitro* and *in vivo*, remarkably preserves dopaminergic neurons and prevents motor deficits in MPTP-treated mice and 6-OHDA-lesioned rats. The compound is effective both when administered concurrently with the toxin and also when initiated up to 14 days after 6-OHDA administration, suggesting both neuroprotective and neurotrophic properties.[85,86] It is not clear whether the protection observed in these models arises from direct effects on neurons, or the enhanced synthesis of neurotrophic factors that may be mediated by AMPA receptors on glia,[86] but this remains an intriguing observation that might provide a novel mechanism for disease modification in PD.

Perampanel (**17**) was recently approved for the treatment of epilepsy, demonstrating that AMPA antagonists are safe and effective in certain diseases, but there appears to be limited further development of AMPA antagonists in PD at this time.

10.3 Recent Progress of Metabotropic Glutamate Receptor Modulators in Parkinson's Disease

10.3.1 Group I mGluRs: Focus on mGluR5 Negative Allosteric Modulators

The abundance of Group I mGluRs expressed within the basal ganglia circuit has led to the postulation that inhibitory modulation of Group I mGluRs could

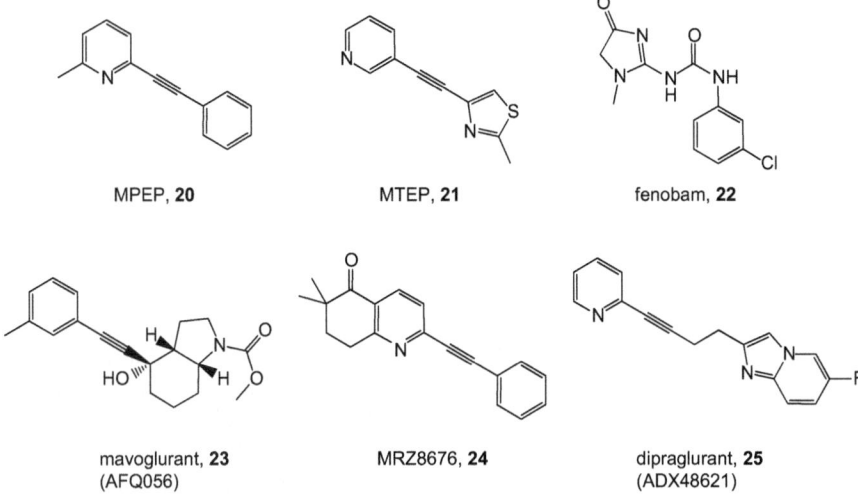

Figure 10.5 Selected mGluR5 negative allosteric modulators (**20–25**).

mediate and alleviate some of the classic symptomatic effects of PD animal models as well as provide a degree of neuroprotective effect against excitotoxicity caused by elevated glutamatergic tone. These hypotheses have been validated through the use of mGluR5 KO animals and the the *in vivo* proof-of-concept in animal models of PD with potent, selective, orally bioavailable mGluR5 NAMs MPEP (**20**), MTEP (**21**), fenobam (**22**), mavoglurant (AFQ056, **23**), MRZ8676 (**24**), and dipraglurant (ADX48621, **25**), represented in Figure 10.5.

The prototypical mGluR5 NAM MPEP (**20**) (5 mg kg^{-1}, i.p.)[87] as well as dipraglurant (**25**) (10 mg kg^{-1}, p.o.)[88] demonstrated a reversal of catalepsy, locomotor impairment and muscle rigidity in HIC in rats, suggesting a potentially beneficial symptomatic impact of mGluR5 blockade in PD. Both MPEP (**20**) (5 mg kg^{-1}, i.p.) and MTEP (**21**) (3 mg kg^{-1}, i.p.)[89,90] reverse haloperidol-induced increase of electromyographic activity during movement in the rat.

Unilateral 6-OHDA lesions in mGluR5 KO mice showed attenuation of the resulting motor impairment and asymmetrical movement compared to wild type.[91] Similar observations were made in akinesia models using chronic administration of MPEP in 6-OHDA-lesioned rats suggesting a significant symptomatic component to this pharmacological intervention.[92] Four-day treatment with MPEP (**20**) (1.5 mg kg^{-1}, i.p.) starting either immediately, one week or four weeks after 6-OHDA lesioning in rats demonstrated immediate improvement of the akinesia effects measured in an adjusting step test, yet no effect of MPEP (**20**) was observed in modifying neuronal survival or neuroinflammation in the nigrostriatal pathway.

Chronic postlesion treatment, but not acute treatment, with MPEP (**20**) (1.5 mg kg^{-1}, i.p.) daily over three weeks completely reversed 6-OHDA-induced akinesia in rats, with moderate improvements observed within the first week and full pre-operative performance restored after three weeks of treatment.[93] Chronic i.p. administration of MPEP (**20**) (3 mg kg^{-1}) has been shown to significantly reverse both the cognitive deficit and anxiogenic aspects of the 6-OHDA model.[94–96] There was no translation of anti-Parkinsonian effects when dosing MTEP (**21**) acutely to MPTP-induced Parkinsonian macaques,[97] however, the combination of MTEP (**21**) (36 mg kg^{-1}, p.o.) and L-DOPA in MPTP-lesioned animals bearing established LIDs significantly reduced the frequency of dyskinesia compared to the non-MTEP treated controls, whilst having no inhibition of the anti-Parkinsonian effects of the L-DOPA. Similarly MPEP (**20**) (10 mg kg^{-1}, p.o.), MTEP (**21**) (10 mg kg^{-1}, p.o.), fenobam (**22**) (10 mg kg^{-1}, p.o.) and mavoglurant (**23**) (25 mg kg^{-1}, p.o.) demonstrated reduction of LIDs in L-DOPA-treated MPTP-lesioned monkeys.[98–100] Additionally, mavoglurant (**23**) in combination with a low dose of L-DOPA was shown to significantly enhance the anti-Parkinsonian effect of L-DOPA whilst keeping the dyskinetic effects of the treatment low. Dipraglurant (**25**) has demonstrated similar beneficial effects on LIDs in MPTP-treated monkeys, with both the chorea and dystonia aspects of LIDs being alleviated by 10–30 mg kg^{-1} dosed orally.[88]

Similar effects on LIDs were observed for MTEP (**20**) (2.5 mg kg^{-1}, i.p.), fenobam (**22**) (30 mg kg^{-1}, p.o.) and MRZ8676 (**24**) (25 mg kg^{-1}, p.o.) in the rat 6-OHDA PD model.[89,99,101,102] Furthermore, chronic dosing of fenobam alongside L-DOPA in drug-naïve 6-OHDA-lesioned rats significantly reduced the time of onset of dyskinetic symptoms caused by L-DOPA treatment, without impacting the anti-Parkinsonian symptomatic effects of L-DOPA.[103]

Various groups have also produced evidence that mGluR5 blockade results in neuroprotective effects. Chronic systemic MPEP (**20**) treatment in 6-OHDA-lesioned rats significantly attenuated the loss of DA neurons in the SNc, thought to be caused by extensive glutamatergic innervation from the STN.[104,105] Similar neuroprotective effects have recently been demonstrated by MPEP (**20**) in the MPTP-lesioned rat model.[106,107]

Taken together, the ability of several mGluR5 NAMs to reduce motor symptoms in preclinical models of PD in rodents and primates, and the potential involvement of mGluR5 in the pathology of LIDs, has provided a strong rationale for investigating their clinical efficacy in PD patients suffering from LIDs. This led to the recent successful clinical trials of mavoglurant (**23**) and dipraglurant (**25**) in PD patients (see Section 10.5.3).

10.3.2 Group II mGluR Modulators

10.3.2.1 mGluR2/3 Agonists

As Group II mGluRs are involved in glutamateric tone regulation, with activation of both mGluR2 and mGluR3 being shown to inhibit presynaptic striatal glutamate release, it can be hypothesized that modulation of these targets may prove useful in treating aspects of PD. A set of dual-selective mGluR2/3 agonists (Figure 10.6) has been evaluated in a classical rodent model

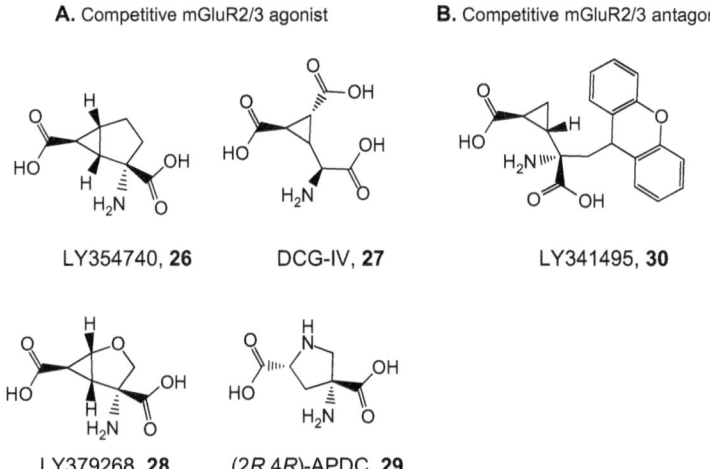

Figure 10.6 Selected mGluR2/R3 orthosteric agonists (**26–29**) and an antagonist (**30**).

of PD. LY354740 (**26**; 10 mg kg^{-1}, i.p.) dose dependently reduced haloperidol-induced muscle rigidity,[108] while DCG-IV (**27**) alleviated akinesia in the rat RIA when administered intranigrally (0.125 nM).[109] Surprisingly, changing the route of administration (i.p. *vs.* i.c.v.) of the mGluR2/3 agonist LY379268 (**28**) resulted in opposite effects in the rat,[110] such that i.p. administration of LY379268 (**28**) (2.5 mg kg^{-1}) in the unilateral 6-OHDA-lesioned rat model resulted in a reversion towards non-asymmetric rotation behavior. Additionally, a Group II mGluR agonist has been shown to have no effect on L-DOPA-induced dyskinesia,[102] limiting the potential therapeutic benefit of dual mGluR2/R3 orthosteric agonists.

In recent years, the potential neuroprotective effects of mGluR2/3 activation have received increasing interest, since DCG-IV (**27**) was shown to protect DA neurons against MPP^{+}-induced neurotoxicity, proposing a mechanism linked to induction and release of brain-derived neurotrophic factor BDNF.[111] Additionally, subchronic dosing over seven days of (2R,4R)-APDC (**29**) (10 nM intranigral administration) is neuroprotective in the 6-OHDA-lesioned rat model.[106] A similar neuroprotective effect of this compound in this model when administered for seven days 48 h post lesion has been reported, with significant amelioration in neuronal survival and symmetry-based motor deficits.[112] This suggests a degree of functional recovery as well as neuroprotection may be mediated by Group II mGluR activation. Corti *et al.* have shown that this neuroprotective effect is most likely driven *via* mGluR3 activation.[113] Administration of LY379268 (**28**) was shown to be neuroprotective to striatal neurons after an NMDA-induced excitotoxicity challenge in both wild type and mGluR2 KO mice. The same team has recently shown through the i.p administration of the mGluR2/3 agonist LY379268 (**28**) to mGluR2 and mGluR3 KO mice that mGluR3 activation but not mGluR2 activation drives an up-regulation of glial-cell-line-derived neurotrophic factor (GDNF), and that this up-regulation results in a neuroprotective effect in the MPTP mouse model of PD and explains the previously observed neuroprotective effects towards NMDA excitotoxicity.[114]

10.3.2.2 mGluR2/3 Antagonists

There have been reports of mGluR2/3 blockade *via* LY341495 (**30**, Figure 10.6), an mGluR2/3 antagonist, having a beneficial effect in both the acutely monoamine depleted mouse model of PD and the unilateral 6-OHDA-lesioned rat model of PD, including additive effects with apomorphine and L-DOPA, however the full data on these reports has not, to the best of our knowledge, been released.[115]

10.3.3 Group III mGluRs: Focus on mGluR4 and mGluR8 Activators

10.3.3.1 mGluR4 Orthosteric Agonists and Positive Allosteric Modulators

The therapeutic potential of mGluR4 activators (orthosteric agonist and PAMs) in CNS disorders, including PD has been extensively reviewed.[116–119] The progress towards clinical development of Group III mGluR orthosteric ligands and the first generation of potentiators has been a real challenge due their weak activity at the target, lack of selectivity, poor pharmacokinetics and generally low CNS exposure. However, the past two years of investigation have provided the first selective mGluR4 orthosteric agonists, along with two major chemotypes belonging to the allosteric modulator class reaching preclinical development for CNS disorders, including PD.

The first orthosteric agonists L-AP4 (**31**) and L-SOP (**32**), shown in Figure 10.7, proved to reverse reserpine-induced akinesia (intrapallidal route)[120,121] and showed neuroprotective effects in the 6-OHDA-lesioned rat model.[106,122] Cyclic derivatives such as *meso*-ACPT-1 (**33**)[123] and (1*S*,2*R*)-APCPr (**34**)[124] dose dependently reversed HIC in rats, as well as akinesia in the 6-OHDA-lesioned rat model.[125] More recently, preferential mGluR4 orthosteric agonists have been discovered through a virtual high-throughput screening approach,[126,127] such as LSP1-2111 (**35**)[128], LSP1-3081 (**36**)[129] and LSP4-2022 (**37**), the first potent (EC$_{50}$ 110 nM) and selective mGluR4 agonist reported to date.[130] LSP1-2011 (**35**) and LSP4-2022 (**37**) are brain penetrant and exert their central anti-cataleptic effects by reversing HIC after i.p. administration, and LSP1-3081 (**36**) improved 6-OHDA-induced akinesia upon intrastriatal dosing. Interestingly, while LSP1-2011 (**35**) failed to reduce LIDs through acute dosing, reduction of AIMs was demonstrated upon chronic dosing (30 mg kg^{-1}, i.p.) in combination with L-DOPA (10 mg kg^{-1}, i.p.) over nine days.[131]

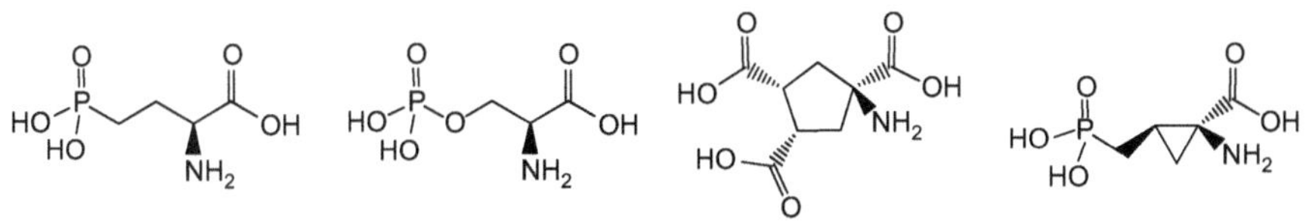

L-AP4, **31** L-SOP, **32** *meso*-ACPT-1, **33** (1*S*,2*R*)-APCPr, **34**

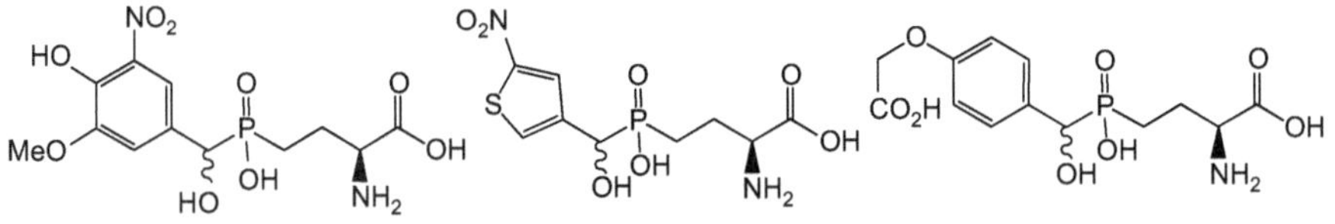

LSP1-2111, **35** LSP1-3081, **36** LSP4-2022, **37**

Figure 10.7 Selected mGluR4 orthosteric agonists (**31**–**37**).

N-Phenyl-7-(hydroxyimino)cyclopropa[*b*]chromen-1*a*-carboxamide PHCCC (**38**; Figure 10.8), was independently characterized in 2003 by Novartis and Merck & Co.[132,133] PHCCC demonstrated anti-Parkinsonian-like effects in several rodent models after i.c.v. administration, as well as neuroprotective effects in MPTP-induced nigrostriatal degeneration in mice.[134] Recently, systemic administration of optimized PHCCC analogues such as chromenone (**39**) from Domain Therapeutics showed anti-catalepic activity in rats (30 mg kg^{-1}, i.p.).[135]

The allosteric agonist VU0155041 (**40**) has demonstrated efficacy in the HIC and RIA models after i.c.v. administration in rat,[136] and neuroprotection against 6-OHDA-induced lesion in rat.[137] The recently disclosed carboxamide analogue LuAF21934 (**41**), successfully demonstrated improved permeability and CSF exposure, compared to VU0155401 (**40**). LuAF21934 (**41**) (10 and 30 mg kg^{-1}; i.p.) does not only exert stand-alone anti-Parkinsonian efficacy in the unilateral 6-OHDA-lesioned rat model, but also produced a synergistic effect at with a subthreshold dose of levodopa, improving akinesia.[138] Finally, LuAF21934 (**41**) (30 mg kg^{-1}, ip) significantly reduced dyskinesia duration whilst not affecting LID severity when dosed with L-DOPA (5 mg kg^{-1}, ip).[138]

Similar combination therapies with L-DOPA, or other anti-Parkinsonian drugs, have been also reported with mGluR4 PAMs belonging to two major chemical classes (picolinamide- and pyrazolyl aminothiazole-based family). The first generation of these picolinamide compounds (**42** and **43**), disclosed in 2008, had poor oral bioavailability and required i.c.v administration to exert their anti-Parkinsonian effect in the HIC model in rats.[139] Structural modifications led to an improved molecule with systemic exposure and efficacy. Co-administered with a threshold dose of L-DOPA, VU0364770 (**44**) showed potentiation and a synergistic effect in the HIC model in rat (30 mg kg^{-1}, s.c.). VU0364770 (**44**) also potentiates the efficacy of the A_{2A} antagonist preladenant at low doses (0.1, 0.3 and 1 mg kg^{-1}) as demonstrated in the HIC model. As a monotherapy, VU0364770 (**44**) produced an anti-cataleptic effect in rat (10, 30

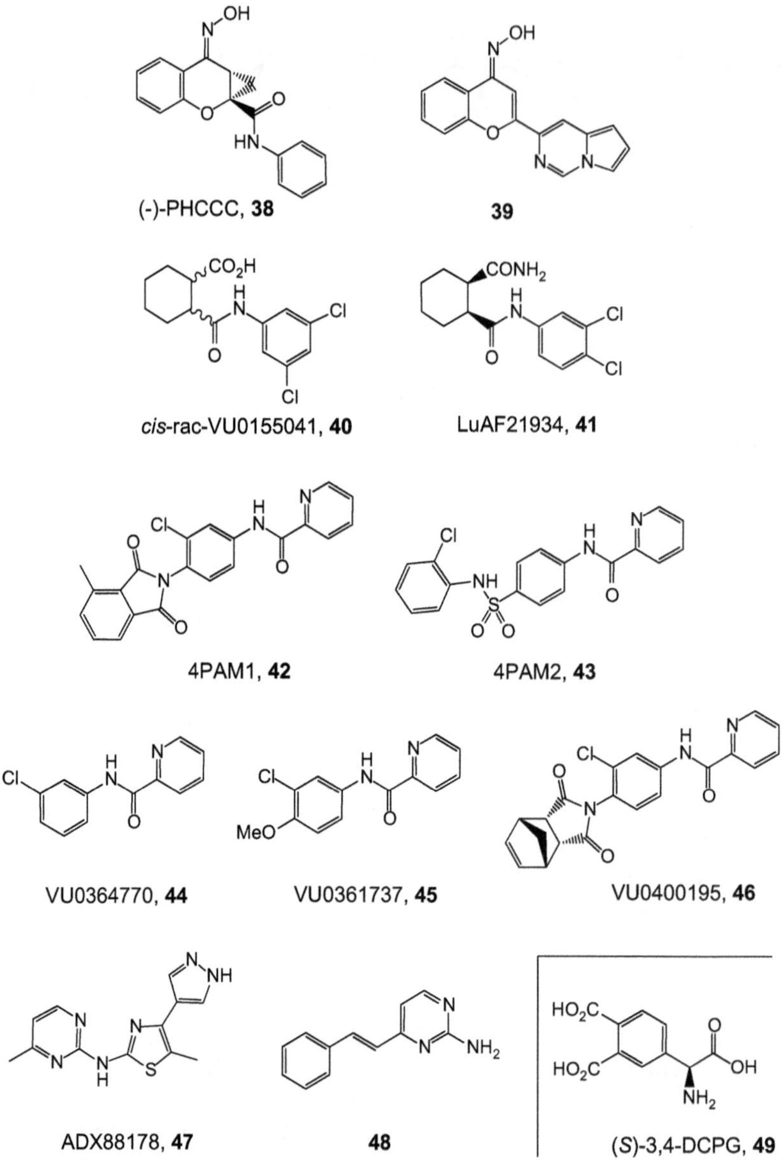

Figure 10.8 Selected mGluR4 positive allosteric modulators (**38–48**) and an mGluR8 orthosteric agonist, DCPG (**49**).

and 56.6 mg kg^{-1}, s.c.), and reversed forelimb akinesia in the unilateral 6-OHDA-lesioned rat model (100 mg kg^{-1}, s.c.) and in the bilateral 6-OHDA-lesioned rat model (56.6 mg kg^{-1}, s.c.).[140] The structurally related VU0361737 (**45**; 10, 30 and 56.6 mg kg^{-1}; i.p.)[141] and VU0400195 (**46**; 0.1–56.6 mg kg^{-1}; p.o.)[142] reverse HIC in rat, further reinforcing this series as a promising anti-Parkinsonian class of compounds.

Recently, ADX88178 (**47**; Figure 10.8), was reported as the first nanomolar, orally bioavailable and brain-penetrant mGluR4 PAM, displaying efficacy in several rodent models of PD.[143] ADX88178 (**47**) (3 and 10 mg kg^{-1} p.o.) reversed HIC in rat as monotherapy, but no statistically significant effect was seen on forelimb akinesia in rats subjected to a bilateral 6-OHDA lesion of the *striatum*. However, ADX88178 (**47**) (3, 10 and 30 mg kg^{-1}, p.o.) dose-dependently reversed forelimb akinesia deficit when combined with a low dose of L-DOPA, without worsening LIDs. Finally, the combination of ADX88178 (**47**) (30 mg kg^{-1}, p.o.) and L-DOPA (15 mg kg^{-1}) produced a significant increase in ambulatory activity over L-DOPA effects alone in the genetic MitoPark mice model. Finally, the chemically divergent mGluR4 PAM TCN238 (**48**; 3 and 10 mg kg^{-1}, p.o.), discovered by Boehringer Ingelheim and Evotec, has also demonstrated efficacy in reversing HIC.[144]

Taken together, all these data suggest that activation and/or modulation of mGluR4 represents a promising non-dopaminergic strategy for the treatment of PD motor symptoms,[145] with additional potential for neuroprotection and disease modification.[146]

10.3.3.2 mGluR8 Receptor Agonists

Racemic (*RS*)-3,4-dicarboxyphenylglycine (DCPG) is a mixed antagonist of AMPA receptors and an agonist of mGluR8. The (*S*)-enantiomer of DCPG (**49**; Figure 10.8) is the most selective mGluR8 agonist known to date, and has recently shown some potentially interesting results in the PD arena. Acute i.c.v. administration of (*S*)-DCPG (**49**) showed mixed results in reversing RIA in rats depending on the study.[147,148] However, (*S*)-DCPG (**49**) (10 nM, i.c.v) demonstrated efficacy upon chronic treatment, first in reversing HIC and RIS, and second in improving forelimb use asymmetry in the unilateral 6-OHDA-lesioned rat model,[148] suggesting selective activation of mGluR8 may be a viable and novel symptomatic approach in PD treatment.

10.4 Clinical Development of Novel Glutamate-Based Therapeutics in Parkinson's Disease

10.4.1 NMDA Receptor Blockers

There is extensive experience with the use of NMDA antagonists in clinical studies, with trials evaluating a range of compounds against stroke, traumatic brain injury, pain, and Alzheimer's disease to mention just a few. However, with respect to PD, there has not been robust success with the NMDA antagonists used so far. This section will describe trials with a few compounds that have been evaluated recently. Two NR2B antagonists, traxoprodil (**6**) and MK0657 (**8**) have been in small studies in PD patients. Nutt and colleagues studied traxoprodil infusion in PD patients and evaluated motor scores and dyskinesias at two doses.[149] They reported that Parkinsonism was not improved but that there was a trend towards improvement of dyskinesia. Although NR2B antagonists in general do not induce the florid behavioral adverse effects of some non-selective antagonists, amnesia and dissociation were noted as adverse effects at the higher dose used. Addy and colleagues[150] treated patients with L-DOPA or MK0657 (**8**) at 7 mg in a cross-over design, and found no improvement in the UPDRS score with the NR2B antagonist while L-DOPA was effective as a positive control. Elevated blood pressure was the dose-limiting adverse effect established in Phase I studies, and was observed in PD patients at this dose. The newest NMDA antagonist in clinical use is CNS5161 (**14**, Figure 10.3). Interestingly, early clinical experience with this compound in a pain study reported that the dose-limiting effect was elevated blood pressure.[151] This will not impede the use of CNS5161 (**14**) for PET imaging NMDA receptors,[152] which is a very exciting development for the study of dyskinesia mechanisms, but it illustrates an important if underappreciated limitation of NMDA antagonists. It appears that elevated blood pressure may be an on-target adverse effect associated with both non-selective agents and NR2B antagonists, which limits the dose of these compounds that can be used. Although there is excellent evidence for the efficacy of NR2B antagonists in rodent and primate studies, it may not be possible to safely use a sufficient dose of compound to produce a robust anti-Parkinsonian effect.

10.4.2 AMPA Receptor Modulators

There have been far fewer clinical trials of selective AMPA inhibitors in PD. Talampanel (**16**) and perampanel (**17**) (Figure 10.4) were advanced into clinical studies in PD based on the interesting preclinical data described above. Talampanel (**16**) was evaluated for efficacy in treating dyskinesias (Clinical Trial Identifier: NCT00108667), but these results have not yet been published. Perampanel (**17**) progressed into pivotal Phase III studies in PD where the key end-point was "off" time. Unfortunately, the end-points in this study were also not met,[153,154] perhaps reflecting the challenges associated with developing agents for "off" time without a well-validated animal model to predict efficacy.

10.4.3 mGluR5 Receptor Negative Allosteric Modulators

mGluR5 receptors are involved in several important CNS functions and have been reported to have a role in a variety of chronic neurodegenerative diseases like Parkinson's and Huntington's; they may also play an important role in psychiatric disorders in which glutamatergic neurotransmission is abnormally regulated.

To date, two compounds have been reported to be in clinical development for PD LIDs. Mavoglurant (**23**) (AFQ056, Figure 10.5), developed by Novartis, has shown activity in three randomized controlled clinical trials. Two Phase IIa randomized, double-blind, placebo-controlled, parallel group, in-patient studies with PD patients with moderate-to-severe LIDs (study 1) and severe LIDs (study 2) on stable dopaminergic therapy were performed.[155]

Patients received 25–150 mg of mavoglurant (**23**) or placebo twice daily for 16 days. Study 2 included a four-day down titration. Mavoglurant-treated patients showed significant improvements in dyskinesia on day 16 *vs.* placebo, but no significant changes were seen from baseline on day 16 in the UPDRS Part III in either studies, suggesting that mavoglurant (**23**) does not affect the anti-Parkinsonian activity of L-DOPA and does not improve or worsen motor signs in these patients. The anti-dyskinetic effect of mavoglurant (**23**) was confirmed in an additional 13-week, double-blind, placebo-controlled, multi-center study in 140 out-patients with moderate-to-severe PD LIDs.[156] In this study the improvement was confirmed to occur without worsening of the underlying Parkinsonian symptoms.

Dipraglurant (**25**) (ADX48621, Figure 10.5), developed by Addex Therapeutics, has completed one double-blind, placebo-controlled randomized Phase IIa study in the USA and Europe, in patients with moderate or severe LIDs on stable levodopa therapy (Addex company website: http://www.addex-therapeutics.com/rd/pipeline/dipra-ir). Patients were titrated from 50 mg o.d. (day 1) through to 50 mg t.i.d. (week 2) and 100 mg t.i.d. (week 4). Dipraglurant (**25**) has demonstrated statistically significant reduction in LID severity with both 50 mg and 100 mg doses without interfering with levodopa efficacy. In addition, during week 4, patients reported a mean reduction in daily "off" time of 50 min, suggesting an effect on Parkinsonian motor symptoms as well as the observed reductions in LIDs.[157,158]

The potential anxiolytic profile of some mGluR5 NAMs, such as fenobam (**22**), MPEP (**20**)[94–96] and dipraglurant (**25**)[159] demonstrated in preclinical rodent models suggest that they might be useful for treating the often under-diagnosed and untreated non-motor symptoms of PD patients.[160]

10.5 Conclusion and Perspectives

Since the introduction of L-DOPA, emergent and promising targets such as ionotropic and metabotropic glutamate receptors have been seriously considered for pharmacological intervention in PD, in order to address clear unmet medical needs.[146] Improving motor symptoms and reducing LIDs during the progression of the disease in PD patients[161–163] whilst providing neuroprotection and disease-modification solutions,[20,164] are among these important needs.

The intense interest in iGluR and mGluR receptor function and pharmacology has resulted in the identification of an arsenal of potential therapeutic agents that modulate glutamate receptor function. There is now increasing preclinical and clinical evidence that these molecules are not equal in their ability to improve motor deficits, treat LIDs and prevent neurodegeneration with an adequate safety profile to become the next generation of anti-Parkinsonian drugs.

Whilst NMDA channel blockers clearly have some symptomatic efficacy and may be more robustly neuroprotective, NR2B antagonists are better tolerated and appear to be efficacious in improving motor symptoms and decreasing dyskinesias in rodent and primate models, but lack evidence for neuroprotection. Together with the AMPA receptor blocker perampanel (**17**), two NR2B antagonists traxoprodil (**6**) and MK0657 (**8**) failed to prove their efficacy in human Phase II clinical trials at doses that were sufficiently well tolerated. Drug-repurposing strategies have had modest success successful so far in tackling LIDs, using amantadine (**2**) and potentially the anti-Alzheimer's drug memantine. However, the potential efficacy of engaging these targets has to be considered in the context of the therapeutic window without compromising on L-DOPA efficacy; typically this has proved to be a critical limitation in the ultimate therapeutic utility of engaging iGluRs.

The more recently emergent glutamate-based therapeutics belong to the mGluR family. The positive Phase II trials using two mGluR5 NAMs, mavoglurant (**23**) and dipraglurant (**25**), belonging to the alkyne-based chemotype drugs, developed by Novartis and Addex Therapeutics respectively, are very promising novel mechanisms for treating LIDs in PD patients, while showing benefits in treating motor and non-motor symptoms in preclinical rodent and primate models. There is no drug approved specifically for the treatment of LIDs. The anti-dyskinetic effects observed with mavoglurant (**23**) and dipraglurant (**25**) have demonstrated that mGluR5 inhibition is one of the most promising potential treatments for LIDs and a clinically relevant therapeutic target.

Lastly, the pharmacology and functions of mGluR3, mGluR4 and mGluR8 subtypes suggest these may be emergent targets for PD. In particular, intensive drug discovery efforts have been made during the past 10 years on improving potency, selectivity and CNS exposure from a wide range of mGluR4 activators (agonists and PAMs). A small set of orally bioavailable mGluR4 PAMs is now advancing towards preclinical development, and, both with stand-alone efficacy, and in addition to L-DOPA, demonstrate a reduction of LIDs and neuroprotection potential in rodent models of PD. Therapeutic efficacy of mGluR4 activators in the gold standard primate studies remains to be seen before such agents reach human clinical testing.

The next few years will be crucial to prove the validity of these new treatments which could address the complex symptomatic panel of symptoms in PD patients, with the potential for modifying the course of the disease through their neuroprotective action.

Acknowledgements

The authors are grateful to Dr J.-P. Rocher and Dr C. Keywood from Addex Therapeutics for having peer-reviewed this manuscript.

References

1. M. R. Delong and T. Wichmann, *Arch. Neurol.*, 2007, **64**, 20.
2. Z. G. Ni, R. Bouali-Benazzouz, D. M. Gao, A. L. Benabid and A. Benazzouz, *Brain Res.*, 2001, **899**, 142.

3. X. Liu, H. L. Ford-Dunn, G. N. Hayward, D. Nandi, R. C. Miall, T. Z. Aziz and J. F. Stein, *Clin. Neurophysiol.*, 2002, **113**, 1667.
4. G. A. Carvalho and G. Nikkhah, *Exp. Neurol.*, 2001, **171**, 405.
5. R. Dingledine, K. Borges, D. Bowie and S. F. Traynelis, *Pharmacol. Rev.*, 1999, **51**, 7.
6. P. J. Conn and J. P. Pin, *Annu. Rev. Pharmacol. Toxicol.*, 1997, **37**, 205.
7. M. J. Marino, O. Valenti and P. J. Conn, *Drugs Aging*, 2003, **20**, 377.
8. D. G. Standaert, C. M. Testa, A. B. Young and J. B. Penney, Jr., *J. Comp. Neurol.*, 1994, **343**, 1.
9. T. Gotz, U. Kraushaar, J. Geiger, J. Lubke, T. Berger and P. Jonas, *J. Neurosci.*, 1997, **17**, 204.
10. E. A. Waxman and D. R. Lynch, *Neuroscientist*, 2005, **11**, 37.
11. J. N. Kew and J. A. Kemp, *Psychopharmacology*, 2005, **179**, 4.
12. B. T. Chatha, V. Bernard, P. Streit and J. P. Bolam, *Neuroscience*, 2000, **101**, 1037.
13. V. Bernard, P. Somogyi and J. P. Bolam, *J. Neurosci.*, 1997, **17**, 819.
14. J. P. Pin and R. Duvoisin, *Neuropharmacology*, 1995, **34**, 1.
15. J. P. Pin and F. Acher, *Curr. Drug Targets*, 2002, **1**, 297.
16. P. J. Conn, G. Battaglia, M. J. Marino and F. Nicoletti, *Nat. Rev. Neurosci.*, 2005, **6**, 787.
17. L. T. May, K. Leach, P. M. Sexton and A. Christopoulos, *Annu. Rev. Pharmacol. Toxicol.*, 2007, **47**, 1.
18. S. Urwyler, *Pharmacol. Rev.*, 2011, **63**, 59.
19. S. Duty and P. Jenner, *Br. J. Pharmacol.*, 2011, **164**, 1357.
20. A. M. Hellman, J. F. Morley and J. E. Duda, *Neurodegener. Dis. Manage.*, 2012, **2**, 379.
21. F. Blandini and M. T. Armentero, *Expert Opin. Invest. Drugs*, 2012, **21**, 153.
22. D. Galter, K. Pernold, T. Yoshitake, E. Lindqvist, B. Hoffer, J. Kehr, N. G. Larsson and L. Olson, *Genes Brain Behav.*, 2010, **9**, 173.
23. M. I. Ekstrand, M. Terzioglu, D. Galter, S. Zhu, C. Hofstetter, E. Lindqvist, S. Thams, A. Bergstrand, F. S. Hansson, A. Trifunovic, B. Hoffer, S. Cullheim, A. H. Mohammed, L. Olson and N. G Larsson, *Proc. Natl. Acad. Sci. USA*. 2007, **104**, 1325.
24. R. M. Santangelo, T. M. Acker, S. S. Zimmerman, B. M. Katzman, K. L. Strong, S. F. Traynelis and D. C. Liotta, *Expert Opin. Ther. Pat.*, 2012, **22**, 1337.
25. C. Beinat, S. Banister, I. Moussa, A. J. Reynolds, C. S. McErlean and M. Kassiou, *Curr. Med. Chem.*, 2010, **17**, 4166.
26. W. J. Schmidt and M. Bubser, *Pharmacol. Biochem. Behav.*, 1989, **32**, 621.
27. P. J. Elliott, S. P. Close, D. M. Walsh, A. G. Hayes and A. S. Marriott, *J. Neural. Transm. Park. Dis. Dement. Sect.*, 1990, **2**, 91.
28. M. Carlsson and A. Svensson, *Life Sci.*, 1990, **47**, 1729.
29. A. R. Crossman, D. Peggs, S. Boyce, M. R. Luquin and M. A. Sambrook, *Neuropharmacology*, 1989, **28**, 1271.
30. W. C. Graham, R. G. Robertson, M. A. Sambrook and A. R. Crossman, *Life Sci.*, 1990, **47**, L91.

31. E. F. Domino and J. Sheng, *J. Pharmacol. Exp. Ther.*, 1993, **264**, 221.
32. H. S. Chen, J. W. Pellegrini, S. K. Aggarwal, S. Z. Lei, S. Warach, F. E. Jensen and S. A. Lipton, *J. Neurosci.*, 1992, **12**, 4427.
33. G. Skuza, Z. Rogoz, G. Quack and W. Danysz, *J. Neural Transm. Gen. Sect.*, 1994, **98**, 57.
34. W. Danysz, C. G. Parsons, J. Kornhuber, W. J. Schmidt and G. Quack, *Neurosci. Biobehav. Rev.*, 1997, **21**, 455.
35. C. G. Parsons, W. Danysz, A. Bartmann, P. Spielmanns, T. Frankiewicz, M. Hesselink, B. Eilbacher and G. Quack, *Neuropharmacology*, 1999, **38**, 85.
36. M. Karcz-Kubicha, B. Lorenz and W. Danysz, *Neuropharmacology*, 1999, **38**, 109.
37. K. L. Nicholson and R. L. Balster, *Psychopharmacology*, 2003, **170**, 215.
38. S. Subramaniam, S. D. Donevan and M. A. Rogawski, *J. Pharmacol. Exp. Ther.*, 1996, **276**, 161.
39. J. T. Greenamyre, R. V. Eller, Z. Zhang, A. Ovadia, R. Kurlan and D. M. Gash, *Ann. Neurol.*, 1994, **35**, 655.
40. T. M. Engber, S. M. Papa, R. C. Boldry and T. N. Chase, *Neuroreport*, 1994, **5**, 2586.
41. M. A. Paquette, A. M. Anderson, J. R. Lewis, C. K. Meshul, S. W. Johnson and B. S. Paul, *Neuropharmacology*, 2010, **58**, 1002.
42. B. Gomez-Mancilla and P. J. Bedard, *J. Pharmacol. Exp. Ther.*, 1991, **259**, 409.
43. F. Bibbiani, J. D. Oh, A. Kielaite, M. A. Collins, C. Smith and T. N. Chase, *Exp. Neurol.*, 2005, **196**, 422.
44. P. J. Blanchet, S. Konitsiotis and T. N. Chase, *Mov. Disord.*, 1998, **13**, 798.
45. L. Turski, K. Bressler, K. J. Rettig, P. A. Loschmann and H. Wachtel, *Nature*, 1991, **349**, 414.
46. F. Blandini, R. H. Porter and J. T. Greenamyre, *Mol. Neurobiol.*, 1996, **12**, 73.
47. G. J. Kress and I. J. Reynolds, *Neurobiol. Dis.*, 2005, **20**, 639.
48. R. Srivastava, E. Brouillet, M. F. Beal, E. Storey and B. T. Hyman, *Neurobiol. Aging*, 1993, **14**, 295.
49. A. Tabatabaei, T. L. Perry, S. Hansen and C. Krieger, *Neurosci. Lett.*, 1992, **141**, 192.
50. A. Kupsch, P. A. Loschmann, H. Sauer, G. Arnold, P. Renner, D. Pufal, M. Burg, H. Wachtel, B. G. Ten and W. H. Oertel, *Brain Res.*, 1992, **592**, 74.
51. P. K. Sonsalla, G. D. Zeevalk, L. Manzino, A. Giovanni and W. J. Nicklas, *J. Neurochem.*, 1992, **58**, 1979.
52. A. Zuddas, G. Oberto, F. Vaglini, F. Fascetti, F. Fornai and G. U. Corsini, *J. Neurochem.*, 1992, **59**, 733.
53. J. Kornhuber, M. Weller, K. Schoppmeyer and P. Riederer, *J. Neural Transm. Suppl.*, 1994, **43**, 91.
54. G. C. Palmer, E. F. Cregan, A. R. Borrelli and F. Willett, *Ann. N. Y. Acad. Sci.*, 1995, **765**, 236.

55. I. Shoulson, J. Penney, M. McDermott, S. Schwid, E. Kayson, T. Chase, S. Fahn, J. T. Greenamyre, A. Lang, A. Siderowf, N. Pearson, M. Harrison, E. Rost, A. Colcher, M. Lloyd, M. Matthews, R. Pahwa, D. McGuire, M. F. Lew, S. Schuman, K. Marek, S. Broshjeit, S. Factor, D. Brown, A. Feigin, J. Mazurkiewicz, B. Ford, D. Jennings, S. Dilllon, C. Comella, L. Blasucci, K. Janko, L. Shulman, W. Wiener, D. Bateman-Rodriguez, A. Carrion, O. Suchowersky, A. L. Lafontaine, C. Pantella, E. Siemers, J. Belden, R. Davies, M. Lannon, D. Grimes, P. Gray, W. Martin, L. Kennedy, C. Adler, S. Newman, J. Hammerstad, C. Stone, P. Lewitt, K. Bardram, K. Mistura, J. Miyasaki, L. Johnston, J. H. Cha, M. Tennis, M. Panniset, J. Hall, J. Tetrud, J. Friedlander, R. Hauser, L. Gauger, R. Rodnitzky, A. Deleo, J. Dobson, L. Seeberger, C. Dingmann, D. Tarsy, P. Ryan, L. Elmer, D. Ruzicka, M. Stacy, M. Brewer, B. Locke, D. Baker, C. Casaceli, D. Day, M. Florack, K. Hodgeman, N. Laroia, R. Nobel, C. Orme, L. Rexo, K. Rothenburgh, K. Sulimowicz, A. Watts, E. Wratni, P. Tariot, C. Cox, C. Leventhal, V. Alderfer, A. M. Craun, J. Frey, L. McCree, J. McDermott, J. Cooper, T. Holdich and B. Read, *Neurology*, 2001, **56**, 455.
56. I. J. Reynolds and R. J. Miller, *Mol. Pharmacol.*, 1989, **36**, 758.
57. K. Williams, *Mol. Pharmacol.*, 1993, **44**, 851.
58. B. L. Chenard, J. Bordner, T. W. Butler, L. K. Chambers, M. A. Collins, D. L. De Costa, M. F. Ducat, M. L. Dumont and C. B. Fox, *J. Med. Chem.*, 1995, **38**, 3138.
59. D. R. Lynch, S. S. Shim, K. M. Seifert, S. Kurapathi, V. Mutel, M. J. Gallagher and R. P. Guttmann, *Eur. J. Pharmacol.*, 2001, **416**, 185.
60. N. J. Liverton, R. A. Bednar, B. Bednar, J. W. Butcher, C. F. Claiborne, D. A. Claremon, M. Cunningham, A. G. DiLella, S. L. Gaul, B. E. Libby, E. A. Lyle, J. J. Lynch, J. A. McCauley, S. D. Mosser, K. T. Nguyen, G. L. Stump, H. Sun, H. Wang, J. Yergey and K. S. Koblan, *J. Med. Chem.*, 2007, **50**, 807.
61. M. E. Layton, M. J. Kelly, III, K. J. Rodzinak, P. E. Sanderson, S. D. Young, R. A. Bednar, A. G. DiLella, T. P. McDonald, H. Wang, S. D. Mosser, J. F. Fay, M. E. Cunningham, D. R. Reiss, C. Fandozzi, N. Trainor, A. Liang, E. V. Lis, G. R. Seabrook, M. O. Urban, J. Yergey and K. S. Koblan, *ACS Chem. Neurosci.*, 2011, **2**, 352.
62. M. Morissette, M. Dridi, F. Calon, T. A. Hadj, L. T. Meltzer, P. J. Bedard and P. T. Di, *Mov. Disord.*, 2006, **21**, 9.
63. Z. L. Zhou, S. X. Cai, E. R. Whittemore, C. S. Konkoy, S. A. Espitia, M. Tran, D. M. Rock, L. L. Coughenour, J. E. Hawkinson, P. A. Boxer, C. F. Bigge, L. D. Wise, E. Weber, R. M. Woodward and J. F. Keana, *J. Med. Chem.*, 1999, **42**, 2993.
64. C. F. Claiborne, J. A. McCauley, B. E. Libby, N. R. Curtis, H. J. Diggle, J. J. Kulagowski, S. R. Michelson, K. D. Anderson, D. A. Claremon, R. M. Freidinger, R. A. Bednar, S. D. Mosser, S. L. Gaul, T. M. Connolly, C. L. Condra, B. Bednar, G. L. Stump, J. J. Lynch,

A. Macaulay, K. A. Wafford, K. S. Koblan and N. J. Liverton, *Bioorg. Med. Chem. Lett.*, 2003, **13**, 697.
65. J. E. Nash, M. P. Hill and J. M. Brotchie, *Exp. Neurol.*, 1999, **155**, 42.
66. K. Steece-Collier, L. K. Chambers, S. S. Jaw-Tsai, F. S. Menniti and J. T. Greenamyre, *Exp. Neurol.*, 2000, **163**, 239.
67. P. A. Loschmann, G. C. De, L. Smith, U. Wullner, G. Fischer, J. A. Kemp, P. Jenner and T. Klockgether, *Exp. Neurol.*, 2004, **187**, 86.
68. P. J. Blanchet, S. Konitsiotis, E. R. Whittemore, Z. L. Zhou, R. M. Woodward and T. N. Chase, *J. Pharmacol. Exp. Ther.*, 1999, **290**, 1034.
69. J. E. Nash, P. Ravenscroft, S. McGuire, A. R. Crossman, F. S. Menniti and J. M. Brotchie, *Exp. Neurol.*, 2004, **188**, 471.
70. F. Calon, A. H. Rajput, O. Hornykiewicz, P. J. Bedard and P. T. Di, *Neurobiol. Dis.*, 2003, **14**, 404.
71. M. J. Hurley, M. J. Jackson, L. A. Smith, S. Rose and P. Jenner, *Eur. J. Neurosci.*, 2005, **21**, 3240.
72. K. R. Leaver, H. N. Allbutt, N. J. Creber, M. Kassiou and J. M. Henderson, *Clin. Exp. Pharmacol. Physiol.*, 2008, **35**, 1388.
73. T. Klockgether, L. Turski, T. Honore, Z. M. Zhang, D. M. Gash, R. Kurlan and J. T. Greenamyre, *Ann. Neurol.*, 1991, **30**, 717.
74. P. A. Loschmann, K. W. Lange, M. Kunow, K. J. Rettig, P. Jahnig, T. Honore, L. Turski, H. Wachtel, P. Jenner and C. D. Marsden, *J. Neural Transm. Park. Dis. Dement. Sect.*, 1991, **3**, 203.
75. H. Wachtel, M. Kunow and P. A. Loschmann, *Neurosci. Lett.*, 1992, **142**, 179.
76. B. Stauch-Slusher, K. C. Rissolo, K. F. Anzilotti, Jr. and P. F. Jackson, *J. Neural Transm. Park. Dis. Dement. Sect.*, 1995, **9**, 145.
77. S. Konitsiotis, P. J. Blanchet, L. Verhagen, E. Lamers and T. N. Chase, *Neurology*, 2000, **54**, 1589.
78. C. Marin, A. Jimenez, M. Bonastre, T. N. Chase and E. Tolosa, *Synapse*, 2000, **36**, 267.
79. C. Perier, C. Marin, M. Bonastre, E. Tolosa and E. C. Hirsch, *Eur. J. Neurosci.*, 2002, **16**, 2236.
80. W. E. Rosenfeld, *Clin. Ther.*, 1997, **19**, 1294.
81. B. Ouattara, D. Hoyer, L. Gregoire, M. Morissette, F. Gasparini, B. Gomez-Mancilla and P. T. Di, *Neuroscience*, 2010, **167**, 1160.
82. C. Kobylecki, M. A. Cenci, A. R. Crossman and P. Ravenscroft, *J. Neurochem.*, 2010, **114**, 499.
83. M. Merino, M. L. Vizuete, J. Cano and A. Machado, *J. Neurochem.*, 1999, **73**, 750.
84. S. F. Traynelis, L. P. Wollmuth, C. J. McBain, F. S. Menniti, K. M. Vance, K. K. Ogden, K. B. Hansen, H. Yuan, S. J. Myers and R. Dingledine, *Pharmacol. Rev.*, 2010, **62**, 405.
85. T. K. Murray, K. Whalley, C. S. Robinson, M. A. Ward, C. A. Hicks, D. Lodge, J. L. Vandergriff, P. Baumbarger, E. Siuda, M. Gates, A. M. Ogden, P. Skolnick, D. M. Zimmerman, E. S. Nisenbaum, D. Bleakman and M. J. O'Neill, *J. Pharmacol. Exp. Ther.*, 2003, **306**, 752.

86. M. J. O'Neill, T. K. Murray, M. P. Clay, T. Lindstrom, C. R. Yang and E. S. Nisenbaum, *CNS Drug Rev.*, 2005, **11**, 77.
87. K. Ossowska, J. Konieczny, S. Wolfarth, J. Wieronska and A. Pilc, *Neuropharmacology*, 2001, **41**, 413.
88. S. Poli, *Curr. Neuropharmacol.*, 2012, **9**(1), 8.
89. A. Dekundy, M. Pietraszek, D. Schaefer, M. A. Cenci and W. Danysz, *Brain Res. Bull.*, 2006, **69**, 318.
90. K. Ossowska, J. Konieczny, S. Wolfarth and A. Pilc, *Neuropharmacology*, 2005, **49**, 447.
91. Y. D. Black, D. Xiao, D. Pellegrino, A. Kachroo, A. L. Brownell and M. A. Schwarzschild, *Neurosci. Lett.*, 2010, **486**, 161.
92. G. Ambrosi, M. T. Armentero, G. Levandis, P. Bramanti, G. Nappi and F. Blandini, *Brain Res. Bull.*, 2010, **82**, 29.
93. N. Breysse, C. Baunez, W. Spooren, F. Gasparini and M. Amalric, *J. Neurosci.*, 2002, **22**, 5669.
94. E. De Leonibus, F. Manago, F. Giordani, F. Petrosino, S. Lopez, A. Oliverio, M. Amalric and A. Mele, *Neuropsychopharmacology*, 2009, **34**, 729.
95. M. H. Hsieh, S. C. Ho, K. Y. Yeh, C. R. Pawlak, H. M. Chang, Y. J. Ho, T. J. Lai and F. Y. Wu, *Pharmacol. Biochem. Behav.*, 2012, **102**, 64.
96. L. Chen, J. Liu, U. Ali, Z. H. Gui, C. Hou, L. L. Fan, Y. Wang and T. Wang, *Brain Res. Bull.*, 2011, **84**, 215.
97. T. H. Johnston, S. H. Fox, M. J. McIldowie, M. J. Piggott and J. M. Brotchie, *J. Pharmacol. Exp. Ther.*, 2010, **333**, 865.
98. N. Morin, L. Gregoire, B. Gomez-Mancilla, F. Gasparini and T. Di Paolo, *Neuropharmacology*, 2010, **58**, 981.
99. D. Rylander, H. Iderberg, Q. Li, A. Dekundy, J. Zhang, H. Li, R. Baishen, W. Danysz, E. Bezard and M. A. Cenci, *Neurobiol. Dis.*, 2010, **39**, 352.
100. L. Gregoire, N. Morin, B. Ouattara, F. Gasparini, G. Bilbe, D. Johns, I. Vranesic, S. Sahasranaman, B. Gomez-Mancilla and T. Di Paolo, *Parkinsonism Relat. Disord.*, 2011, **17**, 270.
101. A. Dekundy, A. Gravius, M. Hechenberger, M. Pietraszek, J. Nagel, C. Tober, M. van der Elst, F. Mela, C. G. Parsons and W. Danysz, *J. Neural Transm.*, 2011, **118**, 1703.
102. D. Rylander, A. Recchia, F. Mela, A. Dekundy, W. Danysz and M. A. Cenci, *J. Pharmacol. Exp. Ther.*, 2009, **330**, 227.
103. D. Rylander, H. Iderberg, Q. Li, A. Dekundy, J. Zhang, H. Li, R. Baishen, W. Danysz, E. Bezard and M. A. Cenci, *Neurobiol. Dis.*, 2010, **39**, 352.
104. L. Chen, Q. J. Zhang, J. Liu, S. Wang, U. Ali, Z. H. Gui and Y. Wang, *Brain Res.*, 2009, **1286**, 192.
105. L. Chen, J. Liu, U. Ali, Z. H. Gui, C. Hou, L. L. Fan, Y. Wang and T. Wang, *Brain Res. Bull.*, 2011, **84**, 215.
106. A. C. Vernon, S. Palmer, K. P. Datla, V. Zbarsky, M. J. Croucher and D. T. Dexter, *Eur. J. Neurosci.*, 2005, **22**, 1799.

107. M. H. Hsieh, S. C. Ho, K. Y. Yeh, C. R. Pawlak, H. M. Chang, Y. J. Ho, T. J. Lai and F. Y. Wu, *Pharmacol. Biochem. Behav.*, 2012, **102**, 64.
108. J. Konieczny, K. Ossowska, S. Wolfarth and A. Pilc, *Naunyn-Schmiedeberg's Arch. Pharmacol.*, 1998, **358**, 500.
109. L. Dawson, A. Chadha, M. Megalou and S. Duty, *Br. J. Pharmacol.*, 2000, **129**, 541.
110. T. K. Murray, M. J. Messenger, M. A. Ward, S. Woodhouse, D. J. Osborne, S. Duty and M. J. O'Neill, *Pharmacol. Biochem. Behav.*, 2002, **73**, 455.
111. E. R. Matarredona, M. Santiago, J. L. Venero, J. Cano and A. Machado, *J. Neurochem.*, 2001, **76**, 351.
112. H. Chan, H. Paur, A. C. Vernon, V. Zabarsky, K. P. Datla, M. J. Croucher and D. T. Dexter, *Parkinson's Dis.*, 2010, 190450.
113. C. Corti, G. Battaglia, G. Molinaro, B. Riozzi, A. Pittaluga, M. Corsi, M. Mugnaini, F. Nicoletti and V. Bruno, *J. Neurosci.*, 2007, **27**, 8297.
114. G. Battaglia, G. Molinaro, B. Riozzi, M. Storto, C. L. Busceti, P. Spinsanti, D. Bucci, L. Di, V, G. Mudo, C. Corti, M. Corsi, F. Nicoletti, N. Belluardo and V. Bruno, *PLoS ONE*, 2009, **4**, e6591.
115. J. A. Feeley Kearney and R. L. Albin, *Exp. Neurol.*, 2003, **184**(1), S30.
116. S. Celanire and B. Campo, *Expert Opin. Drug Discov.*, 2012, **7**, 261.
117. C. W. Lindsley and C. R. Hopkins, *Expert Opin. Ther. Pat.*, 2012, **22**, 461.
118. S. Duty, *Br. J. Pharmacol.*, 2010, **161**, 271.
119. C. W. Lindsley, C. M. Niswender, D. W. Engers and C. R. Hopkins, *Curr. Top. Med. Chem.*, 2009, **9**, 949.
120. N. MacInnes, M. J. Messenger and S. Duty, *Br. J. Pharmacol.*, 2004, **141**, 15.
121. P. Q. Trombley and G. L. Westbrook, *J Neurosci.*, 1992, **12**, 2043.
122. A. C. Vernon, V. Zbarsky, K. P. Datla, D. T. Dexter and M. J. Croucher, *J. Pharmacol. Exp. Ther.*, 2007, **320**, 397.
123. J. Konieczny, J. Wardas, K. Kuter, A. Pilc and K. Ossowska, *Neuroscience*, 2007, **145**, 611.
124. P. Sibille, S. Lopez, I. Brabet, O. Valenti, N. Oueslati, F. Gaven, C. Goudet, H. O. Bertrand, J. Neyton, M. J. Marino, M. Amalric, J. P. Pin and F. C. Acher, *J. Med. Chem.*, 2007, **50**, 3585.
125. S. Lopez, N. Turle-Lorenzo, F. Acher, L. E. De, A. Mele and M. Amalric, *J. Neurosci.*, 2007, **27**, 6701.
126. N. Triballeau, F. Acher, I. Brabet, J. P. Pin and H. O. Bertrand, *J. Med. Chem.*, 2005, **48**, 2534.
127. C. Selvam, N. Oueslati, I. A. Lemasson, I. Brabet, D. Rigault, T. Courtiol, S. Cesarini, N. Triballeau, H. O. Bertrand, C. Goudet, J. P. Pin and F. C. Acher, *J. Med. Chem.*, 2010, **53**, 2797.
128. C. Beurrier, S. Lopez, D. Revy, C. Selvam, C. Goudet, M. Lherondel, P. Gubellini, L. Kerkerian-LeGoff, F. Acher, J. P. Pin and M. Amalric, *FASEB J.*, 2009, **23**, 3619.
129. D. Cuomo, G. Martella, E. Barabino, P. Platania, D. Vita, G. Madeo, C. Selvam, C. Goudet, N. Oueslati, J. P. Pin, F. Acher, A. Pisani,

C. Beurrier, C. Melon, G. L. Kerkerian-Le and P. Gubellini, *J. Neurochem.*, 2009, **109**, 1096.
130. T. Deltheil, N. Turle-Lorenzo and M. Amalric, *Curr. Neuropharmacol.*, 2011, **9**(1), 14–15.
131. S. Lopez, A. Bonito-Oliva, S. Pallottino, F. Acher and G. Fisone, *J. Parkinson's Dis.*, 2011, **1**, 339.
132. M. Maj, V. Bruno, Z. Dragic, R. Yamamoto, G. Battaglia, W. Inderbitzin, N. Stoehr, T. Stein, F. Gasparini, I. Vranesic, R. Kuhn, F. Nicoletti and P. J. Flor, *Neuropharmacology*, 2003, **45**, 895.
133. M. J. Marino, D. L. Williams, Jr., J. A. O'Brien, O. Valenti, T. P. McDonald, M. K. Clements, R. Wang, A. G. DiLella, J. F. Hess, G. G. Kinney and P. J. Conn, *Proc. Natl. Acad. Sci. U. S. A.*, 2003, **100**, 13668.
134. G. Battaglia, C. L. Busceti, G. Molinaro, F. Biagioni, A. Traficante, F. Nicoletti and V. Bruno, *J. Neurosci.*, 2006, **26**, 7222.
135. S. Schann, S. Mayer, C. Morice and B. Giethlen, *Pat.*, WO2011/051478, 2011.
136. C. M. Niswender, K. A. Johnson, C. D. Weaver, C. K. Jones, Z. Xiang, Q. Luo, A. L. Rodriguez, J. E. Marlo, P. T. de, A. D. Thompson, E. L. Days, T. Nalywajko, C. A. Austin, M. B. Williams, J. E. Ayala, R. Williams, C. W. Lindsley and P. J. Conn, *Mol. Pharmacol.*, 2008, **74**(1345).
137. M. J. Betts, M. J. O'Neill and S. Duty, *Br. J. Pharmacol.*, 2012, **166**, 2317.
138. K. E. Bennouar, M. A. Uberti, M. D. Bacolod, M. Cajina and H. N. Jimenez, *Curr. Neuropharmacol.*, 2011, **9**(1), 4.
139. I. J. Reynolds, presented at the 6th International Meeting on Metabotropic Glutamate Receptors, Taormina, 2008.
140. C. K. Jones, M. Bubser, A. D. Thompson, J. W. Dickerson, N. Turle-Lorenzo, M. Amalric, A. L. Blobaum, T. M. Bridges, R. D. Morrison, S. Jadhav, D. W. Engers, K. Italiano, J. Bode, J. S. Daniels, C. W. Lindsley, C. R. Hopkins, P. J. Conn and C. M. Niswender, *J. Pharmacol. Exp. Ther.*, 2011, **340**, 404.
141. C. Niswender, presented at the Allosteric Modulator Drug Discovery Congress, San Diego, 2010.
142. C. K. Jones, D. W. Engers, A. D. Thompson, J. R. Field, A. L. Blobaum, S. R. Lindsley, Y. Zhou, R. D. Gogliotti, S. Jadhav, R. Zamorano, J. Bogenpohl, Y. Smith, R. Morrison, J. S. Daniels, C. D. Weaver, P. J. Conn, C. W. Lindsley, C. M. Niswender and C. R. Hopkins, *J. Med. Chem.*, 2011, **54**, 7639.
143. P. E. Le, C. Bolea, F. Girard, S. Poli, D. Charvin, B. Campo, J. Bortoli, A. Bessif, B. Luo, A. J. Koser, L. M. Hodge, K. M. Smith, A. G. DiLella, N. Liverton, F. Hess, S. E. Browne and I. J. Reynolds, *J. Pharmacol. Exp. Ther.*, 2012, **343**, 167.
144. S. P. East, S. Bamford, M. G. Dietz, C. Eickmeier, A. Flegg, B. Ferger, M. J. Gemkow, R. Heilker, B. Hengerer, A. Kotey, P. Loke, G. Schanzle,

H. D. Schubert, J. Scott, M. Whittaker, M. Williams, P. Zawadzki and K. Gerlach, *Bioorg. Med. Chem. Lett.*, 2010, **20**, 4901.
145. M. Amalric, S. Lopez, C. Goudet, G. Fisone, G. Battaglia, F. Nicoletti, J. P. Pin and F. C. Acher, *Neuropharmacology*, 2013, **66**, 54.
146. Y. Smith, T. Wichmann, S. A. Factor and M. R. Delong, *Neuropsychopharmacology*, 2011, **37**, 213.
147. M. Broadstock, P. J. Austin, M. J. Betts and S. Duty, *Br. J. Pharmacol.*, 2012, **165**, 1034.
148. K. A. Johnson, C. K. Jones, M. N. Tantawy, M. Bubser, M. Marvanova, M. S. Ansari, R. M. Baldwin, P. J. Conn and C. M. Niswender, *Neuropharmacology*, 2013, **66**, 187.
149. J. G. Nutt, S. A. Gunzler, T. Kirchhoff, P. Hogarth, J. L. Weaver, M. Krams, B. Jamerson, F. S. Menniti and J. W. Landen, *Mov. Disord.*, 2008, **23**, 1860.
150. C. Addy, C. Assaid, D. Hreniuk, M. Stroh, Y. Xu, W. J. Herring, A. Ellenbogen, H. A. Jinnah, L. Kirby, M. T. Leibowitz, R. M. Stewart, D. Tarsy, J. Tetrud, S. A. Stoch, K. Gottesdiener and J. Wagner, *J. Clin. Pharmacol.*, 2009, **49**, 856.
151. T. Forst, T. Smith, K. Schutte, P. Marcus and A. Pfutzner, *Br. J. Clin. Pharmacol.*, 2007, **64**, 75.
152. M. R. Walters, A. P. Bradford, J. Fischer and K. R. Lees, *Br. J. Clin. Pharmacol.*, 2002, **53**, 305.
153. A. Lees, S. Fahn, K. M. Eggert, J. Jankovic, A. Lang, F. Micheli, M. M. Mouradian, W. H. Oertel, C. W. Olanow, W. Poewe, O. Rascol, E. Tolosa, D. Squillacote and D. Kumar, *Mov. Disord.*, 2012, **27**, 284.
154. O. Rascol, P. Barone, M. Behari, M. Emre, N. Giladi, C. W. Olanow, E. Ruzicka, F. Bibbiani, D. Squillacote, A. Patten and E. Tolosa, *Clin. Neuropharmacol.*, 2012, **35**, 15.
155. D. Berg, J. Godau, C. Trenkwalder, K. Eggert, I. Csoti, A. Storch, H. Huber, M. Morelli-Canelo, M. Stamelou, V. Ries, M. Wolz, C. Schneider, P. T. Di, F. Gasparini, S. Hariry, M. Vandemeulebroecke, W. Abi-Saab, K. Cooke, D. Johns and B. Gomez-Mancilla, *Mov. Disord.*, 2011, **26**, 1243.
156. F. Stocchi, A. Destee, N. Hattori, R. A. Hauser, A. E. Lang, W. Poewe, O. Rascol, M. Stacy, E. Tolosa, C. Trenkwalder, H. Gao, J. Nagel, B. Gomez-Mancilla, M. Menschhemke, S. Tekin and W. Abi-Saab, presented at the 15th International Congress of Parkinson's Disease and Movement Disorders, Toronto, 2011.
157. F. Tison, F. Durif, J.-C. Corvol, K. Eggert, C. Trenkwalder, M. Lew, S. Isaacson, M. Wakefield, C. Keywood and O. Rascol, presented at the 16th International Congress of Parkinson's Disease and Movement Disorders, Dublin, 2012.
158. O. Rascol, F. Tison, AR. Crossman, E. Pioli, F. Ory-Magnel, W. Meissner, S. Poli, M. Wakefield, C. Keywood and E. Bezard, presented at the Society for Neuroscience Meeting, New Orleans, 2012.

159. M. P. Epping-Jordan, V. Mutel, F. Girard, A.-S. Bessis, M. Kalinichev, C. Bolea, B. Bonnet, F. Derouet, A. Bessif, S. Turin, B. Mingard, S. Barbier and S. Poli, presented at the Society for Neuroscience Meeting, New Orleans, 2012.
160. O. Bernal-Pacheco, N. Limotai, C. L. Go and H. H. Fernandez, *Neurologist*, 2012, **18**, 1.
161. W. Poewe, P. Mahlknecht and J. Jankovic, *Curr. Opin. Neurol.*, 2012, **25**, 448.
162. W. G. Meissner, M. Frasier, T. Gasser, C. G. Goetz, A. Lozano, P. Piccini, J. A. Obeso, O. Rascol, A. Schapira, V. Voon, D. M. Weiner, F. Tison and E. Bezard, *Nat. Rev. Drug Discov.*, 2011, **10**, 377.
163. K. Buck and B. Ferger, *Drug Discov. Today*, 2010, **15**, 867.
164. F. Caraci, G. Battaglia, M. A. Sortino, S. Spampinato, G. Molinaro, A. Copani, F. Nicoletti and V. Bruno, *Neurochem. Int.*, 2012, **61**, 559.

CHAPTER 11

LRRK2 Kinase Inhibitors as New Drugs for Parkinson's Disease?

SANDRA SCHULZ,[a,b,†] STEFAN GÖRING,[c,†]
BORIS SCHMIDT*[c] AND CARSTEN HOPF*[a,b]

[a] Instrumental Analysis and Bioanalytics, Mannheim University of Applied Sciences, Mannheim, Germany; [b] Center for Applied Research in Biomedical Mass Spectrometry ABIMAS, Mannheim University of Applied Sciences, Mannheim, Germany; [c] Clemens Schöpf-Institute of Organic Chemistry and Biochemistry, Technische Universität Darmstadt, Darmstadt, Germany
*Email: c.hopf@hs-mannheim.de; Schmidt_Boris@t-online.de

11.1 Introduction

Parkinson's disease (PD) is a common, age-related, devastating neurodegenerative disease that is characterized by neuronal cell loss, predominantly in dopaminergic neurons of the *substantia nigra pars compacta*, and by formation of fibrillar conglomerates of lipids and proteins (Lewy bodies) in the surviving neurons.[1] PD affects about 2% of the population older than 60 years.[2] The symptoms include motor and cognitive dysfunction such as postural instability, rigidity, tremor, as well as dementia and depression.[3] The underlying molecular mechanisms of the disease remain poorly understood. Presently, established therapies treat symptoms of the disease, *e.g.*, reduce motor manifestations or alleviate non-motor symptoms, but no PD treatment is able to prevent,

---

[†]These authors contributed equally to this work

halt or decelerate disease progression.[2] No treatment of its elusive molecular pathogenic mechanisms exists.

Mutations in leucine-rich repeat kinase 2 (LRRK2) are associated with rare autosomal dominant forms of PD. Several non-selective inhibitors with activity against LRRK2 have been identified.[3–7] Most of them have either not been tested for brain penetration or have been found to not enter the brain. A brain-penetrant, non-selective kinase inhibitor, GW5074, is protective in a mouse model of LRRK2-induced neurodegeneration *in vivo*, suggesting that inhibition of LRRK2 kinase activity could be a new treatment option for PD.[6] However, GW5074 is a very non-selective molecule that, among other things, targets several non-kinases. For instance, it displays activity as an allosteric glutamate dehydrogenase inhibitor[8] and as an anti-viral agent with a RAF1 kinase-independent mode of action.[9]

The molecular biology of LRRK2, most notably the consequences of mutations inside (*e.g.*, G2019S) and outside (*e.g.*, R1441C) the kinase domain on kinase activity and LRRK2-induced toxicity *in vitro*, has been summarized in excellent reviews elsewhere.[1,10] Moreover, the current state of validation of LRRK2 as a therapeutic target as well as the question of whether inhibition of LRRK2 kinase activity is the only viable therapeutic strategy for LRRK2-linked PD have been addressed recently.[2,11] Moreover, a comprehensive review of non-selective (as well as the first selective) LRRK2 inhibitors has been provided very recently.[3] We therefore want to focus this book chapter on the following five aspects that are particularly important for LRRK2 kinase-directed drug discovery:

(1) Recent insight into LRRK2 inhibitor structure–activity relationships (SARs) from structural biology studies and molecular modeling;
(2) Recent medicinal chemistry efforts relating to potent and selective LRRK2 kinase inhibitors;
(3) Recent advances in understanding the role of LRRK2 outside the brain and implications for potential mechanism-based toxicity of LRRK2 kinase inhibitors as drugs;
(4) Recent advances in animal models, both invertebrate and vertebrate, for pharmacological evaluation of LRRK2 kinase inhibitors; and
(5) Pharmacokinetics and pharmacodynamics of LRRK2 kinase inhibitors: the current state-of-the-art and future challenges ahead.

11.2 Insight into LRRK2 Inhibitor SARs from Structural Biology Studies and Molecular Modeling

Kinase activity can be controlled by several modes of inhibition. Type I inhibitors, which compete for the ATP-binding site, are usually rather polar and often associated with selectivity problems. This imposes severe obstacles for the chronic treatment of a neurodegenerative disease. Type II inhibitors stabilize the inactive state of the kinase and generally offer benefits in terms of selectivity and

ATP competition. The DFG [Aspartic acid (D), Phenylalanine (F), Glycine (G)] and the APE [Alanine (A), Phenylalanine (P), Glutamic acid (E)] motifs are typical features in the activation loops of kinases. However, in the case of LRRK2, they depend on the availability of a DYG [Aspartic acid (D), Tyrosine (Y), Glycine (G)]-out motif instead of DFG. This DYG motive is changed to DYS [DYG (Aspartic acid (D), Tyrosine (Y), Serine (S)] in the G2019S-LRRK2 mutant. Hence, this modification could be an approach to develop specific LRRK2 kinase inhibitors.[2] However, this mutation results in constitutively active LRRK2, which imposes severe limitations on druggability of such Type II inhibitors in mutation carriers. Selective Type III and Type IV LRRK2 kinase inhibitors have not been reported yet. Again, Type III inhibitors, which stabilize the inactive state of the kinase, are of limited use in LRRK2 mutations which lack the inactive state. Allosteric Type IV inhibitors have not been identified yet. However, the remarkable C2024-C2025 moiety neighboring the DYG/DYS motif may provide a starting point for Type IV inhibition as this motif contributes to conformational stability. The discovery of kinase inhibitors usually progresses by the improvement of a high-throughput-screening (HTS) derived hit series in an iterative process of rational design and molecular docking analysis. This process requires access to robust enzyme structures. Unfortunately, full-length LRRK2 has escaped crystallization so far. Thus construction and use of LRRK2 homology models is the key.[12–17] Choice of structural templates and sequence alignment are the most important parameters for a homology model. The search for closely related kinases by comparison of sequence homology, usually the first choice, has not resulted in robust homology models yet. This enzyme-based approach can be complemented by a ligand-based approach which utilizes inhibitor fingerprinting to search for kinases with similar selectivities. A recent publication by Genentech compared these two approaches and identified a more robust LRRK2 homology model by comparison of inhibitor profiles. The SAR of JAK2 and JAK3 inhibitors with LRRK2 inhibitors suggested JAK2 and JAK3 as privileged templates for homology modeling. The final LRRK2 homology model was constructed from an unpublished JAK2 cocrystal structure.[12] The N-terminal domain of LRRK2 kinase exhibits predominantly β-sheets and an α-helix, whereas the C-terminal domain is organized in α-helices [Figure 11.1(a)]. Several important amino acids are found around the mostly hydrophobic ATP-binding site, which is located at the interface of the N- and C-terminal lobe. Met1947 represents the gatekeeper and forms the rear of the ATP-binding pocket with Lys1906 and Glu1920, whereas the residues Glu1948, Leu1949 and Ala1950 constitute the backbone of the ATP-pocket [Figure 11.1(b)]. The ceiling of the ATP site is created by Phe1883, Leu1885, Val1893 and Ala1904, whereas the residues Ile1933, Gly1953, Ser1954 and Leu2001 form the floor of the ATP-binding pocket.

The analysis of the number of conserved residues *versus* the total number of kinases sharing these indicates a road to selective kinase inhibitors, as eight identical residues in the ATP-binding site are shared by fewer than 20% of all kinases. In keeping with these findings, a detailed binding site sequence analyses has been completed. The list of the 19 amino acid residues in the ATP pocket

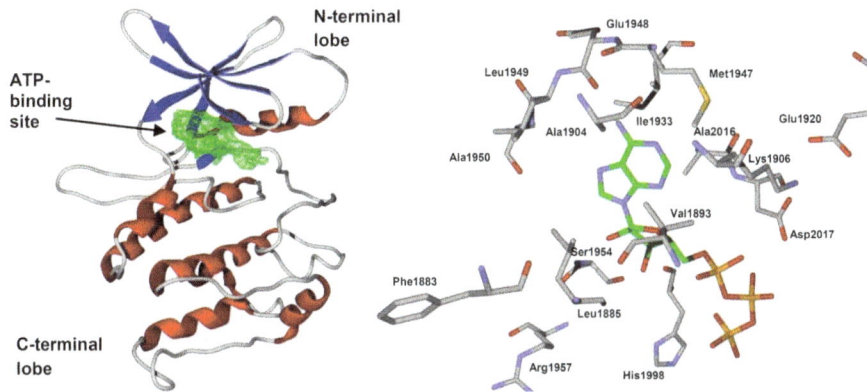

Figure 11.1 (a) Homology model of LRRK2 kinase domain with N- and C-terminal lobes as well as the ATP-binding site (colored in green) at the interface of the two lobes. (b) ATP-binding site of LRRK2 docked with ATP (colored in green) and important amino acids. The homology model was made using the Swiss model server.[18–20] The molecular docking was performed using the software Molegro Virtual Docker 5.

was narrowed down to the most important residues of Phe1883, Leu1949, Ser1954 and Arg1957. Interaction with these four core residues was suggested to provide kinase selectivity of small-molecule inhibitors against LRRK2 regardless of Type I or Type II inhibition.[12] The equilibrium of the active/inactive activation loop is shifted by the G2019S and I2020T mutation towards the active kinase, which imposes limitations on the development of Type II kinase inhibitors, as mutation carriers may not respond to the same dosage regime as non-carriers.

11.3 Medicinal Chemistry for the Design of Potent and Selective LRRK2 Inhibitors

11.3.1 Non-Selective LRRK2 Inhibitors

Almost all LRRK2 inhibitors reported so far have been derived from library screening efforts and act in an ATP-competitive manner. An overview of the different chemotypes is shown in Figure 11.2.

These early LRRK2 hits are all off-target effects of unselective kinase inhibitors and inhibit more than one kinase (Table 11.1). This early data originated from a limited number of *in vitro* assays using wild-type (wt) LRRK2 and G2019S-LRRK2. Unfortunately, these assays vary in the concentration of LRRK2 constructs, substrate and ATP, thus the mere comparison of IC_{50} is misleading. The most sensitive assays utilize radio-isotopes, which allow detection of both autophosphorylation and substrate phosphorylation, but are less suitable for HTS. High-throughput capability was achieved by time-resolved fluorescence resonance energy transfer

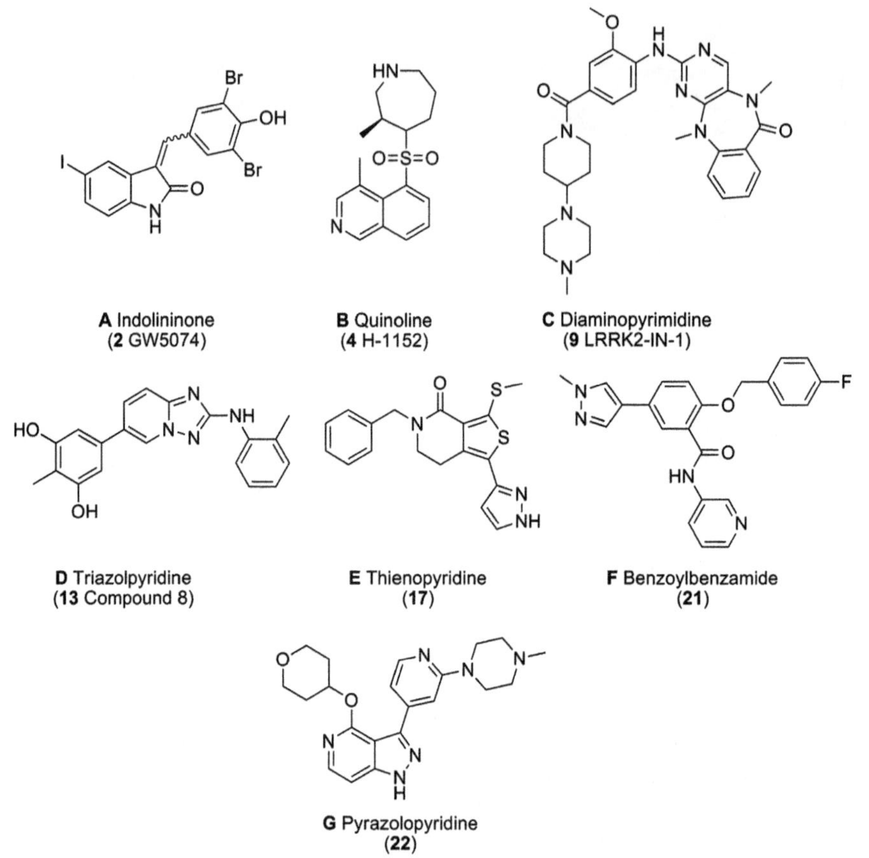

Figure 11.2 Different chemotypes of LRRK2 inhibitors.

(TR-FRET) and the amplified luminescent proximity homogeneous (AlphaScreen) assays.[18] Although truncated LRRK2 and its full-length analog display similar phosphorylation activities, differences have been noted. This may be a result of utilization of different substrates, *e.g.*, LRRKtide and myelin basic protein (MBP).[4,16] Sunitinib (compound **1**), a widely employed kinase inhibitor, inhibits both wt LRRK2 and LRRK2-G2019S (truncated and full-length) with an IC_{50} ranging from 15 to 79 nM (Table 11.1).[4,16,19] The selectivity of sunitinib (**1**) is limited: 12 out of 85 screened kinases are inhibited at 1 μM. However, sunitinib (**1**) is capable of suppressing the activity of full-length LRRK2 expressed from Swiss-3T3 fibroblast cells.[16] The apparent potency of sunitinib (**1**) to inhibit wt LRRK2 drops to an IC_{50} of 370 nM at cellular ATP concentration (1 mM).[20] The indolinones GW5074 (compound **2**) and indirubin-3′-monooxime (compound **3**) are *ca.* three-fold more active against LRRK2-G2019S than the wt kinase, inhibiting the former with IC_{50}s of 880 nM and ~1300 nM, respectively (Table 11.1).[6,21] Closely related LRRK1 and LRRK2 are inhibited in a similar manner, as is phosphorylation of the

Table 11.1 Indolinone derivatives as LRRK2 inhibitors.

No.	Name	Wild-type LRRK2 IC_{50}	LRRK2-G2019S IC_{50}	Substrate	Kinase profile	Literature
1	Sunitinib	79 nM[a]	19 nM[a]	Nictide	12[f] of 85[g]	Refs. 4, 16
		15 nM[b]	26 nM[b]	LRRKtide		and 19
2	Raf-1 kinase inhibitor I (GW5074)	~500 nM[c]	880 nM[d]	LRRKtide MBP[f]	—	Refs. 6 and 21
		3150 nM[d]				
3	Indirubin-3'-monooxime	4830 nM[d]	1310 nM[d]	MBP[e]	—	Ref. 6

[a]GST-LRRK2 (1326-2527; wt/G2019S).
[b]GST-LRRK2 (970-2527; wt/G2019S).
[c]Goat GST-LRRK2.
[d]GST-LRRK2 (wt/G2019S).
[e]Myelin basic protein (MBP).
[f]Number of inhibited kinases, including LRRK2.
[g]Total number of tested kinases.

putative LRRK2 substrate, eukaryotic translation initiation factor 4E-binding protein (4E-BPI). As indolinones and indirubines are well established scaffolds for pan-kinase inhibitors, they are frequently employed in kinase programs and thus provide very limited freedom to operate for new kinase programs.[22]

H-1152 (compound **4**) is a known ROCK2 inhibitor. This compound was profiled against a panel of 85 kinases and found to inhibit Aurora B kinase, BRSK2, wild-type LRRK2 and LRRK2-G2019S. H-1152 (**4**) displayed IC_{50}s ranging from 150 to 600 nM in radioisotope or AlphaScreen phosphorylation assays of wt LRRK2 and LRRK2-G2019S (Table 11.2).[18,19] Docking analysis of H-1152 (**4**) utilized a homology model of LRRK2. The carbon atoms of the LRRK2 sequence were superimposed on the reported ROCK1-H-1152 complex. This analysis indicated a backbone interaction with the NH group of Ala1950 (the PDB code of ROCK1-H-1152 has not been published). Furthermore, the two methyl groups of H-1152 (**4**) were observed to make lipophilic contacts with the ATP-binding site. The amino acid Ala2016 was found to be close to H-1152 (**4**). This could contribute to drug resistance, because the IC_{50} of H-1152 (**4**) for the LRRK2 mutant A2016T increased up to *ca.* 30-fold.[16]

Staurosporine (compound **5**) is a well-established pan-kinase inhibitor. It inhibited both wt LRRK2 and LRRK2-G2019S (truncated and full-length) with an IC_{50} ranging from 0.2 to 40 nM (Table 11.3) in different *in vitro* assays, *e.g.*, radioactive, TR-FRET and AlphaScreen assay.[4,6,18,21] These assays utilized different substrates such as synthetic peptides, *e.g.*, LRRKtide and

Table 11.2 H-1152 as an LRRK2 inhibitor.

H-1152

No.	Name	Wild-type LRRK2 IC_{50}	LRRK2-G2019S	Substrate	Kinase profile	Literature
4	H-1152	244 nM[a]	600 nM[b] 150 nM[b]	GST-Moesin Nictide	12[c] of 85[d]	Refs. 16, 18 and 19

[a]GST-LRRK2 (1326-2527; wt/G2019S).
[b]Full-length Strep-tag LRRK2 (G2019S).
[c]Number of inhibited kinases, including LRRK2.
[d]Total number of tested kinases.

Table 11.3 Staurosporine and derivatives as LRRK2 inhibitors.

Staurosporine   K-252a   K-252b   Gö6976

No.	Name	Wild-type LRRK2 IC_{50}	LRRK2-G2019S IC_{50}	Substrate	Kinase profile	Literature
5	Staurosporine	~1 nM[a] 2 nM[a] 8.2 nM[c] 40 nM[d]	0.2 nM[e] 1.8 nM[b] 40 nM[d]	GST-Moesin LRRKtide MBP[f]	—	Refs. 4, 6, 18 and 21
6	K-252a	~25 nM[a] 3.6 nM[b]	2.8 nM[b]	LRRKtide	—	Ref. 4 and 21
7	K-252b	~50 nM[a]	—	LRRKtide	—	Ref. 21
8	Gö6976	~250 nM[a]	—	LRRKtide	—	Ref. 21

[a]Goat GST-LRRK2.
[b]GST-LRRK2 (970-2527; wt/G2019S).
[c]Full-length LRRK2.
[d]GST-LRRK2 (wt/G2019S).
[e]Full-length Strep-tag LRRK2 (G2019S).
[f]Myelin basic protein (MBP).

derivatives of potential physiological substrates, *e.g.*, GST-Moesin. Staurosporine (**5**) displayed a similar inhibitory profile against LRRK1/LRRK2 autophosphorylation and MBP phosphorylation.[6] Its isoindolinone derivatives

K-252a/b (compounds **6** and **7**) and Gö6976 (compound **8**) also inhibited wt LRRK2 and LRRK2-G2019S with nM potencies of 3.6 to ~25 nM and ~50 nM, respectively (Table 11.3).[4,21]

11.3.2 Potent and Selective LRRK2 Inhibitors

The pyrimidine derivatives LRRK2-IN-1 (compound **9**), CZC-54252 (compound **10**) and CZC-25146 (compound **11**) were optimized to selectively inhibit LRRK2 (Table 11.4).[23,24] LRRK2-IN-1 (**9**) inhibited both truncated wt LRRK2 and LRRK2-G2019S with IC_{50} values of 13 and 6 nM, but LRRK2-A2016T and LRRK2-A2016T + G2019S mutants were found to be ca. 400-fold more resistant to this compound (**9**) (Table 11.4).[23] Molecular docking of LRRK2-IN-1 (**9**) to a homology model of LRRK2-A2016T provided an explanation for resistance of this and the LRRK2-A2016T + G2019S mutants. It revealed an unfavorable steric interaction, which is also present in the H-1152 (**4**) LRRK2 complex.[16] The reversible ATP competitive inhibitor LRRK2-IN-1 (**9**) is selective against a kinase panel of 470 kinases. Surprisingly, LRRK2-IN-1 (**9**) lacked inhibition of LRRK1. The kinase panel revealed additional inhibition of DCLK1, DCLK2 as well as MAPK7 and supported IC_{50}s of greater than 1 μM for AURKB, CHEK2, MKNK2, MYLK (smMLCK), NUAK1, PLK1 and RPS6KA2. LRRK2-IN-1 (**9**) induced a similar dose-dependent Ser910 and Ser935 dephosphorylation and loss of 14-3-3 binding to endogenous LRRK2 in human-derived SHSY5Y neuroblastoma cells and mouse Swiss3T3 cells.

CZC-54252 (**10**) and CZC-25146 (**11**) inhibited the activity of recombinant human wt LRRK2 with an IC_{50} of ~1 to 5 nM. The G2019S mutant was inhibited with an IC_{50} ranging from ~2 to 7 nM in a TR-FRET assay. Potency against the native LRRK2 protein complex in tissue extracts was only slightly lower. In addition, CZC-54252 (**10**) and CZC-25146 (**11**) exhibited good selectivity against 185 native kinases in cell and tissue extracts in a Kinobeads assay (Table 11.4).[24,25] CZC-25146 (**11**) inhibited only five other kinases, PLK4, GAK, TNK1, CAMKK2 and PIP4K2C, with comparable potency, none of them classified as predictors of genotoxicity or hematopoietic toxicity.[26,27] Furthermore, it prevented mutant LRRK2-induced injury of cultured rodent and human neurons with nM potency.

The inhibitor TAE684 (compound **12a**), which fuses structural elements of LRRK-IN-1 (**9**) and the CZC series, inhibited wt LRRK2 and LRRK2-G2019S with IC_{50}s of 7.8 and 6.1 nM, respectively. Although the potencies of TAE684 (**12a**) were similar to LRRK2-IN-1 (**9**), the former compound also inhibited the LRRK2-A2016T and LRRK2-A2016T + G2019S mutants with IC_{50} values of 93.3 and 21.9 nM, respectively (Table 11.4). This finding could be rationalized in a molecular docking study with LRRK2-A2016T. While both TAE684 (**12a**) and LRRK2-IN-1 (**9**) share the aminopyrimidine scaffold, the molecular model suggested that the isopropylsulfone moiety of TAE684 (**12a**) is able to avoid steric contact with the A2016T residue.[28] Furthermore, the selectivity of TAE684 (**12a**) was evaluated against a kinase panel of 442

Table 11.4 Diaminopyrimidine derivatives LRRK2-IN-1, CZC-54252, CZC-25146, TAE684 and HG-10-102-01 as potent and selective LRRK2 inhibitors.

No.	Name	Wild-type LRRK2 IC_{50}	LRRK2-G2019S IC_{50}	LRRK2-A2016T	LRRK2-A2016T + G2019S	Substrate	Kinase profile	Literature
9	LRRK2-IN-1	13 nM[a]	6 nM[a]	2450 nM[a]	3080 nM[a]	Nictide	12[c] of 442[d]	Ref. 23
10	CZC-54252	1.28 nM[b]	1.85 nM[b]	—	—	LRRKtide	10[c] of 184[d]	Ref. 24
11	CZC-25146	4.76 nM[b]	6.87 nM[b]	—	—	LRRKtide	5[c] of 184[d]	Ref. 24
12a	TAE684	7.8 nM[a]	6.1 nM[a]	93.3 nM[a]	21.9 nM[a]	Nictide	6[c] of 102[d]	Ref. 28 and 29
12b	HG-10-102-01	20.3 nM[a]	3.2 nM[a]	153.1 nM[a]	95.9 nM[a]	Nictide	2[c] of 451[d]	Ref. 30

[a]GST-LRRK2 (1326-2527; wt/G2019S/A2016T/G2019S + A2016T).
[b]Human LRRK2 (wt/G2019S).
[c]Number of inhibited kinases, including LRRK2.
[d]Total number of tested kinases.

kinases. This screen revealed that TAE684 (**12a**) is a potent inhibitor of several kinases like CAMKKβ, CHK2, FGF-1R, NUAK1, PBK and TSSK1 in a low nM range.[28,29] The next generation inhibitor in this series is HG-10-102-01 (compound **12b**), which is characterized by improved ligand efficiency at apparently retained selectivity. The inhibitory activity of HG-10-102-01 (**12b**) of wt LRRK2 and LRRK2-G2019S was comparable to LRRK2-IN-1 (**9**) with IC_{50} values of 20.3 and 3.2 nM, respectively. In contrast to LRRK2-IN-1 (**9**) the inhibitor HG-10-102-01 (**12b**) displayed an increased inhibition against the A2016T mutant.[30] Furthermore, HG-10-102-01 (**12b**) was evaluated against panels of 138 and 451 kinases, respectively (Table 11.4). At a concentration of 10 μM HG-10-102-01 (**12b**) displayed good selectivity against the panel of 138 kinases, as only MLK1 and MNK2 were inhibited by greater than 80%. HG-10-102-01 (**12b**) was evaluated further against 451 kinases and revealed excellent selectivity at a concentration of 1 μM, the only exception being the inhibition of one mutant form of c-Kit (L576P).[30]

The triazolopyridine Compound 8 (compound **13**) and the aminopyrimidine **14a** were identified by HTS as potent LRRK2 G2019S inhibitors. The triazolopyridine (**13**) exhibited a K_i value of 10 nM against LRRK2-G2019S (Table 11.5) and had chemical properties suitable for CNS penetration.[12] The aminopyridine **14a** is characterized by a slightly higher ligand efficiency, yet the scaffold is well established in kinase inhibition and thus provides little freedom to operate. Compound 8 (**13**) was tested against a kinase panel of 63 kinases at 1 μM and exhibited good selectivity. Abl was the only kinase which was also inhibited greater than 50% in addition to wt LRRK2 and LRRK2-G2019S. A homology model was developed on the basis of inhibitor fingerprinting and identified JAK3 and to a lesser extent JAK2 as kinases with similar ligand selectivity for a set of 100 triazolopyridines. JAK2 was selected as a template for the LRRK2 model due to the availability of a high-resolution structure determination of JAK2 cocrystals. The docking of Compound 8 (**13**) to the LRRK2 homology model elucidated the binding mode to the ATP pocket, which is indicative of a kinase Type I inhibitor.[12] The 2-aminotriazole establishes strong hydrogen bond contacts with the residue Ala1950. The essential dimethylphenol moiety interacts with the rear of the ATP pocket and forms hydrogen bonds between Lys1906 and Glu1920 with the hydroxyl group and hydrophobic interactions between the phenol and the gatekeeper Met1947 (Figure 11.3). Unfortunately, the dimethylphenol pharmacophore could not be replaced without loss of potency. This lead was abandoned due to the poor drug metabolism and pharmacokinetic (DMPK) properties associated with the phenol moiety.[12]

Compound 15 (compound **14c**) exhibited a K_i value of 3 nM and excellent selectivity against a panel of 63 kinases at 1 μM. FLT-3, the only kinase found to be effected, was inhibited by 53%.[12] The addition of the substituent R^2 (Table 11.5) to the diaminopyrimidines provided improved selectivity of LRRK2 *versus* JAK2 inhibition. This is probably due to the induction of a dihedral twist that destabilizes the JAK2 inhibiting conformer.

Table 11.5 Triazolopyridine and diaminopyridine derivatives as potent and selective LRRK2 inhibitors.

No.	Name	R^1	R^2	Wild-type LRRK2 K_i	LRRK2-G2019S K_i	JAK2 K_i	Substrate	Kinase profile	Literature
13	Compound 8	—	—	—	10 nM	405 nM	LRRKtide	2^a of 63^b	Ref. 12
14a	—	Cl	H	—	6 nM	7 nM	LRRKtide	—	Ref. 12
14b	—	Cl	CH_3	—	7 nM	> 3200 nM	LRRKtide	—	Ref. 12
14c	Compound 15	Cl	OCH_3	—	3 nM	> 3200 nM	LRRKtide	1^a of 63^b	Ref. 12
14d	—	Cl	Br	—	13 nM	> 3200 nM	LRRKtide	—	Ref. 12
14e	—	Br	OCH_3	—	1 nM	3000 nM	LRRKtide	—	Ref. 12
14f	—	F	OCH_3	—	71 nM	> 3200 nM	LRRKtide	—	Ref. 12
14g	—	CF_3	OCH_3	—	1 nM	> 3200 nM	LRRKtide	—	Ref. 12
14h	—	CN	OCH_3	—	13 nM	> 3200 nM	LRRKtide	—	Ref. 12

[a]Number of inhibited kinases, including LRRK2.
[b]Total number of tested kinases.

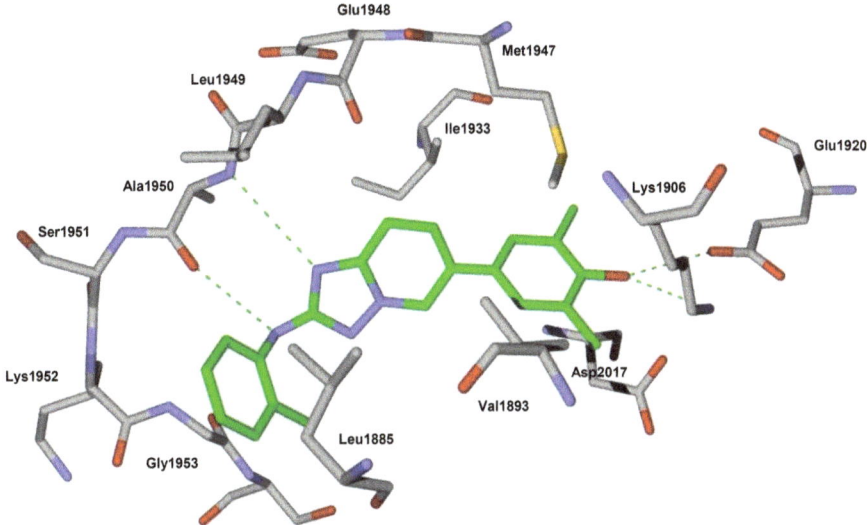

Figure 11.3 Compound 8 (compound 13) (green) in the ATP-binding pocket of LRRK2. Important protein interactions are shown as green dashed lines. The homology model was made using the Swiss model server.[18–20] The molecular docking was performed using the software Molegro Virtual Docker 5.

11.3.3 Examples from Recently Published Patent Applications

Recently, patent applications by GlaxoSmithKline (GSK), the Medical Research Council (MRC) and Genentech were published.[31–34] The cyclic thieno[3,4-c]pyridine derivatives (compounds **15–18**) inhibited LRRK2-G2019S in a low nM range (IC$_{50}$s from 1.3 to 3.2 nM, Table 11.6).[33] LRRK2 inhibitors bearing the benzyloxybenzamide moiety were published by GSK (Table 11.7). Compounds **19–21** exhibited similar potencies (LRRK2-G2019S IC$_{50}$s from 3 to 8 nM) as the thieno[3,4-c]pyridines (**15–18**).[34] Pyrazolo-pyridines (compounds **22–25**) published by Genentech exhibited K_i values of 3 to 35 nM (Table 11.8).[32] Unfortunately, the selectivity of these patented compounds against kinase panels was not disclosed.

11.4 The Role of LRRK2 Outside the Brain and Implications for Potential Mechanism-Based Toxicity of LRRK2 Inhibitors as Drugs

LRRK2 is a large multi-domain phosphoprotein.[35] Several mutations of the catalytic core region of LRRK2 have been unequivocally linked to late-onset autosomal dominant PD[36] and variations around the lrrk2 locus are a risk factor for idiopathic PD[37], with G2019S being the most prevalent in PD patients carrying a LRRK2 mutation (reviewed in ref. 3). Since the G2019S

Table 11.6 Thieno[3,4-c]pyridin-4(5H)-one derivatives as LRRK2 inhibitors.

No.	R^1	Wild-type LRRK2 IC_{50}	LRRK2-G2019S IC_{50}	Substrate	Kinase profile	Literature
15	HO~~~	—	1.7 nM	LRRKtide	—	Ref. 33
16	PhO-(CH2)3-	—	3.2 nM	LRRKtide	—	Ref. 33
17	PhCH2-	—	1.3 nM	LRRKtide	—	Ref. 33
18	Ph-	—	2.0 nM	LRRKtide	—	Ref. 33

Table 11.7 Benzyloxybenzamide derivatives as LRRK2 inhibitors.

No.	R^1	Wild-type LRRK2 $IC_{50}{}^a$	LRRK2-G2019S $IC_{50}{}^a$	Substrate	Kinase profile	Literature
19	-CH2CH2OCH3	—	6 nM	LRRKtide	—	Ref. 34
20	-CH2CH2-morpholine	—	8 nM	LRRKtide	—	Ref. 34
21	Me	—	3 nM	LRRKtide	—	Ref. 34

^{a}6His-Tev-LRRK2 (1326-2527).

mutation increases kinase activity,[38] treating LRRK2-linked PD with kinase inhibitors appears to be a very promising therapeutic approach. However, little is known about the underlying molecular mechanisms of PD pathophysiology.

Table 11.8 Pyrazolopyridine derivatives as LRRK2 inhibitors.

[Core scaffold: pyrazolo[3,4-b]pyridine with R¹ at 4-position and R² at 3-position, N–H on pyrazole]

No.	R¹	R²	K_i^a	Substrate	Kinase profile	Literature
22	tetrahydropyran-4-yloxy	4-(pyridin-2-yl)piperazin-1-yl (via methylene)	6 nM	LRRKtide	—	Ref. 32
23	tetrahydropyran-4-yloxy	cyclopropyl (via methylene)	35 nM	LRRKtide	—	Ref. 32
24	tetrahydropyran-4-yloxy	furan-2-yl	13 nM	LRRKtide	—	Ref. 32
25	tetrahydropyran-4-yloxy	1-isopropyl-N-(oxetan-3-yl)-1H-pyrrole-2-carboxamid-5-yl	3 nM	LRRKtide	—	Ref. 32

^{a}LRRK2 wt (Invitrogen cat# PV4882), LRRK2 G2019S (Invitrogen cat # PV4874).

But also the normal function of this protein and its role outside the brain remain poorly understood. The fact that LRRK2 is not only expressed in various parts of the brain but also in several peripheral tissues including lung, heart and kidney[35] leads to the suspicion that LRRK2 inhibition might cause adverse events which have to be considered early on in on-going attempts to develop LRRK2 selective kinase inhibitors.

The availability of LRRK2 knock-out (KO) mice has recently given insight into novel aspects of the physiological role of LRRK2 and its peripheral biology.[39,40] Tong et al. investigated the normal function of LRRK2 by generating two independent lines of germ-line deletion mice and studying potential effects over a time period of 20 months of mouse development.[40] In the brain, the dopaminergic system remained unaffected and no apparent neurodegeneration was detected. Tyrosine hydroxylase immunostaining of the brain revealed normal numbers and morphology of dopaminergic neurons. Striatal dopamine and α-synuclein levels were not significantly altered in comparison to wt mice. In contrast, kidneys of aging (20-month old) LRRK2 KO mice developed striking changes, and a disturbed protein homeostasis was observed. Loss of LRRK2 led to significant age-dependent kidney abnormalities, including renal atrophy and a 60-fold increased accumulation and aggregation of α-synuclein. Furthermore, protein degradation mediated by the ubiquitin–proteasome system was found to be impaired in the absence of LRRK2, leading to an accumulation of ubiquitinated proteins. In addition, accumulation of lipofuscin granules was observed in kidneys, indicating an impaired autophagy/lysosomal pathway in aging LRRK2 KO mice. The absence of LRRK2, moreover, led to a striking increase in inflammatory responses and apoptosis in mouse kidneys. These dramatic effects in kidney but not in brain may be related to the *ca*. six-fold higher LRRK2 expression in the former.[41] Furthermore, in the brain, LRRK1, a functional homolog of LRRK2 in vertebrates, might compensate for the absence of LRRK2. [35,40] Taken together, LRRK2 plays a crucial role *in vivo* and is involved in regulating protein degradation in an age-dependent manner.[40] For the development of LRRK2 inhibitors as potential treatments of PD, this is critically important, as the full inhibition of LRRK2 may cause mechanism-based renal toxicity.

The essential role of LRRK2 in the kidney was recently confirmed by a team at Novartis.[39] Three different mutant mouse lines were generated, in order to elucidate the essential functions of LRRK2 *in vivo*. The mouse lines were either carrying the pathogenic G2019S mutation inducing an elevated kinase activity (knock-in, KI) or the point mutation D1994S, which inactivates kinase activity (kinase-dead, KD). Additionally, a KO mouse line was generated.[39] As had been observed earlier,[40] LRRK2 mutant brain tissue did not exhibit any neuropathology or quantitative changes compared to wt littermates. However, kidneys were grossly affected in KO and also in KD mice. Both mutant lines developed kidney darkening, tubular dilatation and an elevated lipofuscin accumulation starting at the age of five months. In addition, KO and KD mice presented an early-onset (1.5 month) increase in size and number of secondary lysosomes in the proximal tubule cells. KO mutants, in addition, displayed pathophysiological changes in the lung (microvacuoles and lamellar bodies in lung Type II pneumocytes). These findings further support the essential function of LRRK2 in protein homeostasis. Impaired autophagy was not clearly detected by Herzig *et al*.[39] In contrast to earlier results,[40] LC3 levels, a marker for autophagy, remained unchanged in all mutant lines. mTOR and TCS2 protein levels were reduced in KD and increased in KO and KI mice. In KD mice the level of full-length LRRK2 protein was dramatically reduced in the kidney. The same was noted in wt mice after administration of a LRRK2-selective kinase inhibitor.[39] This may be an important finding for LRRK2 drug discovery, as it illustrates the important role of kinase activity for maintenance of normal LRRK2 protein levels and, hence, normal kidney function. Interestingly, in heterozygous KO mice the reduction of LRRK2 levels was more subtle.[39] In summary, absence of kinase activity is associated with kidney abnormalities and perturbed protein and lysosome homeostasis.[39] Taken together, this extends earlier findings for KO mutants[40] to KD mutants suggesting that protein homeostasis in kidneys depends on LRRK2 kinase activity and not (just) a putative scaffolding role of the LRRK2 protein. Full inhibition of kinase activity might thus not be a suitable therapeutic approach, since it may cause toxic effects in the kidney. However, further investigation, especially of heterozygote animals, is needed to understand this matter properly.

Apart from its essential role in protein homeostasis and kidney morphology and function, LRRK2 has been suggested to be involved in several immunological functions. LRRK2 is expressed in various tissues of the immune system such as the thymus and spleen[35] and in subsets of immune cells.[42,43] LRRK2 mRNA has been detected in human peripheral blood mononuclear cells (hPBMCs), in circulating $CD19^+$ B cells and in $CD14^+$ monocytes. In particular, the monocyte subpopulation $CD14^+CD16^+$ displays high levels of LRRK2 expression, whereas in $CD14^+CD16^-$ cells LRRK2 protein levels are very low. Since $CD14^+CD16^+$ monocytes are augmented in inflammatory processes and considered to be more mature, it is believed that LRRK2 plays a role in monocyte maturation. Additionally, monocyte differentiation experiments revealed that levels of LRRK2 protein remain high in macrophages and dendritic cells after the maturation process is complete.[42] Earlier studies had shown that LRRK2 expression is regulated during B cell differentiation and maturation, as different protein levels of LRRK2 were found in the B-lymphocyte subtypes pre-B, B1 and B2.[43,44] LRRK2 expression in hPBMCs is significantly increased after treatment with IFN-γ, consistent with the fact that the lrrk2 gene contains a consensus binding site for interferon response factors.[42,45] LRRK2 mRNA and protein levels are particularly upregulated by IFN-γ in isolated monocytes. A weaker and more variable response was recently also observed after IFN-β and TNFα treatment,[42] in contrast to earlier results that did not detect up-regulation of LRRK2 upon TNFα stimulation in THP-1 cells.[45] No increase in LRRK2 expression was detected when hPBMCs were treated with interleukins IL-1β, IL-2, IL-15 or with the insulin-like growth factor IGF-1. The fact that LRRK2 is expressed in monocytes, macrophages, dendritic cells and B cells and that its expression is highly regulated by IFN-γ suggest a role in the immune system. Signal transduction cascades that LRRK2 is involved in as well as its precise role in function of the immune system remain to be established. Recent evidence, however, suggests that LRRK2 can be phosphorylated at Ser910 and Ser935 by canonical (IKKalpha and IKKbeta) and IKK-related (IKKepsilon and TBK1) kinases following activation *via* the MyD88 pathway.[46] This finding implies that inhibition of phosphorylation on Ser910 and Ser935 may not be a specific pharmacodynamic biomarker of LRRK2 inhibition. Consequently, Ser910 and Ser935 phosphorylation data needs to be interpreted with care and may not be the quintessential readout for efficacy of LRRK2 inhibitors *in vivo*.

In summary, LRRK2 is involved in B cell and monocyte maturation and differentiation processes. This implies the possibility of mechanism-based toxicity of LRRK2 inhibitors for PD. Kidney, blood and immune function or potentially toxicity need to be monitored carefully during drug discovery and development. Interestingly, in the case of papillary renal and thyroid carcinomas, down-regulation of LRRK2 expression induced increased cell death[47] and in the brain LRRK2 kinase inhibition attenuated microglial inflammatory responses,[48] illustrating once again the complexity of protein function and the consequences of (sometimes only subtle) changes in the expression level.

11.5 Invertebrate and Vertebrate Animal Models for Pharmacological Evaluation of LRRK2 Inhibitors

Pharmacological validation of LRRK2 as a druggable target for PD and testing of efficacy of possible PD therapeutics requires adequate animal models with a PD-like phenotype. For (mutant) LRRK2-induced PD a limited number of relevant *in vivo* models has been described. The lrrk2 gene is highly conserved across species, providing the opportunity to study LRRK2-related diseases in invertebrate and vertebrate animal models.

Invertebrate LRRK2-linked PD animal models have been published for *Drosophila melanogaster* and for *Caenorhabditis elegans*. *Drosophila* possesses a single ortholog (CG5483) of human LRRK2. Several research groups have generated transgenic loss-of-function *Drosophila* mutants using the GAL4/UAS system, taking advantage of the yeast GAL4 transcription factor which binds specifically to an upstream activation sequence. Under the control of the promoter-GAL4 it is therefore possible to express transgenes in specific cells or various tissues.[49] LRRK loss-of-function mutants exhibited impaired locomotive activity and degeneration of dopaminergic (DA) neurons in *Drosophila*.[50] Overexpression of human wt LRRK2 or the LRRK2-G2019S mutant in fly neurons produced late-onset selective loss of DA neurons in the brain, locomotor deficiencies and early mortality with LRRK2-G2019S causing increased autophosphorylation and a more severe Parkinsonism-like phenotype in comparison to wt LRRK2.[51]

C. elegans contains LRK-1, the ortholog of human LRRK2. In deletion mutants, LRK-1 was identified to be involved in polarized sorting of synaptic vesicle proteins to axons by excluding these proteins from the dendrite-specific Golgi Transport machinery.[52] Overexpression of human wt LRRK2 or the mutants R1441C and G2019S in *C. elegans* DA neurons causes age-dependent neurodegeneration, a decrease of dopamine levels, locomotion impairment and behavioral deficits *in vivo*.[53] As observed for *Drosophila*, the mutant worms display a more severe phenotype than wt, providing additional evidence of the importance of appropriate GTPase and kinase activity in PD pathogenesis.

Drosophila and *C. elegans* invertebrate LRRK2 PD models resemble several key features of human PD and might thus offer suitable short-term models of PD pathogenesis and underlying molecular mechanisms. As many genetic tools are available for these model organisms and large-scale genetic screens are feasible, *Drosophila* and *C. elegans* provide the possibility of carrying out inexpensive and rapid compound screens. However, both organisms lack a mammal-like blood–brain barrier. Furthermore, LRRK2 protein–protein interactions that may contribute to the function of human LRRK2, and thus to PD pathogenesis, may not be evolutionarily conserved.

In zebrafish, XM_682700 has been identified as ortholog of human LRRK2. The blockage of zebrafish LRRK2 (zLRRK2) protein by morpholino reagents led to embryonic lethality and severe developmental deficiencies such as growth retardation and loss of neurons. Interestingly, deletion of only the WD40 domain of zLRRK2 by morpholino-directed targeting of the splicing machinery induces Parkinsonism-like phenotypes such as loss of DA neurons in the diencephalon, and locomotor dysfunction.[54] It may therefore provide a useful small vertebrate disease model for PD.

The most common vertebrate models for LRRK2-linked Parkinsonism are rodent models. LRRK2 transgenic mice have been generated that show some neurochemical and behavioral abnormalities yet mostly lack specific PD characteristics like DA neuronal cell loss in the *substantia nigra*.[40,55–58] One exception is a transgenic mouse model expressing human LRRK2 with the mutation G2019S which demonstrated a modest age-dependent degeneration of DA neurons.[59]

Conditional expression of wt LRRK2 or LRRK2-G2019S in double transgenic mice under the control of the calcium/calmodulin-dependent protein kinase II (CaMK II) promoter failed to trigger neurodegeneration of DA neurons and PD-like behavioral phenotypes.[56] Bacterial artificial chromosome (BAC) transgenic mice expressing wt LRRK2 or the mutants R1441G and G2019S show evidence of an age-dependent decrease of dopamine release and axonal pathology of the nigrostriatal dopaminergic projection[55] or a decrease in striatal dopamine content and release.[58] However, no loss of DA neurons in the *substantia nigra* was observed. Human LRRK2 wt and LRRK2 mutant G2019S BAC transgenic mice generated by Melrose *et al.* revealed abnormal dopamine neurotransmission in both wt and mutant transgenic mice demonstrated by a reduction in extracellular dopamine levels and augmented phosphorylation of the microtubule binding protein tau in G2019S mutant mice.[57] Li *et al.* reported hyperphosphorylation of tau in brain tissue and age-dependent motor activity deficits in mutant LRRK2 R1441G BAC transgenic mice.[55] A possible pathophysiological interaction between the PD-related genes LRRK2 and α-synuclein was shown by Lin *et al.* who generated double transgenic mice co-expressing human LRRK2 and α-synuclein.[56] The overexpression of either wt or PD-related G2019S LRRK2 in A53T α-synuclein transgenic mice significantly accelerated the progression of neuropathological abnormalities, while the presence of excess LRRK2 alone did not exacerbate neurodegeneration. Furthermore, LRRK2 promoted the accumulation and aggregation of α-synuclein in A53T mice. Conversely, the silencing of LRRK2 suppressed the accumulation and aggregation of α-synuclein and by this means delayed the progression of neuropathology in A53T mice.[56] These findings suggest that LRRK2 inhibition is a potential therapeutic option for attenuating α-synuclein-induced neurodegeneration, but some controversy remains.[60] In summary, whereas various LRRK2 mouse models have shed some light on pathobiology of wt and mutant LRRK2 in PD, the observed phenotypes, in particular those relating to neurodegeneration, may prove too subtle for disease models of PD that can be used for robust measurement of the efficacy of LRRK2 inhibitors. Moreover, onset of phenotypic changes is usually late, indicating the need for long-term inhibitor studies, which are not a preferred option – at least not in early drug discovery.

A promising, rapid-onset rodent model, however, was generated by transient expression of the LRRK2-G2019S mutant in mouse brain using herpes simplex virus (HSV) amplicons or adeno-associated virus (AAV) vectors for LRRK2 gene delivery into the *striatum* and the *substantia nigra*.[6] Overexpression of LRRK2 mutant induced a significant loss of tyrosine hydroxylase (TH) positive neurons in the *substantia nigra*. LRRK2 wt and, interestingly, the KD mutant LRRK2-G2019S, D1994A did not trigger any signs of neurodegeneration. An *in vivo* efficacy study with non-selective LRRK2 inhibitors in this animal model revealed that the application of moderately potent kinase inhibitors, GW5074 and indirubin-3'-monooxime, were effective in attenuating the neuronal cell loss suggesting that LRRK2 kinase inhibition may be a viable therapeutic strategy for PD.[6]

The AAV mouse model offers an *in vivo* model for LRRK2-induced loss of DA neurons in the *substantia nigra*, a hallmark of PD, only three weeks after stereotactic injection into the *striatum*.[6] It may therefore be suitable for subchronic evaluation of LRRK2 inhibitor drug candidates, since other rodent models usually take months (if ever) to produce PD-like phenotypes. However, because of the known non-selectivity of the tool compounds used in this study, the results are still a matter of debate.[61] Rapidly induced neurodegeneration could also be a drawback, since it might not be a relevant model for human PD as a late-onset disease. Nevertheless, the AAV mouse model is currently the only rodent model where degeneration of DA neurons is robustly observed and may therefore be considered the model of choice. If the described animal models are relevant for idiopathic PD remains unclear, but currently no animal model is able to fully recapitulate a disease.

11.6 Pharmacokinetics and Pharmacodynamics of LRRK2 Inhibitors: The Current State-of-the-Art

Few potent and selective LRRK2 inhibitors have been reported to date and little is known about their pharmacokinetic properties (Table 11.9). Oral bioavailability of these leads is generally 50% and higher indicating that oral administration is a possible route for future target validation and efficacy studies. However, both the reported Novartis[39] and Genentech[12] LRRK2 inhibitors have plasma half-lives of less than one hour, which significantly restricts their use in *in vivo* studies. The structure of the Novartis/NIBR compound has not been disclosed. LRRK2-IN-1[23] and TAE684[28] display extended plasma half-lives of 4.5 and 11.3 h, respectively, as well as significantly higher compound exposure expressed as (AUC blood h^{-1} ng mL^{-1}; Table 11.9) after oral administration than other compounds. CZC-25146,[24] although it cleared more rapidly than the former two compounds, is extensively distributed in tissue and reaches a maximal plasma concentration of 1357 ng mL^{-1} at an oral dose of 5 mg kg^{-1} *vs.* 1618 ng mL^{-1} at an oral dose of 10 mg kg^{-1} for LRRK2-IN-1. Furthermore, compound exposure to CZC-25146 at a 5 mg kg^{-1} dose is about half that of TAE684 at twice that dose suggesting that CZC-25146 may also be useful as a probe compound for

Table 11.9 Pharmacokinetics (PK) of current potent and selective LRRK2 inhibitors. n.r. = not reported.

		CZC-25146	LRRK2-IN-1	NIBR Inhibitor	Cpd 15	TAE684
Mouse PK (1 mg kg^{-1} i.v.)	Clearance/L^{-1} h^{-1} kg^{-1}	2.3	0.3	3.0	—	1.0
	Volume of distribution V_{ss}/L kg^{-1}	5.4	1.7	1.7	—	14.5
	$t_{1/2}$/h	1.6	4.5	0.4	0.2	11.3
	C_{max}/ng mL^{-1}	154	—	—	—	—
	AUC$_{last}$/h ng^{-1} mL^{-1}	419	2974	—	—	772
Mouse PK (p.o.)	Oral dose/mg kg^{-1}	5	10	3	—	10
	$t_{1/2}$/h	1				
	t_{max}/h	0.25	1			
	AUC$_{last}$ blood/h ng^{-1} mL^{-1}	2878	14758	573	—	6374
	C_{max} blood/ng mL^{-1}	1357	1618	312	—	—
	C_{max} brain/ng g^{-1}	—	—	672	—	—
	Oral bioavailability (F)/%	133	49	57	—	83
	Wt mouse total brain-to-plasma ratio	~0.04	n.r.^{a}	—	1.4	2.3^b
	p-gp/BCRP double KO mouse total brain-to-plasma ratio	—	—	—	2.9	—
Mouse PK (30 mg kg^{-1} i.p.)	Total plasma concentration at 1 h/μM	—	—	—	8.4	—
	Total brain concentration at 1 h/μM	—	—	—	6.2	—
Literature		Refs. 66 and 24	Ref. 23	Ref. 39	Ref. 12	Ref. 28

^{a}Following administration of LRRK2-IN-1, no dephosphorylation of LRRK2 Ser910 or Ser935 was observed in the brain, suggesting that very little or none of the compound penetrated into the brain.
^{b}Despite high total brain-to-plasma ratio, suggesting good penetration into the brain, no dephosphorylation of LRRK2 Ser910 or Ser935 was observed in the brain; the reasons for this unexpected result are currently unknown.

future *in vivo* studies. However, whereas CZC-25146 and LRRK2-IN-1 may be useful in *in vivo* studies addressing LRRK2 (inhibitor) function in peripheral organs including kidney and the immune system, they lack a key feature of potential PD drugs: brain penetration.

Two recent LRRK2 inhibitors, Genentech's Compound 15 and TAE684, are brain-penetrant. The former features a total brain-to-plasma ratio of 1.4, which is even higher (2.9) in the p-glycoprotein (p-gp)/ATP-binding cassette subfamily G member 2 (ABCG2/BCRP) double KO mice suggesting that the compound is a moderate substrate of brain efflux transporters.[12] However, 1 h after intraperitoneal dosing at 30 mg kg^{-1}, a compound concentration of 6.2 µM was measured in brain, suggesting that sufficient compound exposure could be obtained at least for acute pharmacodynamic biomarker studies. For TAE684 an even higher total brain-to-plasma ratio was measured (2.3). Nevertheless, despite this evidence for good penetration into the brain, dephosphorylation of LRRK2 Ser910 or Ser935, a potential pharmacodynamics biomarker[23] (but see ref. 46), was only observed in the periphery (*e.g.*, kidney), and not in brain. The reasons for this unexpected result are currently unknown.[28] It should be noted that lack of brain penetration of LRRK2-IN-1 had not been demonstrated, only inferred from lack of LRRK2 Ser910 or Ser935 dephosphorylation,[23] a finding that may have to be revisited in light of the TAE684 study. Future LRRK2 inhibitor drug discovery will have to focus on improving the brain penetration and metabolic stability of selective compounds and on defining suitable biomarkers for acute studies of compound pharmacodynamics in the brain. A very recent study, currently in press, reports a pyrimidine analog, HG-10-102-01, as a potent and selective inhibitor of wt LRRK2 and the G2019S mutant. It inhibited LRRK2 Ser910 and Ser935 phosphorylation at submicromolar concentrations in cells and, as the first compound reported to date, in mouse brain, albeit at high intraperitoneal doses of 50 mg kg^{-1}.[30]

11.7 Will LRRK2 Kinase Inhibitors be Developed into Drugs (for the Treatment of Parkinson's Disease)?

The availability of suitable HTS assay formats such as TR-FRET or AlphaScreen has provided initial inhibitors of LRRK2 activity with moderate selectivity. Lead optimization has resulted in first-generation tools and second-generation inhibitors, which display promising pharmacokinetic properties, but are limited by insufficient brain uptake or brain activity. The increase in patent applications (*e.g.*, GSK and Genentech) indicates a "target on the rise". TTT-3002, a staurosporine derivative in clinical development by TauTaTis, allegedly exhibited good results in LRRK2 inhibition. A Phase I clinical trial of TTT-3002 was expected to start in 2011.[62] However, staurosporines are notorious for their lack of selectivity and the precise status of the compound is unknown.[31,34,63–66]

Clinical development of LRRK2 inhibitors is hampered by the lack of (public) data of the relevant pharmacology, biology and even biomarkers.

For example, it is not known at present whether any of the published potent and selective LRRK2 inhibitors (Tables 11.4 and 11.5) display overt toxicity when chronically administered to animals. If so, given the industry's effort, it would seem fair to assume that any compound-based toxicity could eventually be overcome during drug discovery by further lead optimization. Hence, three main questions need to be addressed:

(1) Patient stratification: Would an ideal LRRK2 kinase inhibitor be a potential drug for all PD patients? Or only for those patients carrying LRRK2 mutations? Or only for those patients carrying activating mutations in the kinase domain, e.g., LRRK2 G2019S?
(2) Mechanism-based toxicity: Given the high expression levels of LRRK2 in peripheral organs such as the kidney and in various blood cells, is LRRK2 inhibition expected to be associated with mechanism-based toxicity that would drastically limit its utility as a treatment for any chronic disease? If so, would such toxicity only be observed if LRRK2 was fully inhibited or could it be mitigated by appropriate dosing regimes? What therapeutic index could be expected for a best-in-class LRRK2 inhibitor?
(3) Utility in other therapy areas besides PD: Given the high expression levels of LRRK2 in peripheral organs such as kidney and in various blood cells, could LRRK2 inhibition be a therapeutic option for diseases other than PD?

11.7.1 Patient Stratification

Since an initial report suggesting that PD-associated mutations in LRRK2 link enhanced GTP-binding and kinase activities to neuronal toxicity,[67] it has generally been assumed that mutations inside the kinase domain of LRRK2 are indeed associated with increased kinase activity that may be attenuated by LRRK2 inhibitors as therapeutics. In contrast, the question of whether mutations outside the kinase domain, e.g., within the GTPase domain, are linked to increased kinase activity and, hence, neurotoxicity is still a matter of debate.[1] Interestingly, use of the selective LRRK2 inhibitor CZC-25146 blocked neurotoxicity induced by LRRK2 R1441C, a mutation outside the kinase domain, in cultured human cortical neurons.[24] To complicate matters further, it has recently been shown that a risk factor for PD, the G2385R mutation in LRRK2's C-terminal WD40 domain, is associated with a partial loss of kinase function and that deletion of the entire C-terminus abolishes kinase function altogether.[68] This study therefore suggests that the simple assumption LRRK2 inhibitor = PD treatment or LRRK2 inhibitor = treatment for patients with testable PD mutations is probably not valid and needs to be thoroughly tested experimentally. On the other hand, careful examination of the precise correlation between the type of mutation, increased kinase activity and therefore possible patient benefit, may result in commercial opportunities for pharmaceutical/diagnostics companies in the field of companion diagnostics. Carriers of gain-of-function mutations in LRRK2 may require fundamentally different dosing regimens to other PD patients. One possible avenue for future drug discovery in the field is therefore that of personalized/stratified medicine.

Moreover, the gain-of-function mutation G2019S may impact the potential of Type II kinase inhibitors, as the equilibrium of the active/inactive state is shifted to the former, which may reduce the efficacy of Type II kinase inhibitors in the mutation carriers.

11.7.2 Mechanism-Based Toxicity

Pronounced expression of LRRK2 in kidney and defined immune cells (but also in other organs)[35,39,42–44] invariably leads to the question of whether or not LRRK2 inhibition may be associated with unwanted effects such as immunosuppression or renal toxicity. The verdict is still out, but it can be assumed that the currently available potent, selective and (reasonably) metabolically stable LRRK2 inhibitors[23,24,28] will be instrumental in addressing this question. Brain penetration is not required for most of these studies; hence, several probe compounds could be tested side-by-side.

11.7.3 Utility in Other Therapy Areas Besides Parkinson's Disease

Molecular and cell biology work a few years ago suggested that LRRK2 be involved in the interferon-γ (IFN-γ) response.[45] A recent detailed study by a Merck–Serono team reported that LRRK2 expression in human peripheral blood mononuclear cells (hPBMCs) is increased after treatment with IFN-γ and -β as well as TNFα, but not by interleukins IL-1β, IL-15, IL-2 or insulin-like growth factor IGF-1.[42] More importantly, treatment of purified monocytes with the selective LRRK2 inhibitor LRRK2-IN-1 decreased IFN-γ-induced CD14 and CD16 expression with half-maximal effective concentrations of *ca.* 10 and 1 µM, respectively.[42] The significance of this finding is two-fold: On the one hand does it provide pharmacological evidence for the potential utility of LRRK2 inhibitors as treatment for immunological diseases that are mitigated by these cytokines. From a drug discovery and development perspective, the results suggest that easily accessible immune cells could be useful in the discovery of clinically useful pharmacodynamic biomarkers for LRRK2 inhibitors. Furthermore, mitigating a possible immunological component of PD may be a complementary function that could enhance the clinical efficacy of LRRK2 inhibitors.

Recently, LRRK2 was found to be a negative regulator of the transcription factor NFAT, an important regulator of the immune system, and to control the severity of inflammatory bowel disease (IBD). A LRRK2 KO led to more severe clinical symptoms of IBD and KO mice were more susceptible to experimental colitis.[69] However, Liu *et al.* showed with a KD mutant that

LRRK2-based NFAT regulation is a kinase-independent mechanism and thus may not present difficulties in LRRK2 kinase inhibitor development.

Since LRRK2 expression may be associated with IBD[69] and several cytokines are up-regulated in PD patients,[70] LRRK2 appears to play an important role in the immunological processes of these diseases and its inhibition might have the potential to attenuate progression of these diseases. Indeed, very recently a study using cultured microglia revealed that LRRK2 activity is increased in microglial inflammation and that inhibition of LRRK2 kinase activity by the selective LRRK2-IN-1 inhibitor attenuated inflammatory signaling in cultured microglia, as assessed by decreased TNF release, decreased lipopolysaccharide (LPS)-induced p38 phosphorylation, decreased LPS-induced increase in microglia process length, and decreased microglial chemotaxis.[48] These pharmacological findings are well supported by LRRK2 knock-down studies in cultured murine microglia.[71] As a consequence, brain-penetrant LRRK2 inhibitors may be tested in animal models of neuroinflammation outside of PD.

References

1. M. R. Cookson, *Nat. Rev. Neurosci.*, 2010, **11**, 791.
2. B. D. Lee, V. L. Dawson and T. M. Dawson, *Trends Pharmacol. Sci.*, 2012, **33**, 365.
3. T. Kramer, F. Lo Monte, S. Göring, B. Okala Amombo and B. Schmidt, *ACS Chem. Neurosci.*, 2012, **2**, 151.
4. V. S. Anand, L. J. Reichling, K. Lipinski, W. Stochaj, W. Duan, K. Kelleher, P. Pungaliya, E. L. Brown, P. H. Reinhart, R. Somberg, W. D. Hirst, S. M. Riddle and S. P. Braithwaite, *FEBS J.*, 2009, **276**, 466.
5. G. Drewes, C. Hopf and V. Reader, *US Pat.*, 2009220992, 2009.
6. B. D. Lee, J. H. Shin, J. Vankampen, L. Petrucelli, A. B. West, H. S. Ko, Y. I. Lee, K. A. Maguire-Zeiss, W. J. Bowers, H. J. Federoff, V. L. Dawson and T. M. Dawson, *Nat. Med.*, 2010, **16**, 998.
7. M. Liu, B. Dobson, M. A. Glicksman, Z. Yue and R. L. Stein, *Biochemistry*, 2010, **49**, 2008.
8. M. Li, C. J. Smith, M. T. Walker and T. J. Smith, *J. Biol. Chem.*, 2009, **284**, 22988.
9. M. Arita, T. Wakita and H. Shimizu, *J. Gen. Virol.*, 2008, **89**, 2518.
10. T. Gasser, *Expert Rev. Mol. Med.*, 2009, **11**, e22.
11. I. N. Rudenko, R. Chia and M. R. Cookson, *BMC Med.*, 2012, **10**, 20.
12. H. Chen, B. K. Chan, J. Drummond, A. A. Estrada, J. Gunzner-Toste, X. Liu, Y. Liu, J. Moffat, D. Shore, Z. K. Sweeney, T. Tran, S. Wang, G. Zhao, H. Zhu and D. J. Burdick, *J. Med. Chem.*, 2012, **55**, 5536.
13. M. Liu, S. Kang, S. Ray, J. Jackson, A. D. Zaitsev, S. A. Gerber, G. D. Cuny and M. A. Glicksman, *Biochemistry*, 2011, **50**, 9399.
14. I. Marin, *Mol. Biol. Evol.*, 2006, **23**, 2423.
15. I. F. Mata, W. J. Wedemeyer, M. J. Farrer, J. P. Taylor and K. A. Gallo, *Trends Neurosci.*, 2006, **29**, 286.

16. R. J. Nichols, N. Dzamko, J. E. Hutti, L. C. Cantley, M. Deak, J. Moran, P. Bamborough, A. D. Reith and D. R. Alessi, *Biochem. J.*, 2009, **424**, 47.
17. H. Yun, H. Y. Heo, H. H. Kim, N. DooKim and W. Seol, *Bioorg. Med. Chem. Lett.*, 2011, **21**, 2953.
18. L. Pedro, J. Padros, L. Beaudet, H. D. Schubert, F. Gillardon and S. Dahan, *Anal. Biochem.*, 2010, **404**, 45.
19. N. Dzamko, M. Deak, F. Hentati, A. D. Reith, A. R. Prescott, D. R. Alessi and R. J. Nichols, *Biochem. J.*, 2010, **430**, 405.
20. L. J. Reichling and S. M. Riddle, *Biochem. Biophys. Res. Commun.*, 2009, **384**, 255.
21. J. P. Covy and B. I. Giasson, *Biochem. Biophys. Res. Commun.*, 2009, **378**, 473.
22. N. Hottecke, M. Liebeck, K. Baumann, R. Schubenel, E. Winkler, H. Steiner and B. Schmidt, *Bioorg. Med. Chem. Lett.*, 2010, **20**, 2958.
23. X. Deng, N. Dzamko, A. Prescott, P. Davies, Q. Liu, Q. Yang, J. D. Lee, M. P. Patricelli, T. K. Nomanbhoy, D. R. Alessi and N. S. Gray, *Nat. Chem. Biol.*, 2011, **7**, 203.
24. N. Ramsden, J. Perrin, Z. Ren, B. D. Lee, N. Zinn, V. L. Dawson, D. Tam, M. Bova, M. Lang, G. Drewes, M. Bantscheff, F. Bard, T. M. Dawson and C. Hopf, *ACS Chem. Biol.*, 2011, **6**, 1021.
25. M. Bantscheff, D. Eberhard, Y. Abraham, S. Bastuck, M. Boesche, S. Hobson, T. Mathieson, J. Perrin, M. Raida, C. Rau, V. Reader, G. Sweetman, A. Bauer, T. Bouwmeester, C. Hopf, U. Kruse, G. Neubauer, N. Ramsden, J. Rick, B. Kuster and G. Drewes, *Nat. Biotechnol.*, 2007, **25**, 1035.
26. A. J. Olaharski, H. Bitter, N. Gonzaludo, R. Kondru, D. M. Goldstein, T. S. Zabka, H. Lin, T. Singer and K. Kolaja, *Toxicol. Sci.*, 2010, **118**, 266.
27. A. J. Olaharski, N. Gonzaludo, H. Bitter, D. Goldstein, S. Kirchner, H. Uppal and K. Kolaja, *PLoS Comput. Biol.*, 2009, **5**, e1000446.
28. J. Zhang, X. Deng, H. G. Choi, D. R. Alessi and N. S. Gray, *Bioorg. Med. Chem. Lett.*, 2012, **22**, 1864.
29. M. I. Davis, J. P. Hunt, S. Herrgard, P. Ciceri, L. M. Wodicka, G. Pallares, M. Hocker, D. K. Treiber and P. P. Zarrinkar, *Nat. Biotechnol.*, 2011, **29**, 1046.
30. H. G. Choi, J. Zhang, X. Deng, J. M. Hatcher, M. P. Patricelli, Z. Zhao, D. R. Alessi and N. S. Gray, *ACS Med. Chem.*, 2012.
31. B. Chan, A. Estrada, Z. Sweeney and E. G. McIver, *Pat.*, WO 2011/141756 A1, 2011.
32. B. Chan, H. Chen, A. Estrada, D. Shore, Z. Sweeney and E. McIver, *Pat.*, WO 2012/038743 A1, 2012.
33. J. A. Miccauley, H. A. Rajapakse, T. J. Greshock, J. Sander, B. Kim, V. L. Rada, J. T. Kern, H. H. Stevenson and M. T. Bilodeau, *Pat.*, WO 2012/058193 A1, 2012.
34. P. L. Nichols, A. J. Eatherton, P. Bamborough, K. S. Jandu, O. J. Philips and D. Andreotti, *Pat.*, WO 2011/038872 A1, 2011.

35. M. Westerlund, A. C. Belin, A. Anvret, P. Bickford, L. Olson and D. Galter, *Neuroscience*, 2008, **152**, 429.
36. A. Zimprich, S. Biskup, P. Leitner, P. Lichtner, M. Farrer, S. Lincoln, J. Kachergus, M. Hulihan, R. J. Uitti, D. B. Calne, A. J. Stoessl, R. F. Pfeiffer, N. Patenge, I. C. Carbajal, P. Vieregge, F. Asmus, B. Muller-Myhsok, D. W. Dickson, T. Meitinger, T. M. Strom, Z. K. Wszolek and T. Gasser, *Neuron*, 2004, **44**, 601.
37. International Parkinson Disease Genomics Consortium, *Lancet*, **377**, 641.
38. A. B. West, D. J. Moore, S. Biskup, A. Bugayenko, W. W. Smith, C. A. Ross, V. L. Dawson and T. M. Dawson, *Proc. Natl. Acad. Sci. U. S. A.*, 2005, **102**, 16842.
39. M. C. Herzig, C. Kolly, E. Persohn, D. Theil, T. Schweizer, T. Hafner, C. Stemmelen, T. J. Troxler, P. Schmid, S. Danner, C. R. Schnell, M. Mueller, B. Kinzel, A. Grevot, F. Bolognani, M. Stirn, R. R. Kuhn, K. Kaupmann, P. H. van der Putten, G. Rovelli and D. R. Shimshek, *Hum. Mol. Genet.*, 2011, **20**, 4209.
40. Y. Tong, H. Yamaguchi, E. Giaime, S. Boyle, R. Kopan, R. J. Kelleher and J. Shen, *Proc. Natl. Acad. Sci. U. S. A.*, 2010, **107**, 9879.
41. S. Biskup, D. Moore, A. Rea, B. Lorenz-Deperieux, C. Coombes, V. Dawson, T. Dawson and A. West, *BMC Neurosci.*, 2007, **8**, 102.
42. J. Thevenet, R. Pescini Gobert, R. Hooft van Huijsduijnen, C. Wiessner and Y. J. Sagot, *PLoS One*, 2011, **6**, e21519.
43. M. Kubo, Y. Kamiya, R. Nagashima, T. Maekawa, K. Eshima, S. Azuma, E. Ohta and F. Obata, *J. Neuroimmunol.*, 2010, **229**, 123.
44. T. Maekawa, M. Kubo, I. Yokoyama, E. Ohta and F. Obata, *Biochem. Biophys. Res. Commun.*, 2010, **392**, 431.
45. A. Gardet, Y. Benita, C. Li, B. E. Sands, I. Ballester, C. Stevens, J. R. Korzenik, J. D. Rioux, M. J. Daly, R. J. Xavier and D. K. Podolsky, *J. Immunol.*, 2010, **185**, 5577.
46. N. Dzamko, F. Inesta-Vaquera, J. Zhang, C. Xie, H. Cai, S. Arthur, L. Tan, H. Choi, N. Gray, P. Cohen, P. Pedrioli, K. Clark and D. R. Alessi, *PLoS One*, 2012, **7**, e39132.
47. B. D. Looyenga, K. A. Furge, K. J. Dykema, J. Koeman, P. J. Swiatek, T. J. Giordano, A. B. West, J. H. Resau, B. T. Teh and J. P. MacKeigan, *Proc. Natl. Acad. Sci. U. S. A.*, 2011, **108**, 1439.
48. M. S. Moehle, P. J. Webber, T. Tse, N. Sukar, D. G. Standaert, T. M. DeSilva, R. M. Cowell and A. B. West, *J. Neurosci.*, 2012, **32**, 1602.
49. A. H. Brand and N. Perrimon, *Development*, 1993, **118**, 401.
50. S. B. Lee, W. Kim, S. Lee and J. Chung, *Biochem. Biophys. Res. Commun.*, 2007, **358**, 534.
51. Z. Liu, X. Wang, Y. Yu, X. Li, T. Wang, H. Jiang, Q. Ren, Y. Jiao, A. Sawa, T. Moran, C. A. Ross, C. Montell and W. W. Smith, *Proc. Natl. Acad. Sci. U. S. A.*, 2008, **105**, 2693.
52. A. Sakaguchi-Nakashima, J. Y. Meir, Y. Jin, K. Matsumoto and N. Hisamoto, *Curr. Biol.*, 2007, **17**, 592.

53. C. Yao, R. El Khoury, W. Wang, T. A. Byrd, E. A. Pehek, C. Thacker, X. Zhu, M. A. Smith, A. L. Wilson-Delfosse and S. G. Chen, *Neurobiol. Dis.*, 2010, **40**, 73.
54. D. Sheng, D. Qu, K. H. H. Kwok, S. S. Ng, A. Y. M. Lim, S. S. Aw, C. W. H. Lee, W. K. Sung, E. K. Tan, T. Lufkin, S. Jesuthasan, M. Sinnakaruppan and J. Liu, *PLoS Genet.*, 2010, **6**, e1000914.
55. Y. Li, W. Liu, T. F. Oo, L. Wang, Y. Tang, V. Jackson-Lewis, C. Zhou, K. Geghman, M. Bogdanov, S. Przedborski, M. F. Beal, R. E. Burke and C. Li, *Nat. Neurosci.*, 2009, **12**, 826.
56. X. Lin, L. Parisiadou, X. L. Gu, L. Wang, H. Shim, L. Sun, C. Xie, C. X. Long, W. J. Yang, J. Ding, Z. Z. Chen, P. E. Gallant, J. H. Tao-Cheng, G. Rudow, J. C. Troncoso, Z. Liu, Z. Li and H. Cai, *Neuron*, 2009, **64**, 807.
57. H. L. Melrose, J. C. Dächsel, B. Behrouz, S. J. Lincoln, M. Yue, K. M. Hinkle, C. B. Kent, E. Korvatska, J. P. Taylor, L. Witten, Y. Q. Liang, J. E. Beevers, M. Boules, B. N. Dugger, V. A. Serna, A. Gaukhman, X. Yu, M. Castanedes-Casey, A. T. Braithwaite, S. Ogholikhan, N. Yu, D. Bass, G. Tyndall, G. D. Schellenberg, D. W. Dickson, C. Janus and M. J. Farrer, *Neurobiol. Dis.*, 2010, **40**, 503.
58. X. Li, J. C. Patel, J. Wang, M. V. Avshalumov, C. Nicholson, J. D. Buxbaum, G. A. Elder, M. E. Rice and Z. Yue, *J. Neurosci.*, 2010, **30**, 1788.
59. D. Ramonet, J. P. L. Daher, B. M. Lin, K. Stafa, J. Kim, R. Banerjee, M. Westerlund, O. Pletnikova, L. Glauser, L. Yang, Y. Liu, D. A. Swing, M. F. Beal, J. C. Troncoso, J. M. McCaffery, N. A. Jenkins, N. G. Copeland, D. Galter, B. Thomas, M. K. Lee, T. M. Dawson, V. L. Dawson and D. J. Moore, *PLoS One*, 2011, **6**, e18568.
60. M. C. Herzig, M. Bidinosti, T. Schweizer, T. Hafner, C. Stemmelen, A. Weiss, S. Danner, N. Vidotto, D. Stauffer, C. Barske, F. Mayer, P. Schmid, G. Rovelli, P. H. van der Putten and D. R. Shimshek, *PLoS One*, 2012, **7**, e36581.
61. L. Osherovich, *SciBX: Science–Business eXchange*, 2010, **3**.
62. http://www.tautatis.com/home.html.
63. J. W. Kim, J. Lee, H.-J. Song, Y. Kim, H. K. Lee, J.-S. Choi, S.-H. Lim and S. Chang, *Pat.*, WO 2011/053861 A1, 2011.
64. J. Lee, H.-J. Song, J. S. Koh, L. K. Lee, Y. Kim, S. Chang, H. W. Kim, S. Chang, S.-H. Lim, J.-S. Choi, J.-H. Kim and S.-W. Kim, *Pat.*, WO 2011/060295 A1, 2011.
65. E. G. Mciver, E. Smiljanic, D. J. Harding and J. Hough, *Pat.*, WO 2010/106333 A1, 2010.
66. N. G. Ramsden, *Pat.*, WO2009127642 A2, 2009.
67. A. B. West, D. J. Moore, C. Choi, S. A. Andrabi, X. Li, D. Dikeman, S. Biskup, Z. Zhang, K. L. Lim, V. L. Dawson and T. M. Dawson, *Hum. Mol. Genet.*, 2007, **16**, 223.
68. I. N. Rudenko, A. Kaganovich, D. N. Hauser, A. Beylina, R. Chia, J. Ding, D. Maric, H. Jaffe and M. R. Cookson, *Biochem. J.*, 2012, **446**, 99.

69. Z. Liu, J. Lee, S. Krummey, W. Lu, H. Cai and M. J. Lenardo, *Nat. Immunol.*, 2011, **12**, 1063.
70. B. Brodacki, J. Staszewski, B. Toczyłowska, E. Kozłowska, N. Drela, M. Chalimoniuk and A. Stępien, *Neurosci. Lett.*, 2008, **441**, 158.
71. B. Kim, M. S. Yang, D. Choi, J. H. Kim, H. S. Kim, W. Seol, S. Choi, I. Jou, E. Y. Kim and E. H. Joe, *PLoS One*, 2012, **7**, e34693.

CHAPTER 12

Phosphodiesterase Inhibitors as a New Therapeutic Approach for the Treatment of Parkinson's Disease

ANA MARTINEZ AND CARMEN GIL*

Instituto de Química Médica-CSIC, Madrid, Spain
*Email: cgil@iqm.csic.es

12.1 Introduction

Parkinson's disease (PD) is a progressive neurodegenerative disorder characterized by the inability to initiate, execute and control movements. Common symptoms of PD patients are tremor, rigidity, bradykinesia, akinesia, postural reflex abnormalities, gait disturbance, *etc.*[1] In addition to motor difficulties, Parkinsonian patients experience bouts of mild depression and irritability, as well as cognitive disturbances characterized by memory and attention perturbations.[2] PD is a disease of high incidence, especially in developed countries, increasing age being one of the main risk factors. With the partial exception of juvenile Parkinsonism, development of PD emerges during the fifth and sixth decade, affecting one in 2000 people, and consistently being more prevalent in men than in women. This incidence increases to one in 500 people older than 65 years.[3] Today, there are more than one million PD patients in the USA and double this figure all over the world, representing high social and sanitary costs.[4]

As with all neurodegenerative diseases, PD shares with these devastating pathologies an unknown etiology, the progressive destruction of specific areas of the central nervous system (CNS), and the lack of an effective treatment. The etiology of PD is not well understood, but is likely to involve both genetic and environmental factors.[5] Neuropathologically, PD is characterized by the loss of dopamine-producing neurons in the *substantia nigra pars compacta* of the mid-brain, followed by striatal dopamine depletion and indirectly by cortical dysfunction.[6] These cells project to the dorso-lateral part of the *striatum*, which is the major component of the basal ganglia, a group of subcortical nuclei involved in the control of motor function. According to the neural cell type lost in this pathology, various kinds of symptoms of PD are caused by decreasing dopamine, and it is well documented that dopamine decrease and the development of PD symptoms have a close and direct relationship.[7]

Among the therapeutic agents for the palliative treatment of PD, L-3,4-dihydroxyphenylalanine (L-DOPA) is the drug most frequently used. It ameliorates Parkinsonian-associated motor impediments, however it is relatively inefficient at alleviating the affective and cognitive symptoms of the disease.[8] L-DOPA is the precursor of dopamine, and this therapy aims to supplement this neurotransmitter. As dopamine does not cross the blood–brain barrier (BBB), it will not reach the brain if it is administered into the circulating blood. In contrast, L-DOPA crosses the BBB, and it is metabolized by DOPA-decarboxylase in the brain to give dopamine, which in turn acts on the dopaminergic receptors. However, long-term administration of L-DOPA causes complications such as dyskinesias.[9] Under these circumstances, there has been a need to develop new therapeutic agents for PD or new concomitant agents to avoid the high-dose administration of L-DOPA.[10]

12.2 Dopamine and Cyclic Adenosine Monophosphate

Dopamine plays a central role in the regulation of numerous biological processes, including psychomotor functions. The first evidence for the involvement of a biochemical mechanism in dopamine systems was the observation that dopamine could dose-dependently stimulate the synthesis of the second messenger cyclic adenosine monophosphate (cAMP),[11] and downstream protein phosphorylation by subsequent activation of cAMP-dependent protein kinase (PKA).[12] Thus, the cAMP/PKA signaling cascade is essential for dopamine neurotransmission although today other mechanisms of action able to modulate the signaling function of dopamine receptors are also known.[13]

At presynaptic dopaminergic terminals, the synthesis of dopamine by tyrosine hydroxylase (TH) and its release into the synaptic cleft are regulated by the cAMP/PKA signaling cascade.[14,15]

Once released from presynaptic terminals, dopamine activates members of a family of G-protein-coupled receptors called D_1 to D_5. Dopamine receptor functions have typically been associated with the regulation of cAMP and PKA *via* G-protein-mediated signaling. The D_1 class receptors, D_1 and D_5, are generally coupled to a tissue-specific GTP-binding protein ($G_{s/olf}$) and, through the activation of adenylyl cyclase, stimulate the production of cAMP and consequently, the activity of PKA. In contrast, D_2-class dopamine receptors (D_2, D_3 and D_4) are coupled to $G_{i/o}$ and negatively regulate the production of cAMP, resulting in a decrease of PKA activity.[16] Thus, in postsynaptic dopaminergic *striatum* neurons, dopamine, acting on D_1 receptors, stimulates the production of cAMP *via* activating adenylyl cyclase, which catalyzes the synthesis of the cyclic nucleotide, increasing its levels.[17] At the same time, dopamine, acting on D_2 receptors, inhibits cAMP/PKA signaling *via* G_i-mediated inhibition of the same enzyme, the adenyl cyclase.[18]

The action of dopamine on cAMP production coupled to the D_1 receptor was one of the first neurotransmitter effects to be identified in the brain and has been extensively studied.[19] In fact, cAMP is an important second messenger with a great variety of cellular functions,[20] its major action being binding to the regulatory subunit of PKA, activating this pleiotropic signaling pathway. In addition, cAMP has been strongly associated with different neuronal functions. Thus, cAMP is involved in neuroinflammation,[21] where increased levels of this nucleotide are associated with a decrease in different pro-inflammatory cytokines, generally preserving the neural death rate.[22] cAMP is also involved in neuromodulation and long-lasting changes in gene expression and synaptic function.[23] It can also directly activate cyclic nucleotide-gated cation channels and guanine nucleotide exchange factors (cAMP-GEF or EPACs)[24] potentiating many of their biological activities.[25] Although the importance of these targets has been increasingly recognized, little information is available on their role in the brain and, more precisely, in the *striatum*, which is the CNS area most affected in PD.[26]

As we have summarized above, cAMP levels are crucial for many cell-signaling cascades including the dopamine signal, and these are determined by the balance of synthesis and degradation. Adenyl cyclase is the enzyme that synthesizes this cyclic nucleotide, and its activation through a different mechanism including D_1-mediated receptors, increases cAMP levels. On the other hand, cAMP degradation is performed biologically by the action of phosphodiesterases (PDEs). Therefore, PDEs play an important role in the control of cAMP levels, and their inhibition increases cAMP levels inside the cell (Figure 12.1). In fact, cAMP degradation by PDEs and the precise roles of each PDE isoform in dopaminergic signaling are not yet fully understood. However, as cyclic nucleotide second messengers are involved in many physiological and pathological functions, PDEs have recently emerged as important drug targets for CNS disorders,[27] including PD.[28]

12.3 Phosphodiesterases and their Role in Dopamine Signaling

In mammals, PDEs are a super-family of enzymes that are involved in the regulation of the intracellular second messengers cAMP and cyclic guanosine monophosphate (cGMP) by controlling their rates of hydrolysis. There are 11

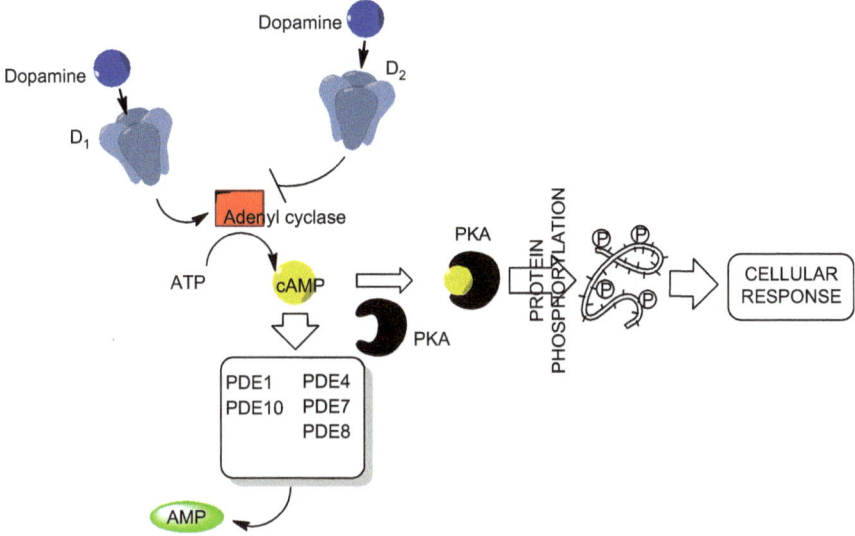

Figure 12.1 Schematic role of cAMP-PDEs in dopamine signaling.

different PDE families and each family typically has multiple isoforms and splice variants expressed in multiple tissues including the brain. Among the 11 PDE families, PDE4, PDE7 and PDE8 are cAMP-specific, whereas PDE5, PDE6 and PDE9 are cGMP-specific; the remaining PDE families hydrolyze both cAMP and cGMP.[29] The expression and subcellular localization of the different PDEs isoforms are tightly regulated.[30]

Within the brain, several PDE families are expressed in the *striatum*. Specifically, PDE10A, PDE1B and PDE7B are enriched in this specific area, while PDE4 (A, B and D isoforms), PDE2A and PDE9A, which are widely distributed in the brain, are also expressed in the *striatum*.[31] These PDEs, all of them able to hydrolyze cAMP, interact with dopamine systems modulating dopamine-mediated behaviors. Moreover, PDE8B, which is expressed mainly in the putamen, is also able to influence dopaminergic neurotransmission in the *striatum*.[32] The inhibition of PDEs induces up-regulation of cAMP levels and subsequent activation of cAMP/PKA signaling in three neuronal subtypes of the *striatum*. The results are: stimulation of dopamine synthesis at dopaminergic terminals, inhibition of dopamine D_2 signaling in pallidostriatal neurons (D_2 antagonist-like, anti-psychotic effect), and stimulation of dopamine D_1 signaling in nigrostriatal neurons (D_1 agonist-like effect).[33] Biochemical and behavioral studies demonstrated that PDE10A inhibition preferentially down-regulates dopamine D_2 signaling, PDE4 inhibition preferentially stimulates dopamine synthesis, and PDE1B inhibition preferentially up-regulates dopamine D_1 signaling, while the inhibition of PDE7 and PDE8 in dopaminergic transmission is not yet fully determined. Although much more information about the specific role of inhibition of *striatum*-expressed cAMP-PDEs in dopamine signaling is needed, there is no doubt about the therapeutic potential of cAMP-PDE inhibitors as new pharmacological treatments for dopamine-related diseases such as PD.[34]

12.4 Phosphodiesterases as Drug Targets Beyond Dopamine

Recent data has indicated that the traditional view of PD as an isolated disorder of the nigrostriatal dopaminergic system alone is an oversimplification of its complex symptomatology. A detailed examination of a substantial number of exemplary preclinical and clinical studies reveals that dopamine deficits may be merely an epiphenomenon of a larger process underlying this disorder, and an increasing interest in new directions for PD research is emerging.[35] Aside from classical motor deficits, various non-motor symptoms including autonomic dysfunction, sensory and cognitive impairments as well as neuropsychiatric alterations and sleep disturbances are common in PD. Some of these non-motor symptoms can even precede the motor problems. Many of them are associated with extranigral neuropathological changes, such as extensive α-synuclein pathology and also neuroinflammatory responses in specific brain regions, which have been implicated in several aspects of PD pathogenesis and progression.[36] Since neuroinflammatory responses are in principle modifiable, such approaches could help to identify new targets or adjunctive therapies for the full spectrum of PD-related symptoms. In this case, it is important to determine the underlying key role of cAMP in neuroinflammation and neuroplasticity.[37] Thus, targeting cAMP-PDEs present in the *striatum* may be a good strategy to decrease microglial and/or astroglial inflammation reaction,[38] and to delay or stop the progression of cell death.[39]

Cognitive disorders, some of them associated with an impaired adult neurogenesis and brain plasticity, are also frequent in PD patients sometimes mediated by α-synuclein accumulation in certain brain regions.[40] The transcription factor cAMP response element binding (CREB) protein plays a crucial role in regulating brain development and neurogenesis in the adult brain, being regulated by PDEs.[41] Thus, modulation of cAMP-PDEs has been recently shown to be a good therapeutic strategy for mood disorders,[42] while their inhibition by small molecules produced cognition and memory enhancement,[43] showing an important role for PDEs in PD patients beyond dopamine signaling.[44]

12.5 Phosphodiesterase Inhibitors as New Drugs for Parkinson's Disease

Cyclic nucleotide signaling is highly compartmentalized within individual cells. The complexity of compartmentalization of PDE signaling is particularly evident in the CNS.[45] For this reason, PDEs as key regulators of cAMP and cGMP bear investigation as means of modulating synaptic plasticity, and their inhibitors could be considered as promising candidates for cognition-enhancing agents.[46] Moreover, as PDE1B, PDE7B, PDE8B and PDE10A are highly expressed in the *striatum*,[27,47] the inhibitors of these enzymes could be considered as promising candidates for the treatment of PD. They can influence dopamine signaling or non-motor symptoms of PD, including the slowdown of dopaminergic neuron death or the differentiation into dopamine neurons of endogenous neural stem cells.[48] As a consequence, the design of selective PDE inhibitors as new drugs for neurodegenerative diseases is currently an area of growing interest.[30]

Regarding PD pharmacological approaches, it is noteworthy that caffeine, a well-known methyl xanthine, has shown neuroprotective effects in animal models of PD.[49,50] The potential mechanism whereby caffeine exerts neuroprotection is not clear yet, and although it could protect neurons by blocking the adenosine receptor A_{2A}, caffeine may be expected to have other effects in the CNS.[51,52] From our point of view, it must be remembered that methyl xanthines in general, and caffeine in particular, also act as non-selective PDE inhibitors as they are able to enhance cAMP levels.[53] In the following sections, inhibitors of cAMP PDEs expressed in brain areas involved in PD will be discussed (Figure 12.2).

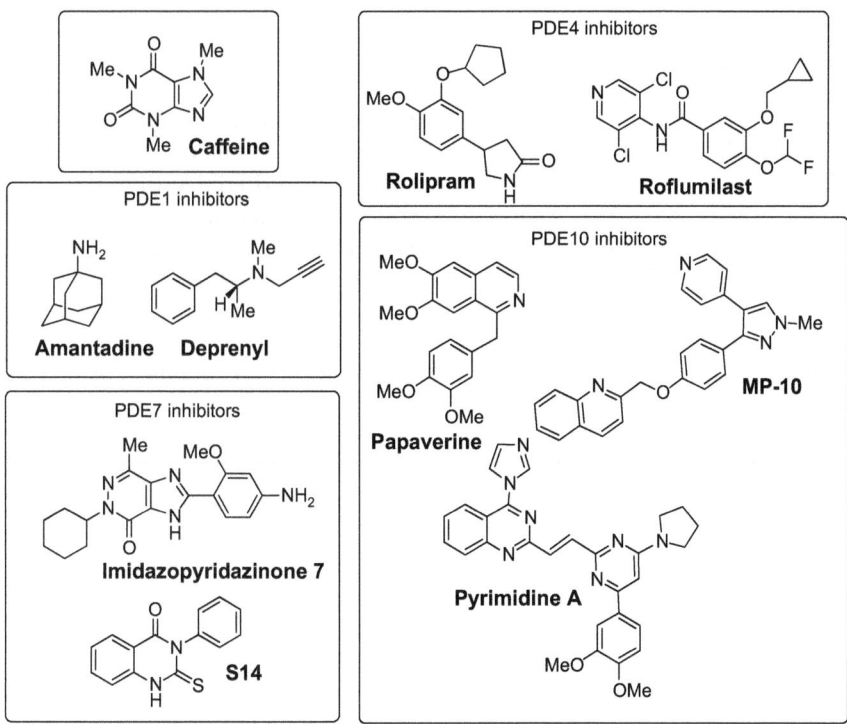

Figure 12.2 Chemical structures of some PDE inhibitors.

12.5.1 PDE1 Inhibitors

The PDE1 family is activated by Ca^{2+}/calmodulin and is capable of acting on both cAMP and cGMP. PDE1 is one of the key enzymes involved in the complex interactions between the cyclic nucleotides and the Ca^{2+} second messenger system.[54,55] This family of enzymes is encoded by three distinct genes, PDE1A, PDE1B, and PDE1C. The high levels of PDE1B and PDE1C mRNAs found in the caudate, the *nucleus accumbens* and the *substantia nigra*, respectively, point to an important role of these PDEs in brain signal transduction. With respect to peripheral tissues, PDE1A is the most prevalent PDE isoenzyme in the kidney, while PDE1C is predominant in the heart.[56] Their unique tissue localization and the fact that in PD patients there is an alteration in cAMP/cGMP levels, suggest a potential role of PDE1 in this neurodegenerative pathology.[57] In fact, the pharmacological use of a PDE inhibitor in the treatment of patients with PD is not completely new. The two classic anti-Parkinson's drugs deprenyl[58] and amantadine,[59] in addition to their activities as MAO B inhibitor and NMDA antagonist respectively, are able to inhibit calmodulin-dependent cyclic nucleotide PDEs. In this way, the protective role exerted by these drugs could also involve a modulation of cAMP levels.

Altogether, PDE1 inhibitors could be considered as compounds able to protect dopaminergic neurons under stress, because of their ability to increase cAMP levels.

12.5.2 PDE4 Inhibitors

The PDE4 family has attracted considerable attention because PDE4-selective inhibitors have potential therapeutic use in a range of major disease areas. PDE4 enzymes are encoded by four distinct genes (PDE4A–PDE4D) that specifically hydrolyze cAMP. Their isoforms show distinct and specific cell-type patterns of expression and distribution, and they play major regulatory roles in different tissues.[60] The involvement of PDE4 in pathological processes associated with these tissues suggests great potential for pharmacological intervention in a variety of disorders, such as inflammatory and neurological disorders, through modulation of cAMP levels.[61,62] A number of PDE4 inhibitors have reached clinical trials for the treatment of inflammatory diseases.[63] Despite the failure of most of them because of side-effects such as emesis and nausea, roflumilast was approved in 2011 for the treatment of chronic obstructive pulmonary disease due to its improved therapeutic window.[64]

Regarding PD, it is interesting to note that the well-known PDE4 inhibitor rolipram, reached clinical trials for PD treatment. The design was a double-blind trial *versus* placebo in 10 treated patients with PD, and no significant deterioration of the therapeutic dopamine agonist lisuride was noted.[65] Moreover, it was shown recently that the same PDE4 inhibitor reduced the toxicity in an MPTP mouse model of PD and increased hippocampal neurogenesis. The survival of dopaminergic neurons[66] and the memory enhancement produced after the administration of rolipram[67] were explained in both experiments by the increase of intracellular cAMP levels.

12.5.3 PDE7 Inhibitors

The PDE7 family is composed of two genes, PDE7A and PDE7B. High mRNA concentrations of both PDE7A and PDE7B are found in rat brains and in numerous peripheral tissues, although the distribution of these enzymes at the protein level has not been reported. Within the brain, PDE7A mRNA is abundant in the olfactory bulb, hippocampus, and several brain-stem nuclei.[68] The highest concentrations of PDE7B transcripts in the brain are found in the cerebellum and dentate gyrus of the hippocampus and *striatum*.[69,70] There is very little information regarding the physiological functions regulated by PDE7 although its transcriptional activation through the D_1 is well documented.[71] It has been shown that PDE7 is involved in pro-inflammatory processes and is necessary for the induction of T-cell proliferation.[72] In addition, the development of inhibitors of PDE7 has been reported as a new approach to be explored for the treatment of neurological disorders due to it ability to increase levels of cAMP,[73] and specifically PDE7 inhibition hinders dopaminergic cell death and glial activation in an animal model of PD. A heterocyclic small molecule inhibitor of phosphodiesterase 7, the quinazoline called S14, conferred significant neuronal protection against different insults both in the human dopaminergic cell line SH-SY5Y and in primary rat mesencephalic cultures. S14 treatment also reduced microglial activation, protected dopaminergic neurons and improved motor function in the lipopolysaccharide rat model of PD. The neuroprotective effects produced by S14 are abolished when a PKA inhibitor is used. This fact proves the relationship between cAMP signaling pathway and its mechanism of action. These results show for the first time that inhibition of the PDE7 enzyme leads to dopaminergic neuronal protection and, therefore, its inhibitors may exert useful therapeutic actions in patients with PD.[74] More recently, the adequate CNS penetration of a new PDE7 inhibitor, belonging to the imidazopyridazinone family, in mice, allowed it to be tested in an MPTP-induced PD model and haloperidol-induced catalepsy model probing the differential pharmacology of PDE7 in the striatal pathway.[75]

Finally, it is noteworthy that cAMP is involved in the differentiation of different cell lines or stem cells into dopamine neurons,[76] which is considered a promising strategy for cell replacement therapy in PD. Recently, we have shown that the quinazoline PDE7 inhibitor S14 is able to promote the differentiation of stem cells into tyrosine-hydroxylase-positive mature neurons isolated from the adult rat hippocampus and from the subventricular zone.[77] Furthermore, S14 promotes new neurogenesis in the *substantia nigra* after the administration of 6-OHDA.

12.5.4 PDE10 Inhibitors

The high level of expression and unique subcellular distribution of PDE10A in the striatal medium spiny neurons suggest that this enzyme has a particular prominent role in the function of these neurons.[78,79] The striatal medium spiny neurons are the principal input site and the first site for information integration in the basal ganglia circuit of the mammalian brain.[80] Significant insight into basal ganglia function has been garnered from human diseases in which dysfunction in this circuit is implicated. These include neuropsychiatric disorders such as schizophrenia and also neurodegenerative disorders such as PD and Huntington's disease (HD).[81–83] The clinical manifestation of PD is caused by the degeneration of the dopaminergic input to the striatal medium spiny neurons. These clinical symptoms are ameliorated early in the disease by the pharmacological restoration of dopamine receptor stimulation of these neurons.[84] Loss of dopamine neurons in experimental Parkinsonism results in altered cyclic nucleotide cAMP and cGMP levels throughout the basal ganglia and inversely regulated PDE10A gene expression in the *striatum*.[85] In HD, the medium spiny neurons degenerate and there are a number of animal behavioral models that have been used to probe the basal ganglia dysfunction that underlies these human diseases, and to predict the therapeutic utility of new approaches to treatment.[86] Manipulation of PDE10A activity was accomplished using the PDE10A inhibitor papaverine and a mouse line in which the gene for PDE10A was disrupted.[87] A prominent phenotype of the PDE10A knock-out mouse is a reduction of motor activity. The effects of pharmacological inhibition of PDE10A are qualitatively similar to those produced by genetic deletion of the enzyme.[88]

Papaverine was the first PDE10A inhibitor reported in the literature,[89] and although there are significant limitations related to its potency and selectivity, its use as a pharmacological tool has been critical to decipher the therapeutic implications of this novel target. Together with papaverine, the related PDE10A inhibitor, named MP-10,[81] which is more potent and selective, has been successfully used to support the hypothesis that PDE10A inhibitors will provide an innovative approach to the treatment of schizophrenia[90,91] and also HD.[82,86,88] More recently, the efficacy of concomitant treatment of L-DOPA with a new heterocyclic PDE10 inhibitor on different models of PD has been reported.[92] An enhancement of dopamine signals in the brain was observed in animals lesioned with 6-OHDA or MPTP, after treatment with the pyrimidine A PDE10 inhibitor together with inducing dopamine signals by L-DOPA or dopamine receptor agonist administration. Simultaneous treatment of PDE10 inhibitors with currently available pharmacological treatments may decrease levels of L-DOPA in PD patients, with the consequent delay in dyskinesias side-effects.

Until now, PDE10A inhibitors have only been considered as promising drugs for the treatment of schizophrenia, but due to the drug profile of the target and its pattern of localization, it is also feasible that this class of compounds could be used in the near future as drugs for other neurological disorders such as PD or HD.

12.6 Conclusions

PD is a devastating neurodegenerative disorder for which the lack of an efficient pharmacological treatment remains in the present century. Primary therapeutic efforts have led to the extensively used dopamine-replacement approach, a palliative therapy that temporarily alleviates clinical motor symptoms but has severe long-term side-effects. Dopamine plays a central role in the regulation of psychomotor functions, and its effect is largely mediated through the cAMP/PKA signaling cascade and, therefore, controlled by PDEs. One of the results from the intensive research into dopamine neurotransmission, in the search for an effective therapy, is the relevance of cAMP-PDEs as new drug targets for PD. Multiple PDEs with different substrate specification and subcellular localization are expressed in the *striatum*. The inhibition of these PDEs up-regulates cAMP/PKA signaling in three neuronal subtypes, resulting in the stimulation of dopamine synthesis at dopaminergic terminals, the inhibition of D_2 signaling in pallidostriatal neurons, and the stimulation of D_1 signaling in nigrostriatal neurons. Furthermore, beyond dopamine, PDEs are involved in other important neuronal processes such neuroinflammation and neuronal plasticity, having a key role in the prevention of dopaminergic neuronal death and in stem cell differentiation. Specific PDE inhibitors with drug-like and safe profiles have emerged as a new avenue in the search for a PD therapy. PDE4 and PDE1B inhibitors have reached the clinic with some positive effects but with a narrow therapeutic window. Currently, some PDE10A inhibitors are in clinical trials for mood disorders, and it will be important to assay them in PD patients. Moreover, PDE7 inhibitors have just shown relevant results, not only in locomotor activity, but also in the neuroprotection and neurogenesis of dopaminergic neurons in animal models. However, clinical trials are still a long way off. Finally, PDE8 has recently been discovered in the brain, and its inhibitors remain to be discovered and tested in PD models.

A vast new horizon in drug research for PD exists and PDEs, specifically cAMP-PDE inhibitors, are emerging as new candidates to be evaluated in human clinical trials as disease-modifying drugs.

Acknowledgements

Financial support from MICINN and MINECO (SAF2009-13015-CO2-01 and SAF2012-33600) is acknowledged.

References

1. S. A. Schneider and J. A. Obeso, *Lancet Neurol.*, 2013, **12**, 10.
2. I. Ferrer, I. Lopez-Gonzalez, M. Carmona, E. Dalfo, A. Pujol and A. Martinez, *Neurobiol. Dis.*, 2012, **46**, 508.
3. C. R. Baumann, *Parkinsonism Relat. Disord.*, 2012, **18**(1), S90.

4. P. Jennum, M. Zoetmulder, L. Korbo and J. Kjellberg, *J. Neurol.*, 2011, **258**, 1497.
5. K. Wirdefeldt, H. O. Adami, P. Cole, D. Trichopoulos and J. Mandel, *Eur. J. Epidemiol.*, 2011, **26**(1), S1.
6. J. Lotharius and P. Brundin, *Nat. Rev. Neurosci.*, 2002, **3**, 932.
7. M. M. Lima, E. F. Martins, A. M. Delattre, M. B. Proenca, M. A. Mori, B. Carabelli and A. C. Ferraz, *CNS Neurol. Disord. Drug Targets*, 2012, **11**, 439.
8. W. J. Weiner, *Mov. Disord.*, 1999, **14**, 716.
9. M. Vidailhet, A. M. Bonnet, R. Marconi, F. Durif and Y. Agid, *Mov. Disord.*, 1999, **14**(1), 13.
10. W. Poewe, P. Mahlknecht and J. Jankovic, *Curr. Opin. Neurol.*, 2012, **25**, 448.
11. J. W. Kebabian, G. L. Petzold and P. Greengard, *Proc. Natl. Acad. Sci. U. S. A.*, 1972, **69**, 2145.
12. B. K. Krueger, J. Forn and P. Greengard, *Psychopharmacol. Bull.*, 1975, **11**, 10.
13. K. N. Boyd and R. B. Mailman, in *Handbook of Experimental Pharmacology*, ed. F. B. Hoffman, Springer, New York, 2012, vol. 213, p. 53.
14. P. R. Dunkley, L. Bobrovskaya, M. E. Graham, E. I. von Nagy-Felsobuki and P. W. Dickson, *J. Neurochem.*, 2004, **91**, 1025.
15. G. Zhu, M. Okada, S. Yoshida, S. Hirose and S. Kaneko, *Neurosci. Lett.*, 2004, **363**, 120.
16. J. M. Beaulieu and R. R. Gainetdinov, *Pharmacol. Rev.*, 2011, **63**, 182.
17. D. Herve, C. Le Moine, J. C. Corvol, L. Belluscio, C. Ledent, A. A. Fienberg, M. Jaber, J. M. Studler and J. A. Girault, *J. Neurosci.*, 2001, **21**, 4390.
18. J. C. Stoof and J. W. Kebabian, *Nature*, 1981, **294**, 366.
19. J. A. Girault, *Prog. Mol. Biol. Transl. Sci.*, 2012, **106**, 33.
20. J. M. Gancedo, *Biol. Rev. Camb. Philos. Soc.*, 2013, doi: 10.1111/brv.12020.
21. M. Peters-Golden, *Sci. Signal.*, 2009, **2**, pe37.
22. S. H. Christiansen, J. Selige, T. Dunkern, A. Rassov and M. Leist, *Neurochem. Int.*, 2011, **59**, 837.
23. D. A. Hoffman and D. Johnston, *J. Neurophysiol.*, 1999, **81**, 408.
24. M. Gloerich and J. L. Bos, *Annu. Rev. Pharmacol. Toxicol.*, 2010, **50**, 355.
25. J. N. Gelinas, J. L. Banko, M. M. Peters, E. Klann, E. J. Weeber and P. V. Nguyen, *Learn. Mem.*, 2008, **15**, 403.
26. S. Hara, M. Kobayashi, F. Kuriiwa, T. Mukai and H. Mizukami, *Free Radical Biol. Med.*, 2012, **52**, 1086.
27. F. S. Menniti, W. S. Faraci and C. J. Schmidt, *Nat. Rev. Drug Discov.*, 2006, **5**, 660.
28. E. Bollen and J. Prickaerts, *IUBMB Life*, 2012, **64**, 965.
29. M. Conti and S. L. Jin, *Prog. Nucleic Acid Res. Mol. Biol.*, 1999, **63**, 1.
30. S. H. Francis, M. A. Blount and J. D. Corbin, *Physiol. Rev.*, 2011, **91**, 651.

31. T. Kleppisch, in *Handbook of Experimental Pharmacology*, ed. H. H. W. Schmidt, F. B. Hofmann and J.-P. Stasch, Springer, New York, 2009, vol. 191, p. 71.
32. S. Appenzeller, A. Schirmacher, H. Halfter, S. Baumer, M. Pendziwiat, V. Timmerman, P. De Jonghe, K. Fekete, F. Stogbauer, P. Ludemann, M. Hund, E. S. Quabius, E. B. Ringelstein and G. Kuhlenbaumer, *Am. J. Hum. Genet.*, 2010, **86**, 83.
33. J. A. Girault, *Adv. Exp. Med. Biol.*, 2012, **970**, 407.
34. A. Nishi and G. L. Snyder, *J. Pharmacol. Sci.*, 2010, **114**, 6.
35. G. L. Willis, C. Moore and S. M. Armstrong, *Rev. Neurosci.*, 2012, **23**, 403.
36. K. J. Doorn, P. J. Lucassen, H. W. Boddeke, M. Prins, H. W. Berendse, B. Drukarch and A. M. van Dam, *Prog. Neurobiol.*, 2012, **98**, 222.
37. T. M. Sanderson and E. Sher, *Neuropharmacology*, 2013, doi: 10.1016/j.neuropharm.2013.1001.1011.
38. A. Suzumura, A. Ito, M. Yoshikawa and M. Sawada, *Brain Res.*, 1999, **837**, 203.
39. A. L. Hebb and H. A. Robertson, *Curr. Opin. Invest. Drugs*, 2008, **9**, 744.
40. P. Desplats, B. Spencer, L. Crews, P. Pathel, D. Morvinski-Friedmann, K. Kosberg, S. Roberts, C. Patrick, B. Winner, J. Winkler and E. Masliah, *J. Biol. Chem.*, 2012, **287**, 31691.
41. S. Dworkin and T. Mantamadiotis, *Expert Opin. Ther. Targets*, 2010, **14**, 869.
42. Y. Xu, H. T. Zhang and J. M. O'Donnell, in *Handbook of Experimental Pharmacology*, ed. S. H. Francis, M. Conti and M. D. Houslay, Springer, New York, 2011, vol. 204, p. 447.
43. O. A. Reneerkens, K. Rutten, H. W. Steinbusch, A. Blokland and J. Prickaerts, *Psychopharmacology*, 2009, **202**, 419.
44. A. Blokland, F. S. Menniti and J. Prickaerts, *Expert Opin. Ther. Pat.*, 2012, **22**, 349.
45. M. P. Kelly and N. J. Brandon, *Prog. Brain Res.*, 2009, **179**, 67.
46. C. J. Schmidt, *Curr. Top Med. Chem.*, 2010, **10**, 222.
47. E. M. Johansson, E. Reyes-Irisarri and G. Mengod, *Neurosci. Lett.*, 2012, **525**, 1.
48. S. Malmersjo, I. Liste, O. Dyachok, A. Tengholm, E. Arenas and P. Uhlen, *Stem Cells Dev.*, 2010, **19**, 1355.
49. A. Kalda, L. Yu, E. Oztas and J. F. Chen, *J. Neurol. Sci.*, 2006, **248**, 9.
50. X. Chen, O. Ghribi and J. D. Geiger, *J. Alzheimer's Dis.*, 2010, **20**(1), S127.
51. R. A. Popat, S. K. Van Den Eeden, C. M. Tanner, F. Kamel, D. M. Umbach, K. Marder, R. Mayeux, B. Ritz, G. W. Ross, H. Petrovitch, B. Topol, V. McGuire, S. Costello, A. D. Manthripragada, A. Southwick, R. M. Myers and L. M. Nelson, *Eur. J. Neurol.*, 2011, **18**, 756.
52. K. Nakaso, S. Ito and K. Nakashima, *Neurosci. Lett.*, 2008, **432**, 146.
53. J. W. Daly, *Cell Mol. Life Sci.*, 2007, **64**, 2153.
54. R. Kakkar, R. V. Raju and R. K. Sharma, *Cell Mol. Life Sci.*, 1999, **55**, 1164.
55. T. A. Goraya and D. M. Cooper, *Cell Signal.*, 2005, **17**, 789.

56. V. Lakics, E. H. Karran and F. G. Boess, *Neuropharmacology*, 2010, **59**, 367.
57. M. Giorgi, V. D'Angelo, Z. Esposito, V. Nuccetelli, R. Sorge, A. Martorana, A. Stefani, G. Bernardi and G. Sancesario, *Eur. J. Neurosci.*, 2008, **28**, 941.
58. R. Kakkar, R. V. Raju, A. H. Rajput and R. K. Sharma, *Life Sci.*, 1996, **59**, PL337.
59. R. Kakkar, R. V. Raju, A. H. Rajput and R. K. Sharma, *Brain Res.*, 1997, **749**, 290.
60. M. Conti, W. Richter, C. Mehats, G. Livera, J. Y. Park and C. Jin, *J. Biol. Chem.*, 2003, **278**, 5493.
61. C. P. Page and D. Spina, in *Handbook of Experimental Pharmacology*, ed. S. H. Francis, M. Conti and M. D. Houslay, Springer, New York, 2011, vol. 204, p. 391.
62. A. Garcia-Osta, M. Cuadrado-Tejedor, C. Garcia-Barroso, J. Oyarzabal and R. Franco, *ACS Chem. Neurosci.*, 2012, **3**, 832.
63. M. D. Houslay, P. Schafer and K. Y. Zhang, *Drug Discov. Today*, 2005, **10**, 1503.
64. M. A. Giembycz and S. K. Field, *Drug Des. Devel. Ther.*, 2010, **4**, 147.
65. M. Casacchia, G. Meco, F. Castellana, L. Bedini, G. Cusimano and A. Agnoli, *Pharmacol. Res. Commun.*, 1983, **15**, 329.
66. P. Hulley, J. Hartikka, S. Abdel'Al, P. Engels, H. R. Buerki, K. H. Wiederhold, T. Muller, P. Kelly, D. Lowe and H. Lubbert, *Eur. J. Neurosci.*, 1995, **7**, 2431.
67. Y. F. Li, Y. F. Cheng, Y. Huang, M. Conti, S. P. Wilson, J. M. O'Donnell and H. T. Zhang, *J. Neurosci.*, 2011, **31**, 172.
68. X. Miro, S. Perez-Torres, J. M. Palacios, P. Puigdomenech and G. Mengod, *Synapse*, 2001, **40**, 201.
69. T. Sasaki, J. Kotera and K. Omori, *Biochem. J.*, 2002, **361**, 211.
70. E. Reyes-Irisarri, S. Perez-Torres and G. Mengod, *Neuroscience*, 2005, **132**, 1173.
71. T. Sasaki, J. Kotera and K. Omori, *J. Neurochem.*, 2004, **89**, 474.
72. A. Nakata, K. Ogawa, T. Sasaki, N. Koyama, K. Wada, J. Kotera, H. Kikkawa, K. Omori and O. Kaminuma, *Clin. Exp. Immunol.*, 2002, **128**, 460.
73. C. Gil, N. E. Campillo, D. I. Perez and A. Martinez, *Expert Opin. Ther. Pat.*, 2008, **18**, 1127.
74. J. A. Morales-Garcia, M. Redondo, S. Alonso-Gil, C. Gil, C. Perez, A. Martinez, A. Santos and A. Perez-Castillo, *PLoS One*, 2011, **6**, e17240.
75. A. Banerjee, S. Patil, M. Y. Pawar, S. Gullapalli, P. K. Gupta, M. N. Gandhi, D. K. Bhateja, M. Bajpai, R. R. Sangana, G. S. Gudi, N. Khairatkar-Joshi and L. A. Gharat, *Bioorg. Med. Chem. Lett.*, 2012, **22**, 6286.
76. R. G. Tremblay, M. Sikorska, J. K. Sandhu, P. Lanthier, M. Ribecco-Lutkiewicz and M. Bani-Yaghoub, *J. Neurosci. Methods*, 2010, **186**, 60.

77. J. A. Morales-García, A. Pérez-Castillo, M. Redondo, S. Alonso-Gil, C. Gil, A. Santos and A. Martínez, *Neurodegener. Dis.*, 2013, **11**(Supp.1).
78. T. F. Seeger, B. Bartlett, T. M. Coskran, J. S. Culp, L. C. James, D. L. Krull, J. Lanfear, A. M. Ryan, C. J. Schmidt, C. A. Strick, A. H. Varghese, R. D. Williams, P. G. Wylie and F. S. Menniti, *Brain Res.*, 2003, **985**, 113.
79. S. Threlfell, S. Sammut, F. S. Menniti, C. J. Schmidt and A. R. West, *J. Pharmacol. Exp. Ther.*, 2009, **328**, 785.
80. T. M. Coskran, D. Morton, F. S. Menniti, W. O. Adamowicz, R. J. Kleiman, A. M. Ryan, C. A. Strick, C. J. Schmidt and D. T. Stephenson, *J. Histochem. Cytochem.*, 2006, **54**, 1205.
81. C. J. Schmidt, D. S. Chapin, J. Cianfrogna, M. L. Corman, M. Hajos, J. F. Harms, W. E. Hoffman, L. A. Lebel, S. A. McCarthy, F. R. Nelson, C. Proulx-LaFrance, M. J. Majchrzak, A. D. Ramirez, K. Schmidt, P. A. Seymour, J. A. Siuciak, F. D. Tingley, 3rd, R. D. Williams, P. R. Verhoest and F. S. Menniti, *J. Pharmacol. Exp. Ther.*, 2008, **325**, 681.
82. C. Giampa, S. Patassini, A. Borreca, D. Laurenti, F. Marullo, G. Bernardi, F. S. Menniti and F. R. Fusco, *Neurobiol. Dis.*, 2009, **34**, 450.
83. T. Chappie, J. Humphrey, F. Menniti and C. Schmidt, *Curr. Opin. Drug Discov. Devel.*, 2009, **12**, 458.
84. T. Nagatsua and M. Sawadab, *Parkinsonism Relat. Disord.*, 2009, **15**(1), S3.
85. M. Giorgi, G. Melchiorri, V. Nuccetelli, V. D'Angelo, A. Martorana, R. Sorge, V. Castelli, G. Bernardi and G. Sancesario, *Neurobiol. Dis.*, 2011, **43**, 293.
86. C. Giampa, D. Laurenti, S. Anzilotti, G. Bernardi, F. S. Menniti and F. R. Fusco, *PLoS One*, 2010, **5**, e13417.
87. J. A. Siuciak, S. A. McCarthy, D. S. Chapin, R. A. Fujiwara, L. C. James, R. D. Williams, J. L. Stock, J. D. McNeish, C. A. Strick, F. S. Menniti and C. J. Schmidt, *Neuropharmacology*, 2006, **51**, 374.
88. R. J. Kleiman, L. H. Kimmel, S. E. Bove, T. A. Lanz, J. F. Harms, A. Romegialli, K. S. Miller, A. Willis, S. des Etages, M. Kuhn and C. J. Schmidt, *J. Pharmacol. Exp. Ther.*, 2011, **336**, 64.
89. J. Siuciak, D. S. Chapin, S. A. McCarthy, J. Harms, C. J. Schmidt, A. Shrikhande, S. Wong, R. Williams, F. Menniti, L. James, C. Strick, J. Stock and J. McNeish, *Schizophenia Res.*, 2003, **60**, 116.
90. S. M. Grauer, V. L. Pulito, R. L. Navarra, M. P. Kelly, C. Kelley, R. Graf, B. Langen, S. Logue, J. Brennan, L. Jiang, E. Charych, U. Egerland, F. Liu, K. L. Marquis, M. Malamas, T. Hage, T. A. Comery and N. J. Brandon, *J. Pharmacol. Exp. Ther.*, 2009, **331**, 574.
91. B. Langen, R. Dost, U. Egerland, H. Stange and N. Hoefgen, *Psychopharmacology*, 2012, **221**, 249.
92. J. Kotera, T. Sasaki, T. Kitazawa, T. Ishii, H. Morimoto and H. Yamada, *US Pat.*, 8338420, 2009.

CHAPTER 13

5-HT$_{1A}$ Receptors as a Therapeutic Target for Parkinson's Disease

SAKI SHIMIZU AND YUKIHIRO OHNO*

Laboratory of Pharmacology, Osaka University of Pharmaceutical Sciences, Osaka, Japan
*Email: yohno@gly.oups.ac.jp

13.1 Introduction

Parkinson's disease (PD) is the most common neurological disorder in the elderly with symptoms characterized by tremor, rigidity and bradykinesia.[1] The general age of onset is from the late 50s to mid 60s and the prevalence is estimated at about 1% among those older than 60 years. Patients with PD typically show progressive extrapyramidal motor disorders including rest tremor, muscle rigidity, hypolocomotion (bradykinesia and akinesia) and postural instability. In addition, various non-motor features are also seen, such as psychiatric changes (depression and anxiety), cognitive impairments (deficits in learning and memory), sleep disturbances and autonomic dysfunctions (orthostatic hypotension and constipation).[1-3]

Pathological findings have revealed that PD is a neurodegenerative disorder, involving the selective loss of nigrostriatal dopaminergic neurons (Figure 13.1). Thus, patients with the disease are responsive to various dopaminergic treatments.[1,4] The dopamine precursor L-3,4-dihydroxyphenylalanine (L-DOPA) is the most effective treatment as it supplements brain dopamine levels and

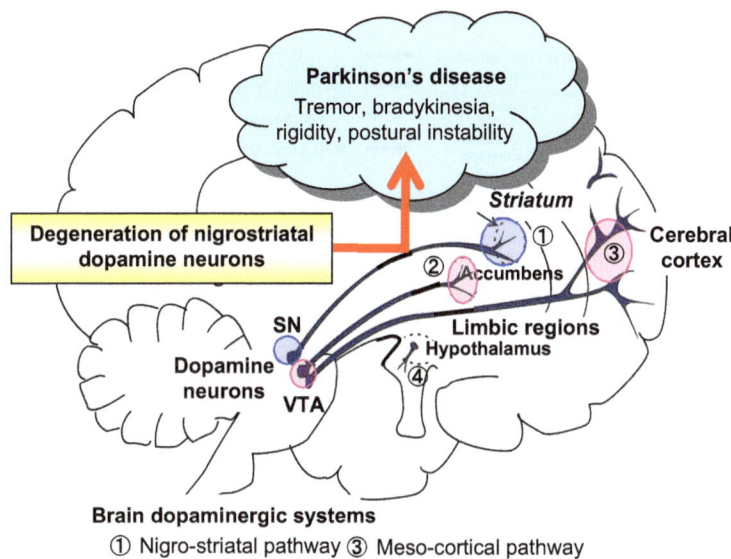

Figure 13.1 Brain dopaminergic systems and PD. Brain dopaminergic pathways consists of four main pathways: (1) nigrostriatal; (2) meso-limbic; (3) meso-cortical; and (4) hypothalamo-pituitary pathways. The nigro-striatal dopaminergic pathway controls extrapyramidal motor functions while the meso-limbic and meso-cortical pathways regulate psychoemotional functions. PD is induced by degeneration of nigrostriatal dopaminergic neurons. SN = Substantia nigra (Pars compacta), VTA = ventral tegmental area.

improves most Parkinsonian symptoms (Figure 13.2).[1] Other dopaminergic drugs include dopamine receptor D_2 (or D_2/D_3) agonists (bromocriptine, pergolide, cabergoline, ropinirole and talipexole), the monoamine oxidase B inhibitor selegiline, the catechol-*O*-methyl-transferase (COMT) inhibitor entacapone and the dopamine releaser/*N*-methyl-D-aspartate (NMDA) antagonist amantadine (Figure 13.2). In addition, antagonists for the muscarinic acetylcholine (mACh) receptor (trihexyphenidyl and biperidene) are also effective against Parkinsonian symptoms, however they frequently cause serious adverse reactions (*e.g.*, intestinal constipation, urinary retention, tachycardia and cognitive impairments).

Although dopaminergic medications are generally effective, there are residual clinical unmet needs, including a lack of drugs suitable for treating neurodegeneration, motor complications (dyskinesia) or psychosis associated with chronic L-DOPA treatments and non-motor symptoms (*e.g.*, mood disturbances and cognitive impairments) in PD (Figure 13.2).[4,5] Specifically, none of the agents currently available can prevent disease development (*i.e.*, neurodegeneration) or restore damaged dopaminergic neurons in PD. In addition, long-term treatments with L-DOPA cause a fluctuation in its efficacy (*e.g.*, wearing-off and on–off phenomena) and serious side-effects (*e.g.*, L-DOPA-induced dyskinesia and psychosis), which often hamper

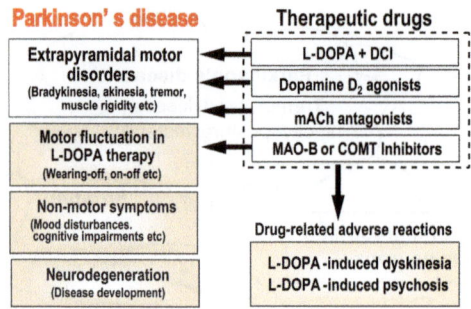

Figure 13.2 Clinical unmet needs in the treatment of PD. Although the dopaminergic agents are generally effective for various symptoms of PD, there are residual clinical unmet needs (shown by shaded boxes), including a lack of effective medications for neurodegeneration, non-motor symptoms (*e.g.*, mood disturbances and cognitive impairments) of PD, a fluctuation in the L-DOPA efficacy (*e.g.*, wearing-off and "on–off" phenomena) and motor complications (*e.g.*, dyskinesia) or psychosis associated with chronic L-DOPA treatments. DCI = aromatic L-amino acid decarboxylase inhibitor, mACh = muscarinic acetylcholine, MAO-B = monoamine oxidase B, COMT = catechol-*O*-methyl-transferase.

continued L-DOPA therapy (Figure 13.2).[1,4,6] Furthermore, new medications and/or substitutes for L-DOPA with pronounced anti-Parkinson actions including efficacy for non-motorsymptoms will be required.

The serotonergic system plays an important role in the control of various physiological processes including psychoemotional, sensorimotor, cognitive and autonomic functions.[7–10] Serotonergic (5-HT) neurons are located in the raphe nuclei and project axons to various brain regions including the cerebral cortex, limbic areas, basal ganglia, diencephalon and the spinal cord. The serotonergic neurotransmission is mediated by multiple 5-HT receptors that are classified into seven families (5-HT$_1$ to 5-HT$_7$) encompassing at least 14 subtypes (*i.e.*, 5-HT$_{1A}$, 5-HT$_{1B}$, 5-HT$_{1D}$, 5-HT$_{1E}$, 5-HT$_{1F}$, 5-HT$_{2A}$, 5-HT$_{2B}$, 5-HT$_{2C}$, 5-HT$_3$, 5-HT$_4$, 5-HT$_{5a}$, 5-HT$_{5b}$, 5-HT$_6$ and 5-HT$_7$ receptors).[8–10] Among these subtypes, 5-HT$_{1A}$ receptors have long been implicated in the pathogenesis and treatment of anxiety and depressive disorders.[11–15] In addition, recent research has revealed new therapeutic roles for 5-HT$_{1A}$ receptors in the alleviation of Parkinsonian motor symptoms,[4,13,15–23] L-DOPA-induced motor disabilities,[6,24–26] and cognitive impairment.[10,27–31] This chapter reviews the functions and pharmacological features of 5-HT$_{1A}$ receptors as a therapeutic target for PD.

13.2 5-HT$_{1A}$ Receptors

5-HT$_{1A}$ receptors are G-protein-coupled receptors with a seven-transmembrane-spanning structure (Figure 13.3).[9,32–34] 5-HT$_{1A}$ receptors are highly expressed in three regions of the brain: the limbic areas (hippocampus

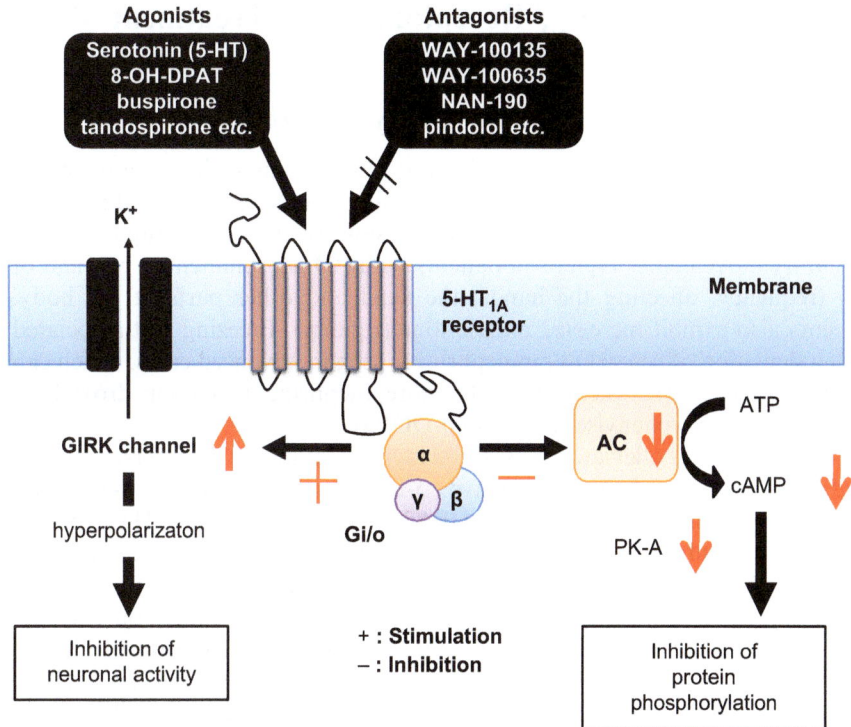

Figure 13.3 Signal transduction pathways of 5-HT$_{1A}$ receptors. GIRK = G-protein-gated inwardly K$^+$ rectifying channel, AC = adenylate cyclase, PK-A = protein kinase A.

and amygdala), septum (lateral septal nucleus) and raphe nuclei (dorsal raphe nucleus and median raphe nucleus).[9,32–34] They also exist at moderate-to-low levels in the cerebral cortex, thalamus, hypothalamus and basal ganglia (*e.g.*, the Substantia nigra (Pars compacta)).[32,33] 5-HT$_{1A}$ receptors function as both presynaptic and postsynaptic receptors. Specifically, 5-HT$_{1A}$ receptors in the raphe nuclei are located on somata and dendrites of 5-HT neurons, where they function as autoreceptors and negatively regulate 5-HT neuron activity.[35] Through this autoreceptor function, 5-HT$_{1A}$ receptors can control the overall activity of the serotonergic system. 5-HT$_{1A}$ receptors are also located on postsynaptic membranes of the neurons or nerve terminals (heteroreceptors) where 5-HT neurons terminate.

5-HT$_{1A}$ receptors are coupled to G$_{i/o}$ protein, inhibit adenylate cyclase and cAMP formation, and reduce protein kinase A (PK-A) activity (Figure 13.3).[4,9,10,32] In addition, 5-HT$_{1A}$ receptors activate G-protein-gated inwardly rectifying K$^+$ (GIRK) channels, which facilitate the efflux of intracellular K$^+$ and hyperpolarize target neurons. Thus, stimulation of 5-HT$_{1A}$ receptors inhibits the firing of target neurons in various regions of the brain including the hippocampus, lateral septum and raphe nuclei.[35–41]

13.3 Role of 5-HT$_{1A}$ Receptors in the Treatment of Parkinson's Disease

13.3.1 Treatment of Parkinsonian Symptoms

Patients with PD show extrapyramidal motor symptoms such as bradykinesia, akinesia, tremor and muscle rigidity.[1–3] Bradykinesia refers to reduced and impaired motor activity and manifests as the slowing and/or interrupted flow of voluntary movements. Tremor in patients with PD is prominent at rest and of low frequency, affecting the hands, the head and other parts of the body. Patients also exhibit increased muscle tone (rigidity). Freezing gait associated with a depletion of central norepinephrine is seen in advanced cases, which can be improved by treatment with the norepinephrine precursor droxydopa (L-*threo*-dihydroxyphenylserine, L-*threo*-DOPS).

5-HT$_{1A}$ agonists alleviate various extrapyramidal motor disorders. A line of studies demonstrated that selective 5-HT$_{1A}$ agonists (*e.g.*, 8-hydroxy-2-(di-*n*-propylamino) tetralin (8-OH-DPAT) and flesinoxan) or 5-HT$_{1A}$ partial agonists (*e.g.*, tandospirone and buspirone) attenuated D$_2$ antagonist-induced catalepsy, bradykinesia and other Parkinsonian symptoms (Figure 13.4 and Table 13.1).[16–22,42] The ameliorative effects of 5-HT$_{1A}$ agonists on extrapyramidal disorders are as potent as those of the anti-Parkinsonian agent trihexyphenidyl (an mACh receptor antagonist) and reversed by 5-HT$_{1A}$ antagonists (*e.g.*, WAY-100135 and WAY-100635) (Figure 13.4).[19,20] In addition, 5-HT$_{1A}$ agonists or partial agonists are also effective against the

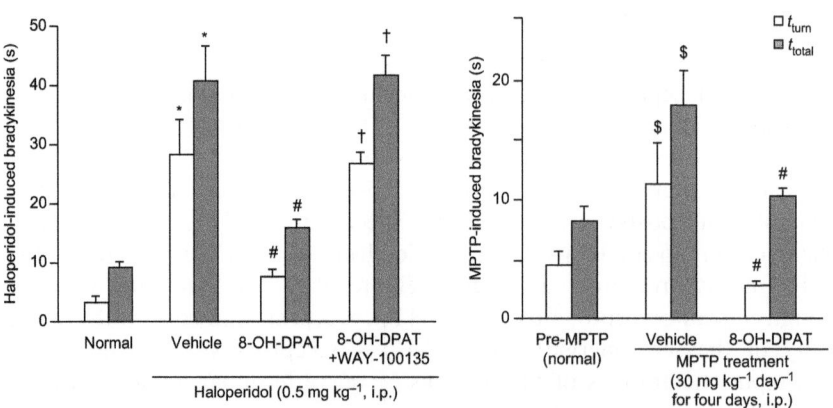

Figure 13.4 Ameliorative actions of the 5-HT$_{1A}$ agonist 8-OH-DPAT on D$_2$ antagonist (haloperidol)- and MPTP-induced bradykinesia and its reversal by the 5-HT$_{1A}$ antagonist WAY-100135. t_{turn} and t_{total} values are the times required for mice to rotate on the pole and to descend to the floor, respectively, in the mouse pole test. Increases in t_{turn} and t_{total} values represent the induction of bradykinesia. * = $p<0.05$ *vs.* normal, # = $p<0.05$ *vs.* vehicle + haloperidol or MPTP, † = $p<0.05$ *vs.* 8-OH-DPAT + haloperidol, $ = $p<0.05$ *vs.* pre-MPTP.

extrapyramidal motor dysfunctions induced by damage to dopaminergic neurons or depletion of central dopamine (Table 13.1).[16,17,22,43] Namely, 5-HT$_{1A}$ agonists reverse hypolocomotion induced by dopaminergic neurotoxins (6-hydroxydopamine (6-OH-DA) and 1-methyl-4-phenyl-1,2,3,6-tetrahydropyridine (MPTP)) or akinesia induced by reserpine which depletes central monoamines. It is also known that 5-HT$_{1A}$ agonists induce contralateral rotation behavior in the 6-OH-DA-hemilesioned rat model of PD and improve motor disability and hypolocomotion in MPTP-treated primate models.[17,18,22,44–49] The anti-Parkinsonian actions of 5-HT$_{1A}$ agonists were further supported by the neurochemical finding that Fos protein expression caused by D$_2$ receptor antagonists in the Substantia nigra (Pars compacta) was significantly reversed by 5-HT$_{1A}$ agonists (*e.g.*, 8-OH-DPAT and tandospirone), indicating that stimulation of 5-HT$_{1A}$ receptors counteracts the blockade of the striatal D$_2$ receptor which is linked to the induction of extrapyramidal motor disorders.[20,21]

Since 5-HT$_{1A}$ agonists effectively alleviate the motor deficits caused by damage to dopaminergic neurons,[17,18,22,44] the anti-Parkinsonian actions of 5-HT$_{1A}$ agonists seem to be mediated by non-dopaminergic mechanisms. Indeed, the alleviating effects of 5-HT$_{1A}$ agonists on akinesia in reserpine-treated rats was antagonized by the 5-HT$_{1A}$ antagonist WAY-100135, but not by the D$_2$ antagonist haloperidol.[17,50,51] It is therefore likely that 5-HT$_{1A}$ agonists produce anti-Parkinsonian actions in an additive fashion with currently used dopaminergic agents.

The anti-Parkinsonian actions of 5-HT$_{1A}$ agonists were not altered by *p*-chlorophenylalanine (PCPA) treatment, which irreversibly inhibits tryptophan hydroxylase and inactivates 5-HT neurons, indicating that 5-HT$_{1A}$ agonists exert anti-Parkinsonian actions at least partly by activating postsynaptic 5-HT$_{1A}$ receptors (Figure 13.5 and Figure 13.6).[19] In addition, microinjection studies demonstrated that local injections of 8-OH-DPAT into the cerebral cortex or Substantia nigra (Pars compacta) were effective in reversing D$_2$-antagonist-induced catalepsy.[22] It is therefore likely that 5-HT$_{1A}$ agonists alleviate Parkinsonism by activating postsynaptic 5-HT$_{1A}$ receptors both in the Substantia nigra (Pars compacta) and in the motor cortex, probably by inhibiting the neuronal activity in these regions (Figure 13.5 and Figure 13.6). Since the activities of striatal neurons are positively regulated by glutamatergic (Glu) neurons derived from the motor cortex,[4] and since NMDA antagonists reduce the neural activity of striatal neurons as well as extrapyramidal motor disorders,[52–55] activation of the cortical 5-HT$_{1A}$ receptors seems to exert anti-Parkinsonian actions by inhibiting the cortico – striatal glutamatergic neurons. On the other hand, the possibility that 5-HT$_{1A}$ autoreceptors in the raphe nuclei also contribute to the anti-Parkinsonian actions of 5-HT$_{1A}$ agonists cannot be ruled out since earlier studies showed that the microinjection of 8-OH-DPAT into the raphe reduced haloperidol-induced catalepsy (Figure 13.5).[56–58] Stimulation of presynaptic 5-HT$_{1A}$ autoreceptors inhibits serotonergic neuron activities, thereby reducing the functions of 5-HT$_{2A/2C}$, 5-HT$_3$ and 5-HT$_6$ receptors which reportedly inhibits extrapyramidal motor disorders.[59,60]

13.3.2 Treatment of L-DOPA-Induced Dyskinesia

The mainstay of treatments for PD is dopamine-replacement therapy with the dopamine precursor L-DOPA, which has been used for over 40 years.[1,4–6] However, long-term treatment with L-DOPA often results in fluctuations in its efficacy (*e.g.*, wearing-off and on–off phenomena) and/or serious side-effects (*e.g.*, dyskinesia and psychosis) (Figure 13.2).[4–6] L-DOPA-Induced dyskinesia is a complex symptom, involving dystonic, choreic and athetotic movements, which affect the limbs, hands, trunk and lingual–facial–buccal muscular systems at different points of an L-DOPA cycle. Although variability exists, the frequency of L-DOPA-induced dyskinesia has been reported at 30–80%.[5] The development of dyskinesia is associated with several factors such as an earlier onset of PD, longer duration of the disease, longer duration of L-DOPA therapy, total L-DOPA exposure and female gender. Since

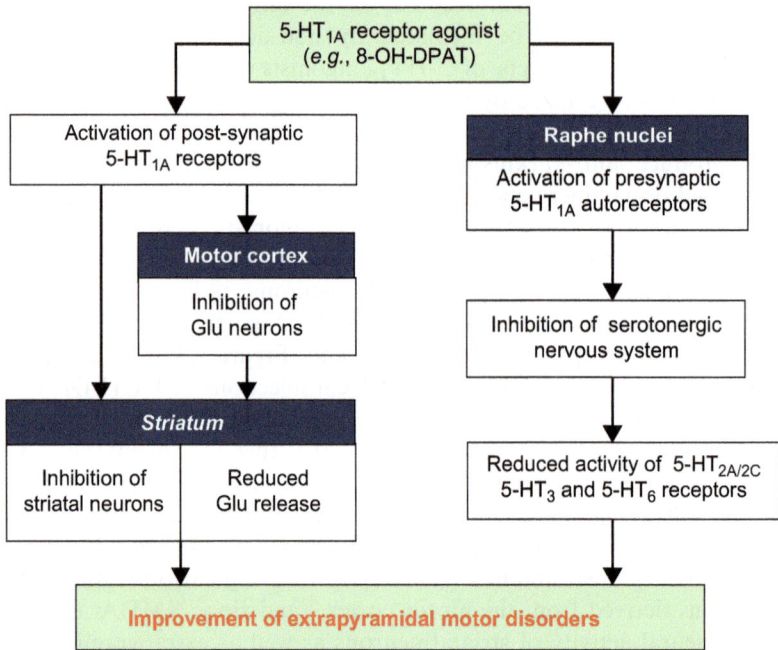

Figure 13.5 Mechanisms of action of 5-HT$_{1A}$ agonists in modulating extrapyramidal motor disorders. The 5-HT$_{1A}$ agonist 8-OH-DPAT improves extrapyramidal motor disorders by stimulating presynaptic and post-synaptic 5-HT$_{1A}$ receptors. Stimulation of postsynaptic 5-HT$_{1A}$ receptors reduces the activity of striatal neurons indirectly by inhibiting Glu neurons in the motor cortex, and directly by hyperpolarizing striatal neurons. In addition, stimulation of presynaptic 5-HT$_{1A}$ autoreceptors inhibits serotonergic neuron activities, thereby reducing the functions of 5-HT$_{2A/2C}$, 5-HT$_3$ and 5-HT$_6$ receptors which reportedly inhibit extrapyramidal motor disorders.[59,60]

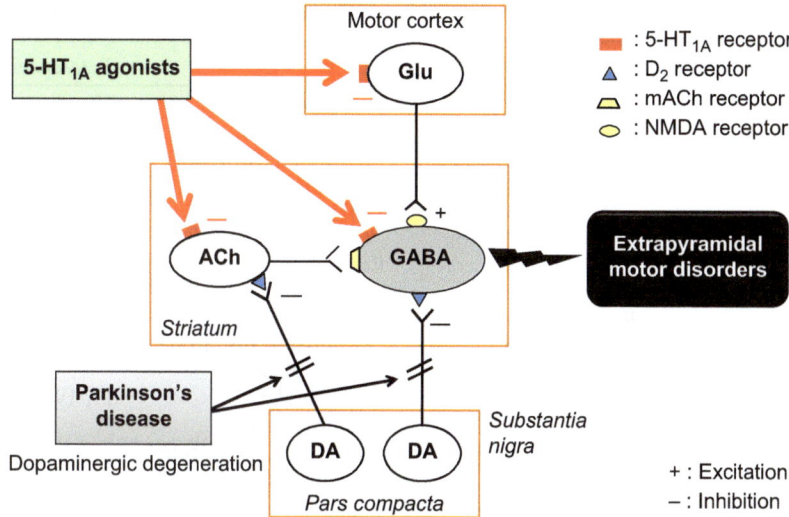

Figure 13.6 Action of 5-HT$_{1A}$ agonists on striatal and cortical 5-HT$_{1A}$ receptors in PD. 5-HT = serotonin, DA = dopamine, ACh = acetylcholine, Glu = glutamate.

L-DOPA-induced dyskinesia are generally difficult to treat and often hampers continued treatment, there is a great clinical need to control it (Figure 13.2).[1,4–6]

Previous studies indicate that dyskinesia is associated with the synchronization of neuronal activity within and outside the basal ganglia.[61–63] Two factors, the primary neurodegenerative process among dopamine neurons in PD and the chronic stimulation of dopamine receptors by L-DOPA therapy, are thought to be important in the pathogenesis of L-DOPA-induced dyskinesia.[5,6] Since L-DOPA, which stimulates both D$_1$ and D$_2$ receptors, is much stronger than the selective D$_2$ receptor agonists in inducing dyskinesia, activation of the D$_1$ receptor and/or an imbalance of D$_1$–D$_2$ receptor functions seem to at least partly be involved in the generation of dyskinesia.[5] Indeed, stimulation of D$_1$ receptors by D$_1$ agonists (*e.g.*, SKF-81297) induces abnormal involuntary movements (vacuous chewing, tongue extrusion, repetitive jaw movements *etc.*) in various animal models.[43] In addition, several neurotransmitter systems including the serotonergic, noradrenergic and glutamatergic systems are also implicated in the development of L-DOPA-induced dyskinesia.[5]

Previous studies have shown that 5-HT$_{1A}$ receptor agonists inhibit dyskinesia in animals primed with chronic L-DOPA treatment (Table 13.1).[24–26] Experiments with MPTP-lesioned primates also showed that 5-HT$_{1A}$ agonists (*e.g.*, 8-OH-DPAT and tandospirone) attenuated dyskinesia without affecting the anti-Parkinsonian efficacy of L-DOPA.[64–69] Although the exact mechanisms involved remain to be clarified, the anti-dyskinetic action of 5-HT$_{1A}$ agonists is mediated at least partly by both striatal and cortical 5-HT$_{1A}$ receptors since the microinjection of 8-OH-DPAT either into the Substantia nigra (Pars compacta) or into the motor cortex reversed L-DOPA-induced dyskinesia.[26,67,70]

Table 13.1 Anti-Parkinsonian and anti-dyskinetic actions of 5-HT$_{1A}$ agonists.

	Animal models	Drugs	References
Anti-Parkinsonian actions	MPTP-induced bradykinesia and motor disturbance	8-OH-DPAT, BAY-639044, repinotan, SLV308, sarizotan	22, 44, 45, 46
	6-OH-DA-induced rotation behavior	8-OH-DPAT, tandospirone	17, 18, 47
	6-OH-DA-induced catalepsy	Buspirone, 8-OH-DPAT	48, 49
	Reserpine-induced hypokinesia	8-OH-DPAT, flesinoxan, buspirone, tandospirone	17, 50, 51
	D$_2$ antagonist-induced catalepsy and bradykinesia	8-OH-DPAT, tandospirone, buspirone, ipsapirone, eptapirone	17, 19, 42
Anti-Dyskinetic actions	L-DOPA-induced dyskinesia	8-OH-DPAT, tandospirone, buspirone, sarizotan	63, 64, 66, 67
	D$_1$ receptor-induced dyskinesia	8-OH-DPAT	43

Specifically, the microinjection of 5-HT$_{1A}$ agonists (*e.g.*, 8-OH-DPAT) into the Substantia nigra (Pars compacta) attenuates L-DOPA-induced dyskinesia as well as D$_1$-agonist-induced dyskinesia,[43] indicating that 5-HT$_{1A}$ receptors counteract the D$_1$ receptors in the Substantia nigra (Pars compacta) regulating L-DOPA-induced dyskinesia. Furthermore, microdialysis studies showed that the local application of 5-HT$_{1A}$ agonists to the cerebral cortex (primary motor areas) reduces glutamate release in the Substantia nigra (Pars compacta), which could inhibit the induction of dyskinesia.[54,71] Therefore, postsynaptic 5-HT$_{1A}$ receptors in the cerebral cortex, which regulate the activity of cortico–striatal glutamatergic neurons, are also likely to be involved in L-DOPA-induced dyskinesia. Thus, stimulation of 5-HT$_{1A}$ receptors is conceivably a potential treatment for L-DOPA-induced dyskinesia in patients with PD.

13.3.3 Treatment of Non-Motor Symptoms in Parkinson's Disease

PD is accompanied by diverse non-motor symptoms including psycho-emotional changes (*e.g.*, anxiety and depression), cognitive impairments, sleep disturbances and autonomic dysfunctions (Figure 13.2).[1–3,13,72] Among these, anxiety and depression are very common, affecting about 50% of individuals with PD, and usually comorbid.[72] It has also been suggested that anxiety and depression are not merely psychological reactions to the illness, but rather, closely linked to the neurobiological basis of PD such as dysfunctions of the dopaminergic and/or serotonergic systems. In addition, cognitive impairments develop in about 40% of patients and most patients (more than 80%) have dementia at the end-stage of PD.[3] Thus, besides the treatment of core motor symptoms, control of non-motor symptoms, especially psychoemotional disturbances and cognitive impairments, is key to improving the quality of life of patients with PD.[1]

13.3.3.1 Anxiety and Depression

5-HT$_{1A}$ receptors are implicated in the pathogenesis and treatment of anxiety and depressive disorders. Numerous studies have shown that 5-HT$_{1A}$ agonists (including partial agonists) produce anti-anxiety effects in various animal models (*e.g.*, Vogel's conflict, elevated-plus maze, social interaction and conditioned-fear freezing tests).[14,15,73–76] The anxiolytic actions of full or partial 5-HT$_{1A}$ agonists (*e.g.*, 8-OH-DPAT, buspirone and tandospirone) are reversed by selective 5-HT$_{1A}$ antagonists (*e.g.*, WAY-100635 and WAY-100135).[74,75] Knock-out (KO) mice lacking 5-HT$_{1A}$ receptors exhibit increased anxiety.[77–79] Conversely, transgenic mice overexpressing 5-HT$_{1A}$ receptors show reduced anxiety behavior, implying that the stimulation of 5-HT$_{1A}$ receptors ameliorates anxiety.[34] In addition, 5-HT$_{1A}$ agonists exhibit significant anti-depressant activity following repeated administration.[80–83]

Figure 13.7 illustrates the mechanisms underlying the anxiolytic and anti-depressant actions of 5-HT$_{1A}$ agonists. As described previously, 5-HT$_{1A}$

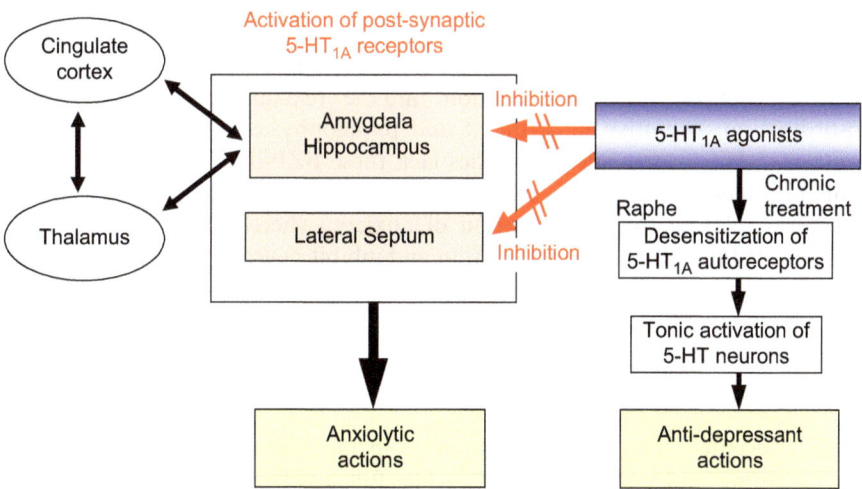

Figure 13.7 Mechanisms underlying the anxiolytic and anti-depressant actions of 5-HT$_{1A}$ agonists. 5-HT$_{1A}$ receptors are mainly expressed in the limbic areas (*e.g.*, the hippocampus and amygdala) and septum (*e.g.*, the lateral septal nucleus), which are involved in the generation and propagation of anxiety. 5-HT$_{1A}$ agonists exert anxiolytic effects by inhibiting the activity of the limbic neurons (*i.e.*, the hippocampus and amygdala) and lateral septal neurons *via* activating GIRK channels. In addition, 5-HT$_{1A}$ agonists induce the desensitization of presynaptic 5-HT$_{1A}$ autoreceptors following chronic treatment, which consequently causes the tonic activation of 5-HT neurons and anti-depressant actions.

receptors are mainly found as postsynaptic receptors in the limbic areas (hippocampus and amygdala) and septum (lateral septal nucleus), which are linked to the generation and propagation of anxiety. Although 5-HT_{1A} receptors also exist in the raphe nuclei as presynaptic autoreceptors, a line of evidence suggests a crucial role for postsynaptic 5-HT_{1A} receptors in regulating anxiety. First, the microinjection of 5-HT_{1A} agonists directly into a limbic region (*e.g.*, the hippocampus) produces significant anxiolytic actions.[74] Second, the denervation of 5-HT neurons, where presynaptic 5-HT_{1A} autoreceptors are located, did not alter the anxiolytic actions of 5-HT_{1A} agonists.[75,77] Finally, by using a tissue-specific conditional rescue strategy, Gross *et al.*[84] showed that the expression of postsynaptic 5-HT_{1A} receptors in the forebrain (*e.g.*, the hippocampus and cerebral cortex) is sufficient to alleviate anxiety in KO mice. It is therefore likely that 5-HT_{1A} agonists exert anxiolytic effects by inhibiting the activity of the limbic neurons and lateral septal neurons *via* the activation of GIRK channels (Figure 13.3 and Figure 13.7).[4,15] On the other hand, desensitization of presynaptic 5-HT_{1A} autoreceptors has been suggested to be involved in the anti-depressant actions of 5-HT_{1A} agonists.[4,15] Repeated treatment of animals with 5-HT_{1A} agonists down-regulates and desensitizes the presynaptic 5-HT_{1A} autoreceptors in the raphe nuclei, releasing 5-HT neurons from 5-HT_{1A} autoreceptor-mediated self-inhibition and resulting in tonic activation of the serotonergic system (Figure 13.7).[4,15,85–87]

It should be noted that 5-HT_{1A} agonistic anxiolytics are superior to conventional benzodiazepine (BZP) anxiolytics, especially in terms of their safety profile. Anxiolytic doses of BZPs (*e.g.*, diazepam) readily caused diverse adverse reactions including sedation, muscle relaxation, impaired motor co-ordination, cognitive impairment and psychophysical dependence.[13,15] In contrast, 5-HT_{1A} agonistic anxiolytics lack these BZP-like side-effects. This is probably due to the specific distribution of 5-HT_{1A} receptors in the limbic structures relating to the generation of anxiety, whereas BZP receptors are widely distributed throughout the brain and inhibit neuronal activities.[88] Thus, the stimulation of 5-HT_{1A} receptors seems to be a favorable strategy for treating anxiety–depressive symptoms in cases of PD.

13.3.3.2 Cognitive Impairment

5-HT_{1A} receptors play an important role in modulating cognitive functions (*e.g.*, learning and memory).[28,31,89–91] The full 5-HT_{1A} agonist 8-OH-DPAT reportedly exhibits biphasic effects on cognitive functions, showing facilitation at low doses and inhibition at high doses.[28,31] These facilitatory and inhibitory effects are now thought to be mediated by the stimulation of presynaptic and postsynaptic 5-HT_{1A} receptors, respectively (Figure 13.8). Thus, the stimulation of postsynaptic 5-HT_{1A} receptors inhibits cognitive functions probably by reducing: (1) the activity of cholinergic or glutamatergic neurons in the basal forebrain regions (*e.g.*, diagonal band of Broca and medial septum) projecting to the cerebral cortex or the hippocampus; (2) the activity of the hippocampal and cortical neurons receiving the above cholinergic or glutamatergic inputs;

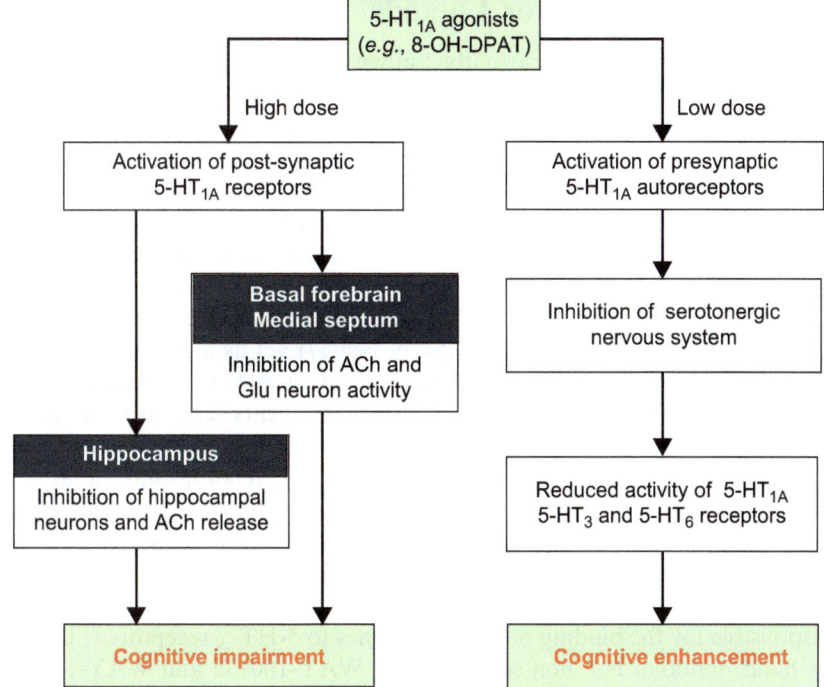

Figure 13.8 Mechanisms of action of 5-HT$_{1A}$ agonists in regulating cognitive function. The 5-HT$_{1A}$ full agonist 8-OH-DPAT facilitates and inhibits cognitive functions by stimulating presynaptic and post-synaptic 5-HT$_{1A}$ receptors, respectively. Stimulation of postsynaptic 5-HT$_{1A}$ receptors inhibits cognitive functions probably by reducing the activity of cholinergic and/or glutamatergic neurons in the basal forebrain regions (e.g., the diagonal band of Broca) and medial septum, which project to the cerebral cortex and hippocampus, and by inhibiting ACh release in the hippocampus. In contrast, stimulation of presynaptic 5-HT$_{1A}$ autoreceptors inhibits serotonergic neuron activities, thereby reducing the functions of 5-HT$_{1A}$, 5-HT$_3$ and 5-HT$_6$ receptors which inhibit cognitive functions.[10]

and (3) the release of ACh into the hippocampus (Figure 13.8). On the other hand, stimulation of presynaptic 5-HT$_{1A}$ autoreceptors inhibits serotonergic neuron activities, thereby reducing the functions of postsynaptic 5-HT$_{1A}$, 5-HT$_3$ and 5-HT$_6$ receptors which impair cognitive functions.[10] In fact, 5-HT$_{1A}$ antagonists (e.g., WAY-100635, WAY-101405, NAD-299 and leozotan) and a partial 5-HT$_{1A}$ agonist (e.g., tandospirone) have been shown to reverse the cognitive impairments induced by mACh receptor antagonists (e.g., scopolamine) or NMDA receptor antagonists (e.g., MK-801).[28,31,89–91] Although several studies suggest that 5-HT$_{1A}$ agonists also improve cognitive impairment,[23,92] it is likely that 5-HT$_{1A}$ antagonists or partial agonists can help alleviate the cognitive deficits in PD.

13.4 5-HT$_{1A}$ Receptor Ligands

The chemical structures of currently known 5-HT$_{1A}$ ligands are shown in Figure 13.9. A three-dimensional model of the 5-HT$_{1A}$ receptor (created by the GPCR modeling program PREDICT) reveals that a binding pocket for 5-HT exists in the extracellular region of the transmembrane (TM) domains of 5-HT$_{1A}$ receptors, where the amine moiety of 5-HT binds to Asp116 of TM3 and its hydroxyl group to Ser199 in TM5 *via* ionic and hydrogen bonds, respectively.[93] 5-HT, 8-OH-DPAT, flesinoxan, and osemozotan (MKC-242) are full 5-HT$_{1A}$ agonists.[13,15,94] Although information is limited, naluzotan (PRX-00023) also acts as a 5-HT$_{1A}$ agonist.[93] Buspirone, tandospirone, ipsapirone, gepirone and piclozotan (SUN N4057) act as a partial agonist at 5-HT$_{1A}$ receptors with intrinsic activities of about 50–80% while buspirone has relatively high D$_2$ blocking activity.[13,15] All these agents seem to be useful in adjunctive therapy for PD as well as monotherapy for anxiety disorders and major depression. Studies of the structure–activity relationships of arylpiperazine derivatives using 3D-QSAR models also showed that the pharmacophore part for binding to 5-HT$_{1A}$ receptors exists in the pyrimidine, piperazine or arylpiperazine part.[93,95,96] In addition, hydrophobic residues, Phe362 in TM6, as well as Asp116 in TM3 and Ser199 in TM5, are suggested to be responsible for the binding of arylpiperazines to 5-HT$_{1A}$ receptors.[93] On the other hand, pindolol is a non-selective, and WAY-100135 and WAY-100635 are selective 5-HT$_{1A}$ antagonists (Figure 13.9).[15]

Several compounds which possess combined D$_2$ and 5-HT$_{1A}$ agonistic actions were recently reported. For instance, SLV308 shows high affinity and full agonistic activity at 5-HT$_{1A}$ receptors and also acts as a partial agonist at D$_2$ receptors (Figure 13.9).[97] It induces contralateral rotation behavior in 6-OH-DA-treated dopaminergic hemi-lesioned rats and alleviates hypolocomotion and L-DOPA-induced motor disability in MPTP-treated common marmosets. Furthermore, a recent double-blind study showed that SLV308 significantly improved motor function in patients with early PD.[98] An aporphine analogue with a D$_2$ and 5-HT$_{1A}$ dual-agonist profile also exhibits anti-Parkinsonian actions and reverses L-DOPA-induced dyskinesia (Figure 13.9).[99] Although further clinical studies are required, these agents may be useful in the treatment of PD.

13.5 Conclusion

This paper reviewed the pharmacological features and functional mechanisms of 5-HT$_{1A}$ receptors in regulating extrapyramidal disorders and updated their therapeutic role in treating PD. Recent studies have revealed new insights into the therapeutic role of 5-HT$_{1A}$ receptors in treating PD. Specifically, 5-HT$_{1A}$ receptors play a crucial role in modulating extrapyramidal Parkinsonian symptoms, L-DOPA-induced dyskinesia, mood disorders (*e.g.*, anxiety and depression) and cognitive impairment (Figure 13.10). Thus: (1) full or partial 5-HT$_{1A}$ agonists improve core Parkinsonian symptoms and L-DOPA-induced

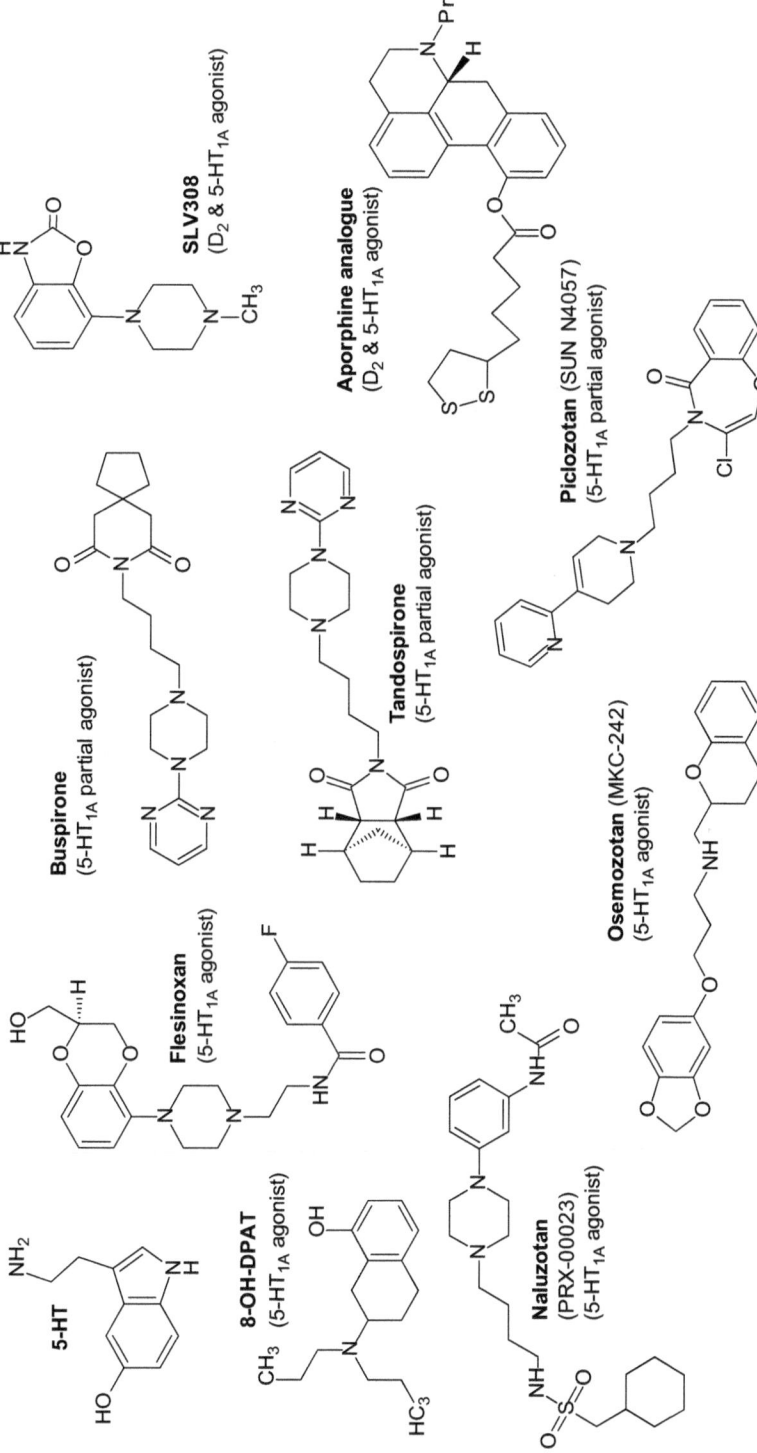

Figure 13.9 Chemical structures of 5-HT$_{1A}$ receptor ligands.

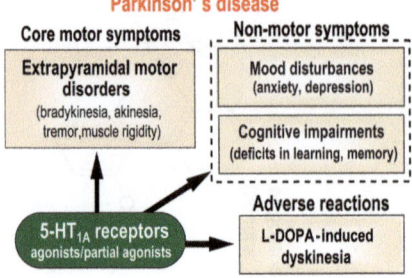

Figure 13.10 5-HT$_{1A}$ receptors as a therapeutic target for PD.

dyskinesia produced by the treatment of PD; (2) 5-HT$_{1A}$ agonists are expected to be effective for mood disorders which are frequently seen in patients with PD; and (3) partial 5-HT$_{1A}$ agonists or antagonists may reduce cognitive deficits in cases of PD. In addition, it should be noted that the amelioration of reserpine-induced akinesia by 5-HT$_{1A}$ agonists is not affected by D$_2$ antagonists,[4,17] but is blocked by 5-HT$_{1A}$ antagonists, implying that the anti-Parkinsonian effects of 5-HT$_{1A}$ agonists are exerted *via* a non-dopaminergic mechanism. Therefore, 5-HT$_{1A}$-receptor-related agents are expected to produce anti-Parkinsonian actions in an additive fashion with currently used dopaminergic agents. These findings should encourage the synthesis of new 5-HT$_{1A}$ agonists with greater potency, higher 5-HT$_{1A}$ selectivity and favorable pharmacokinetic properties. Furthermore, designing 5-HT$_{1A}$ ligands with dopamine-stimulating activity seems to be a promising approach to the treatment of PD. These new ligands for 5-HT$_{1A}$ receptors may overcome limitations of clinical efficacy and/or improve adverse reactions in the current treatment of PD.

References

1. A. Samii, J. G. Nutt and B. Ransom, *Lancet*, 2004, **363**, 1783.
2. D. A. Gallagher and A. Schrag, *Neurobiol. Dis.*, 2012, **46**, 581.
3. J. Meireles and J. Massano, *Front. Neurol.*, 2012, **3**, 88.
4. Y. Ohno, *CNS Neurosci. Ther.*, 2011, **17**, 58.
5. G. Fabbrini, J. M. Brotchie, F. Grandas, M. Nomoto and C. G. Goetz, *Mov. Disord.*, 2007, **22**, 1379.
6. P. A. Cheshire and D. R. Williams, *J. Clin. Neurosci.*, 2012, **19**, 343.
7. B. L. Roth, *Ann. Clin. Psychiatry*, 1994, **6**, 67.
8. H. G. Baumgarten and Z. Grozdanovic, *Pharmacopsychiatry*, 1995, **2**, 73.
9. N. M. Barnes and T. Sharp, *Neuropharmacology*, 1999, **38**, 1083.
10. Y. Ohno, A. Tatara, S. Shimizu and M. Sasa, in *Schizophrenia Research: Recent Advances*, ed. T. Sumiyoshi, Nova Science Publishers, New York, 2012, p. 321.
11. J. P. Feighner and W. F. Boyer, *Psychopathology*, 1989, **22**, 21.

12. R. W. Fuller, *J. Clin. Psychiatry*, 1991, **152**, 52.
13. Y. Ohno, in *Mapping the Progress of Alzheimer's and Parkinson's Disease*, Y. Mizuno, A. Fisher and I. Hanin, Kluwer Academic, New York, 2002, p. 423.
14. E. Akimova, R. Lanzenberger and S. Kasper, *Biol. Psychiatry*, 2009, **66**, 627.
15. Y. Ohno and Cent. Nerv, *Syst. Agents Med. Chem.*, 2010, **10**, 148.
16. L. Mignon and W. A. Wolf, *Psychopharmacology*, 2002, **163**, 85.
17. T. Ishibashi and Y. Ohno, *Biog. Amines*, 2004, **8**, 329.
18. K. Matsubara, K. Shimizu, M. Suno, K. Ogawa, T. Awaya, T. Yamada, T. Noda, M. Satomi, K. Ohtaki, K. Chiba, Y. Tasaki and H. Shiono, *Brain Res.*, 2006, **1112**, 126.
19. Y. Ohno, S. Shimizu, J. Imaki, S. Ishihara, N. Sofue, M. Sasa and Y. Kawai, *Prog. Neuro-Psychopharmacol. Biol. Psychiatry*, 2008, **32**, 1302.
20. Y. Ohno, S. Shimizu, J. Imaki, S. Ishihara, N. Sofue, M. Sasa and Y. Kawai, *Neuropharmacology*, 2008, **55**, 717.
21. Y. Ohno, S. Shimizu and J. Imaki, *J. Pharmacol. Sci.*, 2009, **109**, 593.
22. S. Shimizu, A. Tatara, J. Imaki and Y. Ohno, *Prog. Neuro-Psychopharmacol. Biol. Psychiatry*, 2010, **34**, 877.
23. A. Newman-Tancredi, *Curr. Opin. Investig. Drugs*, 2010, **11**, 802.
24. M. Tomiyama, T. Kimura, T. Maeda, K. Kannari, M. Matsunaga and M. Baba, *Neurosci. Res.*, 2005, **52**, 185.
25. K. B. Dupre, K. L. Eskow, A. Steinige, A. Klioueva, G. E. Negron, L. Lormand, J. Y. Park and C. Bishop, *Psychopharmacology*, 2008, **199**, 99.
26. C. Bishop, D. M. Krolewski, K. L. Eskow, C. J. Barnum, K. B. Dupre, T. Deak and P. D. Walker, *J. Neurosci. Res.*, 2009, **87**, 1645.
27. I. Misane and S. O. Ögren, *Neuropsychopharmacology*, 2003, **28**, 253.
28. M. Lüttgen, E. Elvander, N. Madjid and S. O. Ögren, *Neuropharmacology*, 2005, **48**, 830.
29. N. Madjid, E. E. Tottie, M. Lüttgen, B. Meister, J. Sandin, A. Kuzmin, O. Stiedl and S. O. Ögren, *J. Pharmacol. Exp. Ther.*, 2006, **316**, 581.
30. S. O. Ögren, T.M. Eriksson, E. Elvander-Tottie, C. D'Addario, J. C. Ekström, P. Svenningsson, B. Meister, J. Kehr and O. Stiedl, *Behav. Brain Res.*, 2008, **195**, 54.
31. M. V. King, C. A. Marsden and K. C. Fone, *Trends Pharmacol. Sci.*, 2008, **29**, 482.
32. T. J. Pucadyil, S. Kalipatnapu and A. Chattopadhyay, *Cell. Mol. Neurobiol.*, 2005, **25**, 553.
33. H. Luna-Munguia, L. Manuel-Apolinar and L. Rocha, *Psychopharmacology*, 2005, **181**, 309.
34. H. Kusserow, B. Davies, H. Hörtnagl, I. Voigt, T. Stroh, B. Bert, D. R. Deng, H. Fink, R. W. Veh and F. Theuring, *Brain Res. Mol. Brain Res.*, 2004, **129**, 104.
35. P. Blier and N. M. Ward, *Biol. Psychiatry*, 2003, **53**, 193.
36. A. Hirose, M. Sasa, A. Akaike and S. Takaori, *Neuropharmacology*, 1990, **29**, 93.

37. P. Van den Hooff and M. Galvan, *Br. J. Pharmacol.*, 1992, **106**, 893.
38. V. Hadrava, P. Blier, T. Dennis, C. Ortemann and C. de Montigny, *Neuropharmacology*, 1995, **34**, 1311.
39. Y. Ohno, K. Ishida, T. Ishibashi, H. Tanaka, H. Shimizu and M. Nakamura, in *Serotonin in the Central Nervous System and Periphery*, ed. Y. Takada and G. Curzon, Elsevier, Amsterdam, 1995, p. 159.
40. J. R. Raymond, Y. V. Muklin, T. W. Gettys and M. N. Garnovskaya, *Br. J. Pharmacol.*, 1999, **127**, 1751.
41. K. Tada, K. Kasamo, N. Ueda, T. Suzuki, T. Kojima and K. Ishikawa, *J. Pharmacol. Exp. Ther.*, 1999, **288**, 843.
42. E. P. Prinssen, F. C. Colpaert and W. Koek, *Eur. J. Pharmacol.*, 2002, **453**, 217.
43. K. B. Dupre, K. L. Eskow, C. J. Barnum and C. Bishop, *Neuropharmacology*, 2008, **55**, 1321.
44. C. A. Jones, L. C. Johnston, M. J. Jackson, L. A. Smith, G. van. Scharrenburg, S. Rose, P. G. Jenner and A. C. McCreary, *Eur. Neuropsychopharmacol.*, 2010, **20**, 582.
45. E. Bezard, I. Gerlach, R. Moratalla, C. E. Gross and R. Jork, *Neurobiol. Dis.*, 2006, **23**, 77.
46. F. Bibbiani, J. D. Oh and T. N. Chase, *Neurology*, 2001, **57**, 1829.
47. R. Gerber, C. A. Alter and J. M. Liebman, *Psychopharmacology*, 1988, **94**, 178.
48. A. M. Nayebi, S. R. Rad, M. Saberian, S. Azimzadeh and M. Samini, *Pharmacol. Rep.*, 2010, **62**, 258.
49. J. Mahmoudi, A. Mohajjel Nayebi, M. Samini, S. Reyhani-Rad and V. Babapour, *Daru, J. Fac. Pharm., Tehran Univ. Med. Sci.*, 2011, **19**, 338.
50. S. Ahlenius, V. Hillegaart, P. Salmi and A. Wijkstrom, *Pharmacol. Toxicol.*, 1993, **72**, 398.
51. S. Ahlenius and P. Salmi, *Pharmacol. Toxicol.*, 1995, **76**, 149.
52. E. H. Chartoff, R. P. Ward and D. M. Dorsa, *J. Pharmacol. Exp. Ther.*, 1999, **291**, 531.
53. K. Steece-Collier, L. K. Chambers, S. S. Jaw-Tsai, F. S. Menniti and J. T. Greenamyre, *Exp. Neurol.*, 2000, **163**, 239.
54. N. Hussain, B. A. Flumerfelt and N. Rajakumar, *Neuroscience*, 2001, **102**, 391.
55. S. Yanahashi, K. Hashimoto, K. Hattori, S. Yuasa and M. Iyo, *Brain Res.*, 2004, **1011**, 84.
56. R. W. Invernizzi, L. Cervo and R. Samanin, *Neuropharmacology*, 1988, **27**, 515.
57. M. L. Wadenberg and V. Hillegaart, *Neuropharmacology*, 1995, **34**, 495.
58. M. L. Wadenberg, K. A. Young, J. T. Richter and P. B. Hicks, *Neuropharmacology*, 1999, **38**, 151.
59. Y. Ohno, J. Imaki, Y. Mae, T. Takahashi and A. Tatara, *Neuropharmacology*, 2011, **60**, 201.
60. A. Tatara, S. Shimizu, N. Shin, M. Sato, T. Sugiuchi, J. Imaki and Y. Ohno, *Prog. Neuropsychopharmacol. Biol. Psychiatry*, 2012, **38**, 252.

61. J. A. Brotchie, *Mov. Disord.*, 2005, **20**, 919.
62. M. A. Cenci, *Trends Neurosci.*, 2007, **30**, 236.
63. P. Jenner, *Nat. Rev. Neurosci.*, 2008, **9**, 665.
64. P. J. Elliott, D. M. Walsh, S. P. Close, G. A. Higgins and A. G. Hayes, *Neuropharmacology*, 1990, **29**, 949.
65. F. Bibbiani, J. D. Oh and T. N. Chase, *Neurology*, 2001, **57**, 1829.
66. A. Muñoz, Q. Li, F. Gardoni, E. Marcello, C. Qin, T. Carlsson, D. Kirik, M. Di. Luca, A. Björklund, E. Bezard and M. Carta, *Brain*, 2008, **131**, 3380.
67. C. Y. Ostock, K. B. Dupre, K. L. Jaunarajs, H. Walters, J. George, D. Krolewski, P. D. Walker and C. Bishop, *Neuropharmacology*, 2011, **61**, 753.
68. K. L. Eskow, V. Gupta, S. Alam, J. Y. Park and C. Bishop, *Pharmacol. Biochem. Behav.*, 2007, **87**, 306.
69. L. Grégoire, P. Samadi, J. Graham, P. J. Bédard, G. D. Bartoszyk and T. Di. Paolo, *Parkinsonism Relat. Disord.*, 2009, **15**, 445.
70. P. Huot, T. H. Johnston, J. B. Koprich, L. Winkelmolen, S. H. Fox and J. M. Brotchie, *Neurobiol. Aging*, 2012, **33**, 9.
71. T. Antonelli, K. Fuxe, M. C. Tomasini, G. D. Bartoszyk, C. A. Seyfried, S. Tanganelli and L. Ferraro, *Synapse*, 2005, **58**, 193.
72. I. H. Richard, R. B. Schiffer and R. Kurlan, *J. Neuropsychiatry Clin. Neurosci.*, 1996, **8**, 383.
73. J. P. Feighner and W. F. Boyer, *Psychopathology*, 1989, **22**, 21.
74. Y. Kataoka, K. Shibata, A. Miyazaki, Y. Inoue, K. Tominaga, S. Koizumi, S. Ueki and M. Niwa, *Neuropharmacology*, 1991, **30**, 475.
75. H. Shimizu, T. Tatsuno, H. Tanaka, A. Hirose, Y. Araki and M. Nakamura, *Jpn. J. Pharmacol.*, 1992, **59**, 105.
76. R. Stefański, W. Pałejko, W. Kostowski and A. Płaźnik, *Neuropharmacology*, 1992, **31**, 1251.
77. C. L. Parks, P. S. Robinson, E. Sibille, T. Shenk and M. Toth, *Proc. Natl. Acad. Sci. U. S. A.*, 1998, **95**, 10734.
78. S. Ramboz, R. Oosting, D. A. Amara, H. F. Kung, P. Blier, M. Mendelsohn, J. J. Mann, D. Brunner and R. Hen, *Proc. Natl. Acad. Sci. U. S. A.*, 1998, **95**, 14476.
79. K. C. Klemenhagen, J. A. Gordon, D. J. David, R. Hen and C. T. Gross, *Neuropsychopharmacology*, 2006, **31**, 101.
80. S. Wieland and I. Lucki, *Psychopharmacology*, 1990, **101**, 497.
81. R. Schreiber, M. Brocco, A. Gobert, S. Veiga and M. J. Millan, *Eur. J. Pharmacol.*, 1994, **271**, 537.
82. T. Matsuda, P. Somboonthum, M. Suzuki, S. Asano and A. Baba, *Eur. J. Pharmacol.*, 1995, **280**, 235.
83. W. Koek, B. Vacher, C. Cosi, M. B. Assié, J. F. Patoiseau, P. J. Pauwels and F. C. Colpaert, *Eur. J. Pharmacol.*, 2001, **420**, 103.
84. C. Gross, X. Zhuang, K. Stark, S. Ramboz, R. Oosting, L. Kirby, L. Santarelli, S. Beck and R. Hen, *Nature*, 2002, **416**, 396.

85. P. Blier and C. de Montigny, *Synapse*, 1987, **1**, 470.
86. L. E. Schechter, F. J. Bolaños, H. Gozlan, L. Lanfumey, S. Haj-Dahmane, A. M. Laporte, C. M. Fattaccini and M. Hamon, *J. Pharmacol. Exp. Ther.*, 1990, **255**, 1335.
87. R. Godbout, Y. Chaput, P. Blier and C. de Montigny, *Neuropharmacology*, 1991, **30**, 679.
88. H. Tanaka, H. Shimizu, Y. Kumasaka, A. Hirose, T. Tatsuno and M. Nakamura, *Brain Res.*, 1991, **546**, 181.
89. I. Misane and S. O. Ögren, *Br. J. Pharmacol.*, 2003, **28**, 253.
90. N. Madjid, E. E. Tottie, M. Lüttgen, B. Meister, J. Sandin, A. Kuzmin, O. Stiedl and S. O. Ögren, *J. Pharmacol. Exp. Ther.*, 2006, **316**, 581.
91. S. O. Ögren, T. M. Eriksson, E. Elvander-Tottie, C. D'Addario, J. C. Ekström, P. Svenningsson, B. Meister, J. Kehr and O. Stiedl, *Behav. Brain Res.*, 2008, **195**, 54.
92. R. Depoortère, A.L. Auclair, L. Bardin, F. C. Colpaert, B. Vacher and A. Newman-Tancredi, *Eur. Neuropsychopharmacol.*, 2010, **20**, 641.
93. O. M. Becker, D. S. Dhanoa, Y. Marantz, D. Chen, S. Schachanm, S. Cheruku, A. Heifetz, P. Mohanty, M. Fichman, A. Sharadendu, R. Nudelman, M. Kauffman and S. Noiman, *J. Med. Chem.*, 2006, **49**, 3116.
94. M. Abe, R. Tabata, K. Saito, T. Matsuda, A. Baba and M. Egawa, *J. Pharmacol Exp. Ther.*, 1996, **278**, 898.
95. M. L. López-Rodriguez, D. Ayala, A. Viso, B. Benhamu, R. F. de la Pradilla, F. Zarza and J. A. Ramos, *Bioorg. Med. Chem.*, 2004, **12**, 1551.
96. M. L. López-Rodriguez, M. J. Morcillo, E. Fernandez, B. Benhamu, I. Tejada, D. Ayala, A. Viso, M. Campillo, L. Pardo, M. Delgado, J. Manzanares and J. A. Fuentes, *J. Med. Chem.*, 2005, **48**, 2548.
97. J. C. Glennon, G. Van Scharrenburg, E. Ronken, M. B. Hesselink, J. H. Reinders, M. Van Der Neut, S. K. Long, R. W. Feenstra and A. C. McCreary, *Synapse*, 2006, **60**, 599.
98. J. Bronzova, C. Sampaio, R. A. Hauser, A. E. Lang, O. Rascol, A. Theeuwes, S. V. van de Witte and G. van Scharrenburg, *Mov. Disord.*, 2010, **25**, 738.
99. H. Zhang, N. Ye, S. Zhou, L. Guo, L. Zheng, Z. Liu, B. Gao, X. Zhen and A. Zhang, *J. Med. Chem.*, 2011, **54**, 4324.

CHAPTER 14

Tryptophan Metabolism in Parkinson's Disease: Future Therapeutic Possibilities

ZSÓFIA MAJLÁTH[a] AND LÁSZLÓ VÉCSEI*[a,b]

[a] Department of Neurology, University of Szeged, Szeged, Hungary;
[b] Neurology Research Group of the Hungarian Academy of Sciences and University of Szeged, Szeged, Hungary
*Email: vecsei.laszlo@med.u-szeged.hu

14.1 Introduction

Parkinson's disease (PD) is one of the most common neurodegenerative diseases, with a lifetime risk of 2%.[1] Epidemiological studies have indicated an overall prevalence of 1.5%,[2] with no gender differences;[3] the prevalence and incidence data also reveal an increasing tendency with age.[4] The clinical symptoms were first described by James Parkinson in 1817,[5,6] and the disease has borne his name since the 19th century, when Charcot first referred to the syndrome as the "maladie de Parkinson".[7] The characteristic motor symptoms of PD comprise rigidity, resting tremor, bradykinesia and postural instability.[7] In addition to these features, other non-motor symptoms, such as cognitive decline,[8,9] sleep disturbances[10] or an autonomic dysfunction[11,12] may also develop.[13,14]

The neuropathological background of the disease involves the selective degeneration of dopaminergic neurons in the *substantia nigra pars compacta* (SNc)[15,16] and the presence of Lewy bodies, the exact molecular pathomechanism of which is still not fully understood.[17] The main mechanisms in the pathogenesis of PD are considered to include mitochondrial disturbances,[18] oxidative stress,[19] protein aggregation, excitotoxicity,[20] immunological mechanisms[21] and genetic factors, but evidence is accumulating that implicates an altered tryptophan (TRP) metabolism in PD,[22,23] which may contribute to the development of the disease. TRP is an essential amino acid which, apart from being involved in protein synthesis, is also a precursor of several biologically active molecules. The main routes of the TRP catabolism are the serotoninergic pathway and the kynurenine pathway (KP).[24] The metabolites of the KP, collectively termed "kynurenines", have been demonstrated to be involved in various physiological and pathophysiological processes.[25–28]

14.2 Tryptophan Metabolism

14.2.1 Serotonin Pathway

Serotonin (or 5-hydroxytryptamine, 5HT) was discovered in the early 1940s.[29] Although less than 5% of TRP is catabolized through this pathway, the importance of this metabolic route is outstanding (Figure 14.1). The first step in the pathway (also the rate-limiting step in serotonin biosynthesis) is the hydroxylation of TRP by tryptophan-hydroxylase (TPH) to 5-hydroxytryptophan.[29] Of the two known isoforms of TPH, TPH1 is found in various tissues, including the enterochromaffin cells, leukocytes and pinealocytes, while TPH2 is a brain-specific isoform.[30] The second, and final, step involves the decarboxylation of 5-hydroxytryptophan to yield serotonin.[29] Serotonin can subsequently serve as a precursor in the biosynthesis of melatonin by the action of serotonin-N-acetyl-transferase, furnishing N-acetylserotonin, which is then converted by hydroxyindole-O-methyl-transferase to 5-methoxy-N-acetyltryptamine (melatonin). Serotonin is an important neurotransmitter, which is implicated in the regulation of some of the most important physiological functions including emotional processes, mood,[31] cognition,[32] circadian rhythm[33] and cardiovascular regulation.[34] Alterations in the serotoninergic processes have been demonstrated to be associated with the pathogenesis of a number of disorders, such as irritable bowel syndrome,[35] headache disorders,[36] Alzheimer's disease[37] and depression,[38] among others.

14.2.2 Kynurenine Pathway

In most mammalian tissues, including the human brain, 95% of the TRP is catabolized through the KP.[24,39] This metabolic route is responsible for the synthesis of the essential coenzymes nicotinamide adenine dinucleotide (NAD) and NAD phosphate (Figure 14.1). The central intermediate of this metabolic cascade is L-kynurenine (L-KYN), which is produced from TRP in a two-step process through formylkynurenine. The enzymes that initiate this cascade are indoleamine-2,3-dioxygenase (IDO) and tryptophan-2,3-dioxygenase (TDO).

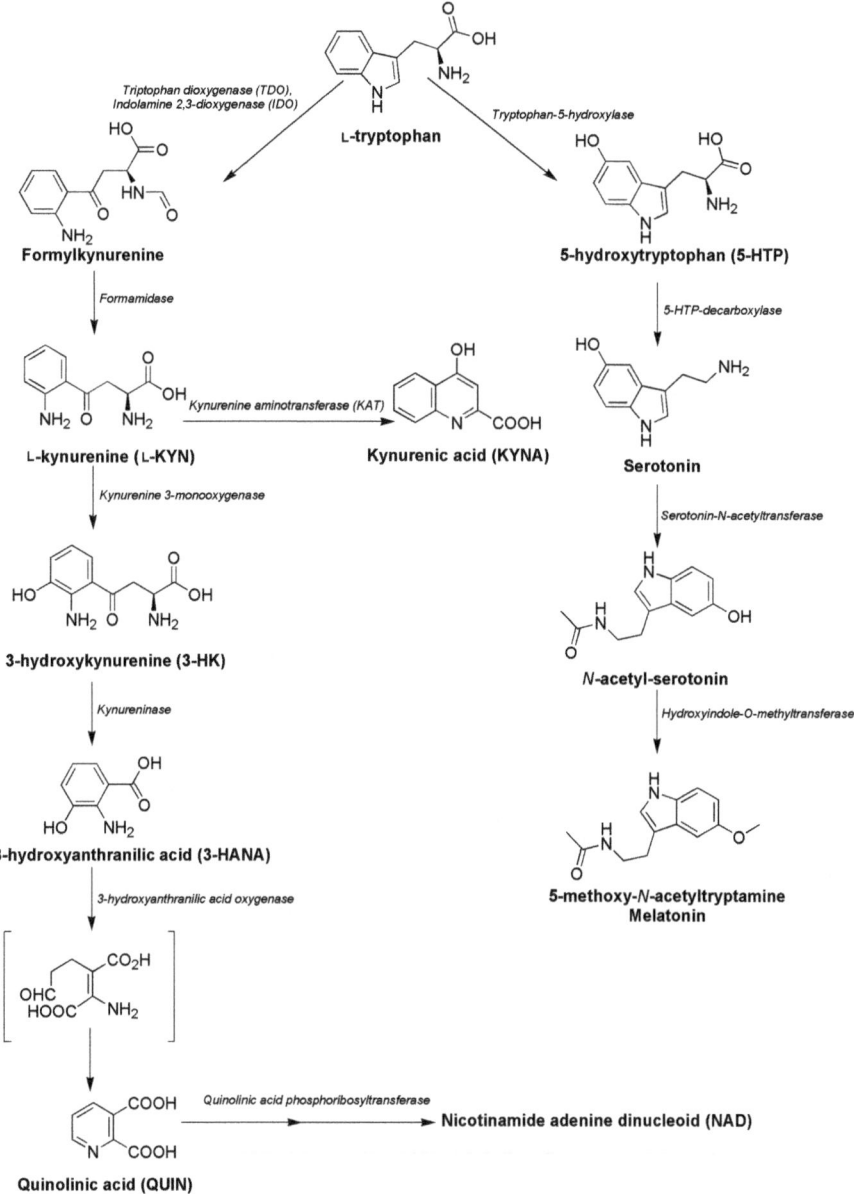

Figure 14.1 Tryptophan metabolism: the kynurenine and serotonin pathways.

L-KYN can be converted *via* two distinct pathways: either through transamination by kynurenine aminotransferases (KATs) to the neuroprotective kynurenic acid (KYNA), or in a sequence of enzymatic steps which produce NAD. The four subtypes of KATs isolated so far[40–42] have slightly different substrate specificities and biochemical properties;[43] they are localized predominantly in the astrocytes within the brain.[44] In the rat brain and the human brain, KYNA is largely produced by KAT-II, whereas in the mouse brain KYNA synthesis can be attributed mainly to the activity of KAT-IV, which is identical to mitochondrial aspartate aminotransferase.[42] KYNA is a broad-spectrum endogenous inhibitor of ionotropic excitatory amino acid receptors,[43] which exerts effects on N-methyl-D-aspartate (NMDA), α-amino-3-hydroxy-5-methyl-4-isoxazolepropionic acid (AMPA) and kainate subtypes of glutamate receptors.[45] It has a high affinity for the strychnine-insensitive glycine-binding site of the NMDA receptors, but with a lower affinity. It is also capable of blocking the glutamate-binding site of the NMDA receptors[46], thereby preventing the deteriorating effects of quinolinic acid (QUIN).[46–49] KYNA is additionally a non-competitive inhibitor of the α7 nicotinic acetylcholine receptor,[50] in this way inhibiting presynaptic glutamate release,[51] which may also attenuate excitotoxic damage. KYNA exhibits a dual effect on the AMPA subtypes of the glutamate receptors in a concentration-dependent manner;[52,53] hence, it may participate in the regulation of neurotransmission. Studies have provided evidence that KYNA may be a ligand for the previously orphan G-protein-coupled receptor GPR35, thus shedding light on new regulating functions of this metabolite.[54] Animal and *in vitro* experiments have demonstrated the neuroprotective capacities of KYNA under different conditions, such as NMDA-, glutamate- and QUIN-induced neurotoxicity or cerebral ischaemia.[55–58]

The other main branch of the metabolic route begins with the formation of 3-hydroxykynurenine (3-HK) from L-KYN *via* the action of kynurenine 3-monooxygenase (KMO). 3-HK is then converted by kynureninase to 3-hydroxyanthranilic acid (3-HANA), and the downstream metabolic cascade yields the NMDA receptor agonist QUIN.[47] A number of studies have afforded evidence that QUIN exerts a neurotoxic effect and causes neuronal damage both in cortical cell cultures and in *in vivo* experiments.[59,60] 3-HK and 3-HANA are potent free radical generators and additionally exhibit neurotoxic properties, contributing to the oxidative damage to proteins and neurons.[61,62] Similarly, QUIN participates in the formation of free radicals; moreover it can decrease anti-oxidant activity and induce lipid peroxidation.[63–65] Besides all these effects, QUIN is implicated in the regulation of glutamate release.[66]

14.3 Pathogenesis of Parkinson's Disease

14.3.1 Some of the Main Aspects of the Pathogenesis of Parkinson's Disease

In spite of the extensive studies that have been conducted, the exact pathogenesis of PD remains ambiguous. However, a growing body of evidence supports the role of mitochondrial impairment,[18,67] oxidative damage[19] and neuroinflammatory processes[68] in the development of PD. The mitochondria play an essential part in energy production and in maintaining the calcium homeostasis of the neurons. Epidemiological data have demonstrated an association between a higher prevalence of PD and exposure to several types of pesticides and environmental toxins capable of mitochondrial damage.[69,70] Although most PD cases are idiopathic, a small proportion of the patients have a family history of this disorder. Wide-ranging genetic investigations have been carried out in recent years, and consequently a number of genetic mutations have been linked to the pathogenesis of PD.[71] Several gene mutations have been proven to exert their effects on mitochondrial function,[72] and a mitochondrial dysfunction has been associated not only with an energy deficit in the cells, but also with oxidative stress and apoptosis induction.[72,73] Mitochondrial dysfunction and oxidative stress have been implicated in the pathogenesis of both sporadic and familial PD forms.[19,67,74,75] Other key factors in the pathogenesis of PD are glutamate excitotoxicity[20] and neuroinflammation,[68] which are likewise closely related to oxidative stress. The elevated levels of certain cytokines and growth factors detected in PD patients[76] points to the association of inflammatory processes with neurodegeneration.[77] The various pathological procedures converge into the final common features of dopaminergic neuronal damage and cell death.[78]

14.3.2 Altered Tryptophan Metabolism in Parkinson's Disease

An altered TRP metabolism has been long implicated in the pathogenesis of neurodegenerative disorders, and imbalances of both the serotoninergic and kynurenine pathways have been detected (Figure 14.2).[27,79–82]

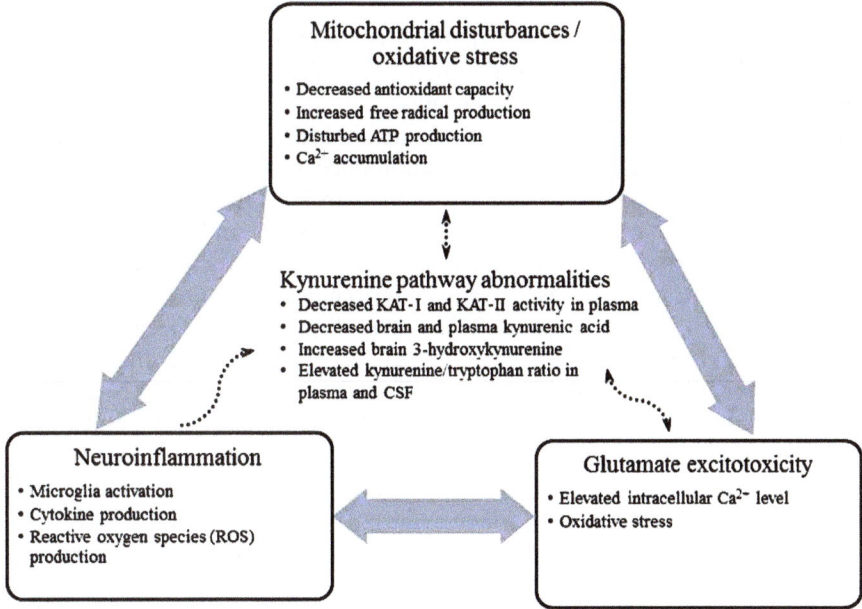

Figure 14.2 Alterations in the KP and their connection with the main pathogenetic pathways of PD.

Human postmortem studies suggested that in neurodegenerative disorders, including PD, the serotoninergic axons are degenerated.[82] Accordingly, decreased serotonin levels have been measured in several brain regions (the *striatum, globus pallidus, substantia nigra*, frontal cortex and thalamus)[83–86] and in the cerebrospinal fluid (CSF) of PD patients.[87] The CSF serotonin levels negatively correlate with the motor symptoms.[87] In addition to these findings, changes have also been observed in the serotonin transporter[88] and serotonin receptor functions[89,90] in the PD brain.[85] Alterations in the serotoninergic system have been attributed to motor and non-motor symptoms, and also to therapy-related complications in PD.[85]

The other branch of TRP catabolism, the KP, is also the focus of research interest, and several important alterations in this metabolic route have been confirmed to be involved in neurodegenerative processes.[23,27,28,91–95]

Impaired mitochondrial Complex I activity has been demonstrated in the *substantia nigra* of the brains of patients with PD.[96] In accordance with the mitochondrial impairment observed in the clinical setting, a number of mitochondrial toxins (some of them found in the environment) have been implicated in the pathogenesis of PD.[97–99] The mitochondrial Complex I inhibitor 1-methyl-4-phenyl-1,2,3,6,-tetrahydropyridine (MPTP)[100,101] is a highly selective dopaminergic neurotoxin that is widely used as an animal model for PD.[102,103] Immunohistochemical methods have revealed that, besides destroying the dopaminergic neurons in the SNc of the mouse brain, MPTP treatment can be associated with decreased KAT-I immunoreactivity in the remaining SNc neurons.[104] Similar effects have been reported following 6-hydroxydopamine treatment.[105]

Luchowski *et al.* demonstrated that mitochondrial inhibitors are able to impair the KAT activity in rat cerebral cortical slices leading to a decreased KYNA concentration.[106] Their results revealed that the 1-methyl-4-phenylpyridinium ion (MPP$^+$), the toxic metabolite of MPTP,[101] is a selective inhibitor of KAT-II in a concentration-dependent manner, while 3-nitropropionic acid diminishes the activity of both KAT-I and KAT-II.[106]

Human postmortem studies indicated decreased KYN and KYNA concentrations in the frontal cortex, putamen and SNc of PD patients as compared with healthy controls, whereas the 3-hydroxykynurenine (3-OH-KYN) levels were increased.[107] Alterations in the TRP metabolism have also been confirmed in the periphery in PD patients: in plasma, decreased KAT-I and KAT-II activity and KYNA[108] and TRP levels,[109] and an increased KYN-to-TRP ratio.[109] Analogous alterations have been detected in the CSF: besides the lower TRP level, and elevated KYN-to-TRP ratio,[109] Molina *et al.* observed an enhanced CSF-to-plasma TRP ratio.[110] Widner *et al.* found elevated serum neopterin levels in the plasma, which reflects cellular immune activation in advanced PD.[109] These results tend to corroborate the possible role of neuroinflammation in the pathogenesis of the disease. The experimental data suggest an increased TRP degradation and a shift in the KP towards the formation of neurotoxic metabolites which contribute to the neurodegenerative process.[22]

Alterations in the KYNA level may also influence dopamine levels in the *striatum via* a negative correlation,[111] and thus there may be a link between dopaminergic, glutamatergic and cholinergic neurotransmission. It has been demonstrated experimentally that elevated KYNA levels are able to inhibit dopamine release; at lower concentrations KYNA is assumed to act *via* the α7 nicotinic acetylcholine receptors,[111] while at higher concentrations the action might involve the blockade of glutamate receptors.[112] On the other hand, L-DOPA treatment has been also found to result in a reduced KYNA level.[113] Decreased KYNA levels may contribute to an enhanced vulnerability to excitotoxic effects by facilitating glutamatergic influences.[114]

14.4 Future Therapeutic Possibilities

As imbalances in the TRP metabolism are strongly involved in the pathogenesis of PD, it seems reasonable to target this metabolic route in a therapeutic approach. Both neuroprotective and neurotoxic metabolites take part in the KP,[26,46] and are additionally implicated in the regulation of neurotransmission;[115,116] consequently, modulation of this pathway could provide a possibility way to afford neuroprotection and delay the progression of PD (Figure 14.3).

One pharmacological approach could be to elevate the level of the neuroprotective KYNA by its systemic administration. It has long been known that the intracerebral injection of KYNA can mitigate MPTP-induced motors symptoms in rats.[117] Regrettably, the systemic administration of KYNA is

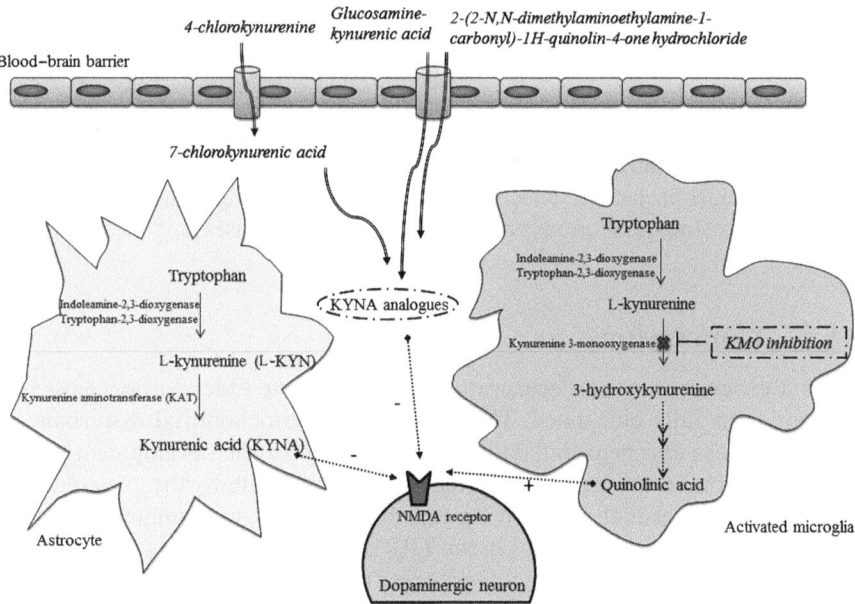

Figure 14.3 Future therapeutic possibilities by modulating the KP.

limited by its poor ability to cross the blood–brain barrier;[118] an alternative method might be the use of KYNA pro-drugs or KYNA analogues. When administered together with L-KYN, probenecid, an inhibitor of KYNA transport from the brain, can effectively raise KYNA levels in the brain.[119] The striatal infusion of 6-hydroxydopamine, a widely used animal model of PD, causes dopaminergic neuronal damage, decreased dopamine levels and motor impairment; all these devastating effects may be prevented by an L-KYN + probenecid treatment.[120]

Another possibility could be modulation of the activity of the enzymes involved in the KP. KMO inhibition appears to be a promising method of neuroprotection by shifting the KP towards the formation of KYNA.[121,122]

KMO inhibition results in an elevated KYNA level in the brain, furthermore, it can also selectively reduce the extracellular glutamate level in the basal ganglia.[123] Miranda et al. demonstrated in 1997, that nicotinylalanine, an inhibitor of kynureninase and KMO, can achieve a significant degree of neuroprotection in the striatal dopaminergic neurons by elevation of the KYNA level, and can prevent NMDA or QUIN-induced neurotoxicity.[57] Moreover, KMO inhibition can also prevent one of the most frequent treatment-related complications, levodopa-induced dyskinesia in an animal model of PD, whereas it did not affect the anti-Parkinsonian effects of levodopa.[124]

Synthetic kynurenine derivatives may also be of importance in the search for therapeutic tools and there is a growing body of experimental data confirming their beneficial properties.[125–127] Some of these kynurenine derivatives have proved to be neuroprotective in the MPTP model of PD by alleviating oxidative stress and improving mitochondrial complex I activity.[128] 4-Chlorokynurenine and the KYNA derivative 7-chlorokynurenate have both been proved to prevent QUIN-induced neurotoxicity.[129,130] Other synthetic KYNA analogues, glucosamine-kynurenic acid and 2-(2-N,N-dimethyl-aminoethylamine-1-carbonyl)-1H-quinolin-4-one hydrochloride, displayed central effects comparable with those of KYNA in in vitro studies,[131,132] and the beneficial neuroprotective effects of the latter compound have also been observed in a transgenic mouse model of Huntington's disease.[126]

14.5 Conclusions

PD is a progressive neurodegenerative disorder whose exact pathogenesis has still not been fully elucidated. It is certain that mitochondrial disturbances, oxidative stress and neuroinflammation contribute to the development of the neurodegenerative process, and in the past decade noteworthy progress has been made as regards the underlying genetic mutations, environmental factors and clinical imaging. Alterations in the TRP metabolism may play a crucial role in this complex pathophysiological process. Targeting the main TRP catabolic route, the KP, may therefore be a promising approach as concerns the development of novel neuroprotective agents.

Acknowledgements

This work was supported by the project TÁMOP-4.2.1/B-09/1/KONV-2010-0005 "Creating the Center of Excellence at the University of Szeged" and by the Neuroscience Research Group of the Hungarian Academy of Sciences and the University of Szeged.

References

1. A. H. Schapira, *Trends Pharmacol. Sci.*, 2009, **30**, 41.
2. M. C. de Rijk, M. M. Breteler, G. A. Graveland, A. Ott, D. E. Grobbee, F. G. van der Meche and A. Hofman, *Neurology*, 1995, **45**, 2143.
3. M. C. de Rijk, C. Tzourio, M. M. Breteler, J. F. Dartigues, L. Amaducci, S. Lopez-Pousa, J. M. Manubens-Bertran, A. Alperovitch and W. A. Rocca, *J. Neurol. Neurosurg. Psychiatry*, 1997, **62**, 10.
4. L. M. de Lau, P. C. Giesbergen, M. C. de Rijk, A. Hofman, P. J. Koudstaal and M. M. Breteler, *Neurology*, 2004, **63**, 1240.
5. J. Parkinson, *An Essay on the Shaking Palsy*, Wiley-Blackwell, Chichester, 1817.
6. P. A. Kempster, B. Hurwitz and A. J. Lees, *Neurology*, 2007, **69**, 482.
7. C. G. Goetz, *Mov. Disord.*, 1986, **1**, 27.
8. D. Aarsland, K. Andersen, J. P. Larsen, A. Lolk, H. Nielsen and P. Kragh-Sorensen, *Neurology*, 2001, **56**, 730.
9. O. Riedel, J. Klotsche, A. Spottke, G. Deuschl, H. Forstl, F. Henn, I. Heuser, W. Oertel, H. Reichmann, P. Riederer, C. Trenkwalder, R. Dodel and H. U. Wittchen, *J. Neurol.*, 2008, **255**, 255.
10. J. P. Larsen and E. Tandberg, *CNS Drugs*, 2001, **15**, 267.
11. O. Appenzeller and J. E. Goss, *Arch. Neurol.*, 1971, **24**, 50.
12. C. G. Goetz, W. Lutge and C. M. Tanner, *Neurology*, 1986, **36**, 73.
13. M. C. Rodriguez-Oroz, M. Jahanshahi, P. Krack, I. Litvan, R. Macias, E. Bezard and J. A. Obeso, *Lancet Neurol.*, 2009, **8**, 1128.
14. J. Massano and K. P. Bhatia, *Cold Spring Harb. Perspect. Med.*, 2012, **2**, a008870.
15. W. R. Gibb and A. J. Lees, *J. Neurol. Neurosurg. Psychiatry*, 1991, **54**, 388.
16. W. R. Gibb, *Eur. Neurol.*, 1991, **31**(1), 48.
17. A. J. Lees, *Neurology*, 2009, **72**, S2.
18. J. T. Greenamyre, G. MacKenzie, T. I. Peng and S. E. Stephans, *Biochem. Soc. Symp.*, 1999, **66**, 85.
19. (a) P. Jenner, *Ann. Neurol.*, 2003, **53**(3), S26; (b) *Ann. Neurol.*, 2003, **53**(3), S36.
20. F. Blandini, *Funct. Neurol.*, 2010, **25**, 65.
21. (a) S. Hunot and E. C. Hirsch, *Ann. Neurol.*, 2003, **53**(3), S49; (b) *Ann. Neurol.*, 2003, **53**(3), S58.
22. N. Szabo, Z. T. Kincses, J. Toldi and L. Vecsei, *J. Neurol. Sci.*, 2011, **310**, 256.

23. D. Zadori, P. Klivenyi, J. Toldi, F. Fulop and L. Vecsei, *J. Neural. Transm.*, 2012, **119**, 275.
24. R. Schwarcz, *Biochem. Soc. Trans.*, 1993, **21**, 77.
25. R. Schwarcz, J. P. Bruno, P. J. Muchowski and H. Q. Wu, *Nat. Rev. Neurosci.*, 2012, **13**, 465.
26. F. Moroni, *Eur. J. Pharmacol.*, 1999, **375**, 87.
27. K. Sas, H. Robotka, J. Toldi and L. Vecsei, *J. Neurol. Sci.*, 2007, **257**, 221.
28. M. P. Heyes, K. Saito, J. S. Crowley, L. E. Davis, M. A. Demitrack, M. Der, L. A. Dilling, J. Elia, M. J. Kruesi, A. Lackner, S. A. Larsen, K. Lee, H. L. Leonard, S. P. Markey, A. Martin, S. Milstein, M. M. Mouradian, M. R. Pranzatelli, B. J. Quearry, A. Salazar, M. Smith, S. E. Strauss, T. Sunderland, S. W. Swedo and W. W. Tourtellotte, *Brain*, 1992, **115**(5), 1249.
29. A. Sirek and O. V. Sirek, *Can. Med. Assoc. J.*, 1970, **102**, 846.
30. D. J. Walther, J. U. Peter, S. Bashammakh, H. Hortnagl, M. Voits, H. Fink and M. Bader, *Science*, 2003, **299**, 76.
31. S. N. Young and M. Leyton, *Pharmacol. Biochem. Behav.*, 2002, **71**, 857.
32. J. A. Schmitt, M. Wingen, J. G. Ramaekers, E. A. Evers and W. J. Riedel, *Curr. Pharm. Des.*, 2006, **12**, 2473.
33. L. P. Morin, *Ann. Med.*, 1999, **31**, 12.
34. Y. Charnay and L. Leger, *Dialogues Clin. Neurosci.*, 2010, **12**, 471.
35. B. Garvin and J. W. Wiley, *Curr. Gastroenterol. Rep.*, 2008, **10**, 363.
36. P. J. Goadsby, *Handbook of Experimental Pharmacology*, ed. Christoph Stein, Springer, New York, 2007, vol. 177, pp. 129–143.
37. P. Newhouse, A. Tatro, M. Naylor, K. Quealey and P. Delgado, *Am J. Geriatr. Psychiatry*, 2002, **10**, 483.
38. C. B. Nemeroff and M. J. Owens, *Clin. Chem.*, 2009, **55**, 1578.
39. H. Wolf, *Scand. J. Clin. Lab. Invest., Suppl.*, 1974, **136**, 1.
40. E. Okuno, M. Nakamura and R. Schwarcz, *Brain Res.*, 1991, **542**, 307.
41. P. Yu, Z. Li, L. Zhang, D. A. Tagle and T. Cai, *Gene*, 2006, **365**, 111.
42. P. Guidetti, L. Amori, M. T. Sapko, E. Okuno and R. Schwarcz, *J. Neurochem.*, 2007, **102**, 103.
43. Q. Han, T. Cai, D. A. Tagle and J. Li, *Cell Mol. Life Sci.*, 2010, **67**, 353.
44. G. J. Guillemin, S. J. Kerr, G. A. Smythe, D. G. Smith, V. Kapoor, P. J. Armati, J. Croitoru and B. J. Brew, *J. Neurochem.*, 2001, **78**, 842.
45. F. Moroni, A. Cozzi, M. Sili and G. Mannaioni, *J. Neural Transm.*, 2012, **119**, 133.
46. T. W. Stone, *Pharmacol. Rev.*, 1993, **45**, 309.
47. T. W. Stone and M. N. Perkins, *Eur. J. Pharmacol.*, 1981, **72**, 411.
48. P. J. Birch, C. J. Grossman and A. G. Hayes, *Eur. J. Pharmacol.*, 1988, **154**, 85.
49. T. W. Stone and J. I. Addae, *Eur. J. Pharmacol.*, 2002, **447**, 285.
50. C. Hilmas, E. F. Pereira, M. Alkondon, A. Rassoulpour, R. Schwarcz and E. X. Albuquerque, *J. Neurosci.*, 2001, **21**, 7463.
51. R. Carpenedo, A. Pittaluga, A. Cozzi, S. Attucci, A. Galli, M. Raiteri and F. Moroni, *Eur. J. Neurosci.*, 2001, **13**, 2141.

52. C. Prescott, A. M. Weeks, K. J. Staley and K. M. Partin, *Neurosci. Lett.*, 2006, **402**, 108.
53. E. Rozsa, H. Robotka, L. Vecsei and J. Toldi, *J. Neural Transm.*, 2008, **115**, 1087.
54. J. Wang, N. Simonavicius, X. Wu, G. Swaminath, J. Reagan, H. Tian and L. Ling, *J. Biol. Chem.*, 2006, **281**, 22021.
55. K. Nozaki and M. F. Beal, *J. Cereb. Blood Flow Metab.*, 1992, **12**, 400.
56. K. Sas, H. Robotka, E. Rozsa, M. Agoston, G. Szenasi, G. Gigler, M. Marosi, Z. Kis, T. Farkas, L. Vecsei and J. Toldi, *Neurobiol. Dis.*, 2008, **32**, 302.
57. A. F. Miranda, R. J. Boegman, R. J. Beninger and K. Jhamandas, *Neuroscience*, 1997, **78**, 967.
58. A. Kumar and G. N. Babu, *Neurochem. Res.*, 2010, **35**, 636.
59. J. P. Kim and D. W. Choi, *Neuroscience*, 1987, **23**, 423.
60. R. Schwarcz, W. O. Whetsell, Jr. and R. M. Mangano, *Science*, 1983, **219**, 316.
61. L. E. Goldstein, M. C. Leopold, X. Huang, C. S. Atwood, A. J. Saunders, M. Hartshorn, J. T. Lim, K. Y. Faget, J. A. Muffat, R. C. Scarpa, L. T. Chylack, Jr., E. F. Bowden, R. E. Tanzi and A. I. Bush, *Biochemistry*, 2000, **39**, 7266.
62. T. W. Stone, *Toxicon*, 2001, **39**, 61.
63. E. Rodriguez-Martinez, A. Camacho, P. D. Maldonado, J. Pedraza-Chaverri, D. Santamaria, S. Galvan-Arzate and A. Santamaria, *Brain Res.*, 2000, **858**, 436.
64. W. M. Behan, M. McDonald, L. G. Darlington and T. W. Stone, *Br. J. Pharmacol,*, 1999, **128**, 1754.
65. C. Rios and A. Santamaria, *Neurochem. Res.*, 1991, **16**, 1139.
66. R. G. Tavares, C. I. Tasca, C. E. Santos, L. B. Alves, L. O. Porciuncula, T. Emanuelli and D. O. Souza, *Neurochem. Int.*, 2002, **40**, 621.
67. H. Bueler, *Exp. Neurol.*, 2009, **218**, 235.
68. M. G. Tansey and M. S. Goldberg, *Neurobiol. Dis.*, 2010, **37**, 510.
69. T. P. Brown, P. C. Rumsby, A. C. Capleton, L. Rushton and L. S. Levy, *Environ. Health Perspect.*, 2006, **114**, 156.
70. J. M. Hatcher, K. D. Pennell and G. W. Miller, *Trends Pharmacol. Sci.*, 2008, **29**, 322.
71. S. Lesage and A. Brice, *Hum. Mol. Genet.*, 2009, **18**, R48.
72. B. Thomas and M. F. Beal, *Hum. Mol. Genet.*, 2007, **16**(2), R183.
73. H. Yang, H. Y. Zhou, B. Li, G. Z. Niu and S. D. Chen, *J. Neuroimmune Pharmacol.*, 2007, **2**, 276.
74. I. P. de Castro, L. M. Martins and S. H. Loh, *Mol. Neurobiol.*, 2011, **43**, 80.
75. J. C. Rochet, B. A. Hay and M. Guo, *Prog. Mol. Biol. Transl. Sci.*, 2012, **107**, 125.
76. M. Mogi, M. Harada, T. Kondo, P. Riederer, H. Inagaki, M. Minami and T. Nagatsu, *Neurosci. Lett.*, 1994, **180**, 147.

77. M. Sawada, K. Imamura and T. Nagatsu, *J. Neural Transm., Suppl.*, 2006, **70**, 373.
78. M. Vila and S. Przedborski, *Nat. Rev. Neurosci.*, 2003, **4**, 365.
79. L. Szalardy, P. Klivenyi, D. Zadori, F. Fulop, J. Toldi and L. Vecsei, *Curr. Med. Chem.*, 2012, **19**, 1899.
80. D. Zadori, P. Klivenyi, L. Szalardy, F. Fulop, J. Toldi and L. Vecsei, *J. Neurol. Sci.*, 2012, **322**, 187.
81. T. W. Stone, G. M. Mackay, C. M. Forrest, C. J. Clark and L. G. Darlington, *Clin. Chem. Lab. Med.*, 2003, **41**, 852.
82. E. C. Azmitia and R. Nixon, *Brain Res.*, 2008, **1217**, 185.
83. B. Scatton, F. Javoy-Agid, L. Rouquier, B. Dubois and Y. Agid, *Brain Res.*, 1983, **275**, 321.
84. M. Guttman, I. Boileau, J. Warsh, J. A. Saint-Cyr, N. Ginovart, T. McCluskey, S. Houle, A. Wilson, E. Mundo, P. Rusjan, J. Meyer and S. J. Kish, *Eur. J. Neurol.*, 2007, **14**, 523.
85. P. Huot, S. H. Fox and J. M. Brotchie, *Prog. Neurobiol.*, 2011, **95**, 163.
86. S. J. Kish, J. Tong, O. Hornykiewicz, A. Rajput, L. J. Chang, M. Guttman and Y. Furukawa, *Brain*, 2008, **131**, 120.
87. H. Tohgi, T. Abe, S. Takahashi, J. Takahashi and H. Hamato, *Neurosci. Lett.*, 1993, **150**, 71.
88. M. Politis, K. Wu, C. Loane, L. Kiferle, S. Molloy, D. J. Brooks and P. Piccini, *Neurobiol. Dis.*, 2010, **40**, 216.
89. P. Huot, T. H. Johnston, T. Darr, L. N. Hazrati, N. P. Visanji, D. Pires, J. M. Brotchie and S. H. Fox, *Mov. Disord.*, 2010, **25**, 1399.
90. B. Ballanger, H. Klinger, J. Eche, J. Lerond, A. E. Vallet, D. Le Bars, L. Tremblay, V. Sgambato-Faure, E. Broussolle and S. Thobois, *Mov. Disord.*, 2012, **27**, 84.
91. T. W. Stone, C. M. Forrest, N. Stoy and L. G. Darlington, *J. Neural Transm.*, 2012, **119**, 261.
92. C. M. Forrest, G. M. Mackay, N. Stoy, S. L. Spiden, R. Taylor, T. W. Stone and L. G. Darlington, *J. Neurochem.*, **112**, 112.
93. E. Gulaj, K. Pawlak, B. Bien and D. Pawlak, *Adv. Med. Sci.*, 2010, **55**, 204.
94. C. Rajda, J. Bergquist and L. Vecsei, *J. Neural Transm., Suppl.*, 2007, **72**, 323.
95. R. Rejdak, A. Junemann, P. Grieb, S. Thaler, F. Schuettauf, T. Choragiewicz, T. Zarnowski, W. A. Turski and E. Zrenner, *Pharmacol. Rep.*, 2011, **63**, 1324.
96. A. H. Schapira, J. M. Cooper, D. Dexter, J. B. Clark, P. Jenner and C. D. Marsden, *J. Neurochem.*, 1990, **54**, 823.
97. P. Jenner, *Trends Neurosci.*, 2001, **24**, 245.
98. D. A. Di Monte, M. Lavasani and A. B. Manning-Bog, *Neurotoxicology*, 2002, **23**, 487.
99. M. A. Collins and E. J. Neafsey, *Neurotoxicol. Teratol.*, 2002, **24**, 571.
100. J. W. Langston, E. B. Langston and I. Irwin, *Acta Neurol. Scand., Suppl.*, 1984, **100**, 49.

101. J. W. Langston, I. Irwin, E. B. Langston and L. S. Forno, *Neurosci. Lett.*, 1984, **48**, 87.
102. J. Bove and C. Perier, *Neuroscience*, 2012, **211**, 51.
103. A. Schober, *Cell Tissue Res.*, 2004, **318**, 215.
104. E. Knyihar-Csillik, B. Csillik, M. Pakaski, B. Krisztin-Peva, E. Dobo, E. Okuno and L. Vecsei, *Neuroscience*, 2004, **126**, 899.
105. E. Knyihar-Csillik, Z. Chadaide, A. Mihaly, B. Krisztin-Peva, R. Fenyo and L. Vecsei, *Acta Neuropathol.*, 2006, **112**, 127.
106. P. Luchowski, E. Luchowska, W. A. Turski and E. M. Urbanska, *Neurosci. Lett.*, 2002, **330**, 49.
107. T. Ogawa, W. R. Matson, M. F. Beal, R. H. Myers, E. D. Bird, P. Milbury and S. Saso, *Neurology*, 1992, **42**, 1702.
108. Z. Hartai, P. Klivenyi, T. Janaky, B. Penke, L. Dux and L. Vecsei, *J. Neurol. Sci.*, 2005, **239**, 31–35.
109. B. Widner, F. Leblhuber and D. Fuchs, *J. Neural Transm.*, 2002, **109**, 181.
110. J. A. Molina, F. J. Jimenez-Jimenez, P. Gomez, C. Vargas, J. A. Navarro, M. Orti-Pareja, T. Gasalla, J. Benito-Leon, F. Bermejo and J. Arenas, *J. Neurol. Sci.*, 1997, **150**, 123.
111. A. Rassoulpour, H. Q. Wu, S. Ferre and R. Schwarcz, *J. Neurochem.*, 2005, **93**, 762.
112. S. Erhardt, H. Oberg and G. Engberg, *Naunyn-Schmiedeberg's Arch. Pharmacol.*, 2001, **363**, 21.
113. H. Q. Wu, A. Rassoulpour and R. Schwarcz, *J. Neural Transm.*, 2002, **109**, 239.
114. B. Poeggeler, A. Rassoulpour, P. Guidetti, H. Q. Wu and R. Schwarcz, *Dev. Neurosci.*, 1998, **20**, 146.
115. J. Banerjee, M. Alkondon and E. X. Albuquerque, *Biochem. Pharmacol.*, 2012, **84**, 1078.
116. M. Alkondon, E. F. Pereira and E. X. Albuquerque, *Biochem. Pharmacol.*, 2011, **82**, 842.
117. W. C. Graham, R. G. Robertson, M. A. Sambrook and A. R. Crossman, *Life Sci.*, 1990, **47**, PL91.
118. S. Fukui, R. Schwarcz, S. I. Rapoport, Y. Takada and Q. R. Smith, *J. Neurochem.*, 1991, **56**, 2007.
119. A. Santamaria, C. Rios, F. Solis-Hernandez, J. Ordaz-Moreno, L. Gonzalez-Reynoso, M. Altagracia and J. Kravzov, *Neuropharmacology*, 1996, **35**, 23.
120. D. Silva-Adaya, V. Perez-De La Cruz, J. Villeda-Hernandez, P. Carrillo-Mora, I. G. Gonzalez-Herrera, E. Garcia, L. Colin-Barenque, J. Pedraza-Chaverri and A. Santamaria, *Neurotoxicol. Teratol.*, 2011, **33**, 303.
121. F. Moroni, A. Cozzi, F. Peruginelli, R. Carpenedo and D. E. Pellegrini-Giampietro, *Adv. Exp. Med. Biol.*, 1999, **467**, 199.
122. D. Zwilling, S. Y. Huang, K. V. Sathyasaikumar, F. M. Notarangelo, P. Guidetti, H. Q. Wu, J. Lee, J. Truong, Y. Andrews-Zwilling, E. W. Hsieh, J. Y. Louie, T. Wu, K. Scearce-Levie, C. Patrick, A. Adame,

F. Giorgini, S. Moussaoui, G. Laue, A. Rassoulpour, G. Flik, Y. Huang, J. M. Muchowski, E. Masliah, R. Schwarcz and P. J. Muchowski, *Cell*, 2011, **145**, 863.
123. F. Moroni, A. Cozzi, R. Carpendo, G. Cipriani, O. Veneroni and E. Izzo, *Neuropharmacology*, 2005, **48**, 788.
124. L. Gregoire, A. Rassoulpour, P. Guidetti, P. Samadi, P. J. Bedard, E. Izzo, R. Schwarcz and T. Di Paolo, *Behav. Brain Res.*, 2008, **186**, 161.
125. T. W. Stone, *Trends Pharmacol. Sci.*, 2000, **21**, 149.
126. D. Zadori, G. Nyiri, A. Szonyi, I. Szatmari, F. Fulop, J. Toldi, T. F. Freund, L. Vecsei and P. Klivenyi, *J. Neural Transm.*, 2011, **118**, 865.
127. K. Nagy, I. Plangar, B. Tuka, L. Gellert, D. Varga, I. Demeter, T. Farkas, Z. Kis, M. Marosi, D. Zadori, P. Klivenyi, F. Fulop, I. Szatmari, L. Vecsei and J. Toldi, *Bioorg. Med. Chem.*, 2011, **19**, 7590.
128. D. Acuna-Castroviejo, V. Tapias, L. C. Lopez, C. Doerrier, E. Camacho, M. D. Carrion, F. Mora, A. Espinosa and G. Escames, *Brain Res. Bull.*, 2011, **85**, 133.
129. A. C. Foster, C. L. Willis and R. Tridgett, *Eur. J. Neurosci.*, 1990, **2**, 270.
130. H. Q. Wu, S. C. Lee and R. Schwarcz, *Eur. J. Pharmacol.*, 2000, **390**, 267.
131. J. Fuvesi, C. Somlai, H. Nemeth, H. Varga, Z. Kis, T. Farkas, N. Karoly, M. Dobszay, Z. Penke, B. Penke, L. Vecsei and J. Toldi, *Pharmacol. Biochem. Behav.*, 2004, **77**, 95.
132. M. Marosi, D. Nagy, T. Farkas, Z. Kis, E. Rozsa, H. Robotka, F. Fulop, L. Vecsei and J. Toldi, *J. Neural Transm.*, 2010, **117**, 183.

CHAPTER 15

Role of P2X7 Receptor Signaling in the Treatment of Parkinson's Disease and Other Neurodegenerative Disorders

TAKATO TAKENOUCHI,*[a] KAZUNARI SEKIYAMA,[b] MASAYO FUJITA,[b] SHUEI SUGAMA,[c] YOSHIFUMI IWAMARU,[d] HIROSHI KITANI[a] AND MAKOTO HASHIMOTO*[b]

[a] Animal Immune and Cell Biology Research Unit, Division of Animal Sciences, National Institute of Agrobiological Sciences, Tsukuba, Japan; [b] Division of Sensory and Motor Systems, Tokyo Metropolitan Institute of Medical Science, Tokyo, Japan; [c] Department of Physiology, Nippon Medical School, Tokyo, Japan; [d] Prion Disease Research Center, National Institute of Animal Health, Tsukuba, Japan
*Email: ttakenou@affrc.go.jp; hashimoto-mk@igakuken.or.jp

15.1 Introduction

Parkinson's disease (PD) is a progressive neurodegenerative disorder of the central nervous system (CNS). PD is more common in the elderly and affects over 6 million people worldwide. Four motor symptoms are generally considered as cardinal signs of PD: resting tremor, rigidity, bradykinesia, and postural instability.[1] These motor symptoms result from a deficiency in dopamine due to degeneration of the dopaminergic neurons in the *substantia* nigra (SN) *pars compacta*. Although levodopa is effective for management of motor symptoms for several years, patients develop motor complications called dyskinesia.[2] In addition, PD patients may suffer from non-motor symptoms, such as dementia and depression, which are progressive during the course of the disease and impair quality of life. Currently, there is no drug or treatment that is curative for PD. Thus, the deleterious effects of PD and related neurodegenerative disorders are serious social problems.

Most PD brains are neuropathologically characterized by formation of eosinophilic inclusions, called "Lewy bodies" (LBs) within surviving neurons. The major proteinous component of LBs is the presynaptic protein α-synuclein, which accumulates in aggregated form. Mechanistically, mature fibrils of α-synuclein sequestered in LBs may be harmless, whereas oligomers and protofibrils of α-synuclein may be causative for diverse neurological alterations such as mitochondrial damage, dysfunction of the ubiquitin–proteasome system, and loss of membrane integrity.[3] Furthermore, a recent study suggests that α-synuclein released *via* an unconventional pathway may propagate to neighboring neurons to stimulate degeneration.[4] In some familial PD cases, a causative link between α-synuclein and PD pathogenesis is clear since three point mutations or multiplications (duplication or triplication) of the α-synuclein gene have been identified.[5] Taken together, these results suggest that α-synuclein may play a central role in the pathogenesis of PD.

Another neuropathological feature of PD is enhancement of neuroinflammation by glial cells, including activated microglia and reactive astrocytes.[6] These glial cells, particularly activated microglia, produce several pro-inflammatory cytokines that are primarily used for neuronal protection against damage such as traumatic injuries, infection, and aging. However, once chronic inflammation occurs, excessive production of such cytokines may exacerbate inflammatory reactions, leading to neurotoxicity. Therefore, inhibition of neuroinflammation is a reasonable strategy to ameliorate neurodegeneration in PD.[7] In this context, since the P2X7 purinergic receptor (P2X7R) is a potent stimulator of neuroinflammation, inhibition of the P2X7R signaling pathway may provide beneficial effects in the treatment of PD.[7] Thus, this chapter focuses on P2X7R as a drug target for immunomodulation in the CNS, and the possible role of this approach as a therapeutic strategy for PD.

15.2 Role of Neuroinflammation in the Progression of Parkinson's Disease

15.2.1 Neuroinflammation in Parkinson's Disease Brains and Animal Models of Parkinson's Disease

Neuroinflammation in PD brains is typified by the accumulation of activated microglia and reactive astrocytes.[8] Microglia are the resident macrophages of the CNS, and their activation plays a pivotal role in induction of neuroinflammation.[7] Significant activation of microglia can be detected in the SN pars compacta in postmortem PD brains,[9,10] and *in vivo* experiments using positron emission tomography show parallel changes in microglial activation and corresponding dopaminergic terminal loss in the affected nigrostriatal pathway in early PD.[11,12] In addition, the appearance of reactive astrocytes has been characterized as another feature of neuroinflammation in PD, and a mild increase in the number of astrocytes immunostained with glial fibrillary acid protein has been observed in the SN *pars compacta* of postmortem PD brains.[13]

Various animal models have been developed to investigate the molecular basis of PD. Since PD is an age-dependent chronic disease, and since the genetic and neuropathological importance of α-synuclein is well established in PD, transgenic mice expressing mutant or wild-type α-synuclein might provide a chronic model to investigate the pathogenesis of PD. In most α-synuclein transgenic mice, astrogliosis occurs concomitantly with accumulation of α-synuclein during the early stage of neurodegeneration.[6] However, neither microglial activation nor dopaminergic neuronal cell death is observed, suggesting that overexpression of α-synuclein (either mutant or wild-type) in neurons may be insufficient to re-capitulate PD pathology in the mouse brain.[6] In contrast, activated microglia are significantly increased in the brains of neurotoxin-induced PD models.[7] Neurotoxins such as 6-hydroxydopamine (6-OHDA), 1-methyl-4-phenyl-1,2,3,6-tetrahydropyridine (MPTP), rotenone, and paraquat have been used in rats or mice to cause acute PD-like syndrome, which mimics the neurological symptoms and exhibits relatively selective dopaminergic neurodegeneration, although aggregation of endogenous α-synuclein is unclear.[14] Furthermore, intranigral injection of lipopolysaccharide (LPS), a component of the Gram-negative bacterial cell wall and a potent inducer of the innate immune response, results in dopaminergic neuronal loss, as seen in PD brains.[15,16] Thus, these results suggest that α-synuclein transgenic mice and neurotoxin models have both advantages and disadvantages for analysis of neuroinflammation in PD. In this context, it is worth noting that LPS models based on α-synuclein transgenic mice have been successfully established.[17]

15.2.2 Role of IL-1β in Neuroinflammation in the Progression of Parkinson's Disease

Interleukin-1β (IL-1β) is one of several powerful pro-inflammatory cytokines that are mainly produced in activated microglia, and to a lesser extent in astrocytes.[18] IL-1β can exert both neuroprotective and neurotoxic effects depending on its concentration, site and duration of action [Figure 15.1(a)].[19] Among the pro-inflammatory cytokines from activated microglia, IL-1β is particularly important since its expression level is elevated in ventricular cerebrospinal fluid and striatal dopaminergic regions in PD patients.[20,21] Notably, chronic expression of IL-1β in the rat SNs achieved with an adenoviral vector elicited most of the characteristics of PD, including progressive dopaminergic cell death, akinesia and glial activation.[22] As further

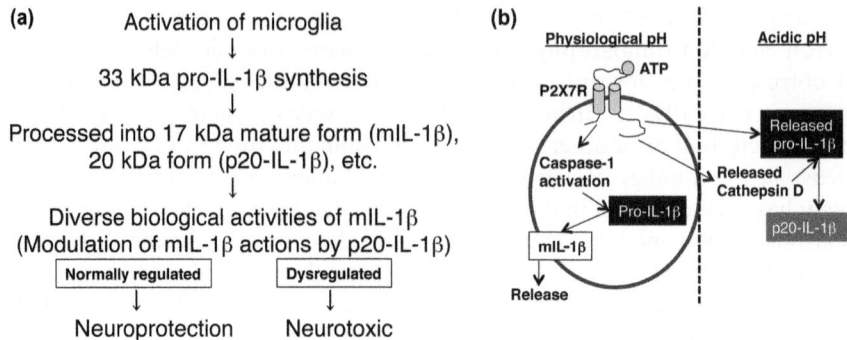

Figure 15.1 (a) Neuroprotective and neurotoxic effects of IL-1β in the CNS. Microglial cells are the major source of IL-1β in the CNS. After activation of microglia by inflammatory agents such as LPS, IL-1β is synthesized as a 33 kDa precursor protein (pro-IL-1β) that is then processed into at least two fragments of 17 or 20 kDa. The 17 kDa fragment is a biologically active mature form of IL-1β (mIL-1β), and the 20 kDa fragment (p20-IL-1β) may modulate the biological actions of mIL-1β. Since mIL-1β possesses numerous and diverse biological activities, its production and release are tightly controlled. When this regulatory system functions normally, IL-1β exerts neuroprotective effects. However, if the system is dysregulated, IL-1β may cause adverse neurotoxic effects. (b) A schematic representation of the proposed mechanisms for processing of pro-IL-1β in microglial cells. At physiological pH, activation of P2X7R by ATP elicits caspase-1 activation followed by processing of pro-IL-1β into mIL-1β. Under acidic extracellular pH, production of mIL-1β decreases and the pro-IL-1β released after ATP stimulation is processed into p20-IL-1β by the released lysosomal enzyme, cathepsin D, in the extracellular space.

support for the relevance of IL-1β to the pathogenesis of PD, a genetic polymorphism in the IL-1β promoter region at position −511 has been associated with the age at onset of Japanese PD patients.[23,24] Similarly, an association of IL-1β gene polymorphisms with PD has been found in Caucasian patients.[25,26] These studies suggest that regulation of IL-1β expression and function might be closely associated with the progression of PD. However, future studies are required to clarify the genetic association of IL-1β gene polymorphisms with PD, since recent studies have failed to show a significant association.[27–30]

IL-1β possesses numerous and diverse biological activities. Therefore, the maturation and release of this cytokine from microglia or other cells of monocyte/macrophage lineage are tightly controlled [Figure 15.1(b)].[31] IL-1β is synthesized as a 33 kDa precursor protein (pro-IL-1β) upon stimulation with inflammatory agents such as LPS. Pro-IL-1β lacks a secretory signal sequence and thus is not transported through the classical endoplasmic reticulum/Golgi-mediated pathway. To exert biological effects, pro-IL-1β must be processed into a 17 kDa mature form (mIL-1β) and released into the extracellular space through unconventional secretion pathways.[32] Activation of caspase-1 (also known as "IL-1β-converting enzyme") is a key event in the maturation of pro-IL-1β. Accumulating evidence indicates the critical involvement of the P2X7 receptor (P2X7R), an adenosine triphosphate (ATP)-gated cation channel, in caspase-1 activation followed by mIL-1β release from macrophage-related cells.[33] Recently, we also demonstrated that P2X7R activation by ATP induces the production and release of an unconventional 20 kDa form of IL-1β (p20-IL-1β) from LPS-primed microglia under acidic extracellular conditions.[34] Since extracellular acidosis develops at sites of inflammation and infection, this finding raises the possibility that p20-IL-1β acts as a novel modulator of biological actions of mIL-1β under some pathophysiological conditions.

To date, P2X7R activation by ATP has been a unique endogenous system that can regulate the processing and release of IL-1β. Given the correlation between IL-1β expression and PD, it is likely that the P2X7R signaling pathway has key roles in the chronic progression of PD through regulation of the production and release of the active form of IL-1β from microglial cells. Thus, modulation of P2X7R functions could serve as a new strategy for PD treatment.

15.3 Expression in the Central Nervous System and Drugs for the Modulation of P2X7R

15.3.1 P2X7R Expression in the Central Nervous System

Two types of P2 purinergic receptors, P2X ionotropic and P2Y metabotropic receptors, have been shown to transduce signals of extracellular ATP. The P2X receptor has seven cloned receptor subtypes (P2X1-P2X7) and intrinsically acts as an ATP-gated ion channel that permits small cations such as K^+, Na^+, and Ca^{2+} to cross the plasma membrane.[35] Unlike other P2X receptor subtypes, P2X7R has a long C-terminal tail, harboring potential motifs for interactions with proteins and lipids in the cytoplasm, and thus activation of this receptor can exert a variety of biological actions *via* the activation of multiple intracellular signaling pathways.[36]

In the CNS, functional expression of P2X7R has mainly been demonstrated in microglia.[37,38] P2X7R activation by ATP induces the secretion of cytokines and chemokines from activated microglia, including mIL-1β, tumor necrosis factor-α (TNF-α), CC-chemokine ligand 3, and CXCL2.[39–42] In intracellular signaling for mIL-1β production, ATP-induced K^+ efflux through P2X7R channels is a critical event in the activation of the inflammasome, a multiprotein complex containing caspase-1, followed by activation of caspase-1 and maturation of pro-IL-1β.[33] P2X7R also mediates the production of reactive oxygen species (ROS) and nitric oxide (NO) in microglia.[43,44] Notably, prolonged activation of P2X7R causes formation of a non-selective membrane pore that is permeable to molecules up to 900 Da in size. Subsequent dilatation of this pore results in apoptotic or necrotic cell death in microglia,[45] whereas, conversely, modest activation of P2X7R causes microglial proliferation.[46]

Functional expression of P2X7R has also been found in other cell types of the CNS, such as astrocytes, oligodendrocytes, and neurons, but to a lesser extent than in microglia.[47,48] As in microglia, P2X7R primarily acts as a cation channel in these cells and transduces extracellular ATP signals to various intracellular signaling pathways with diverse functions. Importantly, activation of P2X7R elicits the release of several neurotransmitters from astrocytes and neurons, such as glutamate, γ-aminobutyric acid, and ATP, which are important mediators in intercellular communication between neurons, astrocytes, and microglia. Therefore, P2X7R expressed in presynaptic sites of neurons and nearby astrocytes plays a number of important roles in the fundamental regulation of synaptic transmission.[49] P2X7R-Mediated responses are less marked in primary neurons compared to primary microglia or astrocytes, but neuroblastoma cell lines such as mouse Neuro2a and human SH-SY5Y cells express relatively high levels of P2X7R and can be used to examine the receptor functions in neural lineage cells.[50,51]

15.3.2 The Dual Neuroprotective and Neurotoxic roles of P2X7R

P2X7R has been postulated to mediate both protective and toxic effects in neurons (Figure 15.2). Under physiological conditions, neuroinflammatory responses accompanied by P2X7R activation primarily have a protective role against neurodegeneration caused by toxic insults. At the affected site, higher levels of extracellular ATP locally released from damaged cells may activate P2X7R in a relatively short period of time, and this may act as a danger signal to alert the immune system and contribute to damage repair.[52] In this regard, P2X7R mediates the production of an endocannabinoid, 2-arachidonoylglycerol, in microglial cells, which has a protective effect against neuronal damage caused by traumatic brain injury.[53] TNF-α released from microglia after P2X7R activation has also been shown to protect neurons against glutamate toxicity.[39] Regarding intracellular signaling pathways, P2X7R-mediated activation of ERK1/2 in cerebellar granule neurons seems to contribute to the protective effects against apoptosis induced by glutamate.[54] Furthermore, P2X7R is coupled to inhibition of glycogen synthase kinase-3 (GSK-3) and confers neuroprotection in cerebellar granule neurons.[55] Collectively, these studies reveal the potential protective roles of P2X7R against neurodegenerative changes.

If insults persistently damage brain cells and danger signals are continuously released, immunocompetent cells are stimulated with endogenous or exogenous inflammatory agents over a longer period of time and cause chronic neuroinflammation that results in exacerbation of neurodegeneration. Under such progressive neurodegenerative conditions, large amounts of ATP released from degenerating neurons may persistently activate P2X7R. Prolonged activation of P2X7R with ATP followed by dilatation of the resulting membrane pore ultimately leads to cell death of both microglia and neurons,[56,57] although astrocytes seem to be relatively resistant to such P2X7R-mediated cell death.[58]

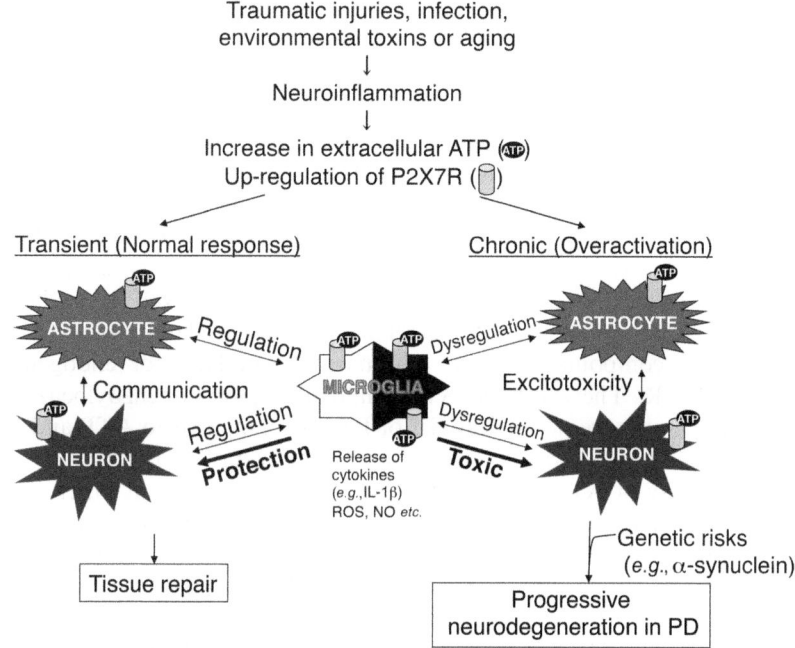

Figure 15.2 Putative roles of P2X7R in the progression of neurodegeneration in PD. Traumatic injuries, infection, environmental toxins and aging may cause neuroinflammation accompanied by an increase in the extracellular ATP concentration and up-regulation of P2X7R expression and function. Although P2X7R is expressed in various cell types of the CNS, immunocompetent microglia play a pivotal role in the induction of neuroinflammatory reactions mediated by P2X7R in the CNS. If the duration of P2X7R activation by ATP is relatively short, modest numbers of cytokines (*e.g.*, IL-1β), ROS, and NO produced from microglia afford protection against neuronal damage and repair the damaged tissue. However, if the neuroinflammation become chronic, P2X7R is persistently activated by ATP released from degenerating cells and overactivated microglia release excessive numbers of neurotoxic factors, which may lead to progressive neurodegeneration, as observed in PD. Expression of genetic risk factors for PD, such as α-synuclein, may facilitate the neurodegenerative process.

In addition, excess numbers of pro-inflammatory cytokines (IL-1β and TNF-α), ROS and NO are produced from microglial cells after sustained P2X7R activation and exert detrimental effects on neurons.[7] Thus, P2X7R appears to mediate acceleration of neurodegeneration under pathophysiological conditions, mainly through activation of microglia.

15.3.3 Modulators of P2X7R Function

Although ATP is considered to be a physiological ligand of P2X7R, its binding affinity and potency for this receptor are relatively low compared to those for other P2X receptor subtypes. In general, mM concentrations of extracellular ATP are required to activate P2X7R to an extent that can be detected *in vitro*. Thus, it is likely that additional physiologically relevant ligands exist for this receptor. In this context, an anti-microbial peptide derived from human cathelicidin (LL37) has been found to act as a direct activator of P2X7R.[59] Administration of adenosine diphosphate (ADP) or adenosine monophosphate (AMP) alone does not activate P2X7R, but brief pretreatment with ATP sensitizes the receptor to stimulation with these ATP metabolites.[60] T-Lymphocytes have a unique mechanism in which exogenous nicotine adenine dinucleotide (NAD) indirectly activates P2X7R *via* ADP-ribosylation of the receptor catalyzed by a toxin-related ADP-ribosylating ectoenzyme, ART2.[61] These compounds may be alternative candidates for endogenous ligands of P2X7R. The synthetic ATP analogue 2'-,3'-O-(4-benzoylbenzoyl)-ATP (BzATP) is a more potent agonist of P2X7R compared to ATP and has been widely used as a pharmacological tool for studying P2X7R.[62]

The low affinity of ATP for P2X7R leads us to speculate that other endogenous factors may be present that can sensitize the receptor to lower concentrations of ATP *in vivo*. Indeed, arachidonic acid (AA), a signaling molecule liberated after plasma membrane phospholipid hydrolysis by phospholipase A2 (PLA2), has been shown to potentiate agonist-induced responses with recombinant and natively expressed P2X7R.[63] Lysophosphatidylcholine (LPC), another molecule produced from the plasma membrane after PLA2 hydrolysis, can also increase the agonist potency at P2X7R.[64,65] Since PLA2 expression is up-regulated at sites of inflammation and P2X7R activation seems to activate PLA2,[66] it is likely that PLA2-catalyzed production of lipid mediators is involved in the positive regulation of P2X7R under physiological conditions.

Drugs that potentiate agonist-induced responses of P2X7R have also been identified. The anti-inflammatory drug tenidap and the anti-biotic polymyxin B enhance the agonist potency at P2X7R[67,68] and the H_1 anti-histamine clemastine has recently been identified as a positive allosteric modulator of recombinant and native human P2X7R.[69] These drugs can be utilized when P2X7R-mediated functions require potentiation.

Regarding inhibition of P2X7R functions, divalent cations including Ca^{2+} and Mg^{2+} and ions such as H^+ and Cl^- are allosteric inhibitors of P2X7R.[70,71] A variety of antagonists have also been developed as pharmacological tools for the study of P2X7R (Figure 15.3), and some are commercially available.[72] Pyridoxalphosphate-6-azophenyl-2',4'-disulfonate (PPADS) was identified as an antagonist for P2X receptor subtypes and has been used to inhibit P2X7R functions.[73] Oxidized ATP (oxATP), a Schiff-base-forming agent, can irreversibly antagonize P2X7R.[74] The isoquinoline KN-62 was first developed as a specific inhibitor of Ca^{2+}/calmodulin-dependent protein kinase II, and additionally characterized as a potent non-competitive antagonist of P2X7R.[75] Brilliant Blue G (BBG) is a relatively selective, non-competitive P2X7R antagonist.[76] However, these antagonists do not have good specificity for P2X7R and also show considerable species variability; *e.g.*, BBG is a more

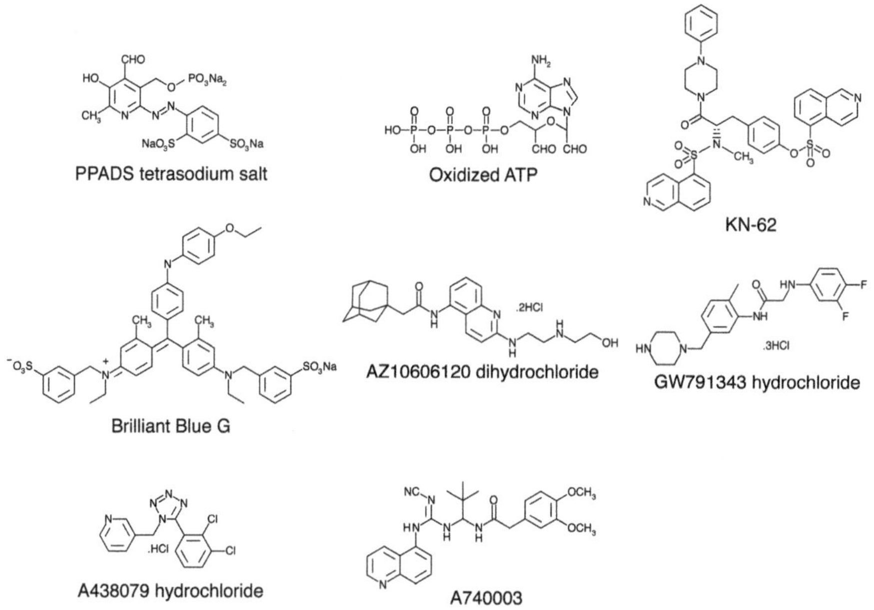

Figure 15.3 Chemical structures of P2X7R antagonists.

effective inhibitor of rat P2X7R than human, whereas KN-62 inhibits human P2X7R but not rat.[77] Similarly, AZ10606120 (also known as "Compound 17") is a negative allosteric modulator of both human and rat P2X7R, whereas GW791343 is a negative allosteric modulator of human P2X7R, but a positive modulator in rat.[78] To overcome these complications, more selective and potent P2X7R antagonists, A438079 and A740003, have been developed by Abbott laboratories.[79] Unlike other P2X7R antagonists, these compounds are reversible and competitive blockers, and tend to show fewer species differences.[77]

15.4 Altered Expression and Function of P2X7R in Parkinson's and Other Neurodegenerative Conditions

The expression and function of P2X7R are up-regulated under various neurodegenerative conditions, which suggests that increased levels of P2X7R are involved in neurodegenerative changes, possibly due to modulation of neuroinflammatory reactions. This also implies that positive or negative modulation of P2X7R may have the potential to prevent progression of neurodegeneration.

LPS administration, which has been used as a model of nigral lesion, may mimic PD-like symptoms in rats or mice because microglia are abundant in the SN and their activation, induced selectively by LPS, elicits degeneration of nigral dopaminergic neurons.[15,16] In line with this finding, LPS injection into the rat *striatum* markedly increases the expression of P2X7R in microglia, suggesting a direct link between LPS-induced neuroinflammation and up-regulation of P2X7R expression.[80] Using toxin-induced PD models, it has also been shown that mRNA expression for P2X7R is up-regulated after MPTP treatment in PC12 cells *in vitro* and in the mouse *striatum* and SN *in vivo*.[81] This study further supports the correlation of increased levels of P2X7R with PD pathology.

In other neurodegenerative disorders, enhanced expression of P2X7R has been shown in adult microglia derived from postmortem Alzheimer's disease (AD) brains, compared to those from non-demented brains.[82] AD is the most common neurodegenerative disorder and is histopathologically characterized by massive neuronal loss, astrocytosis and microglial activation. Accumulation of aggregated amyloid-β (Aβ) in senile plaques is a typical hallmark of the AD brain and Aβ aggregates are known to activate microglia.[83] Intriguing data show that P2X7R expression is required for activation of microglia by Aβ.[84] In a transgenic mouse model of AD, up-regulation of P2X7R expression was found around amyloid plaques where activated microglia and astrocytes are localized.[43] These studies suggest that increased levels of P2X7R are correlated with overactivation of the inflammatory response in microglia, contributing to progression of neuronal loss in AD. Recently, it has been found that activation of P2X7R promotes α-secretase-dependent processing of amyloid precursor protein (APP) in neuronal cells.[85] The stimulatory effect of P2X7R on α-secretase activity is still uncertain because opposite findings that P2X7R activation leads to a reduction of α-secretase activity have also been reported.[86] However, it is an intriguing possibility that P2X7R expressed in neurons and astrocytes is a potential therapeutic target for AD based on modulation of Aβ production.

Expression and function of P2X7R in synaptic terminals are also increased in the brains of Huntington's disease (HD) model mice.[87] HD is a neurodegenerative disorder caused by a CAG triplet-repeat expansion coding for a polyglutamine sequence in the N-terminal region of the Huntington (HTT) protein.[88] ATP-induced synaptic dysregulation and neuronal cell death due to altered expression of neuronal P2X7R may contribute to HD pathogenesis.[87] P2X7R expression is also elevated in apparently normal axon tracts in patients with multiple sclerosis (MS), suggesting the possible involvement of this receptor in the neurodegenerative process in MS.[89]

Altered expression and function of P2X7R in microglia may also be associated with the development of prion disease. Prion disease is caused by an accumulation of abnormal forms of prion protein (PrP^{Sc}) and is histopathologically associated with brain vacuolation, neuronal cell death, astrocytosis, and microgliosis.[90] We have shown that the mRNA level of P2X7R gradually increases in the brain during disease progression in PrP^{Sc}-infected mice.[91] PrP^{Sc}-Infected microglia also exhibited hypersensitivity of P2X7R, as demonstrated by the up-regulation of several receptor functions such as Ca^{2+} influx, membrane pore formation, induction of microglial cell death, and the release of mIL-1β. Thus, it is plausible that alteration of P2X7R expression and function in microglia after PrP^Sc infection may affect the progression of prion disease.

Collectively, these studies suggest that up-regulation of P2X7R is a common mechanism among various neurodegenerative diseases. Further studies are required to clarify the molecular mechanisms affected by altered P2X7R expression and function under neurodegenerative conditions.

15.5 Effects of P2X7R Modulators or Deficiency in Animal Models of Parkinson's and Other Neurodegenerative Diseases

15.5.1 Effects of P2X7R Antagonists or Deficiency

The effects of P2X7R antagonists or P2X7R deficiency have been examined using animal models of various neurodegenerative diseases, as summarized in Table 15.1.

Given the involvement of increased expression of P2X7R in PD, several studies have tested the effects of administration of P2X7R antagonists on disease progression using animal models of PD. Marcellino and colleagues showed that A438079 partially but significantly prevented depletion of striatal dopamine stores in the unilateral 6-OHDA rat model of PD,[92] and proposed that P2X7R antagonists might represent a novel protective strategy for striatal dopamine terminals in PD. This is supported by the finding that oxATP blockade of P2X7R increases neuronal survival after LPS injection in the brain.[80] However, Hracskó et al. reported that P2X7R deficiency or its inhibition by antagonists does not support survival of dopaminergic neurons in experimental models of PD.[81] Although P2X7R antagonists such as BBG and AZ10606120 attenuated the toxic effects of MPTP on PC12 cells *in vitro*, pretreatment with BBG did not show a protective effect against MPTP-induced cell death of primary neurons derived from the SN or against dopaminergic neuron loss after *in vivo* MPTP treatment. Since there is a possibility that P2X7R mediates either neuroprotective or neurotoxic effects depending on the *in vivo* conditions, P2X7R-mediated neuroprotection might still be the dominant effect in such acute toxin-induced PD models. Thus, studies in chronic models of PD, such as transgenic mice models expressing α-synuclein, are needed to evaluate the therapeutic effects of P2X7R antagonists in PD in more detail.

The therapeutic efficacy of P2X7R antagonists has been shown in animal models of other neurodegenerative diseases. Aβ1-42 injection into the rat hippocampus is often used as an AD model that is characterized by increases in P2X7R expression, gliosis, leakiness of the blood–brain barrier (BBB), and hippocampal neuron loss. In this model, BBG reduced the levels of purinergic receptor expression, attenuated gliosis, diminished the leakiness of the BBB, and was neuroprotective.[93] In addition, *in vivo* BBG treatment reduced amyloid plaques in a transgenic mouse model of familial AD through modulation of

Table 15.1 Effects of P2X7R antagonists or P2X7R deficiency in animal models of neurodegenerative diseases. wt = wild type, KO = knock-out.

Antagonist deficiency	Animal	Model	P2X7R expression	Effects	Reference
A438079	Rat	PD (6-OHDA)	Expressed in microglia	Partially prevented striatal dopamine depletion.	92
Deficiency	Mouse (wt, P2X7 KO)	PD (MPTP)	Up-regulation in wt mice	No effect on survival rate and striatal dopamine depletion.	81
oxATP	Rat	PD (LPS injection in the *striatum*)	Up-regulation	Reduced inflammatory responses. Increased neuronal survival.	80
BBG	Rat	AD (Aβ injection)	Up-regulation	Reduced inflammatory responses. Conferred neuroprotection.	93
BBG	Mouse (APP-TG[a])	AD (mutant APP)	No difference between wt and TG mice	Reduced amyloid plaques. Modulated α-secretase and GSK-3 activities.	94
BBG, oxATP	Mouse (HTT-TG)	HD (mutant HTT)	Up-regulation	Prevented neuronal apoptosis. Attenuated body weight loss and motor-coordination deficits.	87
BBG, oxATP	Mouse	MS (MOG[b] injection)	Expressed in oligodendrocyte and myelin	Reduced demyelination. Ameliorated the associated neurological symptoms.	89
Deficiency	Mouse (wt, P2X7 KO)	MS (MOG injection)	—	Reduced incidence of disease, astroglial activation, and axonal damage.	95
oxATP, PPADS, BBG	Rat	Spinal cord injury	Expressed in spinal cord neurons	Reduced inflammatory responses. Conferred neuroprotection. Improved functional recovery.	96, 97
A438079, BBG	Mouse	Seizure (Intra-amygdala injection of kainic acid)	Up-regulation in hippocampus	Reduced seizure duration. Conferred neuroprotection.	98
BBG	Rat	Focal cerebral ischemia	—	Reduced brain damage. Conferred neuroprotection.	99

[a]Transgenic.
[b]Myelin oligodendrocyte glycoprotein-derived peptide.

GSK-3 activity and APP processing.[94] These studies support the beneficial effects of P2X7R antagonists as a treatment strategy for AD.

As with AD, *in vivo* administration of BBG into HD mice ameliorated the pathogenesis of HD by preventing neuronal apoptosis.[87] In animal models of MS, P2X7R deficiency reduced the incidence of the disease, and BBG and oxATP ameliorated the associated neurological symptoms.[89,95] P2X7R antagonists also exhibit beneficial effects in animal models of spinal cord injury, focal cerebral ischemia, and seizure through anti-inflammatory and neuroprotective actions.[96–99] However, the therapeutic effects of P2X7R antagonists in spinal cord injury models failed to be replicated in a recent study, leading to the caution that further preclinical investigations of this potential therapy are required before initiation of clinical trials in patients with severe spinal cord injury.[100]

BBG is derived from a blue food dye called "FD&C Blue No. 1".[101] Besides its P2X7R antagonist action, BBG has been shown to be safe in healthy animals and is approved for use in foodstuffs.[102] Based on its pharmacological properties, which include low toxicity and facilitated transport across the BBB, BBG blockade of P2X7R shows beneficial effects in various animal models of neurodegenerative diseases. The success of BBG also supports the idea that P2X7R could be widely used as a therapeutic target in the treatment of neurodegenerative diseases.

Relatively high concentrations of BBG (~ 50 mg kg^{-1} in rats) have been used for animal studies, but it is yet to be determined if non-toxic doses of BBG are effective for patients with chronic neurodegenerative diseases. Thus, further studies are required to find new compounds that antagonize the P2X7R pathway safely and more effectively than BBG. Besides pharmacological blockade of P2X7R, other potential strategies include knock-down of the expression level of P2X7R through delivery of small RNA molecules (*e.g.*, micro RNA and small interfering RNA) into the brain. A monoclonal antibody to human P2X7R has also been found to functionally block the activation of both recombinant and endogenous P2X7R by ATP.[103] Such alternative strategies could be combined with P2X7R antagonists to suppress the P2X7R signaling pathway.

15.5.2 Possible Effects of Other P2X7R Modulators

To date, both exogenous (*e.g.*, drugs) and endogenous (*e.g.*, lipids) substances have been identified as P2X7R modulators. Among the endogenous agents, lysophospholipids may have roles in the modulation of neurodegenerative progression. We have shown that LPC potentiates intracellular signals mediated by P2X7R activation,[64] while LPC and sphingosylphosphorylcholine (SPC) inhibit P2X7R-mediated maturation and release of IL-1β in microglial cells.[104] Consistent with the latter finding, LPC and SPC exert anti-inflammatory effects and have therapeutic effects in experimental sepsis, despite their intrinsic pro-inflammatory properties.[105–107] Based on our data, it can be speculated that the anti-inflammatory actions of these lysophospholipids appear to be due to their inhibitory effects on P2X7R-mediated inflammation in macrophage-related cells. Since tissue damage following excess inflammation is a key event in the progression of sepsis and neurodegenerative disease, the anti-inflammatory role of lysophospholipids in P2X7R-mediated activation of microglia may also be of potential help in developing therapeutic strategies for the treatment of neurodegenerative diseases. In addition to their anti-inflammatory action, lysophospholipid-induced potentiation of P2X7R-mediated cytoplasmic signals may be beneficial for prevention of neurodegeneration because modest activation of P2X7R leads to neuroprotection.

Up-regulation of autophagy followed by degradation of aggregated α-synuclein has also been suggested as a promising treatment for PD.[108] Autophagy is a fundamental cellular homeostatic process in which cells break down their own components. Indeed, induction of autophagy by rapamycin resulted in decreased expression of α-synuclein in a PD cellular model.[109] In addition, recent evidence has revealed a crucial role for the autophagy pathway in immunity and inflammation.[110] In this context, we and others have reported that P2X7R-mediated signals can control autophagic flux.[111,112] In particular, the autophagy pathway may contribute to the unconventional release of mIL-1β induced by activation of P2X7R.[113–115] Thus, it is expected that modulation of P2X7R may exhibit beneficial effects against PD progression through modulation of the autophagy pathway.

15.6 Conclusion and Perspective

Administration of P2X7R antagonists is effective for treatment of AD and HD in animal models,[87,93] but has failed to ameliorate disease progression in toxin-induced animal models of PD.[81] Similarly, our recent study showed that BBG treatment failed to prevent progression of prion disease, despite inhibitory effects on PrP^{Sc} accumulation.[116] Thus, the results to date show that the effects of P2X7R antagonists are disease-model specific.

There are at least two possibilities for the lack of therapeutic effects in PD and prion disease models following pharmacological antagonism of P2X7R. One explanation is that the effects of P2X7R modulators on protein aggregation, particularly formation of neurotoxic protofibrils, may occur in different ways between disease models, which may lead to differential effects on neurodegeneration. In this context, combined therapy of P2X7R modulators with drugs aimed at the reduction of α-synuclein aggregates, such as autophagy stimulators, may provide an efficient strategy against PD. Alternatively, it is possible that increased levels of P2X7R may participate more in neuroprotective actions rather than neurotoxic actions during the course of the disease, at least in current animal models. If this is the case, stimulation of P2X7R signaling by an agonist may be a promising approach to ameliorate the neurodegenerative changes.

In conclusion, accumulating evidence suggests that the P2X7R pathway plays an important role in the progression of PD and other neurodegenerative diseases. However, the effects of P2X7R antagonists on neurodegenerative diseases remain uncertain. Studies using animal models of PD are required to determine whether modulation of the P2X7R pathway can become a new therapeutic approach for ameliorating PD and other neurodegenerative diseases.

Acknowledgements

This work was supported by a Grant-in-Aid for Science Research on Scientific Research (Category C) and on Innovative Areas ("Brain Environment") from the Ministry of Education, Culture, Sports, Science and Technology of Japan; and by a Grant-in-Aid from the BSE and other Prion Disease Control Project of the Ministry of Agriculture, Forestry, and Fisheries of Japan.

References

1. J. Jankovic, *J. Neurol. Neurosurg. Psychiatry*, 2008, **79**, 368.
2. A. E. Lang and J. A. Obeso, *Lancet Neurol.*, 2004, **3**, 309.
3. J. C. Rochet and P. T. Lansbury, *Curr. Opin. Struct. Biol.*, 2000, **10**, 60.
4. P. Desplats, H. J. Lee, E. J. Bae, C. Patrick, E. Rockenstein, L. Crews, B. Spencer, E. Masliah and S. J. Lee, *Proc. Natl. Acad. Sci. U. S. A.*, 2009, **106**, 13010.
5. T. Gasser, *Expert Rev. Mol. Med.*, 2009, **11**, e22.
6. K. Sekiyama, S. Sugama, M. Fujita, A. Sekigawa, Y. Takamatsu, M. Waragai, T. Takenouchi and M. Hashimoto, *Parkinson's Dis.*, 2012, 271732.
7. C. M. Long, M. Sullivan and Y. M. Nolan, *Prog. Neurobiol.*, 2009, **89**, 277.
8. M. G. Tansey and M. S. Goldberg, *Neurobiol. Dis.*, 2010, **37**, 510.
9. P. L. McGeer, S. Itagaki, B. E. Boyes and E. G. McGeer, *Neurology*, 1988, **38**, 1285.
10. R. B. Banati, S. E. Daniel and S. B. Blunt, *Mov. Disord.*, 1998, **13**, 221.
11. Y. Ouchi, E. Yoshikawa, Y. Sekine, M. Futatsubashi, T. Kanno, T. Ogusu and T. Torizuka, *Ann. Neurol.*, 2005, **57**, 168.
12. Y. Ouchi, S. Yagi, M. Yokokura and M. Sakamoto, *Parkinsonism Relat. Disord.*, 2009, **15**(3), S200.
13. L. S. Forno, L. E. DeLanney, I. Irwin, D. Di Monte and J. W. Langston, *Prog. Brain Res.*, 1992, **94**, 429.
14. W. Dauer and S. Przedborski, *Neuron*, 2003, **39**, 889.
15. A. Castaño, A. J. Herrera, J. Cano and A. Machado, *J. Neurochem.*, 1998, **70**, 1584.
16. A. J. Herrera, A. Castaño, J. L. Venero, J. Cano and A. Machado, *Neurobiol. Dis.*, 2000, **7**, 429.
17. H. M. Gao, P. T. Kotzbauer, K. Uryu, S. Leight, J. Q. Trojanowski and V. M. Lee, *J. Neurosci.*, 2008, **28**, 7687.
18. M. Fujita, S. Sugama, K. Sekiyama, A. Sekigawa, T. Tsukui, M. Nakai, M. Waragai, T. Takenouchi, Y. Takamatsu, J. Wei, E. Rockenstein,

A. R. Laspada, E. Masliah, S. Inoue and M. Hashimoto, *Nat. Commun.*, 2010, **1**, 110.
19. N. J. Rothwell and P. J. Strijbos, *Int. J. Dev. Neurosci.*, 1995, **13**, 179.
20. M. Mogi, M. Harada, T. Kondo, P. Riederer, H. Inagaki, M. Minami and T. Nagatsu, *Neurosci. Lett.*, 1994, **180**, 147.
21. M. Mogi, M. Harada, H. Narabayashi, H. Inagaki, M. Minami and T. Nagatsu, *Neurosci. Lett.*, 1996, **211**, 13.
22. C. C. Ferrari, M. C. Pott Godoy, R. Tarelli, M. Chertoff, A. M. Depino and F. J. Pitossi, *Neurobiol. Dis.*, 2006, **24**, 183.
23. M. Nishimura, I. Mizuta, E. Mizuta, S. Yamasaki, M. Ohta and S. Kuno, *Neurosci. Lett.*, 2000, **284**, 73.
24. M. Nishimura, S. Kuno, R. Kaji, K. Yasuno and H. Kawakami, *Mov. Disord.*, 2005, **20**, 901.
25. P. L. McGeer, K. Yasojima and E. G. McGeer, *Neurosci. Lett.*, 2002, **326**, 67.
26. T. Schulte, L. Schöls, T. Müller, D. Woitalla, K. Berger and R. Krüger, *Neurosci. Lett.*, 2002, **326**, 70.
27. E. Pascale, E. Passarelli, C. Purcaro, A. R. Vestri, A. Fakeri, R. Guglielmi, F. Passarelli and G. Meco, *Acta Neurol. Scand.*, 2011, **124**, 176.
28. G. J. Liu, R. N. Feng, C. Luo and S. Bi, *Neurosci. Lett.*, 2010, **480**, 158.
29. International Parkinson's Disease Genomics Consortium and Wellcome Trust Case Control Consortium 2, *PLoS Genet.*, 2011, **7**, e1002142.
30. M. A. Nalls, V. Plagnol, D. G. Hernandez, M. Sharma, U. M. Sheerin, M. Saad, J. Simón. Schulte, S. Lesage, S. Sveinbjörnsdóttir, K. Stefánsson, M. Martinez, J. Hardy, P. Heutink, A. Brice, T. Gasser, A. B. Singleton, N. W. Wood and the International Parkinson Disease Genomics Consortium, *Lancet*, 2011, **377**, 641.
31. C. A. Dinarello, *Annu. Rev. Immunol.*, 2009, **27**, 519.
32. Y. Qu, L. Franchi, G. Nunez and G. R. Dubyak, *J. Immunol.*, 2007, **179**, 1913.
33. D. Ferrari, C. Pizzirani, E. Adinolfi, R. M. Lemoli, A. Curti, M. Idzko, E. Panther and F. Di Virgilio, *J. Immunol.*, 2006, **176**, 3877.
34. T. Takenouchi, Y. Iwamaru, S. Sugama, M. Tsukimoto, M. Fujita, A. Sekigawa, K. Sekiyama, M. Sato, S. Kojima, B. Conti, M. Hashimoto and H. Kitani, *J. Neurochem.*, 2011, **117**, 712.
35. R. A. North, *Physiol. Rev.*, 2002, **82**, 1013.
36. H. M. Costa-Junior, F. Sarmento Vieira and R. Coutinho-Silva, *Purinergic Signal.*, 2011, **7**, 7.
37. G. Collo, S. Neidhart, E. Kawashima, M. Kosco, A. North and G. Buell, *Neuropharmacology*, 1997, **36**, 1277.
38. D. Ferrari, M. Villalba, P. Chiozzi, S. Falzoni, P. Ricciardi. and Di Virgilio, *J. Immunol.*, 1996, **156**, 1531.
39. T. Suzuki, I. Hide, K. Ido, S. Kohsaka, K. Inoue and Y. Nakata, *J. Neurosci.*, 2004, **24**, 1.

40. M. Shiratori, H. Tozaki, M. Yoshitake, Tsuda and K. Inoue, *J. Neurochem.*, 2010, **114**, 810.
41. A. Kataoka, H. Tozaki, M. Koga, Tsuda and K. Inoue, *J. Neurochem.*, 2009, **108**, 115.
42. T. Takenouchi, S. Sugama, Y. Iwamaru, M. Hashimoto and H. Kitani, *Crit. Rev. Immunol.*, 2009, **29**, 335.
43. L. K. Parvathenani, S. Tertyshnikova, C. R. Greco, S. B. Roberts, B. Robertson and R. Posmantur, *J. Biol. Chem.*, 2003, **278**, 13309.
44. F. P. Gendron, M. Chalimoniuk, J. Strosznajder, S. Shen, F. A. González, G. A. Weisman and G. Y. Sun, *J. Neurochem.*, 2003, **87**, 344.
45. D. Ferrari, P. Chiozzi, S. Falzoni, M. Dal Susino, G. Collo, G. Buell and F. Di Virgilio, *Neuropharmacology*, 1997, **36**, 1295.
46. M. Monif, C. A. Reid, K. L. Powell, M. L. Smart and D. A. Williams, *J. Neurosci.*, 2009, **29**, 3781.
47. B. Sperlágh, E. S. Vizi, K. Wirkner and P. Illes, *Prog. Neurobiol.*, 2006, **78**, 327.
48. C. Matute, *Mol. Neurobiol.*, 2008, **38**, 123.
49. M. R. Bennett, *Aust. N. Z. J. Psychiatry*, 2007, **41**, 563.
50. P. Y. Wu, Y. C. Lin, C. L. Chang, H. T. Lu, C. H. Chin, T. T. Hsu, D. Chu and S. H. Sun, *Cell Signal.*, 2009, **21**, 881.
51. K. P. Larsson, A. J. Hansen and S. Dissing, *J. Neurochem.*, 2002, **83**, 285.
52. A. La Sala, D. Ferrari, F. Di Virgilio, M. Idzko, J. Norgauer and G. Girolomoni, *J. Leukoc. Biol.*, 2003, **73**, 339.
53. A. Witting, L. Walter, J. Wacker, T. Möller and N. Stella, *Proc. Natl. Acad. Sci. U. S. A.*, 2004, **101**, 3214.
54. F. Ortega, R. Pérez-Sen, E. G. Delicado and M. Teresa Miras-Portugal, *Neuropharmacology*, 2011, **61**, 1210.
55. F. Ortega, R. Pérez-Sen, E. G. Delicado and M. T. Miras-Portugal, *Neurotox. Res.*, 2009, **15**, 193.
56. D. J. Jun, J. Kim, S. Y. Jung, R. Song, J. H. Noh, Y. S. Park, S. H. Ryu, J. H. Kim, Y. Y. Kong, J. M. Chung and K. T. Kim, *J. Biol. Chem.*, 2007, **282**, 37350.
57. C. Delarasse, P. Gonnord, M. Galante, R. Auger, H. Daniel, I. Motta and J. M. Kanellopoulos, *J. Neurochem.*, 2009, **109**, 846.
58. S. Duan, C. M. Anderson, E. C. Keung, Y. Chen and R. A. Swanson, *J. Neurosci.*, 2003, **23**, 1320.
59. A. Elssner, M. Duncan, M. Gavrilin and M. D. Wewers, *J. Immunol.*, 2004, **172**, 4987.
60. Y. Chakfe, R. Seguin, J. P. Antel, C. Morissette, D. Malo, D. Henderson and P. Séguéla, *J. Neurosci.*, 2002, **22**, 3061.
61. M. Seman, S. Adriouch, F. Scheuplein, C. Krebs, D. Freese, G. Glowacki, P. Deterre, F. Haag and F. Koch-Nolte, *Immunity*, 2003, **19**, 571.
62. P. G. Baraldi, F. Di Virgilio and R. Romagnoli, *Curr. Top Med. Chem.*, 2004, **4**, 1707.

63. S. Alloisio, R. Aiello, S. Ferroni and M. Nobile, *Mol. Pharmacol.*, 2006, **69**, 1975.
64. T. Takenouchi, M. Sato and H. Kitani, *J. Neurochem.*, 2007, **102**, 1518.
65. A. D. Michel and E. Fonfria, *Br. J. Pharmacol.*, 2007, **152**, 523.
66. E. Alzola, A. Pérez. Kabré, D. J. Fogarty, M. Métioui, N. Chaïb, J. M. Macarulla, C. Matute, J. P. Dehaye and A. Marino, *J. Biol. Chem.*, 1998, **273**, 30208.
67. J. M. Sanz, P. Chiozzi and F. Di Virgilio, *Eur. J. Pharmacol.*, 1998, **355**, 235.
68. D. Ferrari, C. Pizzirani, E. Adinolfi, S. Forchap, B. Sitta, L. Turchet, S. Falzoni, M. Minelli, R. Baricordi and F. Di Virgilio, *J. Immunol.*, 2004, **173**, 4652.
69. W. Nörenberg, C. Hempel, N. Urban, H. Sobottka, P. Illes and M. Schaefer, *J. Biol. Chem.*, 2011, **286**, 11067.
70. P. A. Verhoef, S. B. Kertesy, K. Lundberg, J. M. Kahlenberg and G. R. Dubyak, *J. Immunol.*, 2005, **175**, 7623.
71. C. Virginio, D. Church, R. A. North and A. Surprenant, *Neuropharmacology*, 1997, **36**, 1285.
72. S. A. Friedle, M. A. Curet and J. J. Watters, *Recent Pat. CNS Drug Discov.*, 2010, **5**, 35.
73. I. P. Chessell, A. D. Michel and P. P. Humphrey, *Br. J. Pharmacol.*, 1998, **124**, 1314.
74. M. Murgia, S. Hanau, P. Pizzo, M. Rippa and F. Di Virgilio, *J. Biol. Chem.*, 1993, **268**, 8199.
75. B. D. Humphreys, C. Virginio, A. Surprenant, J. Rice and G. R. Dubyak, *Mol. Pharmacol.*, 1998, **54**, 22.
76. L. H. Jiang, A. B. Mackenzie, R. A. North and A. Surprenant, *Mol. Pharmacol.*, 2000, **58**, 82.
77. D. L. Donnelly, T. Namovic, P. Han and M. F. Jarvis, *Br. J. Pharmacol.*, 2009, **157**, 1203.
78. A. D. Michel, L. J. Chambers and D. S. Walter, *Br. J. Pharmacol.*, 2008, **153**, 737.
79. D. L. Donnelly and F. Jarvis, *Br. J. Pharmacol.*, 2007, **151**, 571.
80. H. B. Choi, J. K. Ryu, S. U. Kim and J. G. McLarnon, *J. Neurosci.*, 2007, **27**, 4957.
81. Z. Hracskó, M. Baranyi, C. Csölle, F. Gölöncsér, E. Madarász, A. Kittel and B. Sperlágh, *Mol. Neurodegener.*, 2011, **6**, 28.
82. J. G. McLarnon, J. K. Ryu, D. G. Walker and H. B. Choi, *J. Neuropathol. Exp. Neurol.*, 2006, **65**, 1090.
83. C. M. Sondag, G. Dhawan and C. K. Combs, *J. Neuroinflammation*, 2009, **6**, 1.
84. J. M. Sanz, P. Chiozzi, D. Ferrari, M. Colaianna, M. Idzko, S. Falzoni, R. Fellin, L. Trabace and F. Di Virgilio, *J. Immunol.*, 2009, **182**, 4378.
85. C. Delarasse, R. Auger, P. Gonnord, B. Fontaine and J. M. Kanellopoulos, *J. Biol. Chem.*, 2011, **286**, 2596.

86. M. León-Otegui, R. Gómez-Villafuertes, J. I. Díaz-Hernández, M. Díaz-Hernández, M. T. Miras-Portugal and J. Gualix, *FEBS Lett.*, 2011, **585**, 2255.
87. M. Díaz-Hernández, M. Díez-Zaera, J. Sánchez-Nogueiro, R. Gómez-Villafuertes, J. M. Canals, J. Alberch, M. T. Miras-Portugal and J. J. Lucas, *FASEB J.*, 2009, **23**, 1893.
88. The Huntington's Disease Collaborative Research Group, *Cell*, 1993, **72**, 971.
89. C. Matute, I. Torre, F. Pérez. Pérez. Alberdi, E. Etxebarria, A. M. Arranz, R. Ravid, A. Rodríguez. Sánchez and Domercq, *J. Neurosci.*, 2007, **27**, 9525.
90. C. Crozet, F. Beranger and S. Lehmann, *Vet. Res.*, 2008, **39**, 44.
91. T. Takenouchi, Y. Iwamaru, M. Imamura, N. Kato, S. Sugama, M. Fujita, M. Hashimoto, M. Sato, H. Okada, T. Yokoyama, S. Mohri and H. Kitani, *FEBS Lett.*, 2007, **581**, 3019.
92. D. Marcellino, D. Suárez, D. Sánchez., A. Aguirre, T. Yoshitake, S. Yoshitake, B. Hagman, J. Kehr, L. F. Agnati, K. Fuxe and A. Rivera, *J. Neural Transm.*, 2010, **117**, 681.
93. J. K. Ryu and J. G. McLarnon, *Neuroreport*, 2008, **19**, 1715.
94. J. I. Diaz-Hernandez, R. Gomez-Villafuertes, M. León-Otegui, L. Hontecillas-Prieto, A. Del Puerto, J. L. Trejo, J. J. Lucas, J. J. Garrido, J. Gualix, M. T. Miras-Portugal and M. Diaz-Hernandez, *Neurobiol. Aging*, 2012, **33**, 1816.
95. A. J. Sharp, P. E. Polak, V. Simonini, S. X. Lin, J. C. Richardson, E. R. Bongarzone and D. L. Feinstein, *J. Neuroinflammation*, 2008, **5**, 33.
96. X. Wang, G. Arcuino, T. Takano, J. Lin, W. G. Peng, P. Wan, P. Li, Q. Xu, Q. S. Liu, S. A. Goldman and M. Nedergaard, *Nat. Med.*, 2004, **10**, 821.
97. W. Peng, M. L. Cotrina, X. Han, H. Yu, L. Bekar, L. Blum, T. Takano, G. F. Tian, S. A. Goldman and M. Nedergaard, *Proc. Natl. Acad. Sci. U. S. A.*, 2009, **106**, 12489.
98. T. Engel, R. Gomez-Villafuertes, K. Tanaka, G. Mesuret, A. Sanz-Rodriguez, P. Garcia-Huerta, M. T. Miras-Portugal, D. C. Henshall and M. Diaz-Hernandez, *FASEB J.*, 2012, **26**, 1616.
99. J. Arbeloa, A. Pérez, Gottlieb and C. Matute, *Neurobiol. Dis.*, 2012, **45**, 954.
100. A. Marcillo, B. Frydel, H. M. Bramlett and W. D. Dietrich, *Exp. Neurol.*, 2012, **233**, 687.
101. J. F. Borzelleca, K. Depukat and J. B. Hallagan, *Food Chem. Toxicol.*, 1990, **28**, 221.
102. S. M. Hess and O. G. Fitzhugh, *J. Pharmacol. Exp. Ther.*, 1955, **114**, 38.
103. G. Buell, I. P. Chessell, A. D. Michel, G. Collo, M. Salazzo, S. Herren, D. Gretener, C. Grahames, R. Kaur, M. H. Kosco and P. Humphrey, *Blood*, 1998, **92**, 3521.
104. T. Takenouchi, Y. Iwamaru, S. Sugama, M. Sato, M. Hashimoto and H. Kitani, *J. Immunol.*, 2008, **180**, 7827.

105. O. Murch, M. Abdelrahman, M. Collino, M. Gallicchio, E. Benetti, E. Mazzon, R. Fantozzi, S. Cuzzocrea and C. Thiemermann, *Crit. Care Med.*, 2008, **36**, 550.
106. O. Murch, M. Collin, B. Sepodes, S. J. Foster, H. Mota-Filipe and C. Thiemermann, *Br. J. Pharmacol.*, 2006, **148**, 769.
107. J. J. Yan, J. S. Jung, J. E. Lee, J. Lee, S. O. Huh, H. S. Kim, K. C. Jung, J. Y. Cho, J. S. Nam, H. W. Suh, Y. H. Kim and D. K. Song, *Nat. Med.*, 2004, **10**, 161.
108. T. Pan, S. Kondo, W. Le and J. Jankovic, *Brain*, 2008, **131**, 1969.
109. J. L. Webb, B. Ravikumar, J. Atkins, J. N. Skepper and D. C. Rubinsztein, *J. Biol. Chem.*, 2003, **278**, 25009.
110. B. Levine, N. Mizushima and H. W. Virgin, *Nature*, 2011, **469**, 323.
111. T. Takenouchi, M. Nakai, Y. Iwamaru, S. Sugama, M. Tsukimoto, M. Fujita, J. Wei, A. Sekigawa, M. Sato, S. Kojima, H. Kitani and M. Hashimoto, *J. Immunol.*, 2009, **182**, 2051.
112. D. Biswas, O. S. Qureshi, W. Y. Lee, J. E. Croudace, M. Mura and D. A. Lammas, *BMC Immunol.*, 2008, **9**, 35.
113. T. Takenouchi, M. Fujita, S. Sugama, H. Kitani and M. Hashimoto, *Autophagy*, 2009, **5**, 723.
114. J. Harris, M. Hartman, C. Roche, S. G. Zeng, A. O'Shea, F. A. Sharp, E. M. Lambe, E. M. Creagh, D. T. Golenbock, J. Tschopp, H. Kornfeld, K. A. Fitzgerald and E. C. Lavelle, *J. Biol. Chem.*, 2011, **286**, 9587.
115. N. Dupont, S. Jiang, M. Pilli, W. Ornatowski, D. Bhattacharya and V. Deretic, *EMBO J.*, 2011, **30**, 4701.
116. Y. Iwamaru, T. Takenouchi, Y. Murayama, H. Okada, M. Imamura, Y. Shimizu, M. Hashimoto, S. Mohri, T. Yokoyama and H. Kitani, *PLoS One*, 2012, **7**, e37896.

Neuroregenerative Strategies

CHAPTER 16

Carotid Body Transplants as a Therapy for Parkinson's Disease

JAVIER VILLADIEGO,*[a,b,c]
ANA BELÉN MUÑOZ-MANCHADO,[a,b,c,d]
SIMÓN MENDEZ-FERRER,[a,b,c,e]
JUAN JOSÉ TOLEDO-ARAL[a,b,c] AND
JOSÉ LÓPEZ-BARNEO[a,b,c]

[a] Instituto de Biomedicina de Sevilla, Hospital Universitario Virgen del Rocío/CSIC/Universidad de Sevilla, Sevilla, Spain; [b] Departamento de Fisiología Médica y Biofísica, Universidad de Sevilla, Sevilla, Spain; [c] Centro de Investigación Biomédica en Red sobre Enfermedades Neurodegenerativas, Madrid, Spain; [d] Present address: Deparment of Medical Biochemistry and Biophysics, Karolinska Institute, Stockholm, Sweden; [e] Present address: Centro Nacional de Investigaciones Cardiovasculares (CNIC), Madrid, Spain and Icanh School of Medicine at Mount Sinai, New York, USA
*Email: fvilladiego@us.es

16.1 Cell Therapy in Parkinson's Disease

Parkinson's disease (PD) is characterized by the progressive degeneration of specific neuronal populations, particularly the dopaminergic neurons in the *substantia nigra* (SN) projecting to the *striatum*. Loss of these neurons leads to a lack of striatal dopamine, which is responsible for the principal motor symptoms characteristic of the disease (tremor, rigidity, slowness of movement and postural instability).[1–3] Current PD pharmacological therapies are based on the administration of pro-dopaminergic drugs, such as levodopa (a dopamine precursor), agonists of dopamine receptors, or inhibitors of dopamine degradation. However, none of these therapeutic strategies can stop disease progression. Moreover, they become less effective with time and can eventually produce motor complications as dyskinesias.[4]

During recent decades intrastriatal transplantation of dopamine-secreting cells has been intensely investigated as a possible treatment to re-establish striatal dopamine levels in PD patients.[5,6] Among the different cell types and transplantation protocols assayed, the intrastriatal graft of fetal mesencephalic neurons has provided the best clinical results.[7-10] However, the clinical efficacy of this procedure has been questioned, since in two double-blind controlled trials[11,12] it showed little clinical benefit and in some patients it induced the appearance of disabling dyskinesias. In addition to these 'neuroreparative' dopamine cell transplants, neurotrophic factors have been shown to have beneficial effects in several preclinical models of PD.[13-17] Based on these promising results cell therapy protocols that deliver trophic factors in the *striatum* have also been applied to protect the nigrostriatal neurons affected by the ongoing neurodegenerative process. This 'neuroprotective' cell therapy aims to diminish the progression of PD and even induce a partial reversal. Therefore, the availability of dopaminergic and/or neurotrophic-factor-producing cells is a major limitation in the search for effective novel cell therapies for PD.

For over a decade, our group has studied the anti-Parkinsonian benefits of intrastriatal carotid body (CB) transplants. In this chapter we review the effects and mechanisms of the action of CB grafting on different preclinical models of PD as well as in two open, Phase I/II, clinical trials performed with Parkinsonian patients.

16.2 Anatomical and Physiological Features of the Carotid Body

The CB is a small, paired, organ located at the carotid bifurcation [Figure 16.1(a)]. It is a highly irrigated organ that receives blood through a vessel branch originating from the external carotid artery. The CB is composed of neural-crest-derived parenchyma, formed by the migration of sympathoadrenal progenitors from the superior cervical ganglion during fetal development, and afferent sensory nerve fibers joining the glossofaringeal nerve.[18] The adult CB parenchyma is organized in clusters of cells, called "glomeruli", which are in close contact with capillaries and nerve fibers [Figure 16.1(b)–(d)]. These glomeruli contain two main cell types: the neuron-like glomus, or Type I, cells, surrounded by the glial-like sustentacular, or Type II, cells. Type I cells are highly dopaminergic and can be easily identified by the expression of tyrosine hydroxylase (TH; see Figure 16.1), the limiting enzyme in dopamine biosynthesis. In contrast, Type II cells, which are TH-negative, express classical glial markers such as the glial fibrillary acidic protein [GFAP; see Figure 16.1(d)].

The CB, the principal arterial chemoreceptor, mediates cardiorespiratory homeostatic reflexes in response to changes in the chemical composition of the blood. Hypoxemia, the main stimulus for CB,[18,19] triggers hyperventilation and sympathetic activation. Besides acute hypoxia, other parameters in arterial blood, such as hypercapnia, acidosis or hypoglycemia, can also activate CB cells.[20] It is well established that Type I, or glomus, cells are electrically excitable and function as the chemoreceptive elements of the CB. These cells contain secretory vesicles with, among other neurotransmitters, high dopamine content.[21] They behave as presynaptic-like elements that upon stimulation release neurotransmitters to activate afferent sensory nerve fibers of the IX cranial nerve. Besides its role in sensing acute hypoxia, the CB is also special among other adult neural or paraneural tissues because it can grow in conditions of chronic hypoxemia. It is well known that sustained hypoxia lasting several days stimulates CB cell proliferation and excitability,[18,22] as well as the synthesis of dopamine, due to TH induction.[23] This special sensitivity to hypoxia makes the CB particularly well-suited for intracerebral transplantation due to its particular durability in low oxygen tensions, a normal environmental condition in brain tissue,[24] which is probably accentuated inside intracerebral grafts. Recently, our laboratory has shown that, besides proliferation of TH-positive cells, hypoxic CB growth is produced by the activation of a population of resident stem cells in the adult organ. CB stems cells are the Type II, sustentacular cells (or a subpopulation of them), that under physiological hypoxia can proliferate and differentiate into new CB glomus cells.[25] Type II

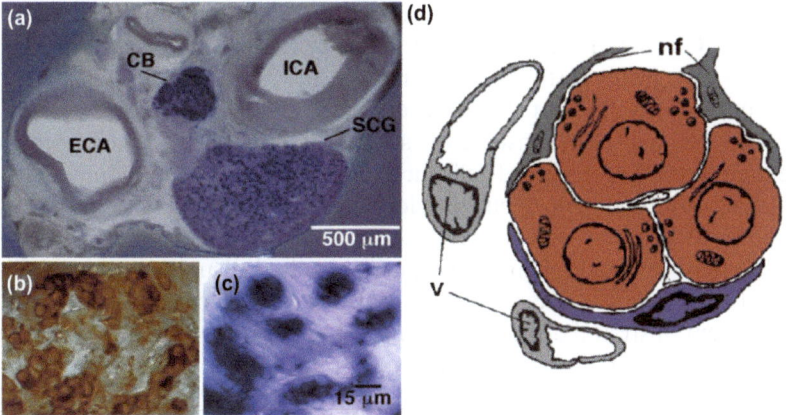

Figure 16.1 Structural organization of the carotid body (CB). (a) Histological section of the rat carotid bifurcation stained by *in situ* hybridization with a probe against tyrosine hydroxylase mRNA (TH; blue). Note the localization of the CB between the internal (ICA) and external (ECA) carotid arteries and near to the superior cervical ganglion (SCG). (b) and (c) Clusters, glomeruli, of TH^+ glomus cells in the CB, revealed by immunohistochemistry of TH [(b); brown] and TH *in situ* hybridization (c). (d) Schematic representation of a CB glomerulus indicating Type I (red) and Type II (purple) cells, blood vessels (V) and sensory nerve fibers (nf). Adapted from refs. 24 and 39.

cells are non-excitable[26,27] and comprise around 15–20% of the cells within the CB. Classically, they had been considered to belong to peripheral glia with a supportive function, but the recent identification of Type II cells as peripheral neural progenitors has generated interest in unraveling the interactions between Type I and Type II cells in the CB.

Another interesting physiological feature of the CB, of special value for its use in neural protection and repair, is that it contains high levels of several neurotrophic factors (brain-derived neurotrophic factor (BDNF), glial cell line neurotrophic factor (GDNF) and artemin, among others).[28–32] During recent years we have shown that the CB is among the tissues with the highest levels of GDNF in the adult rodent nervous system.[30]

16.3 Carotid Body Cell Therapy for Parkinson's Disease

16.3.1 Initial Preclinical Studies: The Carotid Body as a Source of Dopamine Cells

The use of CB grafts in PD animal models was initially proposed as a dopamine cell replacement therapy, based on the high content in dopamine of CB glomus cells.[33,34] The first study describing the use of CB glomus cells implants in PD models was reported by Gash and colleges in the 1980s.[35] The authors performed grafts of enzymatically dispersed CB glomus cells in the hemi-Parkinsonian 6-hydroxydopamine (6-OHDA) rat model, showing a slight behavioral recovery after implantation. However, in this study only a small number of CB glomus cells remained viable four weeks after grafting. In the late 1990s our group performed CB grafts in the same hemi-Parkinsonian rat model but using a different experimental methodology, which consisted of the use of CB cell aggregates instead of isolated glomus cells. CB cell aggregates were used instead of dispersed cell since enzymatically treated glomus cells are known to lose some of their physiological properties.[36] Using this graft procedure, the CB transplants produced a notable behavioral recovery three months after implantation, and subsequent histological and functional analysis revealed numerous clusters of TH-positive CB glomus cells and the improvement of striatal dopamine release.[37] During recent years several groups have confirmed that intrastriatal CB cell transplantation induces a marked recovery of hemi-Parkinsonian rats, as determined by behavioral, histological and neurochemical analyses.[38–41]

The favorable results obtained in the hemi-Parkinsonian rat model prompted us to evaluate the beneficial effects of CB transplantation in a non-human primate PD model. CB grafts were performed in Parkinsonian monkeys, that were previously injected with the neurotoxin 1-methyl-4-phenyl-1,2,3,6,-tetrahydropyridine (MPTP). CB transplantation induced, in this PD model, a long-term (five month) amelioration of the Parkinsonian symptoms that were accompanied by survival of CB glomus cells in the grafted *striata*.[42]

16.3.2 Recent Preclinical Studies: The Carotid Body as a Biological Pump Releasing Dopaminotrophic Factors

As indicated above, the first studies attempting transplantation of CB cells in PD animal models were originally thought of as a dopamine cell replacement therapy. However, they unexpectedly revealed that the CB grafted *striata* showed clear signs of re-innervation, in both rat and monkey PD models, with a high density of striatal TH immunoreactive fibers.[35,37,42] These results posed the question of whether the recovery induced by the CB grafts was due to dopamine release by the transplanted glomus cells or because the transplants secreted trophic substances that induced the re-generation of the host dopaminergic striatal fibers. Thus, we analyzed the long-term recovery after CB grafting (between 5 and 15 months after transplantation) of hemi-Parkinsonian rats with a degree of SN lesions higher than the animals analyzed in the previous studies.[35,37] Interestingly, CB-grafted animals could be clearly differentiated in two distinct groups on the basis of their behavioral and histological characteristics. One group of rats showed a significant and stable behavioral recovery [Figure 16.2(a)] that correlated with an important re-innervation of the grafted *striatum* [Figure 16.2(b)]. The origin of the fibers re-innervating the *striatum* was studied by retrograde labeling experiments, showing that these fibers originated from the remaining ipsilateral dopaminergic SN neurons [Figure 16.2(c)]. In contrast, the other group of rats did not show behavioral recovery despite the fact that they presented a large CB graft, well located in the *striatum* and with numerous highly dopaminergic glomus cells [Figure 16.2(d)]. The histological examination of the nigrostriatal pathway of these animals revealed a complete denervation of the transplanted *striatum* [Figure 16.2(e)] and a total destruction of the ipsilateral SN [Figure 16.2(f)]. Altogether, the behavioral and histological analyses of these animals suggested that the beneficial effect of CB grafts on Parkinsonian animals was due to a trophic effect on dopaminergic nigrostriatal neurons, rather than to the local release of dopamine by the transplant.[39]

The trophic action of CB grafts, suggested by these experiments, could be explained by the fact that the adult CB produces high amounts of the dopaminotrophic factor GDNF.[29,30] GDNF has been shown to induce neuroprotection and fiber outgrowth in several animal models of PD.[13-16] Using different methodologies, including reverse transcription-polymerase chain reaction (RT-PCR), genetically modified animals (GDNF/LacZ) and enzyme-linked immunosorbent assay (ELISA), we have shown that the CB is one of the few areas in the nervous system expressing high levels of GDNF in adult life.[30,39] Interestingly, CB GDNF is produced selectively in the dopaminergic glomus, or Type I, cells and GDNF expression is maintained after intrastriatal transplantation (Figure 16.3).[30]

The preliminary evidence supporting the trophic action of CB implants on the nigrostriatal pathway was obtained using 6-OHDA hemi-Parkinsonian rats.[39] However, this model presents significant limitations to identify and study a trophic effect. Firstly, it lacks an internal control to normalize the

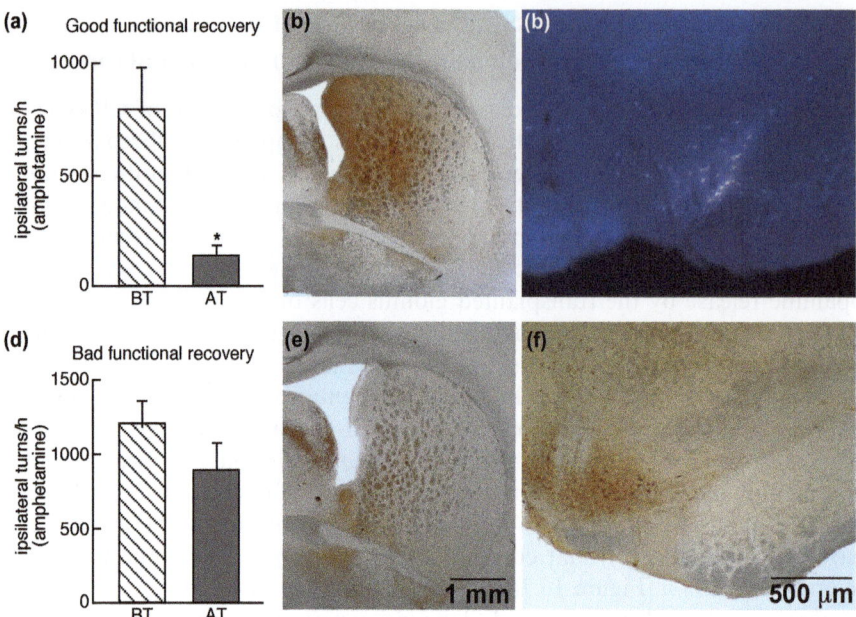

Figure 16.2 Functional and histological analysis of long-term CB-grafted hemi-Parkinsonian rats. (a) Rotational behavior of hemi-Parkinsonian rats with optimal functional recovery, before (BT) and after (AT) CB transplantation. (b) and (c) Histological sections of the *striatum* [(b), after TH immunohistochemistry] and SN [(c) labeled with the fluorescent retrograde tracer fluorogold] of a rat with optimal functional recovery, showing a significant re-innervation of the grafted *striatum* (b) arising from the remaining ipsilateral SN neurons (c). (d) Rotational behavior of hemi-Parkinsonian rats with bad functional recovery, before (BT) and after (AT) CB transplantation. (e) and (f) The histological examination, after TH immunohistochemistry, of the nigrostriatal pathway of these rats showed a complete denervation of the transplanted *striatum* (e) and a total destruction of the ipsilateral SN (f). *t*-Test *$p<0.05$.
Adapted from ref. 37.

variability of the lesion. Moreover, the acute lesion can mask the slow and progressive protective effects of the transplant on the nigrostriatal pathway. Finally, the lesion of the nigrostriatal pathway can be non-uniform, thus it is uncertain if the graft is placed in a region of the *striatum* that preserves the dopaminergic axon terminals necessary for the uptake of the trophic factors released by the transplanted cells. To further investigate if CB implants can trophically protect the nigrostriatal pathway, and thus amieloarate Parkinsonism, we performed CB grafts in a novel systemic and chronic MPTP mouse model. This chronic MPTP model recapitulates better the slow and progressive death of dopaminergic neurons in PD and allowed us to test for the trophic effect of unilateral CB grafts on the nigrostriatal pathway, using for comparison the contralateral sham-grafted *striatum* as a robust internal control. With this experimental procedure we have recently shown that

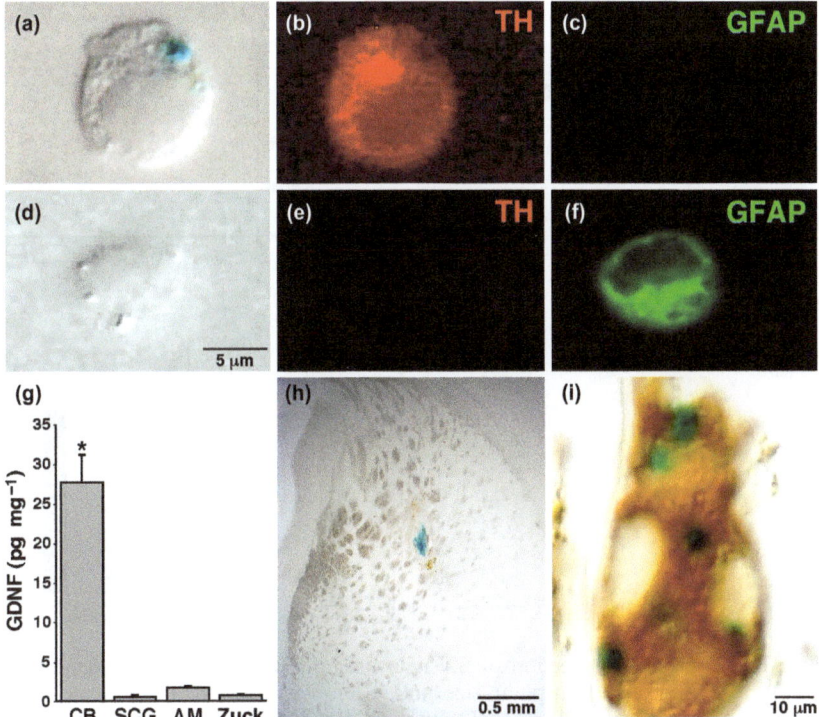

Figure 16.3 GDNF production in CB glomus cells *in situ* and after intrastriatal transplantation. (a)–(c) GDNF expression [blue precipitate, (a)] in a dispersed TH-positive [(b), red fluorescence] and GFAP-negative (c) CB Type I or glomus cell. (d)–(f) Lack of GDNF expression in a Type II or sustentacular CB cell, showing the characteristic GFAP expression [(f), green fluorescence] and absence of TH (e). The cells are representative examples obtained from primary cultures of heterozygous GDNF/LacZ CB, which were stained for Xgal and immunofluorescence (TH and GFAP). (g) GDNF protein content measured by ELISA in CB and other neural or paraneural tissues (SCG = superior cervical ganglion, AM = adrenal medulla, Zuck = Zuckerland's organ). (h) and (i) GDNF expression in intrastriatally grafted CB glomus cells. Note the GDNF expression (blue stain) in an intrastriatal implant of a heterozygous GDNF/lacZ CB (h), counterstaining with TH antibodies the blue GDNF-lacZ dots clearly appeared in the transplanted glomus CB cells (i). *t*-Test $*p < 0.05$.
Adapted from ref. 29.

intrastriatal CB grafts demonstrate a marked protective action on ipsilateral SN neurons projecting to the area of the transplant [Figure 16.4(a) and (b)], and produce fiber outgrowth in the *striatum*. In fair consistency with a classical trophic effect, the trophic protection exerted by the CB graft on the nigrostriatal dopaminergic neurons showed dose–response dependence in relation to the size of the CB transplant [Figure 16.4(c)]. Moreover, the dose-dependent trophic action of CB transplants was also analyzed by performing CB grafts

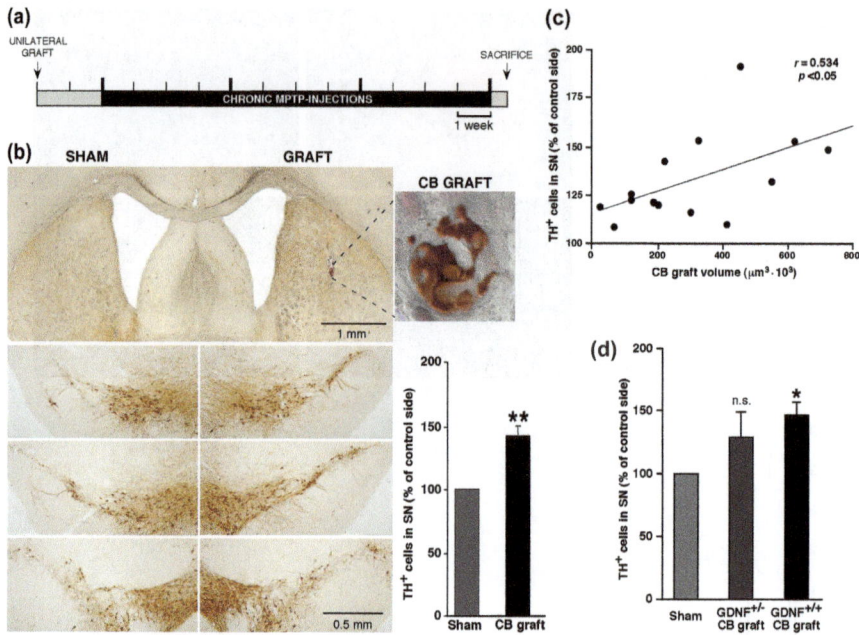

Figure 16.4 Trophic protection of SN neurons by CB grafts and dose-dependence regarding transplant size and GDNF expression. (a) Scheme of the experimental protocol. Briefly, unilateral striatal CB grafts were performed with a sham graft in the contralateral hemisphere, afterwards mice were chronically (20 mg kg^{-1}, three times per week over three months) treated with MPTP. (b) Brain coronal sections after immunochemistry of TH showing the trophic protection exerted by CB graft (inset on right panel) on the nigrostriatal pathway. Note the stereological quantification, expressed as the percentage of sham-grafted side, of TH$^+$ nigral neurons on the right-bottom plot. (c) Graph illustrating the linear regression ($r = 0.534$, $p < 0.05$) established between the nigral protection exerted by the CB transplant (ordinate) and the graft volume (abscissa). (d) Stereological quantification of dopaminergic SN neurons of MPTP-treated mice grafted with GDNF$^{+/-}$ or GDNF$^{+/+}$ CBs. GDNF$^{+/-}$ CB-grafted mice showed a reduced and not significant (n.s.) protection of SN neurons compared with wild type (GDNF$^{+/+}$), which induces a notable protection of these dopaminergic neurons. t-Test *$p < 0.05$, **$p < 0.001$. Adapted from ref. 44.

from heterozygous GDNF/lacZ (GDNF$^{+/-}$) mice, which contained less GDNF than wild-type controls.[43,44] Interestingly, intrastriatal grafting of GDNF$^{+/-}$ CB showed a reduced, non-significant, protection of nigral neurons, compared with wild-type (GDNF$^{+/+}$) CB grafts that, as indicated before, produced a strong preservation of SN neurons [Figure 16.4(d)].[44] Altogether, these results strongly support the view that the beneficial action produced by CB grafting is compatible with a retrograde trophic action of the grafted tissue on the nigrostriatal pathway, and positively correlates with the size and the GDNF expression level of the CB implant.

Based on the high content of trophic factors, especially GDNF, encountered in the adult CB, different authors have proposed the cografting of CB and ventral mesencephalic neurons to improve the survival and the anti-Parkinsonian effects of the grafted fetal dopaminergic neurons.[40,45] Moreover, it have been shown that CB grafts can promote an increase in the number of striatal dopaminergic cells in Parkinsonian monkeys.[46] This effect of CB grafting has been attributed to the release of GDNF by the transplant, because the administration of neurotrophic factors can produce similar effects.[47] In addition, a neuroprotective effect of CB grafts has been also suggested in experimental stroke.[48,49]

16.4 Clinical Studies of Carotid Body Autotransplantation on Parkinson's Disease Patients

The significant improvement induced by the CB graft in the different PD animal models encouraged the evaluation of the efficacy of CB transplantation in PD patients. In addition to their dopaminergic nature and their high content of neurotrophic factors, especially GDNF, the CB presents an important clinical advantage because its unilateral surgical re-section has no significant side-effects.[50] Two pilot Phase I/II open trials were performed to test the feasibility, safety and clinical efficacy of CB autotransplantation in Parkinsonian patients.[51,52] The experimental procedure in these trials consisted of unilateral removal of the CB and subsequent preparation of CB cell aggregates with fine scissors that were bilaterally placed in each putamen. Thirteen patients with advanced PD assessed before and up to 1–3 years after surgery were included in the two clinical trials. The primary outcome measure was change in motor ability [motor subscale (Part III) of the Unified Parkinson's Disease Rating Scale (UPDRS)] in the "off" medication state evaluated by an independent neurologist in a blinded fashion from masked and randomly presented video sequences.

Clinical improvement in the primary outcome measure was observed in 10 of 12 patients, being maximal at 6–12 months after transplantation (5–74%; 23% of mean improvement). Although a long-term trend towards presurgical clinical status was generally observed, a sustained improvement was detected in three of the six patients evaluated three years after grafting. In seven patients ^{18}F-DOPA positron emission tomography (PET) scans were performed before and one year after transplantation.[52] In these patients we observed a trend towards a 5% increment in intraputaminal ^{18}F-DOPA uptake instead of the expected yearly decrement, characteristic of advanced PD patients, estimated in approximately 10% of patients.[53] Because of technical problems one patient was successfully transplanted in only one hemisphere and received a needle track in the contralateral hemisphere. Interestingly, this patient only showed motor improvement in the contralateral hemi-body.[51] The results obtained in this patient indicated that the clinical effect of CB transplant is mediated by a specific action of the grafted cells, as it occurs in the different animal models of PD, rather than by placebo or some unspecific effect of the surgical procedure.

The clinical outcome obtained by autotransplantation of CB is similar to the effects reported by grafting fetal dopaminergic neurons.[11,12] However, CB-grafted patients did not develop dyskinesias unlike those patients subjected to fetal neuron transplantation. This could be related to the fact that the number of dopaminergic CB cells transplanted was markedly smaller than in the case of fetal grafts. Hence, it seems that the main action of CB-grafted cells is a neuroprotective effect on nigrostriatal neurons rather than solely dopamine cell replacement. Additionally, the most significant predictive factors for motor improvement in the patients analyzed were the histological integrity of the CB (an estimation of the number of dopaminergic-GDNF-secreting cells) and a milder disease severity, which further indicated that the main action of the CB transplant is a trophic neuroprotective effect.

16.5 Conclusions and Perspectives

During recent years several studies have demonstrated that intrastriatal transplantation of CB cells produces a significant histological and functional recovery in various preclinical models of PD.[37,39,42,44] These beneficial actions of CB grafting have been independently confirmed by several authors.[38,40,46] Detailed analyses of the mechanism underlying the anti-Parkinsonian action of CB transplants have revealed that their effect is mainly due to a trophic stimulation of the nigrostriatal pathway, producing both striatal fiber outgrowth and protection of the SN neurons.[39,44] This trophic action is mediated, at least in part, by the release of GDNF by the CB graft.[30,44] Thus, dopaminergic glomus CB cells appear to be ideally suited for the endogenous delivery of neurotrophic factors, especially GDNF, in PD and probably in others neurological disorders.

Two pilot Phase I/II open trials have shown that CB autotransplantation is a safe and feasible procedure with potential clinical applicability to treat PD, producing a clinical improvement[51,52] similar to that obtained after fetal mesencephalic transplantation.[11,12] However, the effectiveness of CB cell therapy observed in clinical trials is considerably lower than in experimental models. This differential efficacy had led us to revaluate experimentally putative limitations that can affect the clinical outcome of CB transplantation, such as the severity of the disease, patient age and the amount (and integrity) of CB tissue grafted. The influence of these variables on anti-Parkinsonian CB cell therapy is currently under investigation. We are analyzing the effect of the donor and receptor age on the trophic protection exerted by the CB graft on the nigrostriatal pathway. Moreover, we are testing the putative efficacy of *in vitro* expanded CB cells, based on the recent identification of adult CB stem cells that can proliferate and differentiate into new dopaminergic and GDNF-producing CB glomus cells.[25] The results obtained in these on-going scientific projects would help to design new clinical approaches that could eventually improve the outcome of CB transplantation.

Acknowledgements

This work was supported by grants from the Spanish Government (FIS, Red TERCEL, CIBERNED and MEC), the Andalusian Government (Excellence projects) and the Marcelino Botín Foundation.

References

1. A. E. Lang and A. M. Lozano, *N. Engl. J. Med.*, 1998, **339**, 1130.
2. W. Dauer and S. Przedborski, *Neuron*, 2003, **39**, 889.
3. H. Braak, K. Del Tredici, U. Rüb, R. A. I. de Vos, E. N. H. Jansen Steur and E. Braak, *Neurobiol. Aging*, 2003, **24**, 197.
4. T. Deierborg, D. Soulet, L. Roybon, V. Hall and P. Brundin, *Prog. Neurobiol.*, 2008, **85**, 407.
5. D. M. Yurek and J. R. Sladek, *Annu. Rev. Neurosci.*, 1990, **13**, 415.
6. S. B. Dunnett and A. Björklund, *Nature*, 1999, **399**, A32.
7. O. Lindvall, P. Brundin, H. Widner, S. Rehncrona, B. Gustavii, R. Frackowiak, K. L. Leenders, G. Sawle, J. C. Rothwell and C. D. Marsden, *Science*, 1990, **247**, 574.
8. C. R. Freed, R. E. Breeze, N. L. Rosenberg, S. A. Schneck, E. Kriek, J. X. Qi, T. Lone, Y. B. Zhang, J. A. Snyder and T. H. Wells, *N. Engl. J. Med.*, 1992, **327**, 1549.
9. H. Widner, J. Tetrud, S. Rehncrona, B. Snow, P. Brundin, B. Gustavii, A. Björklund, O. Lindvall and J. W. Langston, *N. Engl. J. Med.*, 1992, **327**, 1556.
10. P. Piccini, D. J. Brooks, A. Björklund, R. N. Gunn, P. M. Grasby, O. Rimoldi, P. Brundin, P. Hagell, S. Rehncrona, H. Widner and O. Lindvall, *Nat. Neurosci.*, 1999, **2**, 1137.
11. C. R. Freed, P. E. Greene, R. E. Breeze, W. Y. Tsai, W. DuMouchel, R. Kao, S. Dillon, H. Winfield, S. Culver, J. Q. Trojanowski, D. Eidelberg and S. Fahn, *N. Engl. J. Med.*, 2001, **344**, 710.
12. C. W. Olanow, C. G. Goetz, J. H. Kordower, A. J. Stoessl, V. Sossi, M. F. Brin, K. M. Shannon, G. M. Nauert, D. P. Perl, J. Godbold and T. B. Freeman, *Ann. Neurol.*, 2003, **54**, 403.
13. E. Arenas, M. Trupp, P. Akerud and C. F. Ibáñez, *Neuron*, 1995, **15**, 1465.
14. A. Tomac, E. Lindqvist, L. F. Lin, S. O. Ogren, D. Young, B. J. Hoffer and L. Olson, *Nature*, 1995, **373**, 335.
15. J. H. Kordower, M. E. Emborg, J. Bloch, S. Y. Ma, Y. Chu, L. Leventhal, J. McBride, E. Y. Chen, S. Palfi, B. Z. Roitberg, W. D. Brown, J. E. Holden, R. Pyzalski, M. D. Taylor, P. Carvey, Z. Ling, D. Trono, P. Hantraye, N. Déglon and P. Aebischer, *Science*, 2000, **290**, 767.
16. D. Kirik, B. Georgievska and A. Björklund, *Nat. Neurosci.*, 2004, **7**, 105.
17. A. Pascual, M. Hidalgo-Figueroa, J. I. Piruat, C. O. Pintado, R. Gómez-Díaz and J. López-Barneo, *Nat. Neurosci.*, 2008, **11**, 755.

18. J. López-Barneo, P. Ortega-Sáenz, R. Pardal, A. Pascual and J. I. Piruat, *Eur. Respir. J.*, 2008, **32**, 1386.
19. E. K. Weir, J. López-Barneo, K. J. Buckler and S. L. Archer, *N. Engl. J. Med.*, 2005, **353**, 2042.
20. J. López-Barneo, *Curr. Opin. Neurobiol.*, 2003, **13**, 493.
21. C. A. Nurse, *Auton. Neurosci.*, 2005, **120**, 1.
22. K. H. McGregor, J. Gil and S. Lahiri, *J. Appl. Physiol.*, 1984, **57**, 1430.
23. M. F. Czyzyk-Krzeska, D. A. Bayliss, E. E. Lawson and D. E. Millhorn, *J. Neurochem.*, 1992, **58**, 1538.
24. M. Erecińska and I. A. Silver, *Respir. Physiol.*, 2001, **128**, 263.
25. R. Pardal, P. Ortega-Sáenz, R. Durán and J. López-Barneo, *Cell*, 2007, **131**, 364.
26. M. R. Duchen, K. W. Caddy, G. C. Kirby, D. L. Patterson, J. Ponte and T. J. Biscoe, *Neuroscience*, 1988, **26**, 291.
27. J. Ureña, J. López-López, C. González and J. López-Barneo, *J. Gen. Physiol.*, 1989, **93**, 979.
28. J. T. Erickson, T. A. Brosenitsch and D. M. Katz, *J. Neurosci.*, 2001, **21**, 581.
29. M. L. Leitner, L. H. Wang, P. A. Osborne, J. P. Golden, J. Milbrandt and E. M. Johnson, *Exp. Neurol.*, 2005, **191**(1), S68.
30. J. Villadiego, S. Méndez-Ferrer, T. Valdés-Sánchez, I. Silos-Santiago, I. Fariñas, J. López-Barneo and J. J. Toledo-Aral, *J. Neurosci.*, 2005, **25**, 4091.
31. A. Izal-Azcárate, S. Belzunegui, W. San Sebastián, P. Garrido-Gil, M. Vázquez-Claverie, B. López, I. Marcilla and M. A. R. Luquin, *Respir. Physiol. Neurobiol.*, 2008, **161**, 95.
32. A. Porzionato, V. Macchi, A. Parenti and R. De Caro, *Int. Rev. Cell. Mol. Biol.*, 2008, **269**, 1.
33. J. T. Hansen and D. S. Christie, *Life Sci.*, 1981, **29**, 1791.
34. S. Fidone, C. González and K. Yoshizaki, *J. Physiol.*, 1982, **333**, 81.
35. G. Y. Bing, M. F. Notter, J. T. Hansen and D. M. Gash, *Brain Res. Bull.*, 1988, **20**, 399.
36. J. López-Barneo, R. J. Montoro, P. Ortega-Sáenz and J. Ureña, in *Oxygen Regulation of Ion Channels and Gene Expression*, ed. J. López-Barneo and K. Weir, Wiley-Blackwell, New York, 1998, pp. 127–144.
37. E. F. Espejo, R. J. Montoro, J. A. Armengol and J. López-Barneo, *Neuron*, 1998, **20**, 197.
38. G. Hao, Y. Yao, J. Wang, L. Zhang, N. Viroonchatapan and Z. Z. Wang, *Physiol. Behav.*, 2002, **77**, 519.
39. J. J. Toledo-Aral, S. Méndez-Ferrer, R. Pardal, M. Echevarría and J. López-Barneo, *J. Neurosci.*, 2003, **23**, 141.
40. S. Shukla, A. K. Agrawal, R. K. Chaturvedi, K. Seth, N. Srivastava, C. Sinha, Y. Shukla, V. K. Khanna and P. K. Seth, *J. Neurochem.*, 2004, **91**, 274.
41. J. J. Toledo-Aral, S. Méndez-Ferrer, R. Pardal and J. López-Barneo, *Brain Res. Bull.*, 2002, **57**, 847.

42. M. R. Luquin, R. J. Montoro, J. Guillén, L. Saldise, R. Insausti, J. Del Río and J. López-Barneo, *Neuron*, 1999, **22**, 743.
43. H. A. Boger, L. D. Middaugh, P. Huang, V. Zaman, A. C. Smith, B. J. Hoffer, A. C. Tomac and A.-C. Granholm, *Exp. Neurol.*, 2006, **202**, 336.
44. A. B. Muñoz-Manchado, J. Villadiego, N. Suárez-Luna, A. Bermejo-Navas, P. Garrido-Gil, J. L. Labandeira-García, M. Echevarría, J. López-Barneo and J. J. Toledo-Aral, *Neurobiol. Aging*, 2013, **34**, 902.
45. J. Rodriguez-Pallares, B. Joglar, A. B. Muñoz-Manchado, J. Villadiego, J. J. Toledo-Aral and J. L. Labandeira-García, *Regen. Med.*, 2012, **7**, 309.
46. W. San Sebastián, J. Guillén, M. Manrique, S. Belzunegui, E. Ciordia, A. Izal-Azcárate, P. Garrido-Gil, M. Vázquez-Claverie and M. R. Luquin, *Brain*, 2007, **130**, 1306.
47. S. Palfi, L. Leventhal, Y. Chu, S. Y. Ma, M. Emborg, R. Bakay, N. Déglon, P. Hantraye, P. Aebischer and J. H. Kordower, *J. Neurosci.*, 2002, **22**, 4942.
48. G. Yu, L. Xu, M. Hadman, D. C. Hess and C. V. Borlongan, *Brain Res.*, 2004, **1015**, 50.
49. G. Yu, C. Fournier, D. C. Hess and C. V. Borlongan, *Neurosci. Biobehav. Rev.*, 2005, **28**, 803.
50. Y. Honda, *J. Appl. Physiol.*, 1992, **73**, 1.
51. V. Arjona, A. Mínguez-Castellanos, R. J. Montoro, A. Ortega, F. Escamilla, J. J. Toledo-Aral, R. Pardal, S. Méndez-Ferrer, J. M. Martín, M. Pérez, M. J. Katati, E. Valencia, T. García and J. López-Barneo, *Neurosurgery*, 2003, **53**, 321.
52. A. Mínguez-Castellanos, F. Escamilla-Sevilla, G. R. Hotton, J. J. Toledo-Aral, A. Ortega-Moreno, S. Méndez-Ferrer, J. M. Martín-Linares, M. J. Katati, P. Mir, J. Villadiego, M. Meersmans, M. Pérez-García, D. J. Brooks, V. Arjona and J. López-Barneo, *J. Neurol. Neurosurg. Psychiatr.*, 2007, **78**, 825.
53. R. Hilker, A. T. Portman, J. Voges, M. J. Staal, L. Burghaus, T. van Laar, A. Koulousakis, R. P. Maguire, J. Pruim, B. M. de Jong, K. Herholz, V. Sturm, W.-D. Heiss and K. L. Leenders, *J. Neurol. Neurosurg. Psychiatr.*, 2005, **76**, 1217.

CHAPTER 17

Stem-Cell-Based Cell-Replacement Therapy in Parkinson's Disease

JAN TØNNESEN[†] AND MERAB KOKAIA*

Experimental Epilepsy Group, Division of Neurology, Lund University Hospital, Lund, Sweden
*Email: merab.kokaia@med.lu.se

17.1 Parkinson's Disease and Cell-Replacement Therapy

In Parkinson's disease (PD) degeneration of dopaminergic neurons in the *substantia nigra pars compacta* (SNc) causes denervation of target striatal areas.[1,2] As a result, excitatory D_1 and inhibitory D_2 dopamine (DA) receptors on distinct subsets of striatal medium spiny neurons (MSNs) are not sufficiently activated, ultimately leading to a reduced excitatory drive onto the motor cortex, which is observed clinically as the hallmark symptoms of PD, including rigidity, tremor at rest, akinesia and postural instability, however, cognitive symptoms such as dementia may also occur.[3]

Currently, there is no available cure for chronically progressive PD, and therapies aim merely at symptom relief. Treatment typically consists of administration of the dopamine precursor Levodopa (L-DOPA), which is

---

[†]Present address: Interdisciplinary Institute for Neuroscience, University of Bordeaux Segalen, Bordeaux, France.

RSC Drug Discovery Series No. 34
Emerging Drugs and Targets for Parkinson's Disease
Edited by Ana Martinez and Carmen Gil
© The Royal Society of Chemistry 2013
Published by the Royal Society of Chemistry, www.rsc.org

taken up by surviving dopaminergic neurons, promoting increased release of DA, or sometimes administration of a D_2 agonist.[4] Both strategies, however, offer only partial relief and come with severe side-effects. The most prevalent side-effect of L-DOPA administration is motor dyskinesias, but detrimental cognitive effects can also occur.[5,6]

At more severe stages of the disease, some symptom relief may be achieved through electrical deep brain stimulation of the subthalamic nucleus, although this therapy has been associated with neuropsychological side-effects such as personality changes.[7]

As neurodegeneration in PD is largely confined to SNc dopaminergic neurons, with the resulting denervation of the *striatum*, dopaminergic cell-replacement therapy appears an attractive therapy option for targeting DA re-innervation of the *striatum*. In essence, cell-replacement therapy in PD aims to replace degenerated SNc dopaminergic neurons, by injecting new dopaminergic cells directly into the *striatum* since nigral grafts do not seem to grow their axons to the *striatum*.[8] While the outline of such a treatment is simple, in practice numerous challenges must be met to realize this scheme clinically.

Breaking down the development of stem-cell-replacement therapy for PD into steps, the major obstacles are to provide proof-of-principle for the treatment strategy, to identify a safe and consistent source of cells with the potential to be used for clinical PD therapy, and to understand how a given therapeutic effect is mediated. As for the latter point, it would of course be necessary to understand how the beneficial effects of cell therapy are mediated, but as experimental work has shown, understanding the mechanisms underlying functional recovery in PD after cell therapy is by no means straightforward.

To date, some of the challenges just outlined above have been met fully or partially, while others are still very real challenges. In the following sections the major steps toward developing cell-replacement therapy for PD are addressed.

17.2 Proof-of-Principle: Fetal-Cell-Replacement Therapy

That dopaminergic cell-replacement therapy can work in PD has been proven in clinical trials using postmitotic fetal dopaminergic neurons as a graft source. Several independent clinical trials have explored the effects of transplanting fetal mid-brain tissue, rich in dopaminergic neurons, into the *striatum* to replace projections from the degenerating dopaminergic SNc neurons.[9-13] In some studies, remarkable improvement in PD patients have been reported,[9-11,14-16] while others have reported lack of improvement or even adverse effects in the form of graft-induced motor dyskinesias.[14,17,18]

The variability in outcomes between clinical trials is likely caused by the lack of standardized procedures for cell therapy. This fact applies to patient selection, tissue derivation procedures, age and preparation of the transplanted fetal tissue, number and co-ordinates of cell-injection sites, the duration of immunosuppression after the transplantation to prevent graft rejection, and more. The lesson to be learned is that cell therapy can indeed work; an important step forward would be to establish a standardized and reliable cell source for transplantation. Currently, using stem-cell-derived neurons for replacement instead of postmitotic fetal cells seems to be the most promising option.

Both from clinical trials and animal models it has emerged that the fraction of dopaminergic neurons that represent a SNc phenotype within the graft is decisive for the outcome.[19–21] Therefore, future stem-cell-based replacement therapy will rely on protocols for differentiating stem cells into this specific cell type in sufficient quantities.

On a synaptic level, the mechanisms underlying the therapeutic effects of fetal-cell-replacement therapy are still being established. It has been shown that grafted dopaminergic neurons survive in the host brain, express markers associated with SNc dopaminergic neurons (*e.g.*, tyrosine hydroxylase, TH, and the G-protein-coupled, inwardly rectifying K^+ Channel 2, GIRK2), and morphologically innervate the surrounding host tissue.[12,20] Furthermore, positron emission tomography (PET) scans have found that grafted neurons are able to take up the radioactively labeled dopamine precursor ^{13}F-DOPA, and bind to the D_2 agonist ^{11}C-raclopride.[14–16] It is currently unknown to what extent synaptic communication between graft and host neurons occurs, and how important synaptic integration is for a positive therapeutic outcome.

Unfortunately, fetal-cell-replacement therapy has been associated with severe side-effects in the form of graft-induced dyskinesias in a subset of patients. These are progressive dyskinesias that are believed to originate from serotonergic-graft-derived neurons innervating the *striatum*,[22] and can be dampened by inhibiting serotonergic signaling activity pharmacologically.[23] Again, the synaptic connectivity of such serotonergic graft neurons is yet to be established. It is likely that a graft containing few or no serotonergic neurons would have a better clinical outcome.

Fetal-cell-replacement therapy typically requires 4–8 fetuses to treat a single patient.[24] Therefore, there are obvious practical and ethical issues related to this strategy, setting major limitations, including tissue availability. These restrictions have spurred extensive efforts to identify a stem cell source for deriving dopaminergic neurons for clinical replacement therapy.

17.3 Candidate Stem Cells for Parkinson's Disease Cell-Replacement Therapy

17.3.1 Fetal Neural Stem Cells

The first stem cells to be explored in the setting of experimental PD were fetal neural precursors, with limited expansion potential. Early studies relying on fetal stem cells provided encouraging outcomes in animal models of PD, though the resulting number of surviving dopaminergic neurons was modest.[25–27]

Consequently, *in vitro* characterizations have reported that neural stem-cell-derived grafted neurons can mature morphologically and electrophysiologically to resemble dopaminergic neurons *in vivo*, and grafted predifferentiated neural stem cells can survive in the host brain, and mature into functional dopaminergic neurons in sufficient numbers to induce symptomatic recovery.[28]

A limiting factor in taking fetal stem cell therapy to the clinic is that such endogenously patterned stem cells have limited expandability *in vitro*, and reportedly lose their phenotypic identity with repeated expansion, so that the fraction of dopaminergic neurons upon differentiation declines.[28] This prevents true large-scale production of cells, and conservation of lines over considerable periods of time. Thus, the requirement to start cultures from a new tissue source for each trial will inevitably lead to batch-to-batch variation, even if using a standardized protocol. In addition, a fetal source of neural stem cells is associated with ethical considerations.

17.3.2 Embryonic Stem Cells

Embryonic stem (ES) cells are derived from the inner cell mass of the blastocyst, can be expanded indefinitely, and hold the potential to differentiate into any cell of the body, including dopaminergic mid-brain neurons.[29] Stable neural stem cell lines can be derived from pluripotent ES cells, which can be patterned towards specific phenotypes *in vitro* and differentiate into functionally mature neurons.[30]

Grafted dopamine neurons derived from ES cells have been shown to morphologically integrate in Parkinsonian animals, and mature into electrophysiologically mature neuronal phenotypes.[31,32] While early studies on human ES cell transplantation in PD models described tumorigenic overgrowth,[33] more recent protocols have reportedly overcome this issue, and found that human ES cells can give rise to dopaminergic neurons in grafts and ameliorate PD symptoms without teratoma formation or overgrowth in rodent models,[34,35] as well as in monkeys (Figure 17.1).[36] To eliminate contamination of the graft cell suspension with undifferentiated potentially tumorigenic cells, ES cells expressing early mid-brain dopaminergic neuronal markers Nurr1 or Pitx3 are genetically marked for a developmental cut-off for selection and purification by cell sorting (FACS) prior to grafting.[37,38]

Experimentally, ES-derived cells are commonly injected into the brain at the neuroblast stage, when they are already postmitotic neurons destined to develop into a certain phenotype, but still highly plastic and with the capacity to migrate and extend neuronal processes out of the graft area. Currently, ES cells are a putative candidate source of stem cells for clinical use for PD treatment.

17.3.3 Induced Pluripotent Stem Cells

Human-induced pluripotent stem (iPS) cells posses the same potential as ES cells for differentiating into any cell type of the body, but are derived by reverting adult somatic cells, *e.g.*, fibroblasts, to a pluripotent ES-like state.[39–41]

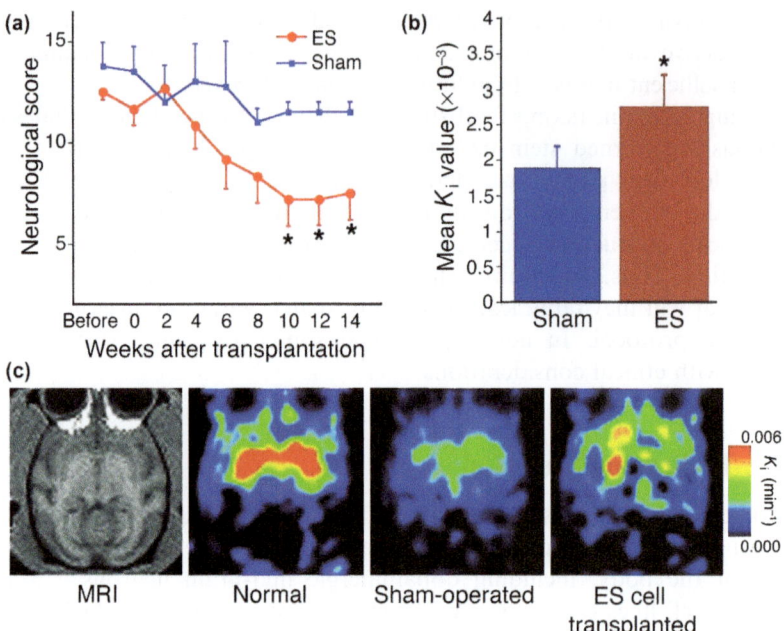

Figure 17.1 Grafting of ES-derived dopaminergic neurons in Parkinsonian monkeys. (a) Grafting gradually improves neurological scores in grafted Parkinsonian monkeys. (b) The metabolic rate constant (K_i) of ^{18}F-DOPA is increased in the lesioned putamen after grafting, confirming increased dopamine metabolism. The K_i value is calculated based on PET images. (c) An MRI image visualizing the putamen and other areas of a monkey brain. Three PET images illustrate normal dopamine metabolism in normal monkey putamen, decreased metabolism in the Parkinsonian sham-operated monkey, and partially recovered metabolism in the Parkinsonian monkey with graft.
Reproduced from ref. 36 with permission. © American Society for Clinical Investigation, 2005.

After somatic cells are reverted to pluripotency, they can be differentiated into neural stem cells, similar to ES cells.

The main advantage of iPS cells is that experimenters or clinicians can potentially derive patient-specific cell lines to target the same individuals, or immunocompatible groups of patients, and thereby minimize the risk of graft rejection by the recipient's immune system, eliminating the need for administering immunosuppressive drugs following cell implantation. Originally the conversion to pluripotency required the genetic manipulations of cells, but currently this can be done epigenetically, thereby increasing the likelihood of iPS cells being clinically relevant for cell-replacement therapy in general.[42] In addition, genetically converted iPS cells can carry the introduced conversion transgenes, which is not the case with epigenetically converted cells that can still be expanded indefinitely and give rise to functionally mature neurons with a mid-brain dopaminergic identity.[42]

iPS-Cell-derived neurons resembling mid-brain dopaminergic neurons have been found to survive, morphologically integrate, become electrophysiologically mature, and offer behavioral recovery in animal models of PD (Figure 17.2), although tumorigenic mitotic cells have also been reported in the graft site.[43,44] Tumorigenic cells were practically eliminated from the grafts when undifferentiated cells were selectively removed from the injected cell

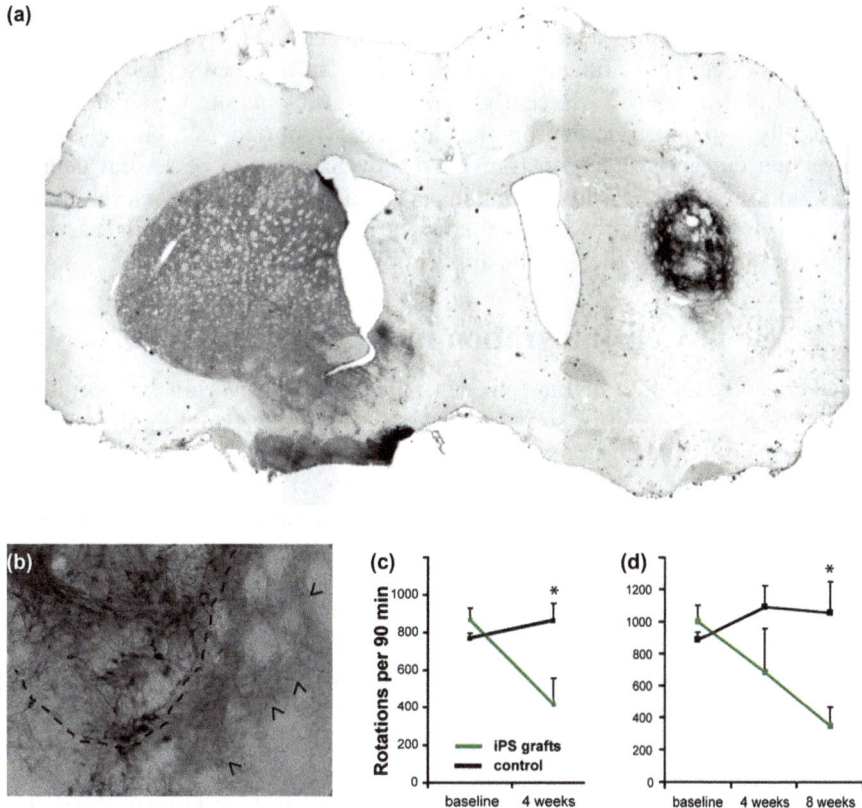

Figure 17.2 Grafting of iPS-cell-derived dopaminergic neurons in Parkinsonian rats. (a) TH immunostaining of a graft site in hemi-Parkinsonian rat brain. In the right, lesioned hemisphere, only the graft shows reactivity to TH, depicted four weeks after injection of cells. The unlesioned hemisphere shows reactivity throughout from innervating nigral dopaminergic fibers. (b) Higher magnification image of the graft, showing TH expressing somas and fibers re-innervating the surrounding striatal tissue (arrowheads). (c) Parkinsonian animals receiving unsorted iPS cell therapy experience rotational recovery compared to controls not receiving the cell therapy. (d) Parkinsonian animals receiving iPS cells where the SSEA-1 expressing mitotic fraction has been eliminated prior to injection, experience increased recovery compared to non-treated controls.
Reproduced from ref. 44 with permission. © National Academy of Sciences, 2008.

population by FACS purification prior to grafting. In contrast to isolating and injecting Nurr1 or Pitx3 positive cells as described above, in this particular study, the selected pluripotency marker stage-specific embryonic antigen 1 (SSEA-1) expressing cells were isolated and discarded, while the remaining cells were injected into the brain.[44]

17.3.4 Directly Induced Neurons

Recent advances in the stem cell field have allowed the conversion of mature fibroblasts directly into functional neurons, without the need to first revert the cells to pluripotency.[45] Recently, direct conversion of fibroblasts into specifically dopaminergic neurons has also been reported.[46,47] Though direct conversion currently relies on transgenetic overexpression of several defined transcription factors, it is likely that an epigenetic protocol will be developed in the near future, which will eliminate the issue of grafted cells carrying transgenes in putative therapeutic applications.

17.4 Stem Cell Integration in a Host Tissue

After grafting, the fate of the cells can be evaluated in various ways and to various degrees of detail. A straightforward evaluation is counting the number of surviving cells at a given timepoint, although more elaborate analyses can be conducted to understand the phenotypic profiles of grafted cells, their transcription activity, morphological and synaptic integration, and the timecourse of phenotypic differentiation.

To date, our knowledge about functional integration of grafted stem cells into the host tissue is rather limited. A commonly used analysis is immunohistochemical postmortem identification of the graft, with emphasis on markers that are classically associated with SNc dopaminergic neurons *in vivo*, such as TH and GIRK2.[48]

Whether morphological and synaptic integration is a necessary prerequisite for behavioral recovery is not quite clear. However, there are few studies addressing this issue in a comprehensive way, and little is known about the extent of functional integration, the factors influencing integration, and the correlation between integration of grafted stem cells and behavioral improvement.

In clinical trials, the fetal cell grafts in patients experiencing symptomatic relief have been shown to be viable and to largely re-innervate the surrounding host striatal tissue and thereby restore dopamine levels in the *striatum*.[49] What is still not fully understood is to what degree the grafted neurons synaptically communicate with the host neurons, and whether given grafted cells extend and receive inputs from other graft neurons or from the host neurons.

Experimental work has reported that Parkinsonian animals may experience some behavioral recovery even from grafts that do not morphologically integrate into the host tissue, and presumably have no synaptic interaction with the host.[50] This observation may be explained by the non-synaptic release of dopamine or neurotrophic factors from the grafted cells, or from the beneficial effect of an immunological response triggered by the graft. Generally, behavioral recovery has been associated with a higher degree of maturation and morphological integration of stem-cell-derived neurons, and with the ability of these cells to release and metabolize dopamine.[28,34,51]

In animals experiencing behavioral recovery, grafts have been found to contain functionally mature dopaminergic neurons.[32] Furthermore, stem-cell-derived dopaminergic neurons within grafts have been described to display functional hallmarks of SNc dopaminergic neurons, including spontaneous firing of action potentials of longer duration (≥ 2 ms), and expression of the inhibitory metabotropic D_2 autoreceptors.[32,52] The extent of synaptic communication between graft cells and host neurons however is largely unknown. This is due to the difficulties of addressing these two interspersed neuronal populations selectively and simultaneously. It is conceivable that when a highly concentrated suspension of immature and plastic cells is injected into the host brain tissue, these cells will primarily connect synaptically with each other, because of their mutual high degree of plasticity, rather than with the host neurons that are less plastic. Such a scenario could render the graft somewhat autonomous and less interactive with host circuitry. To address this question, double whole-cell patch-clamp recordings of grafted and host pairs of neurons have been attempted, but the probability of synaptic connections between graft/host pairs has proven to be too low to document and characterize such connections.[32] Instead, electrical field stimulations have been applied to stimulate areas of grafted neurons or host tissue, respectively, while measuring synaptic currents from individual grafted dopaminergic neurons.[32] The interpretation of such data is rather complicated by the fact that stem-cell-derived dopaminergic neurons are not the only cell population present in the graft, and the other neuronal populations that are glutamatergic may interconnect synaptically with grafted dopaminergic neurons, making it impossible to determine whether excitatory postsynaptic currents recorded in grafted neurons originate from synapses formed by other grafted neurons or by host neurons.[28] Glutamatergic signaling within grafts may originate directly from glutamatergic cells within the graft, or from dopaminergic cells coreleasing glutamate, as has been reported *in vivo* and *in vitro*.[53–55]

Recently, novel optogenetic tools have allowed the bidirectional control of neurons by exposing them to light of defined wavelengths.[56,57] Optical control is achieved by genetically introducing light-activated cell membrane pumps or channels into desired populations of neurons by promoter-specific targeting of the expression. After transduction and expression of the optogenetic probes, cells can be depolarized or hyperpolarized, depending on the chosen probe and respective light wavelength. Recently, this optogenetic technology has been applied to selectively activate or inhibit either host or grafted cells to show bidirectional synaptic connectivity between stem-cell-derived dopaminergic neurons and host neurons in cortico–striatal slice cultures *in vitro*.[52] This study suggested that synaptic afferents from the host onto the grafted neurons in the *striatum* were predominantly excitatory, likely originating from the cortex.

It was further suggested that there was extensive intragraft excitatory glutamatergic synaptic connectivity, while synapses formed by grafted neurons onto the host striatal cells were relatively rare. Although these findings are based on an *in vitro* grafting model, the methodology will be directly transferable to *in vivo* grafting experiments in PD models, and optogenetic probes still hold huge potential in this respect.

Understanding the detailed mechanisms underlying synaptic integration of grafts in PD models is an important step in understanding the significance of such integration for therapeutic outcomes, and perhaps for further optimization of neuronal derivation protocols, as well as for improving transplantation procedures.

One of the fascinating and controversial aspects of stem cell transplantation and integration is the phenomenon of fusion of grafted and host cells. This fusion has been reported to occur at a notable rate, and seems to be facilitated by host microglia cells.[58,59] Fusion between host and grafted cells may be potentially misleading and generate false data for synaptic integration studies, as well as for assessments of the functional properties of grafted neurons. The functional consequences of the fusion of grafted stem cells to host neurons are still poorly understood, but if the phenomenon is proven to be common in transplantation studies, it could be explored in further detail as a potential therapeutic alternative.[60]

17.5 Genetically Enhanced Stem Cells

Although introducing transgenes into stem cells before grafting will pose further concerns and restrictions on their clinical implementation, the abilities offered by expressing certain genes have attracted the interest of researchers in the setting of PD. Genetic manipulation of stem cells may increase the therapeutic potential of grafts. This can be achieved by increasing the number of resulting dopaminergic neurons upon differentiation, *e.g.*, when overexpressing the homeobox transcription factor Lmx1a,[61] or anti-apoptotic protein Bcl-X_L.[62–64] Genetic immortalization of fetal stem cells has been shown to increase their expandability and their phenotypic stability over time,[65,66] while retaining their ability to generate functional neurons *in vitro*.[67,68] However, the use of proto-oncogenes, such as avian *v-myc*, for the immortalization procedure does raise concerns for their clinical implementation. Finally, the optogenetic transduction of stem-cell-derived neurons, outlined above, may not only provide a means for accessing functional integration, but may also be exploited in therapeutic settings, to increase or decrease the activity of a graft. This would, however, require the placement of one or more optical fibers in close vicinity to the graft site.

17.6 Concluding Remarks and Future Perspectives

Stem cell therapy for PD is technically still in its infancy, and much experimental work in animal models still needs to be performed before clinical trials are considered. However, the fact that some PD patients receiving fetal ventral mesencephalic tissue implants, rich in postmitotic dopaminergic cells, have demonstrated remarkable functional recovery and have provided a proof-of-concept for the cell therapy approach, maintains the motivation for stem cell research to strive towards developing clinical-grade dopaminergic neurons from stem cells.

Current protocols for deriving dopaminergic neurons from both iPS and ES cells are encouraging, since these cells provide behavioral recovery when grafted in animal models of PD, seemingly without giving rise to tumors. These protocols need to be further optimized and standardized to be applicable as clinical-grade products. The risk of tumor formation needs to be eliminated, taking into consideration that experiments in rodents are relatively short-term compared to the potential lifetime of a graft residing in a human patient. It is therefore essential to rigorously test for the cell division potential of differentiated stem cells derived *via* a given protocol. The adverse effects of the cell therapy approach, such as dyskinesias observed in some patients receiving fetal tissue grafts, need to be addressed in experimental models and minimized when using stem cell sources for replacement therapy. Part of this process is to increase our understanding of the synaptic integration of the grafted stem-cell-derived neurons into the host tissue, in particular to clarify to what extent synaptic cross-talk between graft and host tissues occurs and is required for functional recovery, as well as how many of the intrinsic properties of these neurons need to resemble those of the endogenous dopaminergic neurons of the SN.

Injection of cells into the brain is not risk-free, and the beneficial effects of putative stem cell therapy must be weighed against other available treatment options, such as pharmacological approaches and deep brain stimulation. Stem cell therapy will need to prove that it is superior compared to already available treatment options in order to be a viable therapeutic alternative.

Nevertheless, stem-cell-based therapy for PD still holds significant potential as a therapeutic approach that needs to be further investigated.

References

1. S. Y. Ma, M. Roytta, J. O. Rinne, Y. Collan and U. K. Rinne, *J. Neurol. Sci.*, 1997, **151**, 83.
2. B. Pakkenberg, A. Moller, H. J. Gundersen, A. Mouritzen Dam and H. Pakkenberg, *J. Neurol. Neurosurg. Psychiatry*, 1991, **54**, 30.
3. J. Jankovic, *J. Neurol. Neurosurg. Psychiatry*, 2008, **79**, 368.
4. R. B. Godwin-Austen, E. B. Tomlinson, C. C. Frears and H. W. Kok, *Lancet*, 1969, **2**, 165.
5. J. E. Ahlskog and M. D. Muenter, *Mov. Disord.*, 2001, **16**, 448.
6. P. Jenner, *Nat. Rev. Neurosci.*, 2008, **9**, 665.
7. D. De Gaspari, C. Siri, A. Landi, R. Cilia, A. Bonetti, F. Natuzzi, L. Morgante, C. B. Mariani, E. Sganzerla, G. Pezzoli and A. Antonini, *J. Neurol. Neurosurg. Psychiatry*, 2006, **77**, 450.

8. I. Mendez, A. Dagher, M. Hong, P. Gaudet, S. Weerasinghe, V. McAlister, D. King, J. Desrosiers, S. Darvesh, T. Acorn and H. Robertson, *J. Neurosurg.*, 2002, **96**, 589.
9. C. R. Freed, R. E. Breeze, N. L. Rosenburg, S. A. Schneck, E. Kriek, J. x. Qi, T. Lone, Y.-b. Zhang, J. A. Snyder, T. H. Wells, L. O. Ramig, L. Thompson, J. C. Mazziotta, S. C. Huang, S. T. Grafton, D. Brooks, G. Sawle, G. Schroter and A. A. Ansari, *N. Engl. J. Med.*, 1992, **327**, 1549.
10. R. A. Hauser, T. B. Freeman, B. J. Snow, M. Nauert, L. Gauger, J. H. Kordower and C. W. Olanow, *Arch. Neurol.*, 1999, **56**, 179.
11. O. V. Kopyov, D. Jacques, A. Lieberman, C. M. Duma and R. L. Rogers, *Cell Transplant.*, 1996, **5**, 327.
12. J. H. Kordower, T. B. Freeman, E. Y. Chen, E. J. Mufson, P. R. Sanberg, R. A. Hauser, B. Snow and C. W. Olanow, *Mov. Disord.*, 1998, **13**, 383.
13. O. Lindvall, P. Brundin, H. Widner, S. Rehncrona, B. Gustavii, R. Frackowiak, K. L. Leenders, G. Sawle, J. C. Rothwell, C. D. Marsden and A. Björklund, *Science*, 1990, **247**, 574.
14. C. R. Freed, P. E. Greene, R. E. Breeze, W. Y. Tsai, W. DuMouchel, R. Kao, S. Dillon, H. Winfield, S. Culver, J. Q. Trojanowski, D. Eidelberg and S. Fahn, *N. Engl. J. Med.*, 2001, **344**, 710.
15. O. Lindvall, G. Sawle, H. Widner, J. C. Rothwell, A. Bjorklund, D. Brooks, P. Brundin, R. Frackowiak, C. D. Marsden, P. Odin and S. Rehncrona, *Ann. Neurol.*, 1994, **35**, 172.
16. P. Piccini, D. J. Brooks, A. Bjorklund, R. N. Gunn, P. M. Grasby, O. Rimoldi, P. Brundin, P. Hagell, S. Rehncrona, H. Widner and O. Lindvall, *Nat. Neurosci.*, 1999, **2**, 1137.
17. C. W. Olanow, C. G. Goetz, J. H. Kordower, A. J. Stoessl, V. Sossi, M. F. Brin, K. M. Shannon, G. M. Nauert, D. P. Perl, J. Godbold and T. B. Freeman, *Ann. Neurol.*, 2003, **54**, 403.
18. C. W. Olanow, J. M. Gracies, C. G. Goetz, A. J. Stoessl, T. Freeman, J. H. Kordower, J. Godbold and J. A. Obeso, *Mov. Disord.*, 2009, **24**, 336.
19. S. Grealish, M. E. Jonsson, M. Li, D. Kirik, A. Bjorklund and L. H. Thompson, *Brain*, 2010, **133**, 482.
20. I. Mendez, R. Sanchez-Pernaute, O. Cooper, A. Vinuela, D. Ferrari, L. Bjorklund, A. Dagher and O. Isacson, *Brain*, 2005, **128**, 1498.
21. L. Thompson, P. Barraud, E. Andersson, D. Kirik and A. Bjorklund, *J. Neurosci.*, 2005, **25**, 6467.
22. T. Carlsson, M. Carta, C. Winkler, A. Bjorklund and D. Kirik, *J. Neurosci.*, 2007, **27**, 8011.
23. M. Politis, K. Wu, C. Loane, N. P. Quinn, D. J. Brooks, S. Rehncrona, A. Bjorklund, O. Lindvall and P. Piccini, *Sci. Transl. Med.*, 2010, **2**, 38ra46.
24. O. Lindvall and A. Bjorklund, *NeuroRx*, 2004, **1**, 382.
25. R. Sanchez-Pernaute, L. Studer, K. S. Bankiewicz, E. O. Major and R. D. McKay, *J. Neurosci. Res.*, 2001, **65**, 284.
26. L. Studer, V. Tabar and R. D. McKay, *Nat. Neurosci.*, 1998, **1**, 290.
27. C. N. Svendsen, M. A. Caldwell, J. Shen, M. G. ter Borg, A. E. Rosser, P. Tyers, S. Karmiol and S. B. Dunnett, *Exp. Neurol.*, 1997, **148**, 135.

28. C. L. Parish, G. Castelo-Branco, N. Rawal, J. Tonnesen, A. T. Sorensen, C. Salto, M. Kokaia, O. Lindvall and E. Arenas, *J. Clin. Invest.*, 2008, **118**, 149.
29. S. H. Lee, N. Lumelsky, L. Studer, J. M. Auerbach and R. D. McKay, *Nat. Biotechnol.*, 2000, **18**, 675.
30. P. Koch, Z. Kokaia, O. Lindvall and O. Brustle, *Lancet Neurol.*, 2009, **8**, 819.
31. L. M. Bjorklund, R. Sanchez-Pernaute, S. Chung, T. Andersson, I. Y. Chen, K. S. McNaught, A. L. Brownell, B. G. Jenkins, C. Wahlestedt, K. S. Kim and O. Isacson, *Proc. Natl. Acad. Sci. U. S. A.*, 2002, **99**, 2344.
32. J. H. Kim, J. M. Auerbach, J. A. Rodriguez-Gomez, I. Velasco, D. Gavin, N. Lumelsky, S. H. Lee, J. Nguyen, R. Sanchez-Pernaute, K. Bankiewicz and R. McKay, *Nature*, 2002, **418**, 50.
33. N. S. Roy, C. Cleren, S. K. Singh, L. Yang, M. F. Beal and S. A. Goldman, *Nat. Med.*, 2006, **12**, 1259.
34. A. Kirkeby, S. Grealish, D. A. Wolf, J. Nelander, J. Wood, M. Lundblad, O. Lindvall and M. Parmar, *Cell Rep.*, 2012, **1**, 703.
35. S. Kriks, J. W. Shim, J. Piao, Y. M. Ganat, D. R. Wakeman, Z. Xie, L. Carrillo-Reid, G. Auyeung, C. Antonacci, A. Buch, L. Yang, M. F. Beal, D. J. Surmeier, J. H. Kordower, V. Tabar and L. Studer, *Nature*, 2011, **480**, 547.
36. Y. Takagi, J. Takahashi, H. Saiki, A. Morizane, T. Hayashi, Y. Kishi, H. Fukuda, Y. Okamoto, M. Koyanagi, M. Ideguchi, H. Hayashi, T. Imazato, H. Kawasaki, H. Suemori, S. Omachi, H. Iida, N. Itoh, N. Nakatsuji, Y. Sasai and N. Hashimoto, *J. Clin. Invest.*, 2005, **115**, 102.
37. Y. M. Ganat, E. L. Calder, S. Kriks, J. Nelander, E. Y. Tu, F. Jia, D. Battista, N. Harrison, M. Parmar, M. J. Tomishima, U. Rutishauser and L. Studer, *J. Clin. Invest.*, 2012, **122**, 2928.
38. E. Hedlund, J. Pruszak, T. Lardaro, W. Ludwig, A. Vinuela, K. S. Kim and O. Isacson, *Stem Cells*, 2008, **26**, 1526.
39. K. Takahashi, K. Okita, M. Nakagawa and S. Yamanaka, *Nat. Protoc.*, 2007, **2**, 3081.
40. K. Takahashi, K. Tanabe, M. Ohnuki, M. Narita, T. Ichisaka, K. Tomoda and S. Yamanaka, *Cell*, 2007, **131**, 861.
41. J. Yu, M. A. Vodyanik, K. Smuga-Otto, J. Antosiewicz-Bourget, J. L. Frane, S. Tian, J. Nie, G. A. Jonsdottir, V. Ruotti, R. Stewart, I. I. Slukvin and J. A. Thomson, *Science*, 2007, **318**, 1917.
42. D. Kim, C. H. Kim, J. I. Moon, Y. G. Chung, M. Y. Chang, B. S. Han, S. Ko, E. Yang, K. Y. Cha, R. Lanza and K. S. Kim, *Cell Stem Cell*, 2009, **4**, 472.
43. J. Cai, M. Yang, E. Poremsky, S. Kidd, J. S. Schneider and L. Iacovitti, *Stem Cells Dev.*, 2010, **19**, 1017.
44. M. Wernig, J. P. Zhao, J. Pruszak, E. Hedlund, D. Fu, F. Soldner, V. Broccoli, M. Constantine-Paton, O. Isacson and R. Jaenisch, *Proc. Natl. Acad. Sci. U. S. A.*, 2008, **105**, 5856.

45. T. Vierbuchen, A. Ostermeier, Z. P. Pang, Y. Kokubu, T. C. Sudhof and M. Wernig, *Nature*, 2010, **463**, 1035.
46. M. Caiazzo, M. T. Dell'Anno, E. Dvoretskova, D. Lazarevic, S. Taverna, D. Leo, T. D. Sotnikova, A. Menegon, P. Roncaglia, G. Colciago, G. Russo, P. Carninci, G. Pezzoli, R. R. Gainetdinov, S. Gustincich, A. Dityatev and V. Broccoli, *Nature*, 2011, **476**, 224.
47. U. Pfisterer, A. Kirkeby, O. Torper, J. Wood, J. Nelander, A. Dufour, A. Bjorklund, O. Lindvall, J. Jakobsson and M. Parmar, *Proc. Natl. Acad. Sci. U. S. A.*, 2011, **108**, 10343.
48. C. Karschin, E. Dissmann, W. Stuhmer and A. Karschin, *J. Neurosci.*, 1996, **16**, 3559.
49. O. Lindvall and Z. Kokaia, *Stroke*, 2004, **35**, 2691.
50. P. C. Baier, J. Schindehutte, K. Thinyane, G. Flugge, E. Fuchs, A. Mansouri, W. Paulus, P. Gruss and C. Trenkwalder, *Stem Cells*, 2004, **22**, 396.
51. J. A. Rodriguez-Gomez, J. Q. Lu, I. Velasco, S. Rivera, S. S. Zoghbi, J. S. Liow, J. L. Musachio, F. T. Chin, H. Toyama, J. Seidel, M. V. Green, P. K. Thanos, M. Ichise, V. W. Pike, R. B. Innis and R. D. McKay, *Stem Cells*, 2007, **25**, 918.
52. J. Tønnesen, C. L. Parish, A. T. Sorensen, A. Andersson, C. Lundberg, K. Deisseroth, E. Arenas, O. Lindvall and M. Kokaia, *PLoS One*, 2011, **6**, e17560.
53. G. D. Stuber, T. S. Hnasko, J. P. Britt, R. H. Edwards and A. Bonci, *J. Neurosci.*, 2010, **30**, 8229.
54. D. Sulzer, M. P. Joyce, L. Lin, D. Geldwert, S. N. Haber, T. Hattori and S. Rayport, *J. Neurosci.*, 1998, **18**, 4588.
55. F. Tecuapetla, J. C. Patel, H. Xenias, D. English, I. Tadros, F. Shah, J. Berlin, K. Deisseroth, M. E. Rice, J. M. Tepper and T. Koos, *J. Neurosci.*, 2010, **30**, 7105.
56. E. S. Boyden, F. Zhang, E. Bamberg, G. Nagel and K. Deisseroth, *Nat. Neurosci.*, 2005, **8**, 1263.
57. F. Zhang, L. P. Wang, M. Brauner, J. F. Liewald, K. Kay, N. Watzke, P. G. Wood, E. Bamberg, G. Nagel, A. Gottschalk and K. Deisseroth, *Nature*, 2007, **446**, 633.
58. C. Cusulin, E. Monni, H. Ahlenius, J. Wood, J. C. Brune, O. Lindvall and Z. Kokaia, *Stem Cells*, 2012, **30**, 2657.
59. Q. L. Ying, J. Nichols, E. P. Evans and A. G. Smith, *Nature*, 2002, **416**, 545.
60. Y. Yu and T. Wen, *Cell Mol. Biol.*, 2011, **57**, OL1528.
61. S. Friling, E. Andersson, L. H. Thompson, M. E. Jonsson, J. B. Hebsgaard, E. Nanou, Z. Alekseenko, U. Marklund, S. Kjellander, N. Volakakis, O. Hovatta, A. El Manira, A. Bjorklund, T. Perlmann and J. Ericson, *Proc. Natl. Acad. Sci. U. S. A.*, 2009, **106**, 7613.
62. I. Liste, E. Garcia-Garcia and A. Martinez-Serrano, *J. Neurosci.*, 2004, **24**, 10786.

63. E. G. Seiz, M. Ramos-Gomez, E. T. Courtois, J. Tonnesen, M. Kokaia, I. Liste Noya and A. Martinez-Serrano, *Exp. Cell Res.*, 2012, **318**, 2446.
64. J. W. Shim, H. C. Koh, M. Y. Chang, E. Roh, C. Y. Choi, Y. J. Oh, H. Son, Y. S. Lee, L. Studer and S. H. Lee, *J. Neurosci.*, 2004, **24**, 843.
65. E. Cacci, A. Villa, M. Parmar, M. Cavallaro, N. Mandahl, O. Lindvall, A. Martinez-Serrano and Z. Kokaia, *Exp. Cell Res.*, 2007, **313**, 588.
66. A. Villa, E. Y. Snyder, A. Vescovi and A. Martinez-Serrano, *Exp. Neurol.*, 2000, **161**, 67.
67. R. Donato, E. A. Miljan, S. J. Hines, S. Aouabdi, K. Pollock, S. Patel, F. A. Edwards and J. D. Sinden, *BMC Neurosci.*, 2007, **8**, 36.
68. J. Tønnesen, E. G. Seiz, M. Ramos, O. Lindvall, A. Martinez-Serrano and M. Kokaia, *Exp. Neurol.*, 2010, **223**, 653.

Subject Index

c-Abl inhibitors 47
Abnormal Involuntary Movement
  Scale (AIMS) 111, 117
abnormal involuntary movements
  (AIMs) 70, 136, 242
ADAGIO clinical trial 76
advanced glycation products
  (AGEs) 180, 189
akinesia and PD symptoms 241, 245,
  294, 308, 312
Alzheimer's disease (AD) 12, 158,
  151, 176, 182, 184, 190–1, 196, 199,
  231, 350, 351
amantadine 18, 113, 115–16, 119,
  160, 242, 299, 309
L-amino acid decarboxylase
  (L-AADC) 84
2-amino-3-(5-methyl-3-oxo-1,2-oxazol-
  4-yl)propanoic acid (AMPA) 239
AMPA receptor modulators
  antagonists 244–6
  KYNA 330
  potentiators 246
  see also NMDA receptor
    blockers
amyloid precursor protein (APP) 350
anosmia 226
anxiety and PD 10, 308, 317–18, 320
apathy 10–11
apomorphine 19, 117–18, 129, 321
arachidonic acid (AA) 189, 348
auranofin 151
autonomic dysfunction 226
autophagy/lysosome pathway
  (ALP) 178, 183

basal ganglia (BG) 7, 38, 41–2, 42–4,
  45, 62, 70, 119, 246
Basque families and hereditary
  Parkinsonism 224
benzodiazepine (BZP) anxiolytics 318
benzoylbenzamide 270
benzyloxybenzamide derivatives as
  LRRK2 inhibitors 278
besonprodil 243
biomarkers 20, 21
bpV(phen) see peroxovanadium
bradykinesia
  5-HT$_{1A}$ receptors 312
  COMT 84
  description 4–5
  dysarthria 4–5
  dysphagia 5–6
  extrapyramidal motor
    symptoms 312
  eye movement abnormalities 6
  gastrointestinal symptoms 14
  hypomimia 4–5
  hypophonia 4–5
  micrographia 4, 6
  PD symptoms 294, 308, 328, 341
  sialorrhea 4, 6
  slow gait 4, 6
brain-derived neurotrophic factor
  (BDNF) 127, 139, 140, 230, 366
Brilliant Blue G (BBG) 348–9, 351–3
bromocriptine 129
buspirone 321

caffeine 299
calcineurin 163

Subject Index

CALM-PD trial 113
CALM-PD-CIT study 75
cAMP response element binding (CREB) 298
Carlsson, Avid 3
carotid body (CB) transplants and PD
    anatomical/physiological features of carotid body 364–6
    cell therapy in PD 363–4
    clinical studies: autotransplantation in PD patients 371–2
    conclusions and perspectives 372
    preclinical studies: biological pump releasing dopaminotrophic factors 367–71
catechol derivatives as COMT inhibitors 85–8
catechol-O-methyl-transferase (COMT) inhibitors
    5-HT_{1A} receptors 309
    conclusions 105–6
    current treatment 18
    dopaminergic treatment of PD 61, 63–5, 70
    human pharmacology 95–100
    introduction 83–4
    mechanism
        in vitro potency 100–3
        in vivo duration of action 103–5
    metabolic profile 93–5
    nitrocatechol COMT inhibitors 88–92
    non-clinical pharmacology 92–3
    pyragallol and catechol derivatives 85–8
    symptomatic treatment of PD 84–5
central nervous system (CNS) and drugs for modulation of P2X7R modulators 347–9

neuroprotection and neurotoxicity 346–7
P2X7R expression 345–6
"channel blockers" (NMDA antagonists) 239, 242–3
chaperone-mediated autophagy (CMA) 179
Charcot, Jean Martin 3, 327
"Classic Lewy bodies" 193–4
clinical Parkinson's disease subtypes
    data-driven subtypes 17
    empirically driven subtypes 17
clozapine 19, 116, 119
co-enzyme Q10 21
coffee consumption and PD 161
cognitive impairment in PD 11–12, 20, 308
continuous drug delivery 76
copper (II) diacetylbis(N(4)-methylthiosemi-carbazonato) 47
"Cortical Lewy bodies" 194
cortico-based degeneration (CBD) 18
Cotzias, George 4
cyclic adenosine diphosphate (cAMP) 295–302, 303
cyto-oxygenase (COX) 187

D_1 receptors 14, 138, 152, 301, 315
D_2 receptors 14, 63–4, 309, 315, 320, 376
D_3 receptor agonists/antagonists as anti-Parkinsonian therapeutics
    conclusions 142
    D_3 receptor-selective ligands 128–34
    introduction 126–8
    localization and distribution in the brain 127–8
    modulation in treatment of L-DOPA-induced dyskinesias 136–8
    neuroprotective actions 138–9
    receptor-independent neuroprotection 139–41

D_3 receptor-selective ligands
　　agonists 129–31
　　antagonists 131–4
　　description 128–9
dardarin 224
dasatinib 47
DATATOP study 75
daytime sleepiness 13, 20
definition of Parkinson's disease (PD) 237
dementia
　　PD 20, 266
　　rivastigmine 19
　　see also Parkinson's disease dementia
dementia with Lewy bodies (DLB) 175–6, 182–4, 191, 195–200
deprenyl 299
depression and PD 9–10, 19, 20, 228, 266, 308, 317–18, 320
diaminopyridine 270
(S)-dicarboxyphenylglycine (DCPG) 253
L-3,4-dihydroxyphenylalanine (L-DOPA)
　　5-HT_{1A} receptor ligands 320
　　5-HT_{1A} receptors and dyskinesias 314–16, 320–2
　　active transport and metabolism 85
　　AMPA antagonists 245
　　anxiety 10
　　chronic treatment 309
　　COMT inhibitors 18, 92–3
　　continuous drug delivery 76
　　CSF 198
　　D_3 receptors
　　　　dyskinesias 134–6, 136–42
　　　　neuroprotection 139
　　　　significance in PD 128
　　dopaminergic treatments for PD 61–4, 69–70, 73–5
　　efficacy 245, 309
　　entacapone 93
　　glia (neuroprotection) 227
　　intraintestinal 117, 118
　　IPX066 (Rytary™) 121–2
　　KYNA 333
　　mavoglurant 255
　　MPTP 248
　　opicapone 91, 93
　　parenteral therapies 117
　　PD treatment 308–9, 376–7
　　PDE10 inhibitors 302
　　phosphodiesterase inhibitors 295, 298
　　pramipexole 113
　　ropinirole 113, 114
　　side-effects 309–10
　　tryptophan metabolism 333
dipraglurant 247, 255
DJ-1 gene 156, 181, 224
dopa decarboxylase see L-amino acid carboxylase
L-DOPA induced dyskinesias (LIDs)
　　5-HT_{1A} receptors 309, 314–16, 320
　　D_3 receptors 135, 136, 138, 142
　　development 315–16
　　dopaminergic-induced dyskinesias 112–13, 115, 116, 120–1, 122
　　glutamate receptors 241, 244, 245–6, 248, 254–5, 256
L-DOPA see also L-3,4-dihydroxyphenylalanine
dopamine (DA)
　　biosynthesis 64
　　catabolism 64
　　central nervous system 128
　　ergoline agonists 129
　　neurons
　　　　D_3 receptor-independent neuroprotection 139
　　　　degeneration 37, 73, 237, 327, 341–2, 363, 377
　　　　inflammatory and immunity channels 32
　　　　metabolic disturbance channel toxins 31
　　　　PDE7 inhibitors 301
　　　　"rejuvenation" 46

Subject Index

  substantia nigra 284
  substantia nigra pars compacta 266
  tauopathy 190
 non-ergoline agonists 129
 psychomotor functions 303
 receptors 70, 77, 138
 reward and addiction 72
"dopamine dysregulation syndrome" (DDS) 72–3
dopamine transporter (DAT) 195
dopamine-replacement therapy (DRT)
 disease progression 73–6
 future directions
  continuous drug delivery 76
  new drugs 76–7
 ICDs 71, 72–3
 motor features of PD
  complications 69–71, 77
  impairment 68–9
 neuroprotection 76
 non-motor features of PD
  addictive-like behavior 72–3
  symptoms 71–2
 strategy for pharmacological management of PD 61–2
dopaminergic systems in brain 308–9
dopaminergic treatments for PD: light and shadows
 conclusions 77
 disease progression 73–7
 drugs 63–7
 future directions 76–7
 introduction 61–3
 motor features of PD 68–71
 non-motor features of PD 71–3
droxydopa 312
drugs used in dopamine-replacement therapy
 adjuncts to L-DOPA: COMT and MAO inhibition 63
 dopaminergic agonists 63–7
 L-DOPA 63

"dual hit hypothesis" (Lewy pathology) 192
dual specificity protein phosphatase 1 (DUSP1, MKP1) 156–7
Duodopa® 118
dysarthria 4–5, 20
dysautonomia
 gastrointestinal symptoms 14–15
 orthostatic hypotension 15
 seborrhea 15
 sexual dysfunction 14
 sweating 15
 urinary bladder symptoms 14, 20
dyskinesias
 AIMs 241
 amantadine 115–16
 AMPA receptors 246
 clozapine 116–17
 D_3 receptors and L-DOPA 136–8
 drug treatment 309
 5-HT_{1A} receptors and L-DOPA 314–16, 320–2
 L-DOPA 377
 pardoprunox 121
 preladenant 120
 see also medical therapies; pharmacologic management
dysphagia 5–6, 20

electron-withdrawing group (EWG) 88
ELEP study 17
ELLDOPA study 75
embryonic stem (ES) cells 379, 385
entacapone 63, 84, 90, 92–3, 95, 309
ergolinic dopaminergic agonists
 cabergoline 65, 66
 dihydroergocriptine 66
 lisuride 65, 66, 129
 pergolide 65, 66
extrapyramidal motor disorders 314
eye movement abnormalities 6

fenobam 255
fibroblast-derived growth factor (FBF) 140

fipamezole (α-2 adrenergic receptor
 antagonist) 119–20
FJORD study 120
flesinoxan 312, 321
Food and Drug Administration
 (FDA) 118, 122
fostreicin 162

G-protein coupled receptors
 (GPCRs) 229
G-protein-gated inwardly rectifying
 K^+ (GIRK) channels 311, 317,
 382
γ-aminobutyric acid (GABA) 195,
 238, 245
gait disturbances
 falls 8
 festination 8
 flexed posture 8, 20
 freezing of gait 7–8, 20, 312
 postural reflexes 8
gallic acid 85
gastrointestinal symptoms 14–15
Gaucher disease 183
GBA gene 183
genome-wide association studies
 (GWAS) 29–30, 182–3
glia (neuroprotective role in PD) 227
glial cell-derived neurotrophic factor
 (GDNF) 34, 36–7, 139, 140, 250,
 366, 367, 369–71, 372
glial cytoplasmic inclusions
 (GCI) 175
globus pallidus 239
globus pallida externa 134
globus pallidus interna (GPi) 5, 7, 134
glucocerebrosidase (GCO) 225
glutamate receptor modulators as
 therapeutic agents for treatment
 of PD
  AMPA receptor modulators
   244–6
  conclusions and perspectives
   255–6
  glutamate-based therapeutics
   253–5

 introduction 237–8
 ionotropic glutamate receptor
  modulators in PD 241–4
 metabotropic glutamate
  receptor modulators in
  PD 246–53
 nomenclature 238–41
glutamate receptors
 allosteric modulators *versus*
  orthosteric ligands 240
 animal models of PD 240–1
 nomenclature
  ionotropic glutamate
   receptors 239
  metabotropic glutamate
   receptors 239–40
glutamate-based therapeutics
 AMPA receptor
  modulators 254
 mGluR5 receptor NAMs 254–5
 NMDA receptor blockers
  253–4
glutathione γ-L-glutamyl-L-
 cysteinylglycine (GSH) 227, 230–1
group II GluR modulators
 MGluR2/3 agonists 248–50
 MGluR2/3 antagonists 250
group III mGluRs
 MGLuR4 orthosteric agonists
  and PAMs 250–3, 256
 MGluR8 receptor agonists 253

haloperidol 68
haloperidol-induced catalepsy
 (HIC) 241, 244, 251, 253
"heat shock protein" 40 180
hereditary Parkinsonism in Basque
 families 224
5-HT_{1A} receptors and treatment
 of PD
  5-HT_{1A} receptor ligands 320
  conclusions 320–1
  description 310–11
  introduction 308–10
  L-DOPA induced
   dyskinesias 314–15

non-motor symptoms in
    PD 316–19
Parkinsonian symptoms 312–13
Huntington protein (HTT) 350
Huntington's disease (HD) 302, 350, 353
6-hydroxydopamine (6-OHDA) 68–70, 74–5, 128, 135, 139, 158, 238, 241–9, 252–3, 302, 313, 320, 334, 343, 351, 366–7
8-hydroxy-2-(di-*n*-propylamino)-tetralin *see* 8-OHDPAT
hypomimia 4–5
hypophonia 4–5

ifenoprodil 153, 243
imidazopyridazinone 7 299
impulse control disorders (ICDs) 11, 71, 72–3
indirubin-3′-monoxime 271
indolininone 270
inflammatory bowel disease (IBD) and LRRK2 gene 288–9
inosine 21
insomnia 13
interferon-γ (IFN-γ) and LRRK2 gene 288
interleukin-1β (IL-1β) 247, 343–5
internal segment *globus pallidus* (ISGP) 134
intraintestinal L-DOPA 118
intratelencephalic (IT) neurons 43
ionotropic glutamate receptors (iGluRs) 238, 241–4
    NMDA receptor blockers 241–4
IPX066 (Rytary™) 121–2
isradipine 21

KTKEGV hexapeptide 177, 222
Kufor–Rakeb syndrome 188
kynurenic acid (KYNA) 329–30, 332–4
kynurenine aminotransferases (KATS) 329
kynurenine pathway (tryptophan metabolism) 328–30, 331–3, 333–4

Lang–Fahn ADL dyskinesia score 113
levetiracetam 120–1
Lewy bodies (LB) 11, 14, 18, 34, 160, 175, 178, 191–2, 193–7, 199, 224, 226, 229, 327, 342
lipopolysaccharide (LPS) 343, 349, 350
lisuride 129
LRRK2 (leucine-rich repeat kinase 2) gene
    immunology 281
    inflammatory bowel disease 288–9
    interferon-γ response 288
    kidney 280–1, 287
    molecular pathogenesis 29–1, 35, 38, 47
    neuroprotection 224–5, 228
    synuclein and PD 180–1, 197
LRRK2 (leucine-rich repeat kinase 2) gene inhibitors (medicinal chemistry)
    patent applications 277
    potent/selective 273–6
    selective 269–73
LRRK2 (leucine-rich repeat kinase 2) gene (inhibitors as new drugs for PD)
    future prospects
        mechanism-based toxicity 288
        other therapy areas besides PD 288–9
        patient stratification 287–8
        summary 286–7
    introduction 266–7
    invertebrate/vertebrate animal models for pharmacological evaluation 282–4
    medicinal chemistry 269–77
    outside of brain and potential mechanism-based toxicity 277–81

LRRK2 (leucine-rich repeat kinase 2) gene (inhibitors as new drugs for PD) (*continued*)
  pharmacokinetics/ pharmacodynamics 284–6
  structure-activity relationships 267–9
lysosomal dysfunction and autophagy 188–9

magnetic resonance imaging (MRI) 18
"maladie de Parkinson" 327
manganese and Parkinsonism 31, 161
MAPT gene 190
Mavoglurant 247, 254–6
medical therapies for dopaminergic-induced dyskinesias
  delaying onset with DA agonists 112–14
    pramipexole 113–14
    ropinirole 114
  new compounds
    fipamezole 119–20
    IPX066 (Rytary™) 121–2
    levetiracetam 120–1
    parduprunox 121
    preladenant 120
  parenteral therapies
    intraintestinal L-DOPA 118
    subcutaneous apomorphine 117–18
  therapeutic strategies for existing dyskinesias 114–16
medium spiny neurons 376
Mendelian inheritance and PD 220, 221, 225
1-methyl-4-phenyl-1,2,3,6-tetrahydropyridine (MPTP) 31, 35, 68–70, 74, 119, 135, 138, 140, 197, 241–2, 246, 248, 302, 312, 332–4, 343, 350, 366
metabolic profile of COMT inhibitors 93–5

metabotropic glutamate receptors (MGluRs)
  description 239–40
  group I mGLuRc: MGLuR5 NAMs 246–8, 254–5
  group II mGLuR modulators 248–50
metformin 46
microglia 189, 342–6, 349–50
micrographia 4, 6
mitochondria and PD 187–8, 330–1, 332
mitochondrial complex I 332, 334
mitochondrial transcription factor A (TFAM) 33
mitogen activating protein kinases (MAPKs) 180, 187
MitoPark mouse 33
molecular pathogenesis and pathophysiology of PD: new targets for new therapies
  introduction 26–7
  molecular pathogenesis
    cell division activation 34
    channels of mitochondrial DNA mutations 32–3
    connecting channels to define pathogenesis of PD 34–6
    description 27–9
    genetic channels 29–31
    inflammatory and immunity channels 32
    metabolic disturbance channel toxins 31
    reactive oxygen and nitrogen species channels 32
    somatic nuclear mutations 33–4
    transcriptional re-programming and epigenetic control 36–8
  pathophysiology of PD
    basal ganglia models 41–2, 42–4, 45

dopamine cells are not
only ones in PD 40–1
dopaminergic neurons are
complex 38–9
targets for New PD
therapies 45–7
monamine oxidase inhibitors
(MAOIs) 18, 220–1, 230–1
monamine oxidase (MAO) 61, 63
motor signs and symptoms
bradykinesia 4–6
current treatment 18
dystonia 9
gait disturbances 7–8
rest tremor 7, 9
rigidity 7
Movement Disorders Society
(MDS) 111
MP-10 299, 302
multiple sclerosis (MS) 350
multiple-system atrophy (MSA) 18,
175, 182, 198

N-methyl-D-aspartate (NMDA) 70,
111, 115–16, 119, 159, 186, 239,
249, 254, 255, 309, 313, 330
see also NMDA receptor
blockers
nafadotride 138
naluzotan 321
NBQX (AMPA antagonist) 244–5
nebicapone 90, 94–5
negative allosteric modulators (NAMs)
240, 245, 246–8, 254–5, 256
neramexane 242
neostriatum see striatum
neuro-opthalmalogical disturbances 16
neuroinflammation in PD progression
brains and animal models 342–3
IL-1β 343–5
neuropathology of Lewy body
disorders
dementia with Lewy bodies and
PD 196–7
sporadic Parkinson's disease
193–6

neuroprotection (new approaches)
in PD
introduction 219–21
neuroprotective therapies:
pathogenesis 129–231
pathogenic mechanisms due to
genetic defects 221–5
strategies 231
temporal pattern of
pathological changes 229
temporal profile of clinical
features and pathological
changes 225–8
neuropsychiatric symptoms
anxiety 10
apathy 10–11
cognitive impairment 11–12
dementia 12
depression 9–10
hallucinations and psychosis 11
impulse control disorders 11
nicotinamide adenosine diphosphate
(NAD) 328
nicotine adenine dinucleotide
(NAD) 348
nitecapone 89–90
nitrocatechol COMT inhibitors 88–92
NMDA receptors (glutamate
receptors)
AMPA receptor
modulators 244
AMPA receptor
potentiators 246
channel blocking NMDA
antagonists 242–3
clinical development 253–4
ionotropic glutamate receptors
239
NR2B-selective antagonists
243–4
non-ergolinic dopaminergic agonists
apomorphine 65, 66
piribidil 67
pramipexole 65, 67
ropinirole 65, 67
rotigotine 65, 67

non-motor symptoms of PD
    5-HT$_{1A}$ receptors 316–19
    anxiety and depression 317–18
    cognitive impairment 318–19
    current treatment 19
    dysautonomia 13–15
    introduction 9
    neuropsychiatric symptoms 9–12
    other symptoms 15–16
    sleep disturbances 12–13
    see also dopamine-replacement therapy
non-selective LRRK2 inhibitors 269–73
nucleus accumbens 40, 127, 131, 300
NURR1 gene 34, 37

8-OH-DPAT 313, 314–16, 317–19, 321
olfactory disturbance 16, 227
opicapone 91–2, 92–3, 95–100, 101–4, 105
orthostatic hypotension 15
osemozotan 321
oxidase inhibitors (MAOI-B) 18

P2X7 receptor signaling in treatment of PD/other neurodegenerative disorders
    altered expression/function 349–51
    CNS and drugs for modulation 345–9
    conclusion and perspective 354–5
    introduction 341–2
    modulators or deficiency in animal models
        antagonists/deficiency 351–3
        other modulators 353–4
    neuroinflammation 342–5
    see also central nervous system
pain 15
papaverine 299, 302
"Paralysis Agitans" 3
paraquat and Parkinsonism 31, 74

parduprunox 121
PARIS (ZNF746) transcriptional repressor 35, 36
PARK genes 164, 176, 180–2, 188, 221
PARKIN gene 30, 32, 35, 38, 47, 154–5, 161, 181–3, 194, 197, 223–4, 226–8
Parkinson, James 3, 327
Parkinson's disease dementia (PDD) 195–7
paroxetine 19
pathogenic mechanisms due to genetic defects in PD
    α-synuclein 221–3
    genes and hereditary Parkinsonism 224–5
    PARKIN gene 223–4
pedunculo-pontine nucleus (PPN) 40, 44
perampanel 245, 256
percutaneous endoscopic gastrostomy (PEG) 118
peroxovanadium 149
PH domain leucine-rich repeat protein phosphatase 163–4
pharmacologic management of dopaminergic-induced dyskinesias in PD
    conclusions 122
    introduction 110–11
    medical therapies 122
    scales for assessment 111
    why do dyskinesias occur? 111
N-phenyl-7-(hydroxyimino)-cyclopropa[*b*]chromen-1*a*-carboxamide (PHCCC) 251–2
phosphodiesterase (PDE) inhibitors and treatment of PD
    conclusions 303
    dopamine and cyclic adenosine monophosphate 295–6
    dopamine signalling 296–8
    drug targets beyond dopamine 298
    introduction 294–5
    new drugs for PD 298–9

PDE1 inhibitors 300
PDE4 inhibitors 300–1
PDE7 inhibitors 301
PDE10 inhibitors 302
piclozotan 321
PINK1 gene 30, 32, 35, 154–6, 180–2, 224
pioglitazone 45
pluripotent stem (iPS) cells 379–82, 385
positive allosteric modulator (PAMs) 240, 250–3, 256
postural reflex impairment and PD 20, 84, 294, 308, 328, 341
postural-instability-gait-disorder (PIGD) 17
potent/selective LRRK2 inhibitors 173–6
pramipexole 113–14, 129, 138–9, 140
  depression 19
pramipexole v. L-DOPA trial on motor complications of PD 113
PREDICT modeling program 320
preladenant (adenosine A2A receptor antagonist) 119, 120
prion disease 350, 354
prion protein (PrP) 192, 350
progressive supranuclear palsy (PSP) 18, 198
PROPARK study 17
proteasome inhibitors 47
protein phosphatases in PD
  future views 164
  introduction 149
  protein serine/threonine phosphatases 157–64
  protein tyrosine phosphatases
    description 149–50
    dual specificity protein phosphatase 1 156–7
    phosphatase and Tensin homolog deleted on chromosome 10 154–6
    protein tyrosine phosphatase PTP-PEST 151

receptor protein tyrosine phosphatases (β/ξ) 150–1
Src homology 2 domain-containing phosphatase 2 153–4
striatum-enriched protein tyrosine phosphatase 150–2
protein serine/threonine phosphatases
  PH domain leucine-rich repeat protein phosphatase 163–4
  protein phosphatase 1 158–60
  protein phosphatase 2A 160–3
  protein phosphatase 3 163
PTEN phosphatase 154, 156, 164
pyrazolopyridine derivatives as LRRK2 inhibitors 279
pyrazolpyridine 270
pyrimidal-tract (PT) neurons 43
pyrimidine A 299
pyrogallol derivatives as COMT inhibitors 85–8

QUIN receptor agonist 330, 334
quinoline 270
quinpirole 131

Raf-1 kinase inhibitor 271
rapamycin 47
rasagiline 76
reactive nitrogen species (RNS) 32
reactive oxygen species (ROS) 32, 33, 139, 156, 180, 189, 347
REAL-PET trials 75
REM sleep behavior disorder (RBD) 11, 12
remacemide 242
reserpine 68, 197
reserpine-induced akinesia (RIA) 241, 251
resting tremor and PD 7, 84, 294, 308, 312, 327, 341
restless legs syndrome (RLS) 13
rigidity and PD symptoms 7, 294, 308, 312, 327, 341

rivastigmine 19
roflumilast 299
rolipram 299
ropinirole 114, 129
rotigotine 19

SCA gene 190
seborrhea 15
serotinergic (5-HT) neurons 310
serotonin pathway (tryptophan
    metabolism) 328, 331–2
sexual dysfunction 14
sialorrhea 4, 6
SINDEPAR study 75
single nucleotide polymorphisms
    (SNP) 29
single photon emission computerized
    tomography (SPECT) 18, 75
sleep disturbances
    daytime sleepiness 13
    insomnia 13
    PD symptoms 327
    REM sleep behavior disorder 12
    restless legs syndrome 13
    vivid dreams 12–13
slow gait 4, 6
SLV308 320–1
SNARE (soluble *N*-ethylmaleimide-
    sensitive factor attachment protein
    receptor) 180, 188
SNCA gene 46, 176, 181, 183
sporadic PD
    mitochondrial dysfunction 331
    α-synuclein/Lewy pathology
        183, 193–6
staurosporine 272
Steele–Richardson syndrome 231
stem cells
    directly induced neurons 381
    embryonic 379
    fetal neural 378–9
    pluripotent 379–82
stem-cell-based cell replacement
    therapy in PD
    candidate stem cells 378–82
    cell-replacement therapy 376–7

conclusions and future
    perspectives 384–5
fetal cell replacement
    therapy 377–8
genetically enhanced stem
    cells 384
stem cell integration in host
    tissue 382–4
striatum
    AMPA receptor modulators
        244, 246
    basal ganglia 42–4
    carotid body 367–8
    cell therapy 363
    COMT inhibitors 84, 93
    D_3 receptors 134–5, 138
    dopamine and cyclic
        AMP 296–7
    dopaminergic neurons 61, 68,
        70, 149, 377
    dyskinesias 111
    fetal cell grafts 382
    glutamate receptor modulators
        237, 239, 241
    LRRK2 inhibitors 284
    P2X7R expression 350
    pathology of PD 38–42
    phosphodiesterases (PDEs) 303
    protein phosphatase 1 158
    α-synuclein 191, 195
    tryptophan metabolism 332
striatum-enriched protein tyrosine
    phosphatase (STEP, PTPN5) 151–3
structure-activity relationships
    (SARs) 89, 90–1, 267–9
subcutaneous apomorphine 117–18
substantia nigra (SN)
    α-synuclein 180, 186
    animal models of PD 241
    basal ganglia 41
    carotid body 367
    D_3 receptors 140–1
    dopaminergic neurons 32, 37,
        46, 284, 363, 372, 377, 385
    hyper-echogenicity 18
    interleukin 1β 343

Subject Index

iron 31, 189
Lewy bodies 195, 229
microglia 44
mitochondrial complex I 332
neuronal loss 83
P2X7R 350, 351
pathological changes 3
PDE1 inhibitors 300
PDE7 inhibitors 301
pedunculo-pontine nuclei 40
pramipexole 138
receptor protein tyrosine phosphatase β/γ 150
serotonin 332
sporadic PD 193
α-synuclein 180, 186, 191
TNFα/TNF receptors 140
substantia nigra pars compacta (SNc)
   5-HT$_{1A}$ agonists 313, 315–16
   6-OHDA Parkinsonism 68
   D$_3$ receptors 141
   dopaminergic areas of brain 9
   dopaminergic neurons 38, 61, 237, 241, 248, 266, 376–7, 377–8, 382–3
   KYNA 332
   motor symptoms 341–2
   neuroinflammation 342–3
   neurons degeneration 61
   PD neuropathology 327
   phosphodiesterase inhibitors 298
substantia nigra pars reticula (SNr)
   D$_3$ receptors 127, 134
   ionotropic glutamate receptors 239
   manganese 31
subthalamic nucleus (STN) 238
sunitinib 270–1, 271
superoxide dismutase (SOD) 189
sweating 15
symptoms, unmet needs and new therapeutic targets
   clinical Parkinson's disease subtypes 16–17
   current diagnosis 18

current treatment 18–19
introduction 3–4
motor signs and symptoms 4–6
non-motor symptoms 9–15
other symptoms 15–16
unmet needs and new therapeutic targets 19–21
α-synuclein
   biochemistry
      genetics 180–3
      localization and regulation 177–9
      neuroinflammation 189–90
      physiological functions 179–80
      structure 176–7
   immunoreactive lesions 227
   Lewy bodies 342
   LPS 343
   neurodegeneration 343
   P2X7R modulators 354
   pathology 228–9
   phosphodiesterases 298
   protein phosphatases 160–1, 164
synuclein and PD: update
   animal models of PD 197
   biochemistry of α-synuclein 176–83
   conclusions and outlook 198–200
   introduction: α-synuclein and disease 175–6
   neuropathology of Lewy body disorders 193–7
   α-synuclein
      biomarker for synucleinopathies 197–8
      interaction with other proteins 190–1
      neurodegradation 183–90
      spread and disease propagation 191–3
   synuclein family 176

talampanel 245
tandospirone 313, 315, 321
tauopathy in PD 190–1
TEMPO clinical trial 76
tezampanel 245
therapeutic drugs and PD 309–10
thieno[3,4-c]pyridin-4(5H)-one
   derivatives as LRRK2
   inhibitors 278
thienopyridine 270
tolcapone 63, 84, 90, 92–3, 94–5, 96,
   101, 103
"toxic-Parkinsonism" 31
toxin-induced
   neurodegeneration 73–4
TRAP1 protein 187
traxoprodil 243
triazolpyridine 270
trophic therapies 21
tropolone 86
tryptophan (TRP) metabolism in PD
   introduction 327–8
   kynurenine pathway 328–30
   pathogenesis of PD
      altered metabolism
         331–3
      conclusions 334

   description 330–1
   future therapeutic
      possibilities 333–4
   serotonin pathway 328
tumor necrosis factor-α (TNF-α) 345

ubiquitin-proteasomal system
   (UPS) 178–9, 186
Unified Parkinson's Disease Rating
   Scale (UPDRS) 69, 75–6, 111, 114,
   116, 121, 254, 255, 371
unmet needs and new therapeutic
   targets
      biomarkers 20
      the cure: cause-directed
         therapy 21
      beyond dopamine 19–20
      neuroprotection 20
      PD treatment 309–10
urinary bladder symptoms 14, 20

venlafaxine 19
Ventral Tegmental Area (VTA) 37
vivid dreams 12–13

walkabout 71, 72
weight changes 16